The Dynamic Earth

Lightning crackles around the column of hot gas and volcanic ash rising from Mount Pinatubo in the Philippines. Several hundred people died as a result of the eruption. This time exposure was taken on the evening of July 1, 1991. Because of the Earth's rotation, stars appear as bright streaks.

The Dynamic Earth
an introduction to physical geology

Second Edition

Brian J. Skinner
Yale University

Stephen C. Porter
University of Washington

John Wiley & Sons, Inc.

New York   Chichester   Brisbane   Toronto   Singapore

ACQUISITIONS EDITOR / Barry Harmon
DEVELOPMENTAL EDITORS / Irene Nunes and
  Barbara Heaney
PRODUCTION MANAGER / Katharine Rubin
DESIGNER / Kevin Murphy
PRODUCTION SUPERVISOR / Nancy Prinz
MANUFACTURING MANAGER / Lorraine Fumoso
COPY EDITOR / Elizabeth Swain
PHOTO RESEARCHERS / Pat Cadley and
  Charles Hamilton
PHOTO EDITOR / Elaine Bernstein
ILLUSTRATION / Edward Starr
PART OPENING CARTOGRAPHY / Alice Thiede /
  Shaded Relief by Hildegard B. Groves
COVER PHOTO / Larry J. Pierce / StockPhotos, Inc.
FRONTISPIECE / Reuters / Bettmann Archive

Recognizing the importance of preserving what has been written, it is a policy of John Wiley & Sons, Inc. to have books of enduring value published in the United States printed on acid-free paper, and we exert our best efforts to that end.

Library of Congress Cataloging-in-Publication Data

Skinner, Brian J., 1928–
    The dynamic earth : an introduction to physical geology / Brian J.
Skinner, Stephen C. Porter.
        p.   cm.
    Includes bibliographical references and index.
    ISBN 0-471-53131-6
    1. Physical geology.   I. Porter, Stephen C.   II. Title.
QE28.2.S55   1992
550—dc20   91-32943
                                                  CIP

Printed and bound by Von Hoffmann Press, Inc.

10 9 8 7 6 5 4 3 2

Preface

The Earth is a giant machine, a machine that never stops. How the Earth-machine works is the story told in this book, and it is a fascinating story.

The science that studies the Earth is geology, and those involved in the studies are geologists. It is a unique science because its laboratory is the world in which we live. Geologists find it very difficult to carry out controlled experiments in their "geological laboratory"—the scales of space and time needed for such experiments are simply too large, and even if the space and time problems could be handled, there is always the chance that the experiments would cause the environment we live in to change in some unfortunate way. Geologists must study the Earth as it exists. From their assembled observations they draw conclusions about the processes that are shaping the Earth today and events that have shaped the Earth over the past 4.5 billion years. Increasingly, geologists are called upon to use their understanding of the Earth-machine to suggest what changes might be expected in the future.

A SCIENCE IN FERMENT

Revolutionary advances in the breadth and depth of our knowledge of the Earth and the other planets of the solar system have occurred during the past 40 years. Never before have so many dramatic discoveries been made in such a short time, and the revolutionary discoveries continue unabated. For example, as this book was being written a spacecraft called *Magellan* went into orbit around Venus and started sending back remarkable radar images of that planet's cloud-obscured surface. Venus is fascinating because it is the nearest and most similar to Earth of the other planets. The radar images make it clear that Venus, like the Earth, has had a long and complex geological history. In the years ahead, with the help of *Magellan* and other spacecraft, geologists will try to determine why Venus and the Earth are so similar but also in some ways so different.

Another example of an intriguing discovery, this time on the Earth, concerns a vast submarine lava field in the western Pacific Ocean. Ocean-going geologists mapped and then drilled into the lava field. The size of the field and the discovery that the lavas were erupted with great rapidity 120 million years ago, just when the Earth's climate became very warm, has led geologists to suggest that gigantic volcanic "burps" released enough carbon dioxide into the atmosphere every now and then to completely change the climate. The most remarkable part of the hypothesis is that the volcanic burps seem to be controlled by the Earth's molten core, implying that the core must play a role in determining the climate. The observations behind these extraordinary concepts are discussed in chapter 19.

Geology is a science in ferment, a science laced with challenging excitement. New discoveries, new insights, and new theories heighten the excitement almost every day. We, the authors of this book, have been privileged to work as geologists all of our professional lives. We have attempted in this book to share the excitement and wonder we feel as we study the intricacies of our planet Earth. In writing this book we have drawn extensively on our own experience, as well as the experience of numerous fellow geologists whose collective geological knowledge spans the field of physical geology. Brian Skinner's research has focused on the physical properties of minerals and on the genesis of base metal deposits. He has worked in Australia, Africa, and North America, and with students in Asia and Europe. Stephen Porter's professional career has been largely concerned with studies of glaciation in many of the world's major mountain systems and with the history of the climatic changes that their deposits record. He also has studied the evolution of midocean and continental volcanoes, the products of their prehistoric eruptions, and how volcanic eruptions may have influenced the Earth's climate. With foreign colleagues he has studied the hazards of large rockfalls in the Alps and the thick, extensive deposits of wind-blown dust in central China that provide one of the longest continuous records we have of climatic change during the last several million years.

Between us, we have carried out geologic field work on all of the continents. This global perspective is emphasized in the book by examples and illustrations from around the world, for we think it is important to emphasize that geology is a global science, a science that recognizes no political boundaries. Only by studying the Earth in its entirety can we hope to understand how our amazing planet works.

KNOWING THE EARTH

The science of geology is like all other sciences in that it is based on observations. We have tried to write this book so that readers can sense first the fun of making their observations and then the challenge of drawing conclusions; the reader does not need a geologic background to read the text and apply the principles it introduces.

We believe that everyone, when given a chance to do so, quickly becomes intrigued to learn how the Earth works. *The Dynamic Earth* was written to provide that chance. It is a book for everyone.

The organization of the second edition is similar to, but not identical with, that of the first edition. The book is divided into three parts. To emphasize the global coverage, each part opens with a map on which the photos used in that part are located. Part 1 covers the Earth's overall features and its materials. Part 2 covers all of the processes that shape the Earth's surface, and Part 3 concerns the evolving Earth, mineral resources, and planetology. Each chapter in the book was written to stand alone. Teachers who prefer to teach a subject in an order that differs from the text can easily do so.

We stress four major themes throughout the book. The first is plate tectonics—the slow, lateral motions of fragments of the Earth's outermost 100 km (called plates) at rates up to 20 cm/y. The shapes and locations of continents and oceans, the locations of mountains and volcanoes, and the violence of earthquakes are all determined by plate tectonics. We have used plate tectonics as a framework within which to integrate most geological processes. Plate tectonics is the link between the Earth's internal and external activities and as a result the topic appears in most chapters. The second theme concerns the influence that the human race is having on the Earth's external processes. We humans are now so numerous (about 5.2 billion and increasing by approximately 3 people a second) that our daily activities are having measurable effects on such things as rainfall, climate, and rates of erosion. To understand how the Earth works today we must appreciate the part played by humans in geological processes. The third and fourth themes also concern the human race. The third is the mitigation of natural disasters. In one way or another all of us are at risk.

We can't stop volcanic eruptions, earthquakes, hurricanes, or floods, but by knowing how the Earth works we can warn populations of pending danger and reduce the threat by taking appropriate actions. The fourth theme is the human use of natural resources. Each of us uses, on average, about 10 tons of mineral resources each year. Finding where to dig those resources is one of the greatest challenges faced by geologists. Understanding the consequences of using mineral resources at the rate we do is another great challenge for geologists. We have, therefore, integrated the topic of natural resources throughout the book, while chapter 17, a short chapter devoted to mineral and energy resources, covers those economic and technical issues of resources that do not fit readily into other chapters.

FEATURES OF THE SECOND EDITION

We have added a new final chapter (chapter 19), titled "Our Changing Planet." In this chapter we cover a theme that has received little attention in physical geology textbooks, namely how human activity is changing the Earth's climate. We emphasize again that the Earth is a complex, interactive, and dynamic system in which a change in one part is likely to change other parts, often in unanticipated ways. As we burn fuels to run society, we pollute the atmosphere and alter its chemical composition. In doing so, we unwittingly have contributed to changes in the Earth's system that may have serious effects on all of us in the years ahead. Geology provides a record of billions of years of natural environmental change on our planet. It therefore has an extremely important role to play in our attempt to understand how human-induced changes may affect our planet and its inhabitants. In chapter 19 we introduce the ozone hole, the greenhouse effect, and other global environmental changes in which each of us plays a role.

Several features of the first edition of *The Dynamic Earth* that drew favorable comments from users and reviewers have been expanded. As with the first edition, each chapter closes with a short essay discussing a topic on which research is continuing. Many of the essays in the second edition are new. Chapter 7, for example, on weathering and soils, closes with an essay on the intriguing problem of the relationship between plate tectonics and global weathering rates, while the chapter discussing the solar system, chapter 18, has a closing essay on the possibility of life existing on a planet in another solar system.

Each chapter now opens with a short essay concerning one of the four themes of the book as it relates to the subject of that chapter. Chapter 3, for example, which deals with volcanism, opens with an essay on the eruptions of Krakatau and Mount St. Helens, while chapter 12, on wind action and deserts, opens with an essay on the dust-bowl years.

Pedagogical material such as chapter summaries, questions for review, and short lists of important words and phrases have been retained, but the number of questions has been increased and the number of technical terms considerably reduced.

The artwork has been revised and simplified; many pieces were redrawn so that most of the art is three dimensional. Photographic research has been intensive. The first edition was commended by users for the fine photos; this second edition has 33 percent more photographs.

SUPPLEMENTS

A full range of supplements to accompany the second edition of *The Dynamic Earth* is available to assist both the instructor and student.

Study Guide. By Michael Kimberley and Susan Kimberley, both of North Carolina State University. This revised guide has been expanded to include detailed chapter introductions and summaries, key terms referenced in the text, and self-testing questions that encourage the student to reinforce concepts covered in the text.

Instructor's Manual and Test Bank. By C. Frederick Lohrengel of Southern Utah University. This supplement contains lecture outlines, teaching suggestions, and a summary of additional resources available to the instructor. The lecture outlines are also available on disk in Macintosh format only. All test questions, including multiple-choice, fill-in, and essay questions, are new to this edition.

Computerized Test Bank. IBM and Macintosh versions are available with full editing features to help you customize tests.

Full-Color Overhead Transparencies. Over 60 full-color transparencies from the text are provided in a form suitable for projection in the classroom. These illustrations will also be available as slides.

Supplementary Slide Set. A full description of this slide set, which contains photos used in the text, many from the authors' personal collections, is included in the Instructor's Manual.

ACKNOWLEDGMENTS

The first edition of *The Dynamic Earth* came to fruition under the guidance of Clifford Mills, then the Earth Sciences editor at John Wiley and Sons. The book was produced by reducing the length, the level, and the range of material covered in a higher-level text, *Physical Geology* (1987). Although the first edition of *The Dynamic Earth* was received enthusiastically by users, it still carried the flavor of its parent. The second edition has therefore been written as a separate book. The Earth Sciences editor, Barry Harmon, and the publisher, Kaye Pace, have been very supportive and encouraging as the planned 30 percent revision grew to a 60 percent revision. The second edition is much the better both for the revision and for their support.

The professional skills and competence of the staff at John Wiley and Sons and of the free-lance experts they found to work with us are outstanding. Barbara Heaney, Nancy Prinz, Katherine Rubin, Stella Kupferberg, Elaine Bernstein, Pat Cadley, Kevin Murphy, Elizabeth Swain, Joan Kalkut, Catherine Faduska, Cynthia Michelson, and everyone else at Wiley were always cordial and always helpful, no matter how badly our travel schedules upset book schedules. John Woolsey, the artist, was a pleasure to work with as he tried to change our crude sketches to finished drawings. As maps were changed from Mercator to Robinson projections, we came increasingly to appreciate the skillful work of Alice Thiede. Above all, the patient, insightful, and very helpful editing of Irene Nunes must be mentioned. Through Irene's eyes we managed to understand why readers have sometimes had trouble with our words. Both this text and our own writing skills are greatly improved, and we are much indebted for her guidance.

We are also indebted to the many people who provided elegant colored photographs that appear in the book. Most of the photographers are geologists, and their discerning eyes can be sensed through the beautiful photos they took. Their names are listed in the Photo Credits at the back of the book, but being so placed is no reflection on their importance. Several people deserve additional thanks for their help in suggesting new sources for photos and for helping with illustrations and data; these include Brian Atwater, Darrell Cowan, Richard Fiske, William Graustein, and Elaine Padovani.

We especially thank the thoughtful and dedicated teachers who commented on the previous text or reviewed the present text. These fine people not only helped us keep a reasonable balance to the book, they also helped us keep the volume as up to date as possible without downplaying the great geological discoveries of the past. They are:

Gary C. Allen
University of New Orleans

J.C. Allen
Bucknell University

N.L. Archbold
Western Illinois University

Philip Brown
University of Wisconsin

Collete D. Burke
Wichita State University

Alan Cain
University of Rhode Island

Robert A. Christman
Western Washington University

Nicholas K. Coch
Queens College

Kristine J. Crossen
Anchorage Community College

John Diemer
University of North Carolina at Charlotte

Grenville Draper
Florida International University

M. Ira Dubins
State University of New York at Oneonta

John Ernissee
Clarion University of Pennsylvania

Stewart Farrar
Eastern Kentucky University

Mike Follo
University of North Carolina at Chapel Hill

Ann G. Harris
Youngstown State University

Robert L. Hopper
University of Wisconsin

Robert Horodyski
Tulane University

Peter L. Kresan
University of Arizona

Albert M. Kudo
University of New Mexico

Judith Kusnick
California State University

Nancy Lindsley-Griffin
University of Nebraska

William W. Locke
Montana State University

David N. Lumsden
Memphis State University

Gerald Matisoff
Case Western University

Robert McConnell
Mary Washington College

Bruce Nocita
University of South Florida

Anne Pasch
University of Alaska-Anchorage

Gary Peters
California State University, Long Beach

John Renton
West Virginia University

Mary Jo Richardson
Texas A&M University

Donald Ringe
Central Washington University

Len Saroka
St. Cloud State University

Fred Schwab
Washington & Lee University

Charles P. Thornton
Pennsylvania State University

Christian Teyssier
University of Minnesota

Graham Thompson
University of Montana

James B. Van Alstine
University of Minnesota

Neil Wells
Kent State University

Monte Wilson
Boise State University

Margaret Woyski
California State University Fullerton

Anne Wyman
University of Nevada-Las Vegas

A CLOSING THOUGHT

After digesting a beginning geology textbook, the reader may well come away with the feeling that we geologists have all the answers, that the major principles are known, and the significant challenges have been resolved. We hope this will not happen with this book. We have tried to show that we do not have all the answers. In fact, it is because we have so many important and challenging questions remaining before us that geology is such a dynamic and exciting science in which to work. Those of us who have seen the remarkable advances of the past 40 years have no reason to doubt that the next 40 will produce even more startling discoveries about our dynamic Earth.

Brian J. Skinner

Stephen C. Porter

Brief Contents

Contents

Fishing boats stranded in the desert by shrinkage of the Aral Sea.

Introduction

T he Aral Sea in Soviet central Asia is shrinking so rapidly that once-prosperous fishing villages are now 50 km from the shore. Human activities caused the change. The story of that change is a lesson for us all: when we alter the balance of nature, unforeseen side effects almost always crop up.

Thirty years ago, the Aral Sea was the fourth largest lake in the world after the Caspian Sea, Lake Superior, and Lake Victoria. The sea covered 68,000 km², had an average depth of 16 m, and yielded 45,000 tons of fish a year. Today the sea is only the sixth largest lake. It now covers 40,000 km², has an average depth of 9 m, is so salty the fishing industry is dead, and is disappearing so fast it will be a waterless desert by 2010.

The Aral Sea is fed by two large rivers, the Amu Dar'ya and the Syr Dar'ya, which carry meltwater across the desert from the snowy mountains of northern Afghanistan. Water leaves the sea by evaporation, so the size of the sea is a balance between evaporation and river inflow. The sea is shrinking because inflow has declined.

A small part of the problem is climatic; there were a number of years in the 1970s when snowfalls were light. The largest part of the problem, however, is the irrigation that has been practiced in the river valleys for millennia. In modern times, the extent of irrigation increased dramatically. By 1960, so much irrigation water was taken from the two rivers that inflow to the Aral Sea had declined to a trickle. The sea has been shrinking steadily ever since.

The people who planned the irrigation systems expected the Aral Sea to shrink. What they did not anticipate were the side effects. The sea, it is now realized, exerts a major influence on the local climate. Because it is shrinking, local rainfall is declining, the average temperature is rising, and wind velocities are increasing. Most of the newly exposed sea bottom is covered with salt. The wind blows the salt around and has created withering salt storms. Potable water supplies have declined, and various diseases, especially intestinal diseases, are afflicting the local population at alarming levels.

The situation could, of course, be reversed by simply reducing the amount of irrigation. The problem is that the irrigated area is now one of the Soviet Union's most prosperous, so the ultimate solution will probably have to be somewhere between returning the sea to its original size and keeping all the irrigated land.

GEOLOGY AND GEOLOGISTS

The Earth is always changing. Small, slow changes are continuous, while massive but rapid changes, like those produced by hurricanes, are sporadic. Fast or slow, large or small, continuous or sporadic, the key word is *change*. The Earth is never still.

The discipline that studies these changes, past and present, is **geology**. The word comes from two Greek roots: *geo-*, meaning of the Earth, and *-logia*, meaning study or science. Scientists who study the Earth are called **geologists**.

Geologists work in every corner of the world, from ice-covered peaks and active volcanoes to the depths of the ocean. They seek to understand all the processes that operate on the Earth and to document the Earth's long, complex history. In their work they study directly any place they can reach. To investigate places they cannot reach, they drill deep holes in solid rock. Beyond the reach of drill holes, geologists must rely on indirect observations. As a doctor listens for noise inside your body with a stethoscope, so a geologist employs sensitive measuring devices to "listen" to the rumbles of distant earthquakes and sense the pulse of activities deep inside the Earth. From their observations geologists try to understand the history of Planet Earth and the origin of its complex landscapes. They try, too, to predict where new oil fields lie, whether a well will strike water, and where rich ore deposits are hidden.

The processes geologists study all obey the fundamental laws of nature discovered by physicists, chemists, and mathematicians. In a sense, then, one might call geology a derivative science. But geology is also a special and very practical science because it is the science of the planet on which we live, the science of our own environment. Geologists investigate our environment using the scientific method. The core of the scientific method of investigation is evidence that can be seen and tested. Based on that tested evidence, geologists draw conclusions and so advance our understanding of the way the Earth works.

Geology is traditionally divided into two broad topic areas with related but differing aims. **Physical geology** is concerned with understanding (1) the *processes* that operate at or beneath the surface of the Earth, and (2) the *materials* on which those processes operate. What causes volcanoes to erupt or how earthquakes, landslides, and floods happen are examples of processes. Examples of materials are soils, sands, rocks, air and seawater. **Historical geology** has as its goal the chronology of the *events*, both physical and biological, that have occurred in the past. Historical geology seeks to resolve questions such as when the oceans formed, when dinosaurs first appeared, when the Rocky Mountains rose, and when and where the first trees appeared.

Physical geology, the subject of this book, serves as a starting point for studying the Earth. Most people seek to know about the Earth because we humans are inherently interested in the things around us. But there is also a very practical reason for studying the Earth. If we are to understand the environment in which we live and be able to make predictions about changes that might lie ahead, we must understand how the Earth works. In order to obtain that understanding, and especially to understand how we humans may be affecting the Earth, we need to examine both materials and processes.

The word *rock* is an important one, and because it is used frequently in this book, we must define it carefully. **Rock** is any naturally formed, nonliving, firm, and coherent aggregate mass of solid matter that constitutes part of a planet. Note that the definition specifies a coherent aggregate, which means that all the rock particles are locked together to make a solid mass. A pile of loose sand grains is not rock because the grains are not locked together—not coherent. A tree is not a rock, even though it is solid, because it is living. But coal, which is a compressed and coherent aggregate of twigs, leaves, and other bits of dead plant matter, is a rock.

LIFE AND THE EARTH

Hurricane Hugo left a trail of death and destruction when it roared through the Caribbean and South Carolina in 1989 (Fig. I.1). Even though people knew the hurricane was coming, there was nothing anyone could do to stop or deflect it. Nor was there any way of stopping Mount St. Helens from erupting in 1980 or preventing the Loma Prieta earthquake from shaking San Francisco in October 1989 (Fig. I.2). Each event was a disaster for those who lived in the area, and each reminded us that we humans do not control nature. It is nature that controls us.

Someday, perhaps, we may learn how to deflect hurricanes or stop earthquakes, but at present the best we can do is try to predict when and where natural disasters might happen and be prepared if they do occur. In order to make predictions as reliable as possible, we must monitor the Earth's continuous changes, and we must learn which changes signal trouble ahead.

Not all the changes to be monitored are natural. Some are the result of our own human activities. We are only beginning to realize just how immense our

FIGURE I.1 Destruction caused by Hurricane Hugo. Remains of a house on the edge of deeply eroded cliffs near Charleston, South Carolina.

collective human activities are. Consider the quantity of mineral resources we take from the ground and use to heat our homes, make our automobiles, build our roads, and produce bottles, radios, and all the rest of the myriad objects we use. On the average, 10 metric tons* of material a year are dug up and used one way or another for every man, woman, and child on the Earth. Since there are 5.2 billion people on the Earth, the total yearly amount dug up is 52 billion tons. Compare this enormous figure with the 16.5 billion tons of dissolved and suspended matter carried each year to the sea by all the rivers of the world.

The human population is now so large, and our collective activities so pervasive, that there is practically nothing left on the Earth we haven't changed. We will go almost everywhere to seek the resources we use (Fig. I.3). We have made rainfall more and more acidic, we have caused fertile top soil to blow away, and we have changed the composition of the soil that remains (Fig. I.4). We have made deserts expand, and changed the composition of the atmosphere, the ocean, streams, and lakes. We have, in short, changed our own environment, and we continue to do so at ever faster rates.

Most human activities have made the world a nicer and friendlier place in which to live. No one could deny that building cities and clearing land for farms causes large changes in the environment. But who

FIGURE I.2 The aftermath of the Loma Prieta earthquake. Damaged houses in the Marina district, San Francisco.

---

*A metric ton is 1000 kg; SI units are used throughout this book. There is a discussion of SI units in appendix A.

FIGURE I.3 Exploring for oil beneath arctic seas. A huge drilling platform in the Beaufort Sea, fifteen kilometers off the northern coast of Alaska. The platform is home to geologists and drillers who are testing possible oil-bearing rocks beneath the seafloor. The jet of seawater coming from the platform (left) is building an ice island to protect the drilling platform from icebergs.

would argue that a beautiful city like Paris is not a proud achievement? Think, too, of the abundant food that flows from modern agriculture. Our ancestors had a much harder time feeding themselves than we do today. The dangerous changes to the environment happened accidentally because we don't understand the Earth's processes sufficiently well. When we started burning coal 300 years ago, no one had the slightest idea that someday the atmosphere might be

changed as a result. But now we do know, although we still do not know what all the consequences will be. For example, if the climate becomes warmer because of changes to the atmosphere, ice in Antarctica might melt, the sea level might rise, and cities might be flooded. Surely those are hazardous consequences we must consider. But note that in each case we say the consequence *might* happen; might, because we do not yet understand the Earth processes in enough detail to be sure. Just as we seek to understand and monitor natural hazards like volcanoes and earthquakes, therefore, so we seek to understand the consequences of our collective human activities.

ENERGY AND THE DYNAMIC EARTH

FIGURE I.4 Corn fields being plowed in Michoacan, Mexico. The wind blows away some of the topsoil when the surface of the soil is disturbed by plowing. Season after season, the rich soil is lost and the fertility slowly declines.

Rain and wind are things that happen in the atmosphere, so are hurricanes. It doesn't take much imagination to realize that everyday wind and rain must somehow be related to hurricanes. Wind, rain, and hurricanes take place because energy in the form of heat is absorbed by the atmosphere. Hurricanes require a huge amount of heat energy. Weather, including hurricanes, changes with the season, and the seasons, we know, are controlled by the Earth's position with respect to the Sun. This straightforward line of reasoning leads to the conclusion that the energy that makes winds blow, rain fall, and hurricanes form must somehow come from the Sun.

Energy from the Sun reaches the Earth as heat rays. You are feeling the effect of heat rays when you sit in front of a fire. When heat rays hit a solid surface or pass through a gas or liquid, some of the heat is absorbed and the temperature of the absorbing body rises.

Approximately 60 percent of the Sun's heat that reaches the Earth is absorbed by the land, the sea, or the atmosphere. (The remaining 40 percent is simply reflected back into space.) The heat rays absorbed by the sea warm the water and cause evaporation. The resulting water vapor forms clouds and eventually rain, snow, sleet, or hail. The heat rays absorbed by the land warm the exposed rocks and soil. Warm seawater, warm rocks, and warm soil heat the air. Warmed air expands, becomes less dense, and rises. Then cool air flows in to take the place of the rising air. Flowing air is wind, and as winds blow over the sea, they create waves. Thus, the familiar everyday processes that happen at the Earth's surface—rain, streams, winds, waves, even glaciers—are produced by the Sun's energy.

Volcanic eruptions and earthquakes, unlike winds, are unrelated to the Sun's energy output. No matter how hot it gets on a summer's day, the Sun's heat is insufficient to melt rocks, and even in frigid Antarctica there are active volcanoes. Earthquakes can happen anywhere on the Earth, no matter how hot or cold the climate may be. We therefore conclude that the energy to form the molten lava that spews from a volcano and the energy that shakes the Earth during an earthquake must come from somewhere inside the Earth.

This conclusion is not difficult to test. If you were to go down into a mine and measure rock tempera-tures, you would find that the deeper you went, the higher the temperature. The Earth's internal heat is the source of the energy that causes earthquakes and volcanoes. It is also the source of the energy that causes mountains to rise and continents to move. You know that if you let a hot cup of coffee stand on the table, it will slowly cool down. The reason it cools is because heat rays from the hot cup radiate out and warm the cool air around the cup. One of the fundamental laws of nature is that heat always flows from a hot place to a cold one. Thus, heat must flow outward from the hot interior of the Earth toward the cool surface. Careful measurements made in mines and drill holes around the world show that the rate of temperature increase with depth (called the **geothermal gradient**) varies from place to place, ranging from 15° to 75°C/km. By extrapolation, we calculate that the temperature at the center of the Earth must be at least 5000°C.

The process by which heat can move through solid rock, or any other solid body, without deforming the solid, is called **conduction**. Heat conduction is a familiar process. It is the way heat moves along the handle of a hot saucepan. Conduction does not cause the movement of hot material from one place to another. But because volcanoes and earthquakes obviously involve movement of hot material, we have to conclude that heat can move in the Earth by another process in addition to conduction. **Convection**, unlike conduction, does cause movement. Convection is the process by which hot, less-dense materials rise upward and are replaced by cold, downward- and sideways-flowing materials to create a **convection current** (Fig. I.5). Wind is an example of convection.

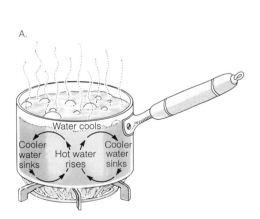

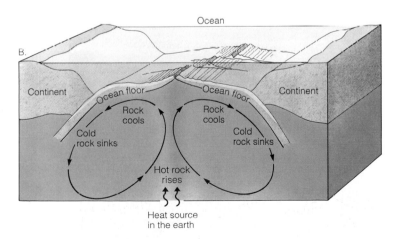

FIGURE I.5 Convection shapes the Earth's surface. A. Convection in a saucepan full of water. Water that is heated expands and rises. As it rises, it starts to cool, flows sideways, and sinks, eventually to be reheated and pass again through the convection cell. B. Convection as it is thought to occur in the Earth. Though much slower than convection in a saucepan, the principle is the same. Hot rock rises slowly from deep inside the Earth, cools, flows sideways, and sinks. The rising hot rock and sideways flow are believed to be the factors that control the positions of ocean basins and continents.

repeated but sporadic events, there is absolutely no evidence to suggest that similar events will not occur again. Nor is there any evidence to suggest when another might occur.

A fascinating but frightening suggestion has been made that a disaster of a different kind may already be happening. The suggestion is that our collective human activities may be changing the Earth so rapidly, and so massively, that we may be causing a change similar in magnitude to some of the major ones in the geological record. At present the suggestion is only a hypothesis. It remains to be tested, and thereby proved or disproved. Nevertheless, the very fact that serious scientists are concerned that the hypothesis might be true emphasizes an important fact: Geology and the welfare of the human race are indissolubly linked.

ABOUT THIS BOOK

The scope of this text is all of physical geology, but four main themes are emphasized:

1. The connective link between internal convection and the Earth's external features through plate tectonics.
2. The influence of the human race on the Earth's external processes and environments.
3. The mitigation of natural disasters.
4. The need for humans to be wise in their use of the Earth's limited store of natural resources.

The first theme, plate tectonics, emphasizes the most important scientific theory to arise from geological investigations in the twentieth century. The second, third, and fourth themes all concern the human race. The second focuses on the well-being of our environment, the third on our protection from natural disasters, and the fourth on our needs and the supplies of resources.

Each chapter ends with three features designed to help you assimilate the material that has been covered: a brief summary, a list of key terms, and a series of questions based on material in the chapter. The following are examples of these features for this Introduction.

SUMMARY

1. Geology is the study of changes, past and present, that happen to the Earth.
2. The Sun's heat energy drives Earth's external processes.
3. The collective effects of human activities influence the Earth's external processes in many ways.
4. The Earth's internal heat energy drives internal processes.
5. Internal heat reaches the Earth's surface by both conduction and convection. The slow convective motions inside the Earth cause plate tectonics and determine the shapes and locations of the Earth's surface features.
6. The Principle of Uniformitarianism states that the internal and external processes operating today have been operating throughout Earth's history.
7. Random massive disasters, such as gigantic meteorite impacts, likely have played an important role in the Earth's history.

IMPORTANT TERMS TO REMEMBER

catastrophism (p. 7)
conduction, of heat (p. 5)
convection, of heat (p. 5)
convection current (p. 5)

erosion (p. 6)

geology (p. 2)
geologists (p. 2)
geothermal gradient (p. 5)

historical geology (p. 2)

physical geology (p. 2)
plate tectonics (p. 6)
Principle of Uniformitarianism (p. 7)

rock (p. 2)

QUESTIONS FOR REVIEW

1. Suggest three human activities that affect the Earth's external processes in a detectable way.

2. Identify three human activities in the area where you live that are causing big changes to the environment.

3. How does the Earth's internal heat energy influence the Earth's surface features?

4. How does the Principle of Uniformitarianism help us understand the history of the Earth?

5. Explain the conflict between the Earth's two gigantic energy sources.

PART I

Coal mine, Wyodak, Wyoming

The Earth's Materials

Stratovolcanoes, Ecuador

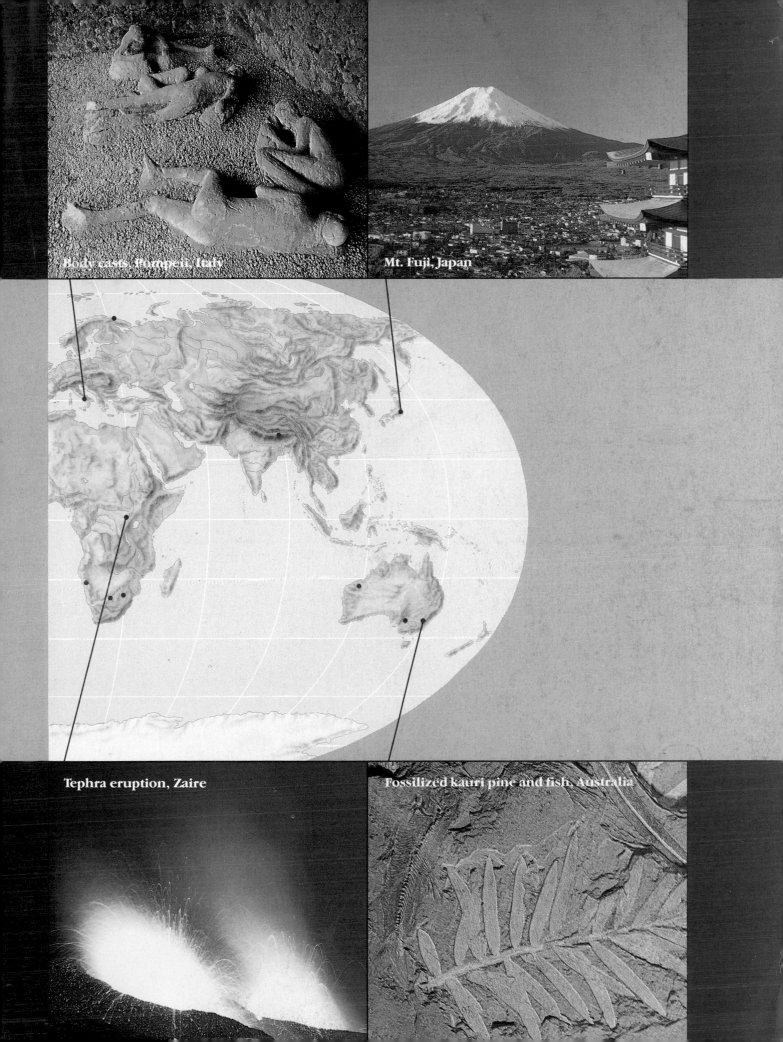

Body casts, Pompeii, Italy

Mt. Fuji, Japan

Tephra eruption, Zaire

Fossilized kauri pine and fish, Australia

A time exposure of stars viewed over Mount Everest, Earth's mightiest mountain, which was formed as a result of the plate tectonic collision between India and Asia. Star trails are curved due to the Earth's rotation.

CHAPTER 1

Planet Earth

N icolaus Copernicus (1473–1543) was one of the boldest thinkers who has ever lived. He was a sixteenth-century Catholic priest who was active at a time when the official doctrine of the Catholic Church was that the Sun and planets revolved around the Earth. Copernicus challenged church doctrine and postulated that the planets revolve around the Sun. He used the crude astronomical measurements of his time to prove the theory. To avoid open conflict with his church superiors, Copernicus delayed publication of his great discovery until just before his death. His work aroused great interest among European intellectuals because his conclusions were based on measurements that others could repeat. Looking back, we can now see that by using the scientific method to test his theory, Copernicus sowed the seeds of the scientific revolution that continues today.

Among those impressed by the Copernican theory was an Italian mathematics and astronomy teacher, Galileo Galilei (1564–1642). Galileo was the first person to view the planets and our Moon through a telescope. The year was 1609, and his homemade device was crude by comparison even with a toy telescope of today. Galileo was astonished to see mountains on the Moon and large flat areas that looked to him like seas. He observed several moons circling Jupiter, and he saw disc-shaped rings around Saturn. The discoveries electrified Galileo's contemporaries. His discoveries, Galileo realized, were at odds with the doctrine that all objects in the solar system circled the Earth. If the doctrine was right, how could moons be circling Jupiter? Galileo realized that Copernicus was right and the theologians were wrong. Galileo, like Copernicus, used the scientific method of observation and measurement to draw his conclusions, but even so he was twice called before the Inquisition. Eventually, in 1635, he was forced to renounce his beliefs.

Today we know that Copernicus and Galileo were correct. The Earth and the planets orbit the Sun, and moons orbit the planets.

THE SPECIAL PLANET

Visits to the Moon by astronauts and images of distant planets sent back by unmanned spaceships have astonished the present generation, just as Galileo's discoveries astonished his contemporaries.

But surely the most remarkable spaceship image brought back by the astronauts is of the Earth itself. For the first time people can see our planet in one sweeping view: see its clouds, oceans, polar ice caps, and continents at the same time. Planet Earth, we can see, is just a small planet in orbit around an ordinary, medium-sized star. But the Earth is a special planet, and that too can be seen. First, the Earth has an overall blue and white hue because it is surrounded by an **atmosphere** of gases, predominantly nitrogen, oxygen, carbon dioxide, and water vapor. No other planet in the solar system has such an atmosphere. In the Earth's atmosphere are white clouds of condensed water vapor. The clouds form because water evaporates from the **hydrosphere** (the "water sphere"), which consists of the world's oceans, lakes, streams, underground water, and all the snow and ice, including glaciers. The hydrosphere is a second reason the Earth is a special place. Planets farther from the Sun are too cold for water to exist as both a liquid and a gas (water vapor), and planets closer to the Sun are too hot. Other planets have hydrospheres, but only the Earth has a hydrosphere consisting of water, ice, and water vapor.

A third reason the Earth is special is the **biosphere** (the "life sphere"), which is the totality of the Earth's living matter. Viewed from space, the biosphere is most dramatically revealed by blankets of green plants on some of the land masses (Fig. 1.1). The biosphere embraces innumerable living things, large and small, grouped into millions of different species. We humans are one of the species. In addition, the biosphere includes dead plants and animals that have not yet been completely decomposed.

A fourth reason the Earth is special concerns its solid surface. Regions that are not green because of dense plant cover appear brown and weather-beaten. The reason is **weathering**, which is the chemical alteration and mechanical breakdown of rock during exposure to the atmosphere, hydrosphere, and biosphere. As a result of weathering the Earth is covered by an irregular blanket of loose rock debris termed the **regolith**. Soils, muds in river valleys, sands in the desert, and all the other friable rock debris are part of the regolith. Most of the world's plants and animals live either on or in the regolith (we live on it, worms live in it), or in the hydrosphere (fish and frogs).

What an image from space makes so obvious is that the environment in which we live is a very thin, very tenuous envelope around Planet Earth. It is an envelope that depends on the involvements of all the external spheres—atmosphere, hydrosphere, biosphere, and regolith. We humans tamper with this envelope at our peril.

THE SOLAR SYSTEM

The solar system consists of the Sun, nine planets, 61 known moons, a vast number of asteroids, millions of comets, and innumerable small fragments of rock and dust called meteoroids. All of the objects in the solar

FIGURE 1.1   The color of the Earth as seen from space. A composite of numerous satellite images showing densely vegetated regions in green, dry deserts in yellow or brown, and ice-covered regions in white.

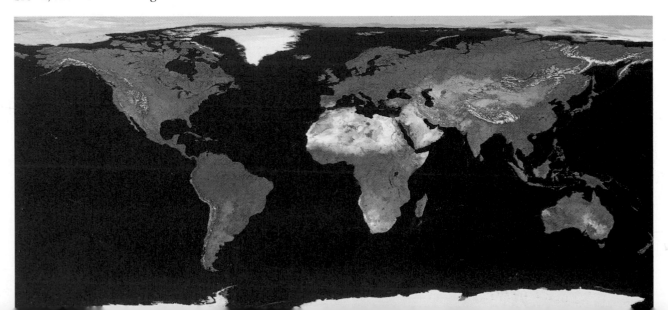

system move through space in smooth, regular orbits, held in place by gravitational attraction. The planets, asteroids, and meteoroids all circle the Sun, while the moons circle the planets.

The distances between the planets are so immense it is difficult to comprehend them. To put the solar system into perspective, think of the Sun as a basketball. The nearest planet, Mercury, would be a speck of dust about 12 m away. The Earth would be a grain of sand about l mm in diameter and 30 m away, Saturn a grape nearly 300 m away, and Pluto, the most distant planet, another grain of sand 1200 m away.

Terrestrial Planets

The planets can be separated into two groups based on their densities and closeness to the Sun (Fig. 1.2A).

The innermost planets, Mercury, Venus, Earth, and Mars, are small, rocky, and dense (Fig. 1.2B). Each has a density of 3 g/cm^3 or more. They are similar in composition and are called the *terrestrial planets* because they are similar to *terra* (the Latin word for Earth).

Jovian Planets

The planets farther from the Sun than Mars are much larger than the terrestrial planets (with the exception of Pluto), yet much less dense. The masses of Jupiter and Saturn, for example, are 317 and 95 times the mass of the Earth, but their densities are only 1.3 and 0.7 g/cm^3, respectively. These *jovian planets*—Jupiter, Saturn, Uranus, Neptune, and Pluto—take their name from *Jove*, an alternate designation for the Roman god

FIGURE 1.2 The planets and their properties. A. The planets, shown in their correct relative sizes and in the correct order outward from the Sun. The Sun is 1.6 million km in diameter, 13 times larger than Jupiter, the largest planet. B. Numerical data concerning the orbits and properties of the planets.

A.

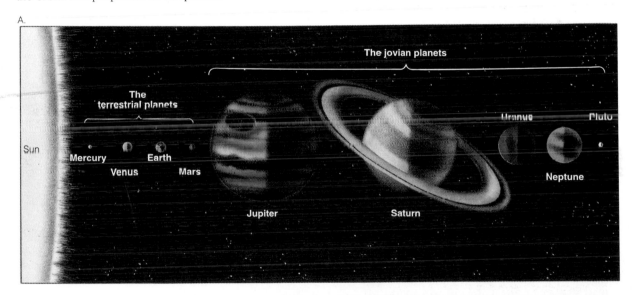

B.

	Mercury	Venus	Earth	Mars	Jupiter	Saturn	Uranus	Neptune	Pluto
Diameter (km)	4880	12,104	12,756	6787	142,800	120,000	51,800	49,500	6000
Mass (Earth=1)	0.055	0.815	1	0.108	317.8	95.2	14.4	17.2	0.003
Density, g/cm^3 (water=1)	5.44	5.2	5.52	3.93	1.3	0.69	1.28	1.64	2.06
Number of moons	0	0	1	2	16	18	15	8	1
Length of day (in Earth hours)	1416	5832	24	24.6	9.8	10.2	17.2	16.1	154
Period of one revolution around Sun (in Earth years)	0.24	0.62	1.00	1.88	11.86	29.5	84.0	164.9	247.7
Average distance from sun (millions of kilometers)	58	108	150	228	778	1427	2870	4497	5900

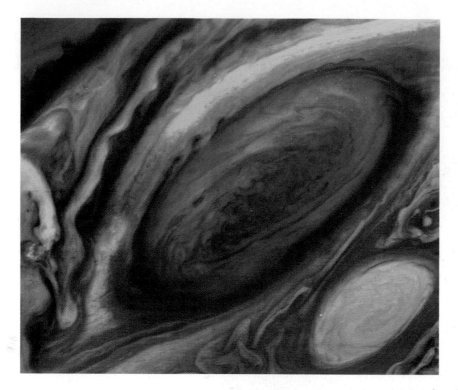

FIGURE 1.3 A gigantic hurricane-like storm has raged for centuries in the atmosphere of Jupiter. First reported by Galileo, the Great Red Spot, as the storm is called, has a diameter twice that of Earth. No adequate explanation exists to explain either the huge size or the long-lived duration of the storm. This extraordinary image was recorded by the *Voyager 2* on July 6, 1979, from a distance of 2,633,000 km.

Jupiter. They all probably have solid centers that resemble terrestrial planets, but, with the exception of Pluto, most of their planetary mass is contained in a thick atmosphere of hydrogen, helium, and other gases. It is the thick atmospheres that we can see and which keep the densities of the jovian planets low (Fig. 1.3).

The Origin of the Solar System

How did the solar system form? We may never know the exact answer to this question, but we can discern the outlines of the process from evidence obtained by astronomers, from our knowledge of the solar system today, and from the laws of physics and chemistry.

The birth throes of our Sun and its planets were similar to those of many other suns. Birth began with space that was not entirely empty. Space was not empty because an earlier star had exploded in what astronomers call a supernova. The explosion scattered atoms of various elements everywhere through a huge volume of space. Most of the atoms were hydrogen and helium, but small percentages of all the other chemical elements were present too. Even though thinly spread, the atoms formed a tenuous, turbulent, swirling cloud of cosmic gas. Over a very long period the gas thickened as a result of a slow gathering of all the thinly spread atoms. The gathering force of the gas was gravity, and as the atoms slowly moved closer together, the gas became hotter and denser. Near the

center of the gathering cloud of gas, hydrogen and helium atoms eventually became so tightly pressed and so hot that they began to fuse to form heavier elements. When, in the gas cloud that formed the solar system, fusion of hydrogen and helium commenced, the Sun was born. The time is estimated to have been about 6 billion years ago.

At some stage the cool outer portions of the cosmic gas cloud became compacted enough to allow solid objects to condense, in the same way that ice condenses from water vapor to form snow (Fig. 1.4). The solid condensates eventually became the planets, moons, and the other solid objects of the solar system.

Planets and moons nearest the Sun, where the temperatures were highest, contain only compounds that can condense at high temperatures. Those compounds consist of chemical elements such as iron, silicon, magnesium, and aluminum; we call them refractory elements, which are elements that form solid compounds at high temperatures. Planets and moons distant from the sun, where temperatures were lower, contain not only refractory elements but also large quantities of volatile elements, such as hydrogen and sulfur, which do not condense at high temperatures and tend to form compounds that are solid only at low temperatures (Fig. 1.5A). The farther away from the Sun the condensation occurred, the lower was the temperature and the greater the fraction of volatile elements. One striking demonstration of this fact is the large amount of hydrogen, present mainly as ice, in the moons of the jovian planets (Fig. 1.5B).

FIGURE 1.4 The gathering of atoms in space created a rotating cloud of dense gas. The center of the gas cloud eventually became the sun; the planets formed by condensation of the outer portions of the gas cloud.

Condensation of the gas cloud is only the first part of the planetary birth story. Condensation formed innumerable small rocky fragments, but the fragments had still to somehow be joined together to form a planet. This happened by impacts between fragments drawn together by gravitational attraction. The largest masses slowly swept up more and more of the condensed rocky fragments, grew larger, and became the planets. Meteorites, such as the one in Figure 1.6, still fall on the Earth, proving that even now some ancient, condensed rocky fragments still exist in space. Some meteorites resemble rocks formed on the Earth, and these meteorites are believed to have been ejected from a planet or a moon as a result of a large meteorite impact, such as the one that formed Meteor Crater

(see Fig. 1.9). Meteorites and the scars of ancient impacts provide evidence of the way the terrestrial planets grew to their present sizes. The growth process—a gathering of more and more bits of solid matter from surrounding space—is called *planetary accretion*.

Planetary Accretion: Still Going On!

Has planetary accretion ceased? The answer, clearly, is no, because a large (but not gigantic) impact formed Meteor Crater only 20,000 years ago, and in 1912 a very large explosion occurred in a remote area of Siberia called Tunguska—it was the largest explosion due to impact in historic times, and so it is fortunate

A.

B.

FIGURE 1.5   Two moons of Jupiter that are rich in volatile elements. A. Io is red colored because it is rich in sulfur. The image shows a volcanic eruption on Io. The volcanic plume is mostly gas, but small solid particles are also distributed by the gas. The plume rises to a height of 100 km above the surface of Io and is believed to be largely sulfur dioxide (SO_2). Several sites of active volcanism have been discovered on Io.

B. The surface of Europa, smallest of the four large moons of Jupiter, is mantled by ice to a depth of 100 km. The fractures indicate that some internal process, probably convection, must be renewing the surface. The dark material in the fractures apparently rises up from below. The cause of the fracturing is not known. The image was taken by *Voyager 2* in July 1979.

FIGURE 1.6   A messenger from space carrying some of the history of the earliest days of the solar system. This stony meteorite fell to the Earth at Pueblito de Allende, Mexico, in 1969.

that it occurred in a relatively uninhabited corner of the world. What such impacts prove is that the Earth and the other planets even today grow larger, albeit exceedingly slowly. Accretion in the early history of the Earth was probably a much more rapid affair.

The velocities of small meteorites entering the Earth's atmosphere have been measured between 4 and 40 km/s. If a large meteorite had such a velocity, the amount of energy released on impact would be enormous. It has been calculated, for example, that a meteorite 30 m in diameter and traveling at a speed of 15 km/s would, on impact, release as much energy as the explosion of 4 million tons of TNT. The resulting impact crater would be the size of Meteor Crater in Arizona—1200 m across and 200 m deep (Fig. 1.9). Cratering is a very rapid geological process; the Meteor Crater event is estimated to have lasted about 1 min. We will discuss impact cratering in greater detail in chapter 19.

THE INTERNAL STRUCTURE OF THE EARTH

Scientists believe that as the terrestrial planets grew larger, their temperatures must have risen. The reason is straightforward: Energy can be changed from one form to another (from electricity to heat, for example), but it cannot be destroyed. A moving object has energy of motion (called *kinetic energy*), and when a meteorite impacts a planet, the kinetic energy is transformed to heat. As planet Earth grew larger and larger, the continual impacts would necessarily have raised its temperature.

In addition, heat must continually have been added from another source. Among the many chemical elements in the Earth are several that are naturally radioactive; that is, they spontaneously transform to another element. Examples are uranium, thorium, and potassium, which transform to lead, lead, and argon respectively. But every time a radioactive transformation occurs, a tiny amount of heat is also produced. Therefore, radioactivity continued to heat the Earth even as the frequency of impacts declined. Eventually, the Earth began to melt. Lighter melted materials, rich in silicon, aluminum, sodium, and potassium, rose toward the surface. Rocks at the Earth's surface are still rich in these elements. Denser melted materials, such as molten iron, sank to the center of the planet. The melting released volatile elements, and these escaped as gases through volcanoes. It was the escaped gases, mainly water vapor, carbon dioxide, methane, and possibly ammonia, that gave rise to the Earth's atmosphere. From the same source came the water we now find in the Earth's oceans. Partial melting changed the Earth from an originally homogeneous planet to a compositionally layered one.

Layers of Differing Composition

Planet Earth contains three compositional layers (Fig. 1.7). At the center is the densest of the three layers, the **core**. The core is a spherical mass, composed largely of metallic iron, with lesser amounts of nickel and other elements.

The thick shell of dense, rocky matter that surrounds the core is called the **mantle**. The mantle is less dense than the core but denser than the outermost layer. Above the mantle lies the thinnest and outermost layer, the **crust**, which consists of rocky matter that is less dense than the rocks of the mantle below.

It is apparent in Figure 1.7 that the core and the mantle have nearly constant thicknesses. The crust is far from uniform, though, and differs in thickness from place to place by a factor of nine. The crust beneath the oceans, the **oceanic crust**, has an average thickness of about 8 km, whereas the **continental crust** averages 45 km and ranges in thickness from 30 to 70 km.

We cannot see and sample either the core or the mantle, and so it is valid to ask how we know anything about their composition. The answer is that indirect measurements are used. One way to determine composition is to measure how the density of rock changes with depth below the Earth's surface. We can do this by measuring the speeds with which earthquake waves pass through the Earth because they move more quickly through dense rocks than through less-dense rocks (see chapter 15). From such measurements, we know that density increases with depth, but not smoothly. At some depths abrupt velocity increases indicate sudden increases in density. From the sudden increases, we can infer that the solid Earth does not have a uniform composition but must instead consist of distinct layers with different densities. Knowing these different densities, we can estimate what the composition of the different layers must be.

Slight compositional variations probably exist within the mantle, but we know little about them. We can see and sample the crust, however, and the sampling shows that even though it is quite varied in composition, the crust's overall composition and density are very different from those of the mantle, and the boundary between them is distinct.

The composition of the core presents the most difficulty. The temperatures and pressures in the core are so great that materials there probably have unusual properties. Some of the best evidence concern-

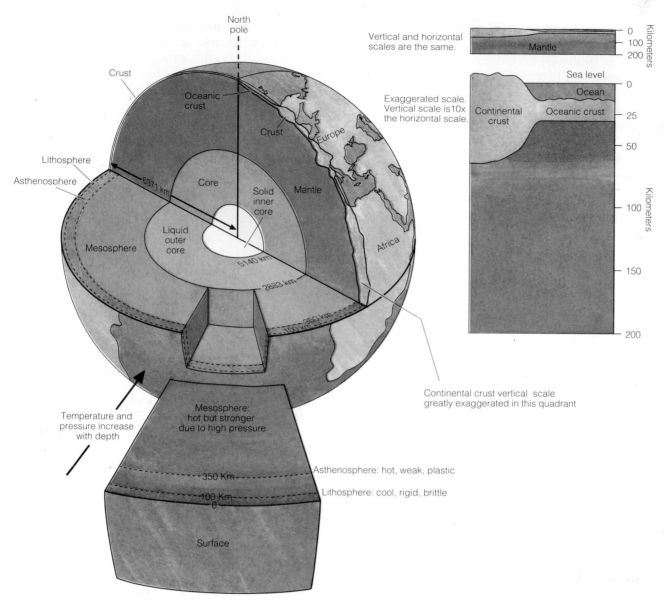

FIGURE 1.7 A sliced view of the Earth reveals layers of different composition and zones of differing rock strength. The compositional layers, starting from the inside, are the core, the mantle, and the crust. Note that the crust is thicker beneath the continents than under the oceans. Note too, that boundaries between zones of differing strength— lithosphere (outermost), asthenosphere, mesosphere—do not coincide with compositional boundaries.

ing core composition comes from iron meteorites. Such meteorites are believed to be fragments from the core of a small terrestrial planet that was shattered by a gigantic impact early in the history of the solar system. Scientists presume that this now-shattered planet must have had compositional layers similar to those of the Earth and the other terrestrial planets.

Layers of Differing Physical Properties

In addition to compositional layering, other changes occur within the Earth. Most important, there are changes of physical properties such as rock strength and solid versus liquid. Changes in physical properties are largely controlled by temperature and pressure rather than rock composition. The places where physical properties change do not coincide exactly with the compositional boundaries between the crust, mantle, and core shown in Figure 1.7.

The Inner and Outer Core

Within the core an inner region exists where pressures are so great that iron is solid despite its high temperature. The solid center of the Earth is the **inner core**. Surrounding the inner core is a zone where temperature and pressure are so balanced that the iron is molten and exists as a liquid. This is the **outer core**. The difference between the inner and outer cores is not one of composition (the compositions are believed to be the same). Instead, the difference lies in the physical states of the two: one is a solid, the other a liquid.

The Mesosphere

The strength of a solid is controlled by both temperature and pressure. When a solid is heated, it loses strength. When it is compressed, it gains strength. Differences in temperature and pressure divide the mantle and crust into three distinct strength regions. In the lower part of the mantle, the rock is so highly compressed that it has considerable strength even though the temperature is very high. Thus a solid region of high temperature but also relatively high strength exists within the mantle from the core-mantle boundary (at 2883 km depth) to a depth of about 350 km and is called the **mesosphere** ("intermediate, or middle, sphere") (Fig. 1.7).

The Asthenosphere

Within the upper mantle, from 350 to about 100 km below the Earth's surface, is a region called the **asthenosphere** ("weak sphere"), where the balance between temperature and pressure is such that rocks have little strength. Instead of being strong, like the rocks in the mesosphere, rocks in the asthenosphere are weak and easily deformed, like butter or warm tar. So far as geologists can tell, the compositions of the mesosphere and the asthenosphere are the same. The difference between them is one of physical properties; in this case the property that changes is strength.

The Lithosphere

Above the asthenosphere, and corresponding approximately to the outermost 100 km of the Earth, is a region where rocks are cooler, stronger, and more rigid than those in the plastic asthenosphere. This hard outer region, which includes the uppermost mantle and all of the crust, is called the **lithosphere**

("rock sphere"). It is important to remember that despite the fact that the crust and mantle differ in composition, it is rock strength, not rock composition, that differentiates the lithosphere from the asthenosphere.

The boundary between the lithosphere and the asthenosphere is again caused by differences in the balance between temperature and pressure. Rocks in the lithosphere are strong and can be deformed or broken only with difficulty; rocks in the asthenosphere below can be easily deformed. One analogy is a sheet of ice floating on a lake. The ice is like the lithosphere, the lake water is like the asthenosphere.

As we will see shortly, the lithosphere is not a continuous layer. Instead, like ice on a lake during a thaw, the lithosphere is broken into a number of huge plates.

PLATE TECTONICS AND THE EXTERNAL STRUCTURE OF THE EARTH

The Earth's internal convection is always moving the plates of lithosphere and changing the Earth's surface. Mountains like the Alps or Appalachians that seem changeless to us are only transient wrinkles when viewed in geological time. Mountain ranges grow when fragments of moving lithosphere collide and heave masses of twisted and deformed rock upward, then the ranges are slowly worn away, leaving only the eroded roots of an old mountain range to record the ancient collision. The Earth's long dynamic history is recorded in rocky scars such as those in Figure 1.8.

A remarkable story that has emerged from geological studies of the Earth's processes is that the continents themselves are still slowly moving. They are drifting sideways at rates up to 10 cm/yr, sometimes bumping into each other and creating a new mountain range by collision, and sometimes splitting apart so that a new ocean basin forms. The Himalaya is a range of geologically young mountains that began to form when the Indian subcontinent collided with Asia about 45 million years ago. The Red Sea is a young ocean that started forming about 30 million years ago when a split developed between the Arabian Peninsula and Africa as the two land masses began to move apart.

But it is not just the continents that move, it is the entire lithosphere. The continents, the ocean basins, and everything else on the surface of the Earth are moving along like passengers on large rafts; the rafts are huge plates of lithosphere that float on the asthenosphere. As a result, all the major features on the Earth's surface, whether submerged beneath the sea or exposed on land, arise as either a direct or indirect

FIGURE 1.8 The scar of an ancient collision. Layers of rock, once horizontal, were twisted and contorted as a result of a collision between two plates. These eroded roots of an ancient mountain range north of Adelaide, South Australia, were recorded in a LANDSAT image in September 1983.

result of the lithosphere drifting on the asthenosphere. Such motions involve complicated events, both seen and unseen, all of which are embraced by the term *tectonics*.

Tectonics

The word *tectonics* is derived from a Greek word, *tekton*, which means carpenter or builder. **Tectonics** is the study of the movement and deformation of the lithosphere.

The special branch of tectonics that deals with the processes by which the lithosphere is moved laterally over the asthenosphere is called **plate tectonics**. The term *plate* is used because the lithosphere moves as a number of separate, platelike pieces. The plates range from several hundred to several thousand kilometers in width (Fig. 1.9).

Plate tectonics was proposed as a theory only in the 1960s. Many details are still inadequately understood and still being investigated. But the discoveries and new understanding that have already come from studies made to test the plate tectonics theory are so profound that the theory has sparked a modern geological

revolution. Necessarily, a lot of attention is paid to plate tectonics in this book, and so it is helpful at this point to introduce the concept and to indicate briefly how it controls the Earth's surface features. First, let's consider the Earth's major surface features.

Continents and Ocean Basins

The ocean covers some 71 percent of the world's surface, and its average depth is 3.7 km. The depth is very irregular, however; the greatest depth—11 km—is near the island of Guam, in the western Pacific.

The remaining 29 percent of the world's surface is occupied by land, which has an average height of 0.8 km above mean sea level. If it were possible to remove all the water from the ocean and then view the dry Earth from a spaceship, we could contrast the ocean basins and the continents. We would see that the continents stand, on average, about 4.5 km above the floor of the ocean basins (Fig. 1.10). The continents stand higher than the ocean basins because continental crust is relatively light (density 2.7 g/cm^3), while oceanic crust is relatively heavy (density close

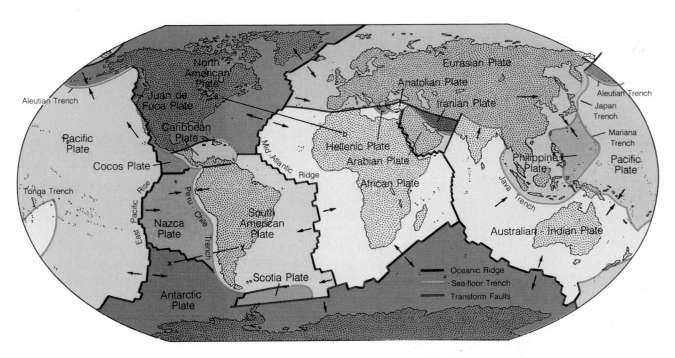

FIGURE 1.9 Six large plates of lithosphere and several smaller ones cover the Earth's surface and move steadily in the directions shown by the arrows. The profile shown in Fig. 1.11 lies along the line a–b, that in Fig. 1.13 lies along the line x–y.

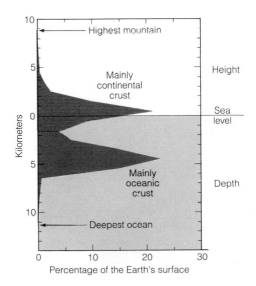

FIGURE 1.10 Distribution of the areas of the Earth's solid surface above and below sea level, expressed as a percentage. Note that areas underlain by continental crust are considerably higher than areas underlain by oceanic crust.

to 3.2 g/cm³). Because the lithosphere is floating on the asthenosphere, those portions of the lithosphere capped by light continental crust stand high while those capped by heavy oceanic crust sit lower.

The Shape of Ocean Basins

Modern shorelines don't coincide exactly with the boundaries between continental crust and oceanic crust. This is so because some ocean water spills out of the ocean basin onto the continent (Fig. 1.11). The boundaries between continental and oceanic crust are therefore covered by water, and today's shorelines are actually on the continents. As a result, each continent is surrounded by a flooded margin of variable width known as the **continental shelf**. The geological edge of the ocean basin is not the shoreline: rather, it is the place where oceanic crust joins the continental crust. The geological edge is at the bottom of the **continental slope**, a pronounced slope beyond the seaward margin of the continental shelf. If, instead of today's shoreline, we take the bottom of the continental slope to be the boundary of the continents, only 60 percent

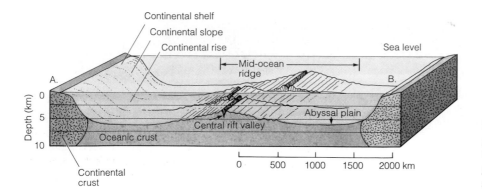

FIGURE 1.11 Simplified diagram of a portion of the Atlantic Ocean showing the major topographic features. The profile is along the line a–b in Fig. 1.9.

of the Earth's surface is occupied by ocean basins, while 40 percent is occupied by continents. Thus 25 percent of the continental crust is covered by seawater (Fig. 1.12).

The **continental rise** lies at the base of the continental slope. It is a region of gently changing slope where the floor of the ocean basin meets the margin of the continent. The rise is actually part of the floor of the ocean basin, but it is a distinctive part because it is underlain by oceanic crust and covered by a thick pile of erosional debris shed from the adjacent continent.

Some continental margins coincide with the edges of tectonic plates. Other continents sit in the middle

of plates and their margins are far from plate edges. Regardless of today's configurations, the margins of all continents have, at some time in the geological past, coincided with plate margins.

Beyond the continental slope and rise lies the strange, rarely seen world of the deep ocean floor. Teams of oceanographers and seagoing geologists, using new devices, have sounded and sampled the ocean bottom during submarine dives of limited duration. As a result of this work, today we know almost as much about the seafloor as we know about the land surface.

The large, flat areas known as the **abyssal plains** are a major topographic feature of the seafloor and

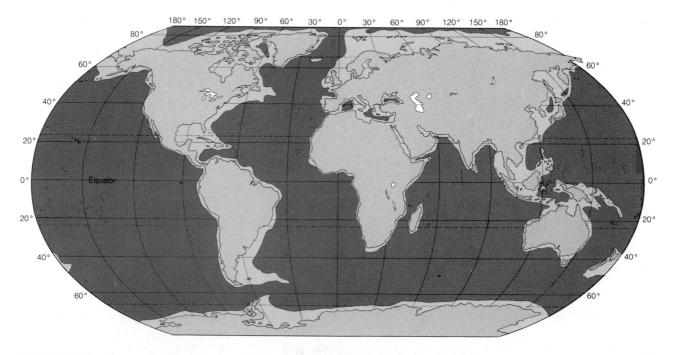

FIGURE 1.12   The continental shelves and slopes (shown in light blue) together form about 25 percent of the continents.

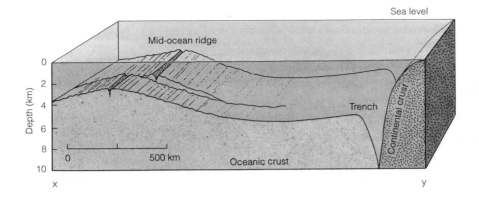

FIGURE 1.13 Simplified diagram of a section across a portion of the Pacific margin of South America. The profile is along the line x–y in Fig. 1.9. Note that the side of the trench adjacent to the continent (to the right) is steeper than the oceanic side.

lie adjacent to the continental rise (Fig. 1.11). They generally are found at depths of 3 to 6 km below sea level and range in width from about 200 to 2000 km. Plains are most common in the Atlantic and Indian oceans, which have large, mud-laden rivers entering them. Abyssal plains form as a result of the mud settling through the ocean water and burying the original seafloor topography beneath a blanket of fine debris.

Two particularly prominent features of the ocean floor are (1) **oceanic ridges** (also referred to as **mid-ocean ridges** or **oceanic rises**), which are rocky ridges on the ocean floor, tens of thousands of kilometers long, many hundreds of kilometers wide, and standing at heights of 0.6 km or more above the seafloor; and (2) **trenches**, which are long, narrow, deep basins in the seafloor (Fig. 1.13).

The oceanic ridge system is a chain of mountains some 84,000 km long that twists and branches in a complex pattern through the ocean basins. This great mountain chain would be one of the most impressive features we would see if we could view a dry Earth from out in space.

A narrow valley, or rift, runs down the center of all oceanic ridges. The rifts are characterized by intense volcanic activity. At several places around the world, the oceanic ridge with its central rift reaches sea level and forms volcanic islands. The largest of these is Iceland, which lies on the center of the Mid-Atlantic Ridge (Fig. 1.14).

Plate Motions

Today the lithosphere is broken into six large plates and numerous smaller ones (Fig. 1.9), all moving at speeds ranging from 1 to 12 cm a year. As a plate moves, everything on it moves too. If the plate is capped partly by oceanic crust and partly by continental crust, then both the ocean floor and the continent move with the same speed and in the same direction.

The hypothesis that the ocean floor might be moving was first proposed in the early 1960s and was one of the key steps that led to the theory of plate tectonics in 1967. But the suggestion that continents move goes

FIGURE 1.14 Long, deep fractures that split Iceland mark the center of a midocean ridge. Iceland is on the Mid-Atlantic Ridge and is one of the few places in the world where the midocean ridge can be seen above sea level.

back to the early years of the present century. The idea of continental movement was most forcefully proposed by a German scientist, Alfred Wegener. The concept came to be called continental drift. When first proposed, the idea did not receive widespread support because at the time no adequate explanation could be offered as to how it could happen. Plate tectonics provided the answer.

The original suggestion for continental drift was that continents must somehow slide across the floor of the ocean. Scientists soon realized, however, that friction would prevent such motions. Rocks on the ocean floor are too rigid and strong for continents to slide over them. Eventually, following the discovery that the oceanic crust on the floor of the ocean also moves, and that the asthenosphere is weak and easily deformed, geologists realized that the entire lithosphere must be in motion, not just the continents, and that plates of hard lithosphere must be sliding across the top of the soft, plastic asthenosphere.

The first clear evidence that seafloor and continent on the same plate of lithosphere move at the same velocity, and in the same direction, came from studies of the magnetic properties of rocks. This evidence is discussed in Chapter 16. Recently, however, a series of remarkable measurements have provided an even more convincing body of evidence.

The new evidence of plate motion comes from satellites. To a close approximation, plates of lithosphere behave as rigid bodies. This means that plates do not stretch and shrink the way rubber sheets do. The distance between, say, New York City and Chicago, both on the North American Plate, remains fixed, even though the plate may flex and warp up and down. Of course, the distances between places on adjacent plates—Los Angeles on the Pacific Plate and San Francisco on the North American Plate, for instance—do change because of plate motions. Figure 1.15 shows the inferred relative motions of plates today, based on past velocities calculated from magnetic measure-

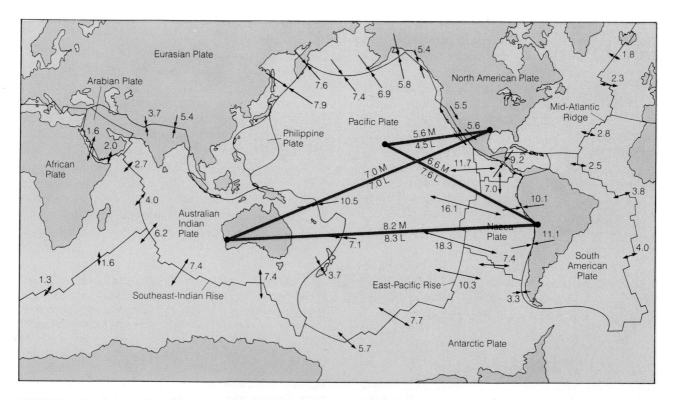

FIGURE 1.15 Present-day plate velocities in centimeters per year, determined in two ways. Numbers along the mid-ocean ridges are mean velocities indicated by magnetic measurements. A velocity of 16.1, as shown for the East Pacific Rise, means that the distance between a point on the Nazca Plate and a point on the Pacific Plate increases, on the average, by 16.1 cm each year in the direction of the arrows. The long red lines connect stations used to determine plate motions by means of satellite laser ranging (L) techniques. The measured velocities between stations are very close to the average velocities estimated from magnetic measurements (M).

ments. The motions recorded by magnetic measurements can be inferred to be today's motions only if actual measurements show the plates really are in motion. The space age has made it possible to get that proof. Using laser beams bounced off satellites, we can measure the distance between two points on the Earth with an accuracy of about 1 cm. By making distance measurements several times a year, therefore, we can measure present-day plate velocities directly. As seen in Figure 1.15, velocities based on these satellite measurements agree very closely with the velocities calculated from magnetic measurements. The agreement implies that the plates move steadily, rather than by starts and stops.

Plate Margins

Plates move as individual units, and interactions between plates occurs along their edges. Plate interactions are most distinctively expressed by earthquakes and volcanism because a majority of the Earth's volcanoes and earthquakes occur along plate margins. It has been through studies of these phenomena, particularly earthquakes, that geologists have been able to decipher the shapes of today's plates.

Plates have three kinds of margins (Fig. 1.16):

1. **Divergent margins**, which are also called **spreading centers** because such margins are fractures in the lithosphere where two plates move apart.

2. **Convergent margins**, where two plates move toward each other. Along convergent margins, one plate must either sink beneath the other, in which case we refer to the margin as a **subduction zone**, or the two plates must collide, in which case we refer to the margin as a **collision zone**.

3. **Transform fault margins**, which are fractures in the lithosphere where two plates slide past each other, grinding and abrading their edges as they do so. Earthquakes are frequent along most transform fault margins.

Spreading Centers

When we examine how a plate moves, a good analogy is a conveyor belt. In a conveyor, the belt appears from below, moves along a certain length, and then turns down and passes temporarily from sight as it completes its circuit. Although broad and irregular rather than long and narrow, a plate of lithosphere acts like the top of a slowly moving conveyor belt.

Each plate moves away from the center just as if it were a continuous belt rising up the fracture from

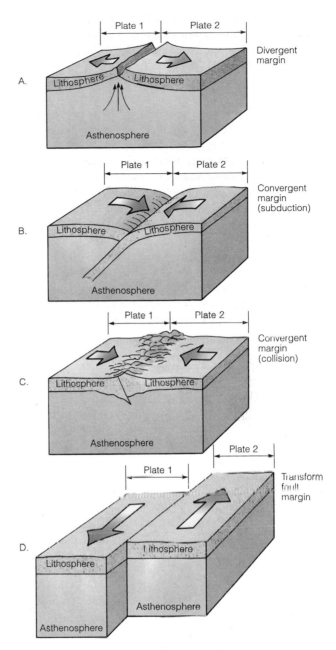

FIGURE 1.16 The various kinds of plate margins shown by schematic diagrams. A. Divergent margin for which the topographic expression is a midocean ridge. B. Convergent subduction margin for which the topographic expression is a seafloor trench. C. Convergent collision margin for which the topographic expression is a mountain range. D. Transform fault margin, which does not produce a consistent topographic expression but is often marked by a long, thin valley due to preferential erosion along the fault.

the mantle below. The analogy is only partly correct because the plate is not rising as a solid ribbon. It is being created by the formation of new crust along the fracture. Another disparity in the analogy is that the two plates are moving away in opposite directions. A more accurate analogy would be two conveyor belts moving in different directions.

When a divergent margin occurs in oceanic crust it coincides with a midocean ridge. We can't see into the mantle beneath the midocean ridges, but it is possible to infer what must be happening. Convection currents bring up hot rock from deep in the mantle, and as a result local portions of the asthenosphere become hot enough to start melting. Molten rock is **magma**. The magma that forms in the asthenosphere beneath the midocean ridge rises upward to the top of the lithosphere, where it cools and hardens to form new oceanic crust (Fig. 1.17).

When a spreading center splits continental crust, an interesting sequence of events occurs. First, a great rift is formed; the African Rift Valley that runs from Ethiopia, through Kenya, Tanzania, and Malawi is a modern example. As the two fragments of continental crust move apart, volcanism commences, as shown in Figure I.6. Continued movement allows the rift to widen and deepen, and eventually the sea enters to

form a long, narrow body of water; the Red Sea is a modern example. Eventually, the fragments of continental crust move far apart, new oceanic crust separates them, and a new ocean, like the Atlantic, has been formed.

Two hundred and fifty million years ago, there was no Atlantic Ocean. Instead, the continents that now border it were joined together into a single huge continent (Fig. 1.18). The place that is now New York was then as far from the sea as central Mongolia is today. About 200 million years ago new spreading centers split the huge continent. We do not yet fully understand why this occurred, but presumably it involved new convection currents in the asthenosphere and mesosphere. The new spreading centers split the lithosphere and in the process broke the ancient continent into the pieces we see today. These fragments, today's continents, then drifted slowly into their present positions. At first the Atlantic Ocean was a narrow body of water that separated North America from Europe and North Africa. As movement continued, the ocean widened and lengthened, splitting South America from Africa and then growing to its present size. The Atlantic is still growing wider by about 5 cm each year.

Evidence is abundant to mark where the torn mar-

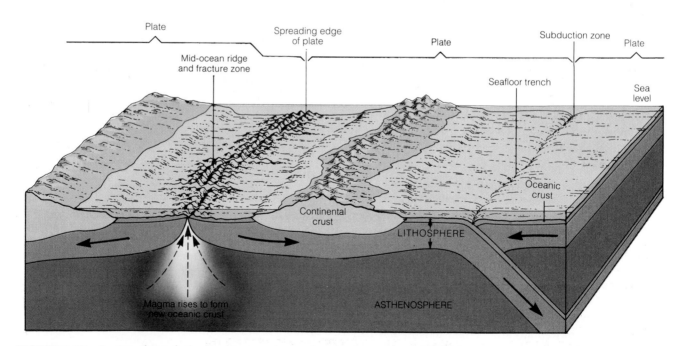

FIGURE 1.17   Section through the Earth's outer layers showing how magma (dashed arrows) moves from the asthenosphere upward into spreading centers in the ocean floor and cools there to form new lithosphere capped by oceanic crust. To accommodate the new material, the lithosphere (solid arrows) moves away from the fracture zone and eventually sinks slowly down into the asthenosphere again, where it is reheated and mixed again with the mantle.

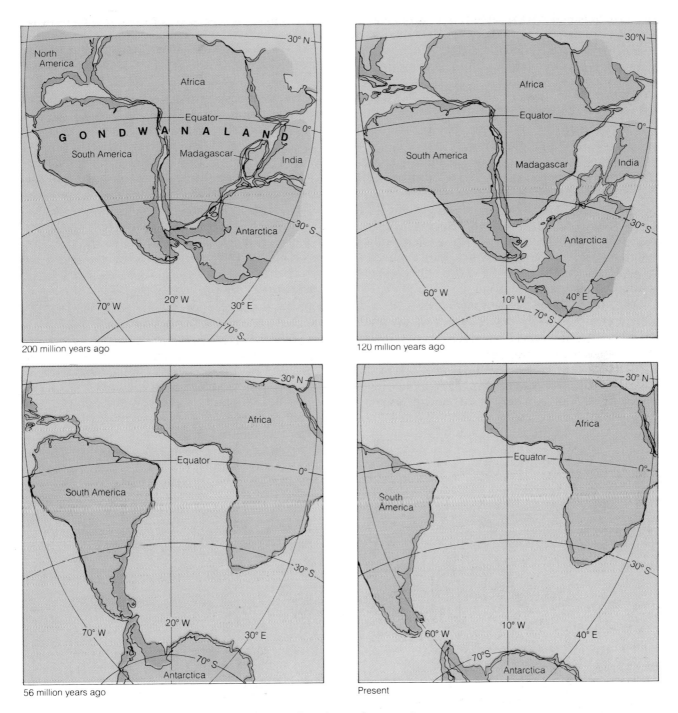

200 million years ago

120 million years ago

56 million years ago

Present

FIGURE 1.18   Breakup of Pangaea. The present southern hemisphere continents were joined together 200 million years ago as the southern half of the supercontinent, Pangaea. Magnetic data obtained from the oceanic crust were used to plot the opening of the southern part of the Atlantic Ocean as South America and Africa drifted apart. When the continents are fitted back together along a line 2000 m below sea level, as shown in the upper left-hand corner, very few overlaps (shaded) or gaps (dark) remain. Notice how the continents move relative to the equator and the way Antarctica slowly moves south.

gins formerly fitted together. If the pieces are reassembled, the continental slopes on each side of the ocean fit like the matched pieces of a jigsaw puzzle (Fig. 1.18). The line of match follows the spreading center, the present Mid-Atlantic Ridge.

Subduction Zones

Near a spreading center, the lithosphere is thin and its boundary with the asthenosphere comes close to the surface (Fig. 1.19). This thinning of the upper mantle and crust happens because magma rising toward the spreading center heats the upper mantle and crust, and only a thin layer near the top retains the hard, rigid strength properties of the lithosphere.

As the lithosphere moves away from the spreading center, it cools and becomes denser. Also, the boundary between the lithosphere and the asthenosphere becomes deeper, and as a result the lithosphere becomes thicker and the asthenosphere thinner. Finally, about 1000 km from the spreading center, the lithosphere reaches a constant thickness and is so cool that it is more dense than the hot, weak asthenosphere below it and starts to sink downward. Like a conveyor belt, old lithosphere with its capping of oceanic crust sinks into the asthenosphere and eventually into the mesosphere. The process by which lithosphere sinks into the asthenosphere is called **subduction**, and the margins along which plates are subducted are called *subduction zones*. They are marked by deep trenches in the seafloor.

As the moving strip of lithosphere sinks slowly through the asthenosphere, it passes beyond the region where geologists can study it directly. Consequently, what happens next is partly conjecture. On one point, however, we can be quite certain: The lithospheric plate does not turn under, as a conveyor belt does, and reappear at the spreading edge; rather, it is heated and slowly mixed with the material of the mantle. The thin layer of oceanic crust on top of the sinking lithosphere melts and becomes magma, and some of this magma reaches the surface to form volcanoes. As a result, subduction zones are marked by an arc of volcanoes parallel to, but about 150 km from, the trench that marks the plate margin (Fig. 1.20).

Collision Zones

Continental crust is not recycled into the mantle; it takes a shorter trip that ends more suddenly.

Continental crust is lighter and less dense than even the hottest regions of the mantle. As a result, continental crust is too buoyant to be dragged downward on top of the sinking lithosphere. So, in continent-sized pieces, such crust floats on plates of lithosphere from place to place on the Earth's surface. Movement stops when two fragments of continental crust collide. Such collisions can happen only when

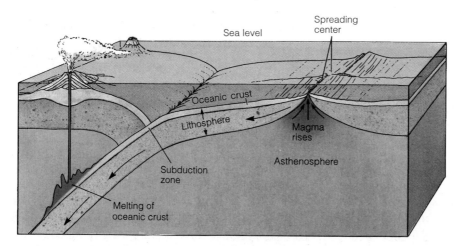

FIGURE 1.19 Schematic diagram of a spreading edge and subduction zone. Where the temperature is high near the spreading edge, because of rising magma, the lithosphere is thin. Away from the spreading edge, the lithosphere cools and becomes thicker so the lithosphere–asthenosphere boundary is deeper. When the lithosphere finally sinks into the asthenosphere at the subduction zone it is reheated. At a depth of about 100 km, the oceanic crust starts to melt, and the magma rises and forms an arcuate belt of volcanoes parallel to the subduction zone.

FIGURE 1.20 This chain of volcanoes in Ecuador sits above the subduction zone where the Nazca Plate sinks below the western edge of the South American Plate. Seven snow-capped volcanoes are visible in this aerial photograph.

subduction of oceanic crust occurs beneath one of the colliding fragments. Because the plate being subducted also carries a fragment of continental crust, a collision will inevitably occur when the two pieces of continental crust meet along the subduction zone (Fig. 1.21). Such collision zones form spectacular mountain ranges. The Alps, the Himalaya, and the Appalachians are the results of continental collisions. Because continental crust cannot sink down into the mantle, much of the evidence concerning ancient plates and their motions is recorded in the bumps and scars of past continental collisions.

Transform Faults

Besides spreading centers and the subduction and collision zones of convergent margins, there is a third kind of plate margin, along which plates simply slip past each other. These margins of slipping are great vertical fractures—or to use a term discussed in more detail in chapter 16, transform faults—that cut right down through the lithosphere. One transform fault much in the public eye because of the threat of earthquakes along it, is the San Andreas Fault in California (Fig. 1.22). This fault, which runs approximately north–south, separates the North American Plate on the east side, on which San Francisco sits, from the Pacific Plate west of the fault, on which Los Angeles sits. The Pacific Plate is moving in a northerly, and the North American Plate in a southerly direction. As the two plates grind and scrape past each other, Los Angeles is slowly moving north and San Francisco is

moving south. At times the plate edges grab and lock, and as they do the rocks on both sides flex and bend. When the locked section breaks free, the flexed rock suddenly snaps back and an earthquake occurs.

INTERACTIONS BETWEEN THE INTERNAL AND EXTERNAL LAYERS

We walk on the regolith, breathe the atmosphere, feel the rain; the evidence is clear that the Earth's external layers are places of intense and continual activity. Water and air penetrate the regolith and far into the crust. Chemical and physical disintegration of rock goes on because the atmosphere, biosphere, and hydrosphere combine to alter and break down the crust.

Cyclic Movements

When rock weathers to form regolith, some of the more soluble constituents dissolve in rainwater. The dissolved salts are transported by streams and eventually concentrated in the ocean. This is the origin of many of the salts in seawater. When raindrops form, they dissolve gases from the atmosphere and carry them down to the Earth's surface. There the gases react to form new minerals in the soil. Since material is constantly being transferred from one of the Earth's spheres to another we can ask some pertinent ques-

— A DRILL HOLE THROUGH THE CRUST? —

A challenging idea was broached during the 1960s—why not drill a hole all the way through the crust and get samples from the mantle? A few drill holes on land for oil and gas had reached depths approaching 9 km, so it seemed to be technically feasible to drill 8 km through the thin oceanic crust. The idea was treated with enthusiasm by many in the geological community, and the project was soon launched. The name Project Mohole was selected. It is derived from moho, a colloquial term for the boundary between the crust and the mantle named for Andrija Mohorovičić, a Yugoslav scientist who first demonstrated the existence of this boundary. After early tests and the drilling of some shallow holes into the deep ocean floor, however, the project was abandoned. The reasons were as much political and financial as they were technical.

In the place of Project Mohole there emerged a less ambitious venture for drilling many shallower holes in order to learn more about the oceanic crust. The Deep Sea Drilling Project (DSDP) was active until 1985, nearly 20 years, and the results it produced were extraordinary. Because of DSDP and its successor, The Ocean Drilling Program (ODP), the geology of the ocean basins has been revealed as never before. Indeed, it is sometimes said we now know the geology of the oceanic crust better than we do that of the continental crust.

The continental crust is thick and contains within it the record of much that has happened on the Earth during the past 4 billion years. Yet we have only scratched the surface of the continental crust, which even at its thinnest points is 30 km thick; the deeper portions remain unsampled and hidden from view. New plans are now afoot for deep-drilling programs, but this time they are programs designed to drill down through the thick layers of igneous and metamorphic rocks of the continental crust.

The first country that successfully mounted a deep-drilling program was the U.S.S.R. While U.S. scientists were planning their oceanic drilling program in the 1960s, the Soviets were planning a continental drilling program. Among the sites they selected was one on the Kola Peninsula, in northwestern Russia, not far from Finland (Fig. B1.1). There, ancient crystalline rocks crop out, and by drilling into them Soviet scientists have found answers to questions such as how rock properties change with depth, how deeply fluids might penetrate, and what differences exist between the geology seen at the surface and the geology predicted at depth. Drilling was slow, but the Kola superdeep well reached 12 km and may eventually be drilled even deeper. Every extra meter will push the hole to a new record and contribute to the development of new technology. The Soviet results are very exciting, and it is hardly surprising that other

---

The more we learn of the Earth's history and the more accurately we determine the timing of past events through radiometric dating (chapter 6), the clearer it becomes that cycle rates have not always been the same. The evidence is strongly against constancy; some rates were once more rapid, others much slower.

One reason the rate of the rock cycle has changed through time is that the Earth is very slowly cooling down as its internal heat is lost. The Earth's internal temperature is maintained, in part, by natural radioactivity. Early in the Earth's history, more radioactive atoms were present than there are today, and so more heat must have been produced than is produced today. Internal processes, which are all driven by the Earth's internal heat, must have been more rapid than they are today. It is possible that 3 billion years ago

oceanic crust was created at a faster rate than it is now, and that continental crust was uplifted and eroded at a faster rate. Either or both actions would cause the rock cycle to speed up.

At the same time, the rates of external processes have also varied. Long-term changes in the rates have occurred because of slow increases in the heat output of the Sun, and also because of the gradual slowing in the rate of rotation of the Earth on its axis—scientists estimate that 600 million years ago there were 400 days in the year, and 2 billion years ago there were 450 days a year, for instance. Short-term effects on external process have also arisen because of changes in the orientation of the Earth's axis of rotation, and changes in the Earth–Sun distance (these changes are discussed more fully in chapter 12). It is clear, therefore, that even though the cycles have been continu-

countries are planning to follow their lead. Germany and France already have drilling tests underway and the United States has a deep-drilling project in advanced stages of planning.

The thickness of continental crust in places where drilling projects are planned is 45 km. The 12-km-deep Soviet hole is less than a third of the way through. Will it ever be possible to drill as deep as 45 km? The answer is probably yes, but there are many technical difficulties to be overcome. Temperatures as high as 600°C will be encountered, pressures vastly in excess of those previously encountered in drilling must be managed, and somehow a way must be found to put tools down the hole without having to stop drilling and pull out all of the equipment every time a change or a new drilling bit is needed. But the scientific returns from a drill hole through the crust and into the mantle will be so great that the technical difficulties will surely be overcome.

FIGURE B1.1 Derrick housing for the world's deepest drill hole. The hole is 250 km north of the Arctic circle near Murmansk in the U.S.S.R. The drilling derrick is enclosed to protect it from the Arctic winter. Started in 1970, the drill hole had reached a depth of 12 km by 1990.

ous, none has maintained a constant rate through time.

The conclusion that rates of geological processes have differed in the past from today's rates is an important one. It means that the relative importance of different geological processes have probably differed in the past. For example, just because glaciation is an important process today, we cannot assume it has been equally important through geological time. But we can assume that when glaciation did affect the Earth in geologically remote times, the processes and effects were the same as the processes and effects of glaciation we observe in Antarctica today.

SUMMARY

1. There are nine planets in orbit around the Sun. The four innermost, Mercury, Venus, Earth, and Mars, are small, dense, rocky bodies. The five outer planets, Jupiter, Saturn, Uranus, Neptune, and Pluto, are larger, less dense objects with thick atmospheres of hydrogen and helium.

2. The planets formed by a two-step process. First, small rocky fragments condensed from a disc-shaped envelope of gas that rotated around the Sun. Then the rocky fragments started accreting into even larger masses. The largest, today's planets, had all formed by 4.6 billion years ago.

3. The heat released by decay of naturally radioactive chemical elements was sufficient, early in the Earth's history, to cause a fraction of the Earth to melt. Heavy materials sank to the center, and lighter ones rose, giving the Earth a compositionally three-layered structure: core, mantle, and crust.

4. The crust consists of two parts: oceanic crust with an average thickness of 8 km, and continental crust with an average thickness of about 45 km.

5. The Earth is also layered with respect to its physical properties, in particular strength. The lithosphere, approximately the outer 100 km of the solid Earth, consists of rock that is strong and relatively rigid. Beneath the lithosphere, down to a depth of 350 km, is the asthenosphere, a region where high temperatures make rock weak and easily deformed. Beneath the asthenosphere is the mesosphere, where rocks become gradually stronger. Within the core there are also two regions differing in physical properties but with the same composition: the inner core is solid, the outer core molten.

6. The lithosphere consists of six large and many small plates that slide slowly over the asthenosphere at rates up to 12 cm/yr.

7. Plates of lithosphere have three kinds of margins; divergent (called spreading centers), convergent (called subduction zones or collision zones), and transform faults.

8. The internal and external layers continually interact in cycles, in which materials flow both from one place to another within a layer and from one layer to another.

9. The hydrological cycle, driven by heat energy from the Sun, is the cyclic movement of water between ocean, air, and land. The movement occurs through evaporation, wind transport, precipitation of snow and rain, stream flow, and percolation.

10. The rock cycle arises from the interactions of the Earth's internal and external processes. Igneous rock is eroded, creating sediment, which is deposited in layers that become sedimentary rock. Burial may lead to changes in temperature and pressure, forming metamorphic rock. Eventually temperatures and pressures may become so high that rock melts and forms new magma. The magma rises, forms new igneous rock and the cycle is repeated.

11. The rock cycle in the oceanic crust interacts with that in the continental crust through the agency of plate tectonics.

IMPORTANT TERMS TO REMEMBER

abyssal plain (p. 26)
asthenosphere (p. 23)
atmosphere (p. 16)

biosphere (p. 16)

collision zone (p. 29)
continental crust (p. 21)
continental rise (p. 26)
continental shelf (p. 25)
continental slope (p. 25)
convergent margin (of a plate)
  (p. 29)
core (p. 21)
crust (p. 21)

divergent margin (of a plate)
  (p. 29)

hydrologic cycle (p. 34)
hydrosphere (p. 16)

igneous rock (p. 35)
inner core (p. 23)

lithosphere (p. 23)

magma (p. 30)
mantle (p. 21)
mesosphere (p. 23)
metamorphic rock (p. 35)
midocean ridge (p. 27)

oceanic crust (p. 21)
oceanic ridge (p. 27)
oceanic rise (p. 27)
outer core (p. 23)

regolith (p. 16)
rock cycle (p. 36)

sediment (p. 35)
sedimentary rock (p. 35)
spreading center (p. 29)
subduction (p. 32)
subduction zone (p. 29)

tectonics (p. 24)
transform fault (p. 33)
trench (p. 27)

weathering (p. 16)

QUESTIONS FOR REVIEW

1. Briefly describe the steps by which scientists believe planets form from a huge cloud of gaseous material.

2. Describe the Earth's compositional layers. Discuss how the layers developed from an originally homogeneous Earth.

3. The Earth is layered with respect to its physical properties. Describe the major physical property layers and discuss how they arise.

4. What are the relationships between the crust, the mantle, and the lithosphere?

5. Why are ocean basins low spots on the Earth's surface and continents high places?

6. Briefly describe the three kinds of plate margins.

7. Describe what happens when two plates topped by oceanic crust converge. Compare your description with what happens when the converging is between two plates capped by continental crust.

8. Identify the major topographic features of the ocean floor, and state how they are related to tectonic plates.

9. Briefly describe the hydrologic cycle. Where does the energy that drives the cycle come from?

10. What is the rock cycle? How does oceanic crust interact with continental crust through the rock cycle?

11. Identify two ways the hydrologic cycle interacts with the rock cycle.

12. Why are sedimentary rocks so abundant at the Earth's surface when igneous rocks make up most of the crust?

The Oppenheimer diamond, a rare yellow stone weighing 253 carats, in the collection of the Smithsonian Institution. The octahedral shape is the characteristic crystal form of diamond. The specimen is 4.5 cm long.

CHAPTER 2

Minerals

D iamond is remarkable. It is the hardest substance known. Diamond saws and diamond drills can cut and drill through all known rocks and metals. Diamond is also the most valuable of all gems. A polished diamond has a sparkle and brilliance that is unique.

A few diamonds are so beautiful they are legendary. The most famous object in the Smithsonian Institution, for example, is the extraordinary Hope diamond; in 1990 alone, more than 6 million people viewed this unique blue stone. Across the Atlantic Ocean 6 million people visit the Tower of London each year to see the British Crown Jewels. They seek especially to see two huge, perfect gems cut from the Cullinan diamond, at 0.605 kg (3025 carats) the largest diamond ever found. Each of the two cut stones is more than 5 cm across!

Few of us will ever have a chance to find a diamond in nature because diamonds are very rare. They are rare because they form only in the mantle at depths of 150 km or more and the only way diamonds reach the Earth's surface is in an unusual kind of igneous rock called kimberlite. How and why kimberlite magma forms in the mantle is still a puzzle, but when it does form it rises explosively upward, punching a circular hole through the crust, carrying broken fragments of mantle rock upward as it does so. Diamonds are among the fragments.

Even the largest kimberlite holes, or pipes as they are usually called, are no more than a few hundred meters in diameter, and the quantity of diamond present is never large. For example, the rich diamond pipes at Kimberley, South Africa, yield, on average, only one carat of diamond for every 5 cubic meters of kimberlite mined. Considering that only a small percentage of the diamonds found are fine enough to be gems it is small wonder that diamonds are so expensive!

MINERALS AND THEIR CHEMISTRY

The word **mineral** has a specific connotation in geology; it is any naturally formed, solid, chemical substance having a specific composition and a characteristic crystal structure. Diamond is a mineral. It is naturally formed, it is a solid, it is pure carbon so it has a specific composition, and the atoms of carbon are packed together in a regular geometric array called its crystal structure. Coal, on the other hand, is not a mineral. Coal is mostly carbon but it contains many different chemical compounds. The composition of coal varies from sample to sample so it does not have a specific composition. Nor does coal have a characteristic crystal structure. Coal is a rock.

Rocks are aggregates of minerals; they are nature's books and in them is recorded the story of the way the Earth works. Rocks tell such stories as the way continents move, how mountains form and slowly erode away, and why volcanoes are located where they are. The words used in nature's books are minerals, and in order to be able to read the words we must investigate the branch of geology that deals with the properties and distribution of minerals. The easiest way to introduce the subject of minerals is through an examination of the two most important characteristics of minerals:

1. **Composition**, which is the kinds of chemical elements present and their proportions; and

2. **Crystal structure**, which is the way in which the atoms of the chemical elements are packed together in a mineral.

Because most minerals contain several chemical elements, it is helpful to commence our discussion by briefly reviewing the way in which chemical elements combine to form compounds.

Elements and Atoms

Chemical Elements

If you were a chemist and you were asked to analyze a mineral or rock, you would report your findings as the kinds and amounts of the chemical elements present. *Chemical elements* are the most fundamental substances into which matter can be separated by chemical means. For example, table salt is not an element because it can be separated into sodium and chlorine. But neither sodium nor chlorine can be further broken down chemically, so each is an element.

Each element is identified by a symbol, such as H for hydrogen and Si for silicon. Some symbols, such as that for hydrogen, come from the element's name in English. Other symbols come from other languages. For example, iron is Fe from the Latin *ferrum*, copper is Cu from the Latin word *cuprum* which in turn comes from the Greek *kyprios*, and sodium is Na from the Latin word *natrium*. The 88 naturally occurring elements and their symbols are listed in appendix B.

A piece of a pure element—even a tiny piece not bigger than a pin's head—consists of a vast number of identical particles of that element called atoms. An **atom** is the smallest individual particle that retains all the properties of a given chemical element. Atoms are so tiny they can only be seen by using the most powerful microscopes ever invented, and even then the image is imperfect because individual atoms are only about 10^{-10} m in diameter.

Atoms and Ions

Atoms are build up from *protons* (which have positive electrical charges), *neutrons* (which, as their name suggests, are electrically neutral), and *electrons* (which have negative electrical charges that balance exactly the positive charges of protons). Protons and neutrons are dense but very tiny particles and they join together to from the core, or *nucleus*, of an atom. Electrons are even tinier particles than protons or neutrons; they move, like a distant and diffuse cloud, in orbits around the nucleus (Fig. 2.1).

Protons give a nucleus a positive charge and the number of protons in the nucleus of an atom is called the *atomic number*. The number of protons in the nucleus is what gives the atom its special physical characteristics, and what makes it a specific element. Elements are catalogued by atomic number, beginning with hydrogen, which has an atomic number of 1 because it has one proton. Hydrogen is followed by helium, which has 2 protons, and so on. All atoms having the same atomic number are atoms of the same element.

Energy-Level Shells

Electrons are confined to specific shells, or orbits which are arranged at predetermined distances from the nucleus. Because the electrons in each shell have a specific amount of energy characteristic for that shell, the shell distances are commonly called **energy-level shells**. The maximum number of electrons that can occupy a given energy-level shell is fixed. As shown in Figure 2.1, shell 1, closest to the nucleus, is small and can accommodate only 2 electrons; shell 2, however, can accommodate up to 8 electrons; shell 3, 18; and shell 4, 32.

A.

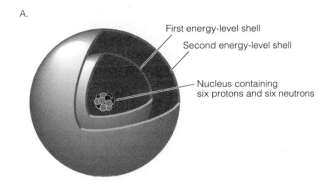

First energy-level shell

Second energy-level shell

Nucleus containing
six protons and six neutrons

B.

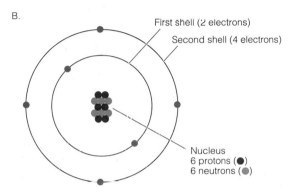

First shell (2 electrons)

Second shell (4 electrons)

Nucleus
6 protons (●)
6 neutrons (●)

FIGURE 2.1 Schematic diagram of an atom of carbon. The nucleus contains six protons and six neutrons. Electrons orbit the nucleus but are confined to specific orbits called energy-level shells. A. Three-dimensional representation showing the first two shells. The first shell can contain two electrons, the second eight. B. Two-dimensional representation of the carbon atom. The first energy level shell is full because it contains two electrons. The second shell contains four electrons and so is half full.

Ions

When an energy-level shell is filled with electrons, it is very stable, like an evenly loaded boat. To fill their energy-level shells and so reach a stable configuration, atoms share or transfer electrons among themselves. An atom is electrically neutral because it has the same number of protons and orbiting electrons. But when transfer of an electron occurs the balance of electrical forces is upset. An atom that loses an electron has lost a negative electrical charge and is left with a net positive charge. An atom that gains an electron has a net negative charge. An atom that has excess positive or negative charges caused by electron transfer is called an **ion**. When the charge is positive (meaning that the atom gives up electrons), the ion is called a **cation**; when negative (meaning an atom adds electrons), an **anion**.

The convenient way to indicate ionic charges is to record them as superscripts. For example, Li^{1+} is a cation (lithium) that has given up an electron, while F^{1-} is an anion (fluorine) that has accepted an electron.

Compounds

Chemical compounds form when one or more anions combine with one or more cations in a specific ratio. For example, two cations of H^{1+} combine with one anion of O^{2-} to make the compound H_2O. In a compound, the sum of the positive and negative charges must be zero.

The formula of a compound is written by putting the cations first, anions second. The numbers of cations or anions are indicated by subscripts, and for convenience the charges of the ions are usually omitted. Thus, we write H_2O rather than $H_2^{1+}O^{2+}$.

An example of the way electron transfer leads to formation of a compound is shown in Figure 2.2 for lithium and fluorine. A lithium atom has energy-level shell 1 filled by two electrons, but has only one electron in shell 2 even though shell 2 can accommodate eight electrons. The lone outer electron in shell 2 can easily be transferred to an element such as fluorine, which already has seven electrons in shell 2 and needs only one more to be completely filled. In this fashion, if the lithium and fluorine are in close proximity, both a lithium cation and fluorine anion finish with filled shells, and the resulting positive charge on the lithium and negative charge on the fluorine draw, or bond, the two ions together.

Lithium and fluorine form the compound lithium fluoride, which is written LiF to indicate that for every Li atom there is a counterbalancing F ion. A combined pair of Li and F ions is called a molecule of lithium fluoride. A *molecule* is the smallest unit that retains all the properties of a compound. Properties of molecules are quite different from the properties of their

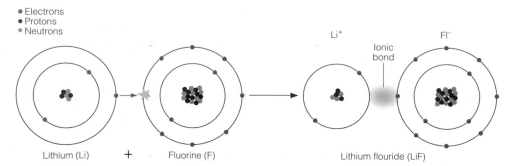

● Electrons
● Protons
● Neutrons

Li⁺

Fl⁻

Ionic bond

Lithium (Li)   +   Fluorine (F)

Lithium flouride (LiF)

FIGURE 2.2   To form lithium fluoride, an atom of lithium combines with an atom of fluorine. The lithium atom transfers its lone outer-shell electron to fill the fluorine atom's outer shell, creating an Li^{1+} cation and a F^{1-} anion in the process. The electrostatic force that draws the lithium and fluorine ions together is an ionic bond.

constituent elements. The elements sodium (Na) and chlorite (Cl) are highly toxic, for example, but the compound sodium chloride (NaCl, table salt) is essential for human health. The manner in which electrons are transferred or shared between atoms leads to several distinctive ways by which ions bond together.

Bonds

Ionic Bonds

Cations and anions formed by electron transfer can exist as free entities, but an electrostatic attraction draws these negatively and positively charged ions together and is called an **ionic bond**. LiF is an example of an ionically bonded compound, as are the minerals halite (NaCl) and fluorite (CaF$_2$).

Covalent Bonds

Some atoms share rather than transfer electrons. The force between two atoms that have filled their energy-level shells by sharing one or more electrons is called **covalent bonding**. Atoms that share electrons, like those that transfer, are called ions. One common substance in which covalent bonds occur is water (H$_2$O). The outer shell of an oxygen atom has six electrons but requires eight for maximum stability. A hydrogen atom has one electron but requires two for maximum stability. Hydrogen is the cation, oxygen the anion. How bonding by the ions is accomplished by a sharing of electrons is shown in Figure 2.3.

Diamond (C) also contains covalent bonds. There are four electrons in the second shell of carbon. By sharing one electron with each of four other carbons,

a stable shell of eight electrons is obtained, as shown in Figure 2.4A. Note that Figure 2.4A is only diagrammatic. It looks as if the carbon ions lie in a plane. In actual fact, they arrange themselves in three dimensions so that each carbon ion is surrounded by four others, which are equally spaced, as if they were at the corner of a tetrahedron as shown in Figure 2.4B. The covalent bond is very strong because it involves shared electrons and that is the reason diamond is such a tough, hard, substance. Covalent bonds also give diamond special optical properties. The reason that diamonds have the sparkle that makes them attractive gems is due to the covalent bonding.

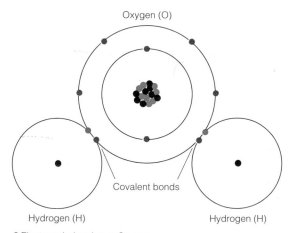

Oxygen (O)

Covalent bonds

Hydrogen (H)   Hydrogen (H)

● Electrons belonging to Oxygen
● Electrons belonging to Hydrogen

FIGURE 2.3   Two atoms of hydrogen form covalent bonds with an oxygen atom through sharing of electrons. The oxygen atom thereby has its most stable configuration with eight electrons in the outer shell, and each of the hydrogen atoms fills its outer shell with two electrons, making the composed H$_2$O.

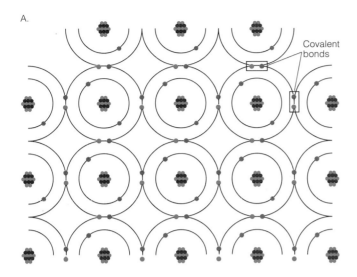

A.

Covalent
bonds

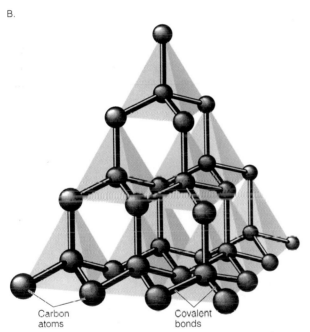

B.

Carbon
atoms

Covalent
bonds

FIGURE 2.4 Covalent bonding in diamond. A. Schematic diagram showing how each carbon atom shares its four outer-shell electrons with four other carbon atoms so that all atoms have stable outer shells of eight electrons. B. The three-dimensional geometric arrangement of carbon atoms in diamond. Note that each atom is surrounded by four others. The actual covalent bonding in diamond is three dimensional, not two dimensional, as shown for simplification in A.

Metallic Bonds

A small number of minerals, each of them metals, have a kind of bonding called the **metallic bond**. In metallic bonding, the ions are held together by covalent bonds but are closely packed because the electron sharing occurs with inner energy-level shells rather than the outermost shell. As a result, electrons in the outermost shell are loosely held and they can move readily from one atom to another. If a little bit of energy is available to move them, the outer shell electrons are free to diffuse through the structure, sometimes drifting, sometimes replacing an electron that is forming a covalent bond. The drifting electrons are the reason that metals have such special properties as high electrical and thermal conductivity and opacity.

Van der Waals Bonds

Lastly, a fourth kind of bond exists called the **van der Waals bond**. It does not involve the transfer or sharing of electrons but is instead a weak electrostatic attraction between certain ions that have already transferred or shared electrons and formed compounds. The van der Waals bond is a weak bond—much weaker than ionic, covalent, or metallic bonds—but it plays an important role in the structure of certain minerals, of which graphite is a good example. Graphite contains only atoms of carbon and it has a sheetlike structure in which each carbon atom has three nearest neighbors at the corners of an equilateral triangle (Fig. 2.5). Carbon atoms are bonded covalently within the sheets and as a result the sheets are very strong and flexible. The graphite used in a tennis racket or a golf club makes use of the strong covalent bonds in the sheets. Adjacent sheets of graphite are held together by weak van der Waals bonds. These van der Waals bonds are so weak they are easily broken. Graphite feels slippery when you rub it between your fingers because the rubbing breaks the van der Waals bonds and the sheets slide easily past each other. Graphite is used as a lubricant for very high temperature purposes, in which case it is the property of weak van der Waals bond that is being employed. Talcum powder is another example of a compound with van der Waals bonds. Talc, the mineral in talcum powder, has a sheet structure analogous to a graphite sheet, and just like graphite the sheets are held together by van der Waals bonds. Talcum powder feels smooth and slippery because the van der Waals bonds are so easily broken.

Complex Ions

Sometimes two kinds of ions form such strong bonds that the combined ions act as if they were a single ion.

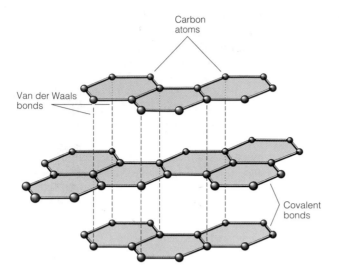

Carbon atoms

Van der Waals bonds

Covalent bonds

FIGURE 2.5 Geometric arrangement of atoms in graphite. Bonding within the sheets of atoms is covalent; bonding between sheets is van der Waals.

Such a strongly bonded pair is called a **complex ion**. Complex ions act in the same way as single ions, forming compounds by bonding with other elements. For example, carbon and oxygen combine to form the complex carbonate anion $(CO_3)^{2-}$. Other important examples of complex ions are the sulfate $(SO_4)^{2-}$, nitrate $(NO_3)^{1-}$ and silicate $(SiO_4)^{4-}$ anions.

CRYSTAL STRUCTURE AND STATES OF MATTER

States of Matter

Compounds and elements can exist in any of the three states of matter—solid, liquid, or gas. H_2O is a familiar example of a compound that can occur in each state under suitable conditions—water (a liquid), ice (a solid), and water vapor (a gas). All minerals are solids. Thus, whether a given compound or element can be called a mineral is controlled by its state. Ice in a glacier is a mineral, but water in the ocean and water vapor in the atmosphere are not minerals.

The state of matter is controlled by temperature and pressure. An example is provided by H_2O, as shown in Figure 2.6. It is clear that at high temperatures or low pressures water vapor is the stable state for H_2O, while ice forms at low temperatures or high pressures. The pressures at sea level and the top of Mount Everest, 101,300 pa and 35,000 pa, respectively, bound the range of pressures at the Earth's surface;

surface temperatures range from about $-100°C$ to $+50°C$. When the limits of temperature and pressure are plotted on Figure 2.6, it is apparent that H_2O can exist in each of the three states on the Earth. On Mars, however, the ranges of temperature and pressure are such that H_2O can only exist as a solid or a vapor.

The temperatures and pressures over which different compounds change from one state to another differ greatly, but the same general statement is true for all substances—low temperature and high pressure favor the solid state, high temperature and low pressure favor the gaseous state, while the liquid state occurs in intermediate ranges of temperature and pressure. There are no minerals in the Sun, for example, because temperatures are too high.

Crystal Structure

Whereas the atoms in gases and liquids are randomly jumbled, the atoms in most solids are organized in regular, geometric patterns, like eggs in a carton, as shown in Figure 2.7. The geometric pattern that atoms

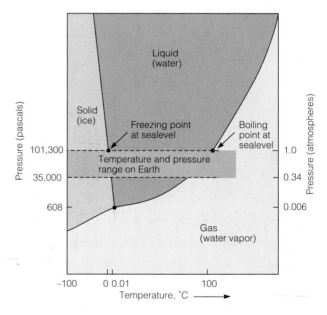

FIGURE 2.6 The controls of temperature and pressure on the state of H_2O. At temperatures and pressures defined by the boundaries between regions, two states can coexist. For example, along the boundary separating ice and water, the solid and liquid states coexist—that is, along the line, water coexists with ice. At one specific temperature and pressure the three lines of coexistence meet, and at this triple point ice, water, and water vapor coexist. The triple point for H_2O is at 0.010°C and 608 Pa. The pressures at sea level and at the top of Mount Everest bound the range of pressures at the Earth's surface.

assume in a solid is called the **crystal structure**, and solids that have a crystal structure are said to be **crystalline**. Solids that lack crystal structures are *amorphous* (Greek for without form). Glass and amber are examples of solids that are amorphous. All minerals are crystalline and the crystal structure of a mineral is a unique property of that mineral. All specimens of a given mineral have identical crystal structure.

Ionic Substitution

The crystal structure of the mineral galena, PbS, a common lead mineral, is shown in Figure 2.8. The bonding is ionic—lead is the cation (Pb^{2+}), sulfur the anion (S^{2-}). Note that the ions in Figure 2.8 are arranged in a cubelike grid in which each sulfur is surrounded by six leads and each lead by six sulfurs.

The anions in Figure 2.8 are larger than the cations. The size of ions is commonly stated in terms of the **ionic radius**, which is the distance from the center of the nucleus to the outermost shell of orbiting electrons. Anions tend to have large radii because addition of an extra electron to fill an energy level shell means that the pull exerted on each orbiting electron by the protons in the nucleus is slightly reduced. As a result, the electron orbits of anions expand a little. By contrast, cations tend to be small because they lose electrons and their remaining electrons are more tightly held. Most of the volume of a crystal structure is taken up by the largest ions, the anions, and as a result the crystal structures of minerals are determined largely by the packing arrangements of anions. The radii of some common ions are shown in Figure 2.9. It is apparent from Figure 2.9 that certain ions have the same electrical charge and are nearly alike in size. For example, Fe^{2+} has a radius of 0.074 nm while Mg^{2+} has

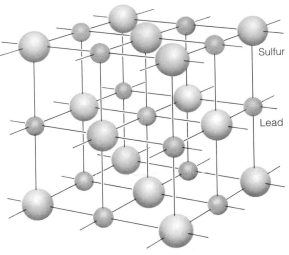

FIGURE 2.8 Arrangement of ions in galena (PbS), the most common mineral containing lead. The packing arrangement is repeated continuously through a crystal, and ions are so small that a cube of galena 1 cm on its edge contains 10^{22} ions each of lead and sulfur. The ions are shown pulled apart along the marked lines to demonstrate how they fit together. Pb is a cation with a charge of 2 +, S is an anion with a charge of 2 −. To maintain a charge balance between the atoms, there must be an equal number of Pb and S ions in the structure.

a radius of 0.066 nm. Because of their similarity in size and charge, ions of Fe^{2+} can substitute for ions of Mg^{2+} in magnesium-bearing minerals. The crystal structure of the magnesium mineral is not changed as a result of the substitution. The substitution of one ion for another in a random fashion throughout a crystal structure is **ionic substitution**.

The way ionic substitutions are indicated in chemical formulas can be demonstrated with the mineral olivine, Mg_2SiO_4. When Fe^{2+} substitutes for Mg^{2+} in olivine the formula is written $(Mg,Fe)_2SiO_4$, which indicates that the Fe^{2+} substitutes for the Mg^{2+}, but not for any other atoms in the structure.

DEFINITION OF A MINERAL

Before discussing specific minerals and their properties let's review what is meant by the term *mineral*. To be called a mineral, a substance must meet four requirements:

1. It must be *naturally formed*. This excludes the vast numbers of substances produced in the laboratory.

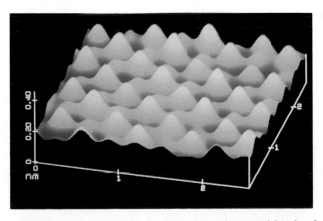

FIGURE 2.7 Atoms can be seen through special kinds of microscopes. Sulfur atoms (large) and lead atoms (small) at the surface of a galena (PbS) crystal are revealed with a scanning-tunneling microscope.

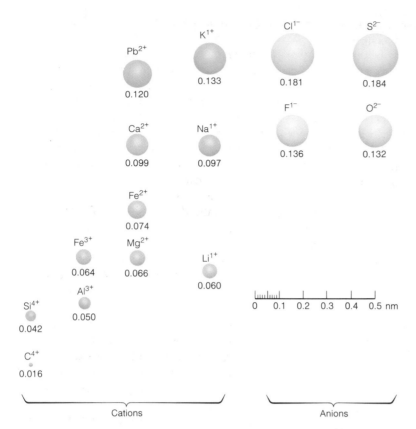

FIGURE 2.9 Ionic radii of some common ions range from C^{4+} at lower left to S^{2-} at upper right. Ions are arranged in vertical groups based on charge, from $^{4+}$ at left to $^{2-}$ at right. Note that the anions tend to have larger ionic radii than cations. Ions in each of the pairs Si^{4+} and Al^{3+}, Mg^{2+}, and Fe^{2+}, and Na^{1+} and Ca^{2+} are about the same size and commonly substitute for each other in crystal lattices. Radii are expressed in nanometers (nm).

2. It must be a *solid*. This excludes all liquids and gases.

3. It must have a *specific chemical composition*. This excludes solids, like glass, that have a continuous composition range that cannot be expressed by an exact chemical formula.

4. It must have a *characteristic crystal structure*. This excludes amorphous materials.

The requirement concerning composition (number 3 above) requires further comment. The term *specific chemical composition* means that a mineral must contain a specific ratio of cations to anions. For example, olivine always contains two 2+ cations, one 4+ cation, and four 2− anions. Even when ionic substitution occurs, the ratio of cations to anions does not change. The term *mineral group* is used to describe a mineral that displays extensive ionic substitution without changing that cation–anion ratio. Special names are given to the ideal, unsubstituted compositions of a mineral group. For example, olivine is a mineral group with the formula $(Mg,Fe)_2SiO_4$. Forsterite is the name given to Mg_2SiO_4, while Fe_2SiO_4 is called fayalite. It is important to remember the common mineral groups because we use them repeatedly throughout this book and because they are commonly used names in science. It is not so important to remember the names used for ideal compositions.

Mineraloids

Some naturally occurring solid compounds do not fulfill the definition of a mineral because they lack either a definite composition, a characteristic crystal structure, or both. Examples are natural glasses and resins, both of which have wide and variable composition ranges and are amorphous. Another example is opal, which has a more or less constant composition, but is amorphous. The term **mineraloid** is used to describe such mineral-like substances.

Polymorphs

Each mineral has a unique crystal structure. Some compounds are known to form two or more different minerals because the ions can be packed to form more than one kind of crystal structure. Diamond and graphite provide an example. Both consist of pure carbon but as shown in Figures 2.4 and 2.5, they have quite different structures. The compound $CaCO_3$ is another example because it forms two different minerals. One is *calcite*, the mineral of which limestone and marble are composed; the other is *aragonite*, most commonly found in the shells of clams, oysters, and snails. Calcite and aragonite have identical compositions, but entirely different crystal structures.

A compound that occurs in more than one crystal structure is called a **polymorph** (many forms). Some common mineral polymorphs are listed in Table 2.1.

THE PROPERTIES OF MINERALS

The properties of minerals are determined by composition and crystal structure. Once we know which properties are characteristic of which minerals, we can use those properties to identify the minerals. It is not necessary, therefore, to analyze a mineral chemically or to determine its crystal structure to discover its identity. The properties most often used to identify minerals are obvious ones such as color, external shape, and hardness—plus some less obvious properties, such as luster, cleavage, and specific gravity. Each property is briefly discussed below and an extensive table of individual mineral properties is given in appendix C.

TABLE 2.1 Examples of Polymorphs, with the Most Common Mineral Listed First

Composition	Mineral Name
C	Graphite
	Diamond
$CaCO_3$	Calcite
	Aragonite
FeS_2	Pyrite
	Marcasite
SiO_2	Quartz
	Cristobalite
	Tridymite

Crystal Form and Growth Habit

Crystal Form

Ice fascinated the ancient Greeks. When they saw glistening needles of ice covering the ground on a frosty morning they were intrigued by the fact that the needles were six-sided and had smooth, planar surfaces. Greek philosophers made many discoveries about the branch of mathematics called geometry but they could not explain how three-dimensional, geometric solids could apparently grow spontaneously. The Greeks called ice *krystallos*; the Romans latinized the name to *crystallum*. Eventually the word **crystal** came to be applied to any solid body that grows with planar surfaces. The planar surfaces that bound a crystal are called **crystal faces**, and the geometric arrangement of crystal faces, called the **crystal form**, became the subject of intense study during the seventeenth century.

Seventeenth-century scientists discovered that crystal form could be used for the identification of minerals. But certain features were difficult for them to explain; for example, why did the sizes of crystal faces differ widely from sample to sample. Under some circumstances a mineral may grow as a long, thin crystal; under others, the same mineral may grow as a short, fat one, as Figure 2.10 shows. Superficially, the two crystals of quartz in Figure 2.10 look very different. It is apparent from Figure 2.10 that the overall size of crystals and crystal faces is not a unique property of a mineral. The person who solved the mystery was a Danish physician, Nicolaus Steno. In 1669 Steno demonstrated that the unique property of crystal of a given

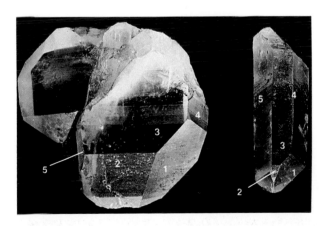

FIGURE 2.10 Two quartz crystals with the same crystal forms. Although the size of the individual faces differ markedly between the two crystals, it is clear that each numbered face on one crystal is parallel to an equivalent face on the other crystal. It is a fundamental property of crystals that, as a result of the internal crystal structure, the angles between adjacent faces are identical for all crystals of the same mineral.

mineral is not the relative face sizes, but rather the angles between the faces. The angle between any designated pair of crystal faces is constant, he wrote, and is the same for all specimens of a mineral, regardless of overall shape or size. Steno's discovery that interfacial angles are constant is made clear by the lettering in Figure 2.10. The same faces occur on both of the quartz crystals. The sets of faces are parallel; therefore, the angle between any two equivalent faces must be the same on each crystal.

Steno and other early scientists suspected that the capacity for crystals to form and for interfacial angles to be constant must depend on some kind of internal order, but the ordered particles—atoms—were too small for them to see, so they could only speculate. Proof that crystal form reflects internal order was finally achieved in 1912. In that year the German scientist, Max von Laue, demonstrated, by use of X rays, that crystals must be made up of atoms packed in fixed geometric arrays, as shown in Figure 2.7.

Crystals can only form when mineral grains can grow freely in an open space. Unfortunately crystals turn out to be uncommon in nature because most minerals do not form in open, unobstructed spaces. Compare Figures 2.10 and 2.11. Figure 2.10 shows crystals of quartz that grew freely into an open space and well developed crystal faces were able to form.

But in the second case, Figure 2.11, quartz grew as irregularly shaped grains in an environment restricted by the presence of other minerals. By making use of von Laue's discovery, it is easy to show that in both a quartz crystal and an irregularly shaped quartz grain all the atoms present are packed in the same strict geometric arrangement; that is, the quartz crystals and the irregular quartz grains are both crystalline. The word *crystalline*, rather than crystal, is therefore used in the definition of a mineral.

Growth Habit

Every mineral has a characteristic crystal form. Some have such distinctive forms we can use the property as an identification tool without having to measure angles between faces. For example, the mineral pyrite (FeS_2) is commonly found as intergrown cubes (Fig. 2.12) with markedly striated faces. A few minerals can even develop distinctive growth habits when they grow in restricted environments. These too can be used as an aid to identification. For example, Figure 2.13 shows asbestos, a variety of the mineral serpentine that characteristically grows as fine, elongate threads.

Cleavage

The tendency of a mineral to break in preferred directions along bright, reflective plane surfaces is called **cleavage.**

FIGURE 2.11 Quartz grains that grew in an environment where other grains prevented development of well-formed crystal faces. Compare with Figure 2.10, which shows crystals that grew in open spaces, unhindered by adjacent grains.

FIGURE 2.12 Distinctive crystal form of pyrite, FeS_2. The characteristic growth habit of pyrite is cube-shaped crystals with pronounced striations on the cube face. The largest crystals in the photograph are 3 cm on an edge. The specimen is from Bingham Canyon, Utah.

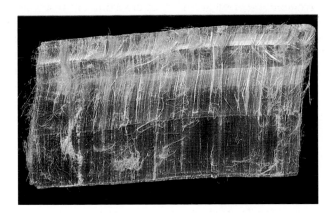

FIGURE 2.13 Some minerals have distinctive growth habits even though they do not develop well-formed crystal faces. The mineral chrysotile sometimes grows as fine, cottonlike threads that can be separated and woven into fireproof fabric, in which case it is referred to as asbestos. Chrysotile is one of several minerals that have asbestiform growth habits and are mined and commercialy processed for asbestos.

If you break a mineral with a hammer, or drop a specimen on the floor so that it shatters, you will probably see that the broken fragments are bounded by surfaces that are smooth and planar, just like crystal faces. In exceptional cases, such as the halite (NaCl) fragments show in Figure 2.14A, all of the breakage surfaces are smooth planar surfaces. Don't confuse crystal faces and cleavage surfaces however, even though the two often look alike. A cleavage surface is a breakage surface, while a crystal face is a growth surface.

The planar directions along which cleavage occurs are governed by the crystal structure (Fig. 2.14B). They are planes along which the bonds between atoms are relatively weak. Because the cleavage planes are direct expressions of the crystal structure the angles between cleavage planes are the same for all grains of a given mineral. Just as interfacial angles of crystals are constant, so are the angles between cleavage planes constant. Cleavage, therefore is a valuable guide for the identification of minerals.

Many common minerals have distinctive cleavage planes. One of the most distinctive is found in mica (Fig. 2.15). Clay minerals also have distinctive cleavage, and it is an easy cleavage direction that makes them feel smooth and slippery when rubbed between the fingers. Other minerals with distinctive cleavages are fluorite (CaF_2), which breaks along four planes into perfect octahedra, and potassium feldspar ($KAlSi_3O_8$), which breaks along two planes that are perpen-

A.

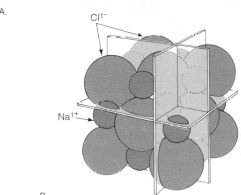

B.

FIGURE 2.14 Relation between crystal structure and cleavage. A. Halite, NaCl, has well-defined cleavage planes; it breaks into fragments bounded by perpendicular faces B. The crystal structure in the same orientation as the cleavage fragments shows that the directions of breakage are planes in the crystal between equal numbers of sodium and chlorine atoms.

FIGURE 2.15 Perfect cleavage of muscovite (a mica), shown by thin, plane flakes into which this specimen is being split. The cleavage flakes suggest leaves of a book, a resemblance embodied in the name "books of mica."

FIGURE 2.16 Two common minerals with distinct cleavages. Fluorite (left) breaks along four planes into octahedral cleavage fragments. Potassium feldspar (right) cleaves along the two planes that are perpendicular. The fluorite is from England, the feldspar from Maine.

dicular, creating fragments that are approximately rectangular (Fig. 2.16).

Luster

The quality and intensity of light reflected from a mineral produce an effect known as **luster**. Two minerals with almost identical color can have quite different lusters. The most important lusters are described as *metallic*, like that on a polished metal surface; *vitreous*, like that on glass; *resinous*, like that of resin; *pearly*, like that of pearl; and *greasy*, as if the surface were covered by a film of oil.

Color and Streak

The color of a mineral is often a striking property, but unfortunately color is not a very reliable means of identification. Color is determined by several factors, but the main cause is chemical composition. Some elements can create strong color effects even when they are present only in very small amounts by ionic substitution. For example, the mineral corundum (Al_2O_3) is commonly white or grayish, but when small amounts of Cr^{3+} replace Al^{3+} by ionic substitution, corundum is blood red, to which the gem name *ruby* is given. Similarly, when small amounts of Fe and Ti are present, the corundum is deep blue and another prized gem, *sapphire*, is the result (Fig. 2.17).

FIGURE 2.17 The many colors of corundum (Al_2O_3), here cut as gemstones, are caused by tiny amounts of trace elements entering the crystal structure by ionic substitution. The red color of ruby is produced by Cr^{3+} substitution for Al^{3+}, and the blue color of sapphire is produced by Fe^{3+} and Ti^{4+}.

FIGURE 2.18 Color contrast between hematite and a hematite streak. Massive hematite is opaque, has a metallic luster, and appears black. On a porcelain plate, hematite gives a red streak.

Streak

Color in opaque minerals with metallic lusters can be very confusing because the color is partly a property of the size of the individual mineral grains. One way to reduce errors of judgment where color is concerned is to prepare a *streak*, which is a thin layer of powdered mineral made by rubbing a specimen on a nonglazed porcelain plate. The powder gives a reliable color effect because all the grains in a powder streak are very small and the grain size effect is reduced. Red streak characterizes hematite (Fe_2O_3) even though the specimen itself looks black and metallic (Fig. 2.18).

Hardness

The term **hardness** refers to the relative resistance of a mineral to scratching. It is a distinctive property of minerals. Hardness, like crystal form and cleavage, is governed by crystal structure and by the strength of the bonds between atoms. The stronger the bonding, the harder the mineral.

Relative hardness values can be assigned by determining the ease or difficulty with which one mineral will scratch another. Talc, the basic ingredient of most body ("talcum") powders, is the softest mineral known, and diamond is the hardest. A scale called the

Moh's relative hardness scale is divided into 10 steps, each marked by a common mineral (Table 2.2). These steps do not represent equal intervals of hardness, but the important feature of the hardness scale is that any mineral on the scale will scratch all minerals below it. Minerals on the same step of the scale are just capable of scratching each other. For convenience, we often test relative hardness by using a common object such as a penny or a penknife as the scratching instrument, or glass as the object to be scratched.

Density and Specific Gravity

Another obvious physical property of a mineral is its density, which in practical terms means how heavy it feels. We know that two equal-sized baskets have different weights when one is filled with feathers and the other with rocks. The property that causes this difference is *density*, or the average mass per unit volume. The units of density are grams per cubic centimeter (g/cm^3). Minerals with a high density, such as gold, have their atoms closely packed. Minerals with a low density, such as ice, have loosely packed atoms.

Minerals are divided into a heaviness or density scale. Gold has a density of 19.3 g/cm^3 and feels very heavy, but many others such as galena (PbS) and mag-

TABLE 2.2 Mohs' Scale of Relative Hardness*

	Relative Number in the Scale	Mineral	Hardness of Some Common Objects
	10	Diamond	
	9	Corundum	
	8	Topaz	
Decreasing	7	Quartz	
	6	Potassium feldspar	
			Pocketknife; glass
	5	Apatite	
	4	Fluorite	
			Copper penny
	3	Calcite	
			Fingernail
	2	Gypsum	
	1	Talc	

*Named for Friedrich Mohs, an Austrian mineralogist, who chose the 10 minerals of the scale.

netite (Fe_3O_4) which have densities of 7.5 g/cm^3 and 5.2 g/cm^3, respectively, also feel heavy by comparison with many common minerals which have densities of 2.5–3.0 g/cm^3.

Density is difficult to measure accurately. We usually measure a property called specific gravity instead. **Specific gravity** is the ratio of the weight of a substance to the weight of an equal volume of pure water. Specific gravity is a ratio of two weights, so it does not have any units. Because the density of pure water is 1 g/cm^3, the specific gravity of a mineral is numerically equal to its density. Specific gravity can be approximated by holding different minerals in the hand and comparing their weights. Metallic minerals feel heavy, whereas nearly all others feel light.

Mineral Properties and Bond Types

Mineral properties depend strongly on the kind of bonding present. Ionic and covalent bonds are strong, and, as shown in Table 2.3, their presence makes minerals hard but brittle. By comparison, metallic and van der Waals bonds are much weaker, and compounds containing them tend to be soft and easily deformed.

COMMON MINERALS

Scientists have identified approximately 3000 minerals. Most occur in the crust, but a few have been identi-

TABLE 2.3 Examples of Mineral Properties that Depend on Bond Type

Bond Type	Mineral	Strength	Hardness	Electrical Conductance	Solubility in Water and Weak Acids
Ionic	Calcite ($CaCO_3$) Halite ($NaCl$)	High	Moderate to high	Very low	High
Covalent	Diamond (C) Sphalerite (ZnS)	Very high	High	Very low	Very low
Mixed Ionic and Covalent	Olivine (Mg_2SiO_4) Muscovite $KAl_2(Si_3Al)O_{10}(OH)_2$	Very high	Moderate to high	Very low	Low
Metallic	Gold (Au) Copper (Cu)	Moderate	Low	High	Very low
van der Waals	Graphite (C) Sulfur (S)	Very low	Very low	Low	Low

TABLE 2.4 The Most Abundant Chemical Elements in the Continental Crust

Element	Percent by Weight
Oxygen (O)	45.20
Silicon (Si)	27.20
Aluminum (Al)	8.00
Iron (Fe)	5.80
Calcium (Ca)	5.06
Magnesium (Mg)	2.77
Sodium (Na)	2.32
Potassium (K)	1.68
Titanium (Ti)	0.86
Hydrogen (H)	0.14
Manganese (Mn)	0.10
Phosphorus (P)	0.10
All other elements	0.77
Total	100.00

fied in meteorites, and two new ones were discovered in the Moon rocks brought back by the astronauts. The total number of minerals may seem large, but it is tiny by comparison with the astronomically large number of ways a chemist can combine naturally occurring elements to form compounds. The reason for the disparity between nature and chemical experiment becomes apparent when we consider the relative abundances of the chemical elements. As Table 2.4 shows, only 12 elements occur in the crust in amounts greater than 0.1 percent. Together these 12—usually referred to as the abundant elements—make up 99.23 percent of the crust mass. The crust is constructed, therefore, of a limited number of minerals in which one or more of the 12 abundant elements is an essential ingredient.

Rather than forming distinct minerals, the many scarce elements tend to occur by ionic substitution. For example, the mineral olivine [$(Mg,Fe)_2SiO_4$], contains, in addition to Mg, Fe, Si, and O, trace amounts of Cu, Ni, Co, Mn, and many other elements as ionic substitutes for the Mg or Fe.

Minerals containing scarce elements certainly do occur, but only in small amounts and those small amounts only form under special and restricted circumstances. A few scarce elements such as hafnium and rhenium, are so rare that they are not known to form minerals under any circumstances—they only occur by ionic substitution.

As Table 2.4 shows, two elements, oxygen and silicon, make up more than 70 percent of the crust. Oxygen forms a simple anion, O^{2-}, and compounds that contain the O^{2-} anion are called oxides. Silicon forms a simple cation, Si^{4+}, but oxygen and silicon together form an exceedingly strong complex ion, the **silicate anion** $(SiO_4)^{4-}$. Minerals that contain the silicate anion are complex oxides, and to distinguish them from simple oxides they are called **silicates**, or **silicate minerals**. For example, MgO is an oxide, but Mg_2SiO_4 is a silicate.

Silicate minerals are the most abundant of all naturally occurring, inorganic compounds, and simple oxides are the second most abundant group. Other mineral groups, all important, but less common than silicates and oxides, are sulfides, which contain the simple anion S^{2-}, carbonates $(CO_3)^{-2}$, sulfates $(SO_4)^{-2}$, and phosphates $(PO_4)^{-3}$.

The Silicate Minerals

The Silicate Tetrahedron

The four oxygen atoms in a silicate anion are tightly bonded to the single silicon cation. The bonding, which is very strong, is largely covalent. Oxygen is a large anion (Fig. 2.9), while silicon is a small cation. In the silicate anion, the oxygen ions pack into the smallest space possible for four large spheres. As can be seen in Figure 2.19, the four oxygens sit at the corners of a tetrahedron and the small silicon cation sits in the space between the oxygens at the center of the tetrahedron. Therefore, the shape of the silicate anion is a tetrahedron. The structures and properties of silicate minerals are all determined by the way the $(SiO_4)^{4-}$ silicate tetrahedra pack together in the crystal structure.

Each silicate tetrahedron has four negative charges. Silicon has four electrons in its outer shell, and oxygen has six. The silicon atom in a silicate tetrahedron shares one of its electrons with each of the oxygens. The four oxygens, in turn, each share one of their electrons with the silicon. This leaves the silicon with a stable outer shell of eight electrons, but each of the four oxygens still requires an additional electron for a stable octet. The oxygens can attain a stable outer shell of eight electrons in two ways:

1. They can accept electrons from, or share electrons with, cations. An example of this is found in olivine (Mg_2SiO_4), in which two Mg atoms give up their two outer shell electrons to the oxygens, forming ionic bonds in the process.

2. They can bond with two Si atoms at the same time. The shared oxygen is covalently bonded to each of two silicons. The outer electron shell of the shared oxygen is now filled because it accepts an electron from each silicon and the two tetrahedral-shaped silicate anions, now joined at a

A.

B.

C.

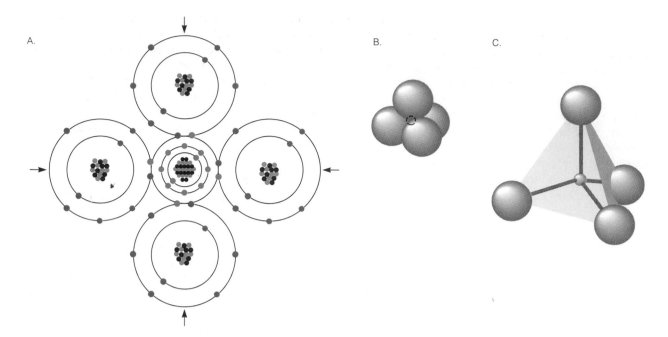

FIGURE 2.19 Bonding and structure of the silicate anion. A. Silicon has four electrons in its outer energy-level shell. Each of the four electrons is shared with one oxygen, and each oxygen in turn shares an electron with the silicon. The silicon finishes with eight electrons in the outer shell, each oxygen with seven electrons. Bonding between the silicate anion and other ions occurs when the four oxygens share or accept an additional electron, thereby filling their outermost energy-level shells. B. Tetrahedral-shaped silicate anion with oxygen ions touching each other in natural positions. Silicon (dashed circle) occupies central space. C. Expanded view showing large oxygen anions at the four corners, equidistant from a small silicon cation. Dotted lines show bonds between silicon and oxygen ions; solid lines outline the tetrahedron.

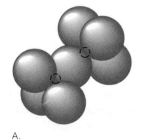

A.

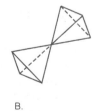

B.

FIGURE 2.20 Two silicate tetrahedra share an oxygen, thereby satisfying some of the unbalanced electrical charges and, in the process, forming a larger and more complex anion group. A. The arrangement of oxygen and silicon ions in double tetrahedra, giving the complex anion $(Si_2O_7)^{6-}$. B. A geometric representation of the double silicate tetrahedra.

common apex, form an even larger complex anion with the formula $(Si_2O_7)^{6-}$, as shown in Figure 2.20. The large $(Si_2O_7)^{6-}$ anion forms compounds by accepting or sharing electrons with cations in the same way that smaller anions do. Only one common mineral, epidote, contains an $(Si_2O_7)^{6-}$ anion formed by the simplest form of oxygen sharing. However, just as more and more beads can be strung on a necklace, so can **polymerization**, the process of linking silicate tetrahedra by oxygen sharing be extended to form huge anions (Fig. 2.21A).

Most common silicate minerals contain very large anions as a result of polymerization. As shown in Figure 2.21B, if a tetrahedron shares more than one oxygen with adjacent tetrahedra, the complex anions can have large circular shapes, endless chains, sheets, and even three-dimensional networks of tetrahedra. No matter how big an anion becomes by polymerization, any oxygen that does not have a stable outer shell of eight electrons can form bonds with cations by accepting or sharing electrons.

Observe in Figures 2.21A and B that an important restriction to the polymerization process is always met—two adjacent tetrahedra never share more than one oxygen. Stated another way, tetrahedra only join at their apexes, never along the edges or faces. The common polymerizations, together with the rock-forming minerals containing them, are illustrated in Figure 2.21C and discussed below in order of increasing complexity of polymerization.

The Olivine Group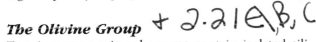

Two important mineral groups contain isolated silicate tetrahedra. The first is the olivine group, which is a glassy-looking group of minerals that are usually pale green in color (Fig. 2.22A). As previously discussed, Fe^{2+} can substitute readily for Mg^{2+} in olivine, giving rise to the group formula $(Mg,Fe)_2SiO_4$. It is the Fe^{2+} that gives olivine its green color. Olivine is one of the most abundant mineral groups in the Earth, being a very common constituent of igneous rocks in the oceanic crust and upper part of the mantle. In rare cases, olivine occurs in such flawless and beautiful crystals that it is used as the gem peridot.

The Garnet Group

The second important mineral group with isolated silicate tetrahedra is the garnet group. As with olivine, ionic substitution gives garnet a range of compositions. But with garnets the range is much wider. The garnet group has the complex formula $A_3B_2(SiO_4)_3$, where A can be any of the cations Mg^{2+}, Fe^{2+}, Ca^{2+},

and Mn^{2+}, or any mixture of them, while B can be either Al^{3+}, Fe^{3+}, Cr^{3+} or a mixture of them. Garnet is characteristically found in metamorphic rocks of the continental crust but it can also be found in certain igneous rocks. One of the most striking features of garnet is its tendency to form beautiful crystals (Fig. 2.22B). An important property of garnet is its hardness, which makes it useful as an abrasive for grinding and polishing.

The Pyroxene and Amphibole Groups

Pyroxenes and amphiboles are two silicate mineral groups that contain long, chainlike anions. Pyroxenes are built from a polymerized chain of tetrahedra, each of which shares two oxygens, so the anion has the general formula $(SiO_3)_n^{2-}$, where n is any very large number. Amphiboles are built from double chains of tetrahedra equivalent to two pyroxene chains in which half the tetrahedra share two oxygens and the other half share three oxygens, giving a general anion formula of $(Si_4O_{11})_n^{6-}$. Both the pyroxene and amphibole chains are bonded together by cations such as Ca^{2+}, Mg^{2+}, and Fe^{2+} which transfer electrons, and thus form bonds with, oxygens in adjacent chains that have unfilled shells.

The general formula for pyroxene is $AB(SiO_3)_2$, where A and B can be any of a number of cations, the most important of which are Mg^{2+}, Fe^{2+}, Ca^{2+}, Mn^{2+}, Na^{1+}, and Al^{3+}. The pyroxenes are most abundantly found in igneous rocks of the oceanic crust and mantle, but they also occur in many igneous and metamorphic rocks of the continental crust. The most common pyroxene is a shiny black variety called *augite*, with the approximate formula $Ca(Mg,Fe,Al)[(Si,Al)O_3]_2$.

Amphiboles have, perhaps, the most complicated formulas of all the common minerals. The general formula is $A_2B_5(Si_4O_{11})_2(OH)_2$, in which A is most commonly either Ca^{2+}, Mg^{2+}, or Na^{1+}, and B is usually Mg^{2+}, Fe^{2+}, or Al^{3+}. Even this complicated formula does not completely describe the composition of the most abundant variety of amphibole, *hornblende*, a dark green to black mineral that looks rather like augite, but can be distinguished from augite because the angles between cleavage surfaces differ in two cases, as shown in Figure 2.23.

The Clays, Micas and Chlorites

Clays, micas, and chlorites are related mineral groups because they all contain polymerized sheets of silicate tetrahedra. The sheets are formed by each tetrahedron sharing three of its oxygens with adjacent tetrahedra to give a general anion formula $(Si_4O_{10})_n^{4-}$. This leaves a single oxygen in each tetrahedron with an unfilled electron shell. Cations such as Al^{3+}, Mg^{2+},

A.

B.

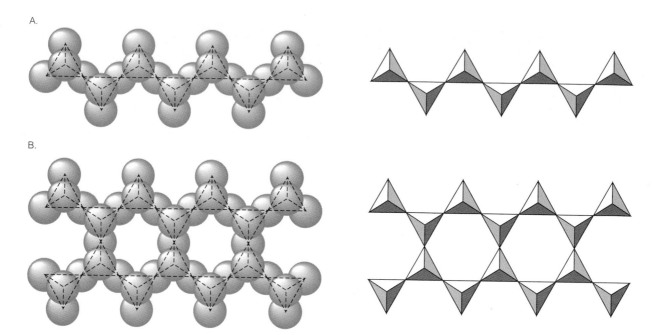

FIGURE 2.21 Formation of complex silicate anions by polymerization. A. Polymerization of silicate anions to form a continuous chain in which each silicate anion shares two oxygens with adjacent anions. A geometric representation of the chain is on right. Formula of the complex anion is $(SiO_3)_n^{2-}$. B. Double chain of a polymerized silicate anions with formula $(Si_4O_{11})_n^{6-}$.

and Fe^{2+} bond with this oxygen and thereby hold the polymerized sheets together in the crystal structure. The Si–O bonds in the polymerized sheets are much stronger than are the bonds between the other cations and oxygen. As a result, the clays, micas, and chlorites all display a very pronounced cleavage parallel to the sheets (Fig. 2.24).

In the case of the clays, the oxygens with unfilled shells bond with Al^{3+} cations, leading to the formula $Al_4Si_4O_{10}(OH)_8$ for the clay mineral *kaolinite*. The clays are among the commonest minerals in the regolith.

With the micas, both Al^{3+} and K^{1+} are the bonding cations. There are two common kinds of mica. The variety called *muscovite* has the formula $KAl_2(Si_3Al) \cdot O_{10}(OH)_2$. Muscovite is a clear, almost colorless mica which takes its name from muscovy, an old term for Russia, which was famous for producing large cleavage sheets of muscovite that could be used for windows. *Biotite*, which has the formula $K(Mg,Fe)_3(Si_3Al) \cdot O_{10}(OH)_2$, is a dark-brown mica whose color is due to the iron. The micas are common minerals in both igneous and metamorphic rocks.

The Chlorite group has the formula $(Mg,Fe,Al)_6 (Si,Al)_4O_{10}(OH)_8$. It is usually green in color and it derives its name from *chloros*, a Greek word meaning

green. Chlorite is a common alteration product from other minerals such as olivine, biotite, hornblende, and augite that contain iron and magnesium. Igneous rocks of the oceanic crust, for example, commonly contain olivine and augite. When such minerals are in contact with seawater, they alter to chlorite. The (OH) in chlorite is one of the ingredients that promotes the melting of oceanic crust during subduction discussed in chapter 1.

Quartz

The only common mineral composed exclusively of silicon and oxygen is quartz, (SiO_2). It provides an example of polymerization filling all unfilled oxygen shells, and in the process forming a three-dimensional network of tetrahedra.

Quartz characteristically forms six-sided crystals (Fig. 2.10), and is found in many beautiful colors. The color comes from minute amounts of iron, aluminum, titanium, and other elements present by ionic substitution. Quartz occurs in igneous, metamorphic, and sedimentary rocks and is one of the most widely used gem and ornamental minerals.

Certain specimens of quartz—those formed by precipitation from cool water solutions—are so fine-grained that they almost appear amorphous, and we

	Arrangement of silica tetrahedra	Formula of the complex ions	Typical mineral	
			Name	Composition
Isolated tetrahedra		$(SiO_4)^{4-}$	Olivine	$(Mg,Fe)_2SiO_4$
Isolated polymerized groups		$(Si_2O_7)^{6-}$	Epidote	$Ca_2Fe_2Al_2O[Si_2O_7][SiO_4](OH)$
	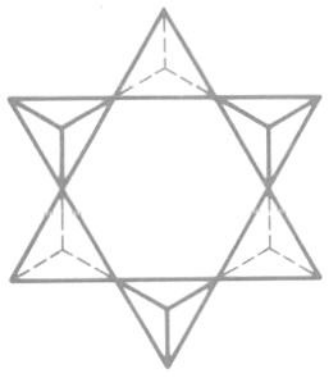	$(Si_6O_{18})^{12-}$	Beryl	$Be_3Al_2Si_6O_{18}$
Continuous chains		$(SiO_3)_n^{2-}$	Pyroxene	$CaMg(SiO_3)_2$ (Variety; diopside)
	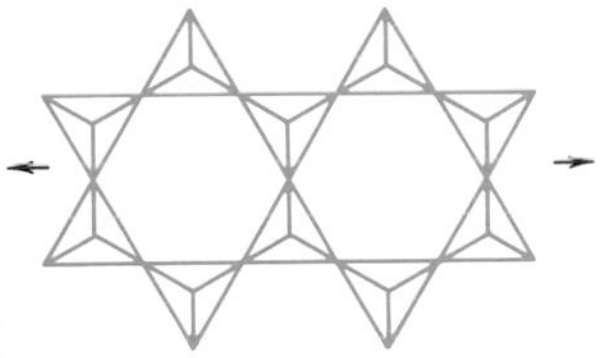	$(Si_4O_{11})_n^{6-}$	Amphibole	$Ca_2Mg_5(Si_4O_{11})_2(OH)_2$ (Variety; tremolite)
Continuous sheets	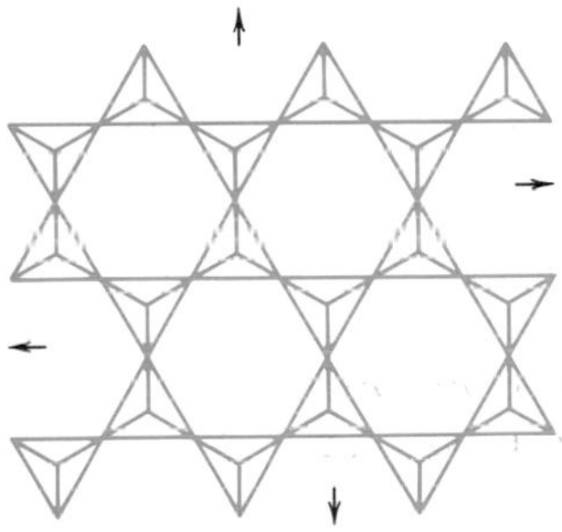	$(Si_4O_{10})_n^{4-}$	Mica	$KAl_2(Si_3Al)O_{10}(OH)_2$ (Variety; muscovite)
Three-dimensional networks	Too complex to be shown by a simple two-dimensional drawing	(SiO_2)	Quartz	SiO_2

FIGURE 2.21 C. Summary of the way silicate anions polymerize to form the complex anions of the common silicate minerals. Polymerizations other than those shown are known but do not occur in common minerals. The most important polymerizations are those that produce chains, sheets, and three-dimensional networks.

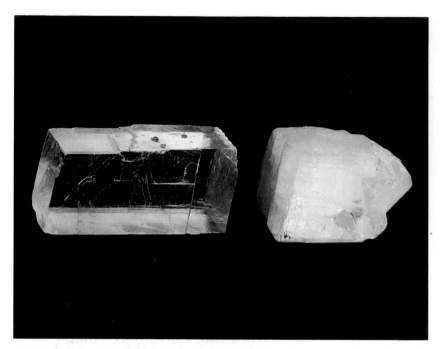

FIGURE 2.27   Calcite ($CaCO_3$) on the left, and dolomite $CaMg(CO_3)_2$ on the right have similar crystal structures and, as a result, similar cleavages. Both cleave along three planes that are not perpendicular, yielding rhombohedral-shaped fragments. The calcite is from Iceland, the dolomite from Traversella, Italy.

nite have the same composition, $CaCO_3$, and are polymorphs. Calcite is much more common than aragonite. Dolomite has the formula $CaMg(CO_3)_2$.

Calcite and dolomite are both very common minerals, and they look very similar. They have the same vitreous luster and the same distinctive cleavage (Fig. 2.27), both are relatively soft, and both are common in sedimentary rocks—calcite in limestone, dolomite in dolostone. One easy way to distinguish between them is with dilute hydrochloric acid (HCl). Calcite reacts vigorously, bubbling and effervescing, while dolomite reacts very slowly with little or no effervescence.

Phosphates

Apatite is by far the most important phosphate mineral. It contains the complex anion $(PO_4)^{3-}$ and has the general formula $Ca_5(PO_4)_3(F,OH)$. Apatite is the substance from which our bones and teeth are made. It is also a common mineral in many varieties of igneous and sedimentary rocks and is the main source of the phosphorus used for making phosphate fertilizers.

Sulfates

All sulfate minerals contain the sulfate anion $(SO_4)^{2-}$. Although many sulfate minerals are known, only two are common, and both are calcium sulfate minerals: *anhydrite,* $CaSO_4$; and *gypsum,* $CaSO_4 \cdot 2H_2O$. Both form when seawater evaporates—anhydrite when temperatures are high, gypsum at lower temperature. Gypsum is the raw material used for making plaster. Plaster of paris got its name from a quarry near Paris where a very desirable, pure white form of gypsum was mined centuries ago.

The Ore Minerals

The term *ore mineral* is used for minerals that are sought and processed for their valuable metal contents. Such minerals tend to be elements, sulfides or oxides. The important ore minerals are listed in appendix D.

Sulfides

The common sulfide minerals all have metallic luster and high specific gravity. The two most common, *pyrite* (FeS_2) and *pyrrhotite* (FeS) are not actually mined for their iron content, but even so they are commonly referred to as ore minerals. Most of the world's lead is won from *galena* (PbS), most of the zinc from *sphalerite* (ZnS) (Fig. 2.28), and most of the copper from *chalcopyrite* ($CuFeS_2$). Other familiar metals won from sulfide ore minerals are cobalt, mercury, molybdenum, and silver.

FIGURE 2.28 Examples of ore minerals. From left to right: Sphalerite (ZnS), the main ore mineral of zinc, from Kapnik, Rumania: cassiterite (SnO_2), the main ore mineral of tin, from Cornwall, England; and uraninite (U_3O_8), the main ore mineral of uranium, from the Goya district, India. Uraninite is black; the yellow-colored mineral is an alteration product.

Oxides

Because iron is one of the most abundant elements in the crust, the iron oxides *magnetite* (Fe_3O_4) and *hematite* (Fe_2O_3) are the two most common oxide minerals. Magnetite takes its name from the Greek word *Magnetis*, meaning stone of Magnesia, an ancient town in Asia Minor. Magnetis had the power to attract iron particles and so the terms *magnet* and *magnetite* eventually joined our vocabulary. Hematite refers to its red color when powdered, the Greek word for red blood being *haima*. Magnetite and hematite are the main ore minerals of iron.

Other oxide ore minerals are *rutile* (TiO_2) the principal source of titanium; *cassiterite* (SnO_2) (Fig. 2.28), the main ore mineral for tin; and *uraninite* (U_3O_8) (Fig. 2.28), the main source of uranium. Other metals won from oxide ore minerals are chromium, manganese, niobium, and tantalum.

MINERALS AS INDICATORS OF THE ENVIRONMENT OF THEIR FORMATION

Minerals should not be regarded merely as objects of beauty or sources of economically valuable materials. Contained within their makeup are the keys to the conditions under which they (and the rocks they are in) formed. The study of minerals, therefore, can provide insight into the chemical and physical conditions in regions of the Earth we cannot observe and measure directly.

Our understanding of the growth environments of minerals has come largely through studying minerals in the laboratory. By suitable experiments, for example, scientists have been able to define the temperatures and pressures at which a diamond will form rather than its polymorph, graphite (Fig. 2.29). Because we can infer how temperature and pressure increase with depth in the Earth, we can state with certainty that rocks in which diamonds are found are samples of the mantle from at least 150 km below the Earth's surface. Another example concerns weathering. The minerals that form in the regolith during weathering are controlled by the climate—cold and wet versus hot and dry, for example. Past climates can therefore be deciphered from the kinds of minerals preserved in sedimentary rocks. The composition of seawater in past ages can also be determined from the minerals formed when the seawater evaporated and deposited its salts. Rather than elaborating many examples at this point, we turn next to a brief examination of rocks and return in later chapters to the way minerals are used to understand past environments.

MINERALS AND SOCIETY

Most minerals that are abundant in the Earth's crust have neither commercial value nor any particular use. Ore minerals that are the raw materials of industry tend to be rare and hard to find. From the ore minerals we get the metals to make our machines and the ingredients for chemicals and fertilizers. Our modern society is totally dependent on an adequate supply of ore minerals. As Figure B2.1 makes clear, each of us uses, directly or indirectly a very large amount of material derived from ore minerals. Without the needed supplies we could not build planes, cars, televisions, or computers. Industry would falter and living standards would decline.

Can the ore minerals in the Earth's crust sustain both a growing population and a high standard of living for everyone? This difficult question has many experts worried. The minerals they worry most about are those used as sources of such important metals as lead, zinc, and copper. Metals, the experts point out, begin the chain of resource use. Without metals, we cannot make machines. Without machines we cannot convert the chemical energy of coal and oil to useful mechanical energy. Without mechanical energy, the tractors that pull plows must grind to a halt; trains and trucks must stop running; and indeed our whole industrial complex must become still and silent.

Experts have no way of telling how long the Earth's supplies of ore mineral will last. Optimistic experts point to the great success our technological society has enjoyed over the past two centuries as ever more remarkable discoveries have been made. If mineral supplies become limited, they suggest, we will find ways to get

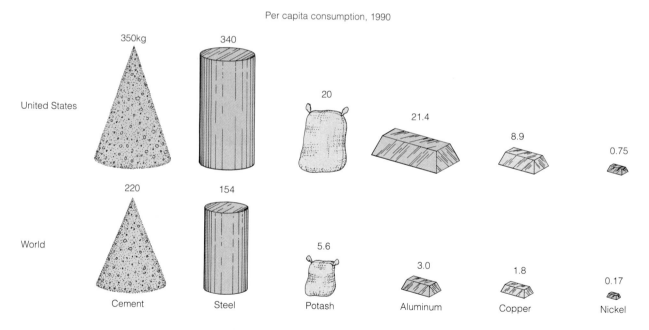

Per capita consumption, 1990

United States: 350kg (Cement), 340 (Steel), 20 (Potash), 21.4 (Aluminum), 8.9 (Copper), 0.75 (Nickel)

World: 220 (Cement), 154 (Steel), 5.6 (Potash), 3.0 (Aluminum), 1.8 (Copper), 0.17 (Nickel)

FIGURE B2.1   The average amount of material consumed per person each year (the per capita consumption) is greater in an industrially advanced country such as the United States than it is for the rest of the world.

around the limits by recycling, by substitution, and through the discovery of new technologies. Many geologists have an opposite and more pessimistic opinion. Technologically advanced societies have faced mineral resource limits in the past, they point out, but the solution has always been to import new supplies from elsewhere rather than trying to develop substitutes or develop effective recycling measures.

England, for instance, was once a great supplier of metals (Figure B2.2). Today, the minerals are mined out, most of her mines are closed, and English industry runs on raw materials imported from abroad. The United States, too, was once self-sufficient in most minerals and an exporter of many minerals. Slowly the situation has changed, so that now the United States is a net importer and has to rely on supplies from such countries as Australia, Chile, South Africa, and Canada. The only large industrial country that can still supply most of its mineral needs is the Soviet Union. But eventually the Soviet mines will be depleted of their minerals, and so too will the mines of Australia and other countries. Where then does society turn?

The answer to the question just posed is not obvious, but it is one that must be answered in the foreseeable future. It is highly likely that within the lifetimes of the people who read this book, mineral limitations may occur. Which minerals, and therefore which metals, would first be in short supply is still an open question. That limitations will eventually happen, however, is no longer an open question. How society will cope and respond, and when it will have to do so, are just two of the great social and scientific issues still to be solved.

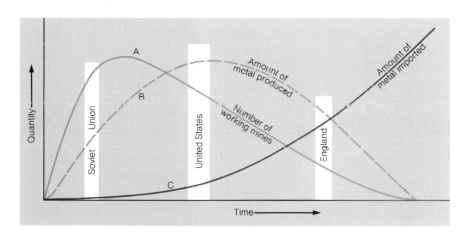

FIGURE B2.2 Changing metal production illustrated by the histories of three countries. The number of mines rises rapidly (curve A) but declines when the rate of mine exhaustion exceeds the discovery rate. The amount of metal produced (curve B) also rises, then falls when mines become exhausted. Curve C represents the growing imports of metal when internal production fails to meet needs. As time passes, the position of a country moves from left to right. In the late nineteenth century, Britain was about where the United States is today, at which time the United States was at about the present position of the Soviet Union.

SUMMARY

1. Minerals are naturally formed, solid chemical elements or compounds having a definite composition and a characteristic crystal structure. Crystal structure is the geometric array of atoms in a crystalline solid.

2. Minerals are formed through the bonding together of cations and anions of different chemical elements. The most important kinds of bonding in minerals are ionic bonds, which involve the transfer of electrons, and covalent bonds which involve the sharing of orbiting electrons.

3. The composition of some minerals varies because of ionic substitution, whereby one ion in a crystal structure can be replaced by another ion having a like charge and like ionic radius.

4. Some compounds have the same composition but different crystal structures. Each structure is a separate mineral. Minerals having the same compositions but differing structures are called polymorphs.

5. The principal properties used to characterize and identify minerals are crystal form, mineral habit, cleavage, luster, color and streak, hardness, and specific gravity.

6. Approximately 3000 minerals are known, but of these about 20 make up more than 95 percent of the Earth's crust.

7. Silicates are the most common minerals, followed by oxides, carbonates, sulfides, sulfates, and phosphates.

8. The basic building block of silicate minerals is the silicate tetrahedron, a complex anion in which an Si^{4+} ion is bonded to four O^{2-} ions. The four O^{2-} ions sit at the apexes of a tetrahedron, with the Si^{4+} at its center. Adjacent silicate tetrahedra can bond together to form larger complex anions by sharing one or more oxygens. The process is called polymerization.

9. The feldspars are the most abundant group of minerals in the Earth's crust, comprising approximately 60 percent of the volume of the crust. Quartz is the second most common mineral in the crust.

10. The differences between rocks can be described in terms of the mineral assemblage, and the texture imparted by the constituent particles present.

IMPORTANT TERMS TO REMEMBER

anion (p. 45)
atom (p. 44)

cation (p. 45)
cleavage (p. 52)
complex ion (p. 48)
covalent bond (p. 46)
crystal face (p. 51)
crystal form (p. 51)
crystal (p. 51)
crystalline (p. 49)
crystal structure (p. 49)

energy-level shell (p. 44)

hardness (p. 55)

ion (p. 45)
ionic bond (p. 46)
ionic radius (p. 49)
ionic substitution (p. 49)

luster (p. 54)

metallic bond (p. 47)
mineral (p. 44)
mineral assemblage (p. 66)
mineraloid (p. 51)

polymerization (p. 59)
polymorph (p. 51)

silicate (p. 57)
silicate anion (p. 57)
silicate mineral (p. 57)
specific gravity (p. 56)

texture (p. 66)

van der Waals bond (p. 47)

MINERAL NAMES TO REMEMBER

agate (p. 63)
albite (p. 63)
amphibole group (p. 59)

anhydrite (p. 64)
anorthite (p. 63)
apatite (p. 64)

aragonite (p. 63)
augite (p. 59)

biotite (p. 60)

graphite (p. 47)
gypsum (p. 64)

plagioclase (p. 63)
potassium feldspar (p. 63)
pyrite (p. 64)
pyroxene group (p. 59)
pyrrhotite (p. 64)

calcite (p. 63)
cassiterite (p. 65)
chalcedony (p. 63)
chalcopyrite (p. 64)
chlorite group (p. 60)
clay group (p. 59)
corundum (p. 54)

halite (p. 53)
hematite (p. 65)
hornblende (p. 59)

quartz (p. 60)

ice (p. 48)

rutile (p. 65)

diamond (p. 46)
dolomite (p. 63)

jasper (p. 63)

sphalerite (p. 64)

kaolinite (p. 60)

talc (p. 47)

feldspar group (p. 63)
flint (p. 63)

magnetite (p. 65)
muscovite (p. 60)

uraninite (p. 65)

galena (p. 49)
garnet group (p. 59)

olivine group (p. 59)
*opal (p. 50)

QUESTIONS FOR REVIEW

1. What is a mineral? Give three reasons why the study of minerals is important.

2. Describe three ways chemical elements bond together to form compounds. Name two minerals that are examples of each bond type.

3. What properties besides composition and crystal structure can be used to identify minerals?

4. What are polymorphs? What are the mineral names of the two polymorphs of $CaCO_3$? ot C?

5. What mineral has a crystal structure in which Pb and S atoms alternate at the corners of a three dimensional, right-angled geometric grid?

6. What is ionic substitution? Illustrate your answer with two examples.

7. Approximately how many common minerals are there? What is the most common one?

8. Can you name five common minerals that are found in the area in which you live?

9. Describe the structure, composition, and ionic charge of the silicate anion.

10. Describe how silicate anions join together to form silicate minerals.

11. Describe the polymerization of silicate anions in the pyroxenes, the micas, and the feldspars.

12. Name five common minerals that are not silicates and name the anion each contains.

13. Are any minerals mined in the area in which you live? What are they and what are they mined for?

14. Can a rock be uniquely defined on the basis of its mineral assemblage? If not, what additional information is needed.

15. Describe the ways by which loose aggregates of sediment can be transformed into rock.

16. What holds together the mineral grains in metamorphic and igneous rocks?

*a mineraloid

MAGMA

We learned in chapter 1 that magma is molten rock. A complete definition for **magma** is molten rock, together with any suspended mineral grains and dissolved gases, that forms when temperatures rise sufficiently high for melting to happen in the crust or mantle. When magma reaches the Earth's surface, it does so through a **volcano,** which is a vent from which magma, solid rock debris, and gases are erupted. The term *volcano* comes from the name of the Roman god of fire, Vulcan, and it immediately conjures up visions of streams of **lava**—magma that reaches the Earth's surface—pouring out over the landscape. Some lava does flow as hot streams, but magma can also be erupted as clouds of tiny red-hot fragments, as was the case at Mount St. Helens. Magma, volcanoes, and eruption processes are much more varied than is commonly realized.

Volcanoes are the only places we can actually see and study magma, so we start this chapter by gaining some insight into volcanoes and the properties of magma, then proceed to a discussion of the kinds of igneous rock formed by the cooling of magma, and finally to a discussion of the way magma is thought to form.

By observing the eruption of lava, we can draw three important conclusions concerning magma.

1. Magma is characterized by a *range of compositions* in which silica (SiO_2) is always predominant.
2. Magma is characterized by *high temperatures*.
3. Magma has the properties of a liquid, including the *ability to flow*. This is true even though some magma is almost as stiff as window glass. Most magma is a mixture of crystals and liquid (often referred to as *melt*).

Composition

The composition of magma is controlled by the most abundant elements in the Earth—Si, Al, Fe, Ca, Mg, Na, K, H, and O. Because O^{2-} is the most abundant anion, it is usual to express compositional variations of magmas in terms of oxides, such as SiO_2, Al_2O_3, CaO, and H_2O. The most abundant component is SiO_2.

Three distinct types of magma are more common than all others. The first type contains about 50 percent SiO_2, the second about 60 percent, and the third about 70 percent. The three magmas are *basaltic, andesitic,* and *rhyolitic* and the names of the common igneous rocks derived from them are **basalt, andesite,** and **rhyolite** (Fig. 3.1). The three magmas are not formed in equal abundance. Approximately 80 percent of all magma erupted by volcanoes is basaltic, while andesitic and rhyolitic magmas are each about 10 percent of the total. Hawaiian volcanoes such as Kilauea and Mauna Loa are basaltic, Mount St. Helens and Krakatau are both andesitic volcanoes, and the now dormant volcanoes at Yellowstone National Park are rhyolitic. As we shall see later in this chapter, the distribution of the different kinds of volcanoes is closely related to plate tectonics.

Gases Dissolved in Magma

Small amounts of gas (0.2 to 3% by weight) are dissolved in all magma, and even though present in only small amounts, these gases strongly influence the properties of magma. The principal gas is water vapor which, together with carbon dioxide, accounts for more than 98 percent of all gases emitted from volcanoes. Other gases include nitrogen, chlorine, sulfur, and argon, which are rarely present in amounts exceeding 1 percent.

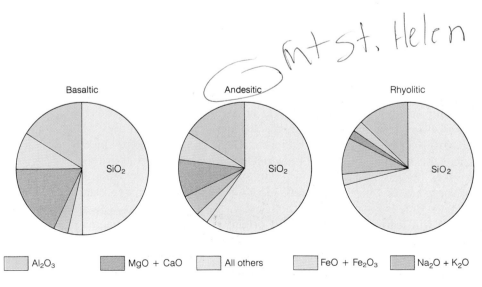

FIGURE 3.1 The average compositions (in weight percent) of the three principal kinds of magma.

Temperature

The temperature of magma is difficult to measure, but it can sometimes be done during volcanic eruptions. As we have seen, volcanoes are dangerous places. Because scientists who study them are not eager to be roasted alive, measurements must be made from a distance using optical devices. Magma temperatures determined in this manner during eruptions of volcanoes, such as Kilauea in Hawaii and Mount Vesuvius in Italy, range from 1000° to 1200°C. Experiments on synthetic magmas in the laboratory suggest that, under some conditions, magma temperatures might even be as high as 1400°C.

Viscosity

Dramatic pictures of lava flowing rapidly down the side of a volcano prove that some magmas are very fluid. Basaltic lava moving down a steep slope on Mauna Loa in Hawaii has been clocked at 16 km/h. Such fluidity is rare, and flow rates are more commonly measured in meters per hour or even meters per day. As seen in Figure 3.2, which shows basaltic lava destroying houses in Hawaii, flow rates are usually slow enough so that people are not endangered. Magma containing 70 percent or more SiO_2 flows so slowly that movement can hardly be detected. The properties of such magma are more akin to those of solids than to liquids.

The internal property of a substance that offers resistance to flow is called **viscosity.** The more viscous a magma, the less fluid it is. Viscosity of a magma depends on temperature and composition (especially the silica and dissolved-gas contents).

Effect of Temperature on Viscosity

The higher the temperature, the lower the viscosity, and the more readily a magma flows. A very hot magma erupted from a volcano may flow readily, but it soon begins to cool, becomes more viscous, and eventually slows to a complete halt. In Figure 3.3, the smooth ropy-surfaced lava, locally called *pahoehoe* in Hawaii, formed from a hot, gas-charged, very fluid lava. The rubbly, rough-looking lava, by contrast, formed from a cooler, gas-poor lava with a high viscosity. Hawaiians call this rough lava *aa*.

Effect of Silica Content on Viscosity

The SiO_4^{4-} anions that occur in silicate minerals (chapter 2) are also present in magmas. Just as they do in minerals, these anions polymerize by sharing oxygens. Unlike the anions in silicate minerals, however, those in magma form irregularly shaped groupings of chains, sheets, and networks. As the average number of silicate tetrahedra in the polymerized groups becomes larger, the magma becomes more viscous—in other words, more resistant to flow—and behaves increasingly like a solid.

The number of tetrahedra in the groups depends on the silica content of the magma. The higher the silica content, the larger the polymerized groups. For this reason rhyolitic magma is always more viscous

FIGURE 3.2 An advancing tongue of basaltic lava setting fire to a house in Kalapana, Hawaii, during an eruption of Kilauea volcano in June 1989. Flames at the edge of the flow are due to burning lawn grass.

FIGURE 3.3 The way lava flows is controlled by viscosity. Two different flows are visible. They have the same basaltic composition. The lower flow, on which the geologist is standing, is a pahoehoe flow formed from a low-viscosity lava like that shown in Figure 3.2. The upper flow (the one being sampled by the geologist), which is very viscous and slow moving, is an aa flow erupted from Kilauea volcano in 1989. The pahoehoe flow was erupted in 1959.

than basaltic magma, and andesitic magma has a viscosity that is intermediate between the two.

ERUPTION OF MAGMA

Magma, like most other liquids, is less dense than the solid rock from which it forms. Therefore, once formed, the lower-density magma will exert an upward push on the enclosing higher density rock and will slowly force its way up. There is, of course, a pressure on a rising mass of magma due to the weight of all the overlying rock. The pressure is proportional to depth. As a magma rises upward, therefore, the pressure must decrease.

Pressure controls the amount of gas a magma can dissolve—more at high pressure, less at low. Gas dissolved in a rising magma acts the same way as gas dissolved in soda water. When a bottle of soda is opened, the pressure inside the bottle drops, gas comes out of solution, and bubbles form. Gas dissolved in an upward-moving magma also comes out of solution and forms bubbles. What happens to the bubbles once they are formed is determined by the viscosity of the liquid.

Non-explosive Eruptions

It is understandable why people tend to regard any volcanic eruption as a hazardous event and active volcanoes as dangerous places that should be avoided. However, geologists have discovered that some volcanoes are comparatively safe and relatively easy to study. Nonexplosive eruptions, like those we can witness in Hawaii, are relatively safe compared to violent, explosive events like the 1980 eruption of Mount St. Helens in Washington and the 1982 eruption of El Chichón in Mexico, each of which caused substantial destruction and loss of life. The differences between nonexplosive and explosive eruptions are largely a function of magma viscosity and dissolved gas content. Nonexplosive eruptions are favored by low-viscosity magmas and low dissolved-gas contents.

Even nonexplosive eruptions may appear violent during their initial stages. Gas bubbles in a low-viscosity basaltic magma will rise rapidly upward, like the gas bubbles in a glass of soda water. If a basaltic magma rises rapidly, which means the pressure drop will be fast, gas can bubble so rapidly out of solution that spectacular fountaining will occur (Fig. 3.4). Bits of falling lava spatter when they strike the ground and can pile up as a *spatter cone* or *spatter rampart* beside the vent. When fountaining dies down, hot, fluid lava emerging from the vent will flow rapidly downslope (Fig. 3.5). Because heat is lost quickly at the top of a flow, the surface forms a crust beneath which the liquid lava will continue to flow downslope along well-defined channels. These enclosed lava tubes inhibit upward loss of heat and enable a low-viscosity lava to move along just below the surface for great distances from the vent. As the lava cools and continues to lose dissolved gases, its viscosity increases and the character of flow changes. The very fluid lava initially forms thin pahoehoe flows, but with increasing viscosity the rate of movement slows and the stickier lava may be transformed into a clinkery aa flow that now moves very slowly. Thus, during a single, nonexplosive, Hawaiian-type eruption, spatter cones, pahoehoe, and aa may be formed from the same batch of magma.

As a basaltic lava cools and the viscosity rises, the gas bubbles find it increasingly difficult to escape. When the lava finally solidifies to rock, the last-formed

FIGURE 3.5 This stream of low-viscosity, basaltic lava moving smoothly away from an eruptive vent demonstrates how fluid and free flowing lava can be. The temperature of the lava is about 1100°C. The eruption occurred in Hawaii in 1983.

bubbles become trapped and their form preserved. These bubble holes are called *vesicles*, and the texture they produce in an igneous rock is said to be *vesicular* (Fig. 3.6). Vesicular basalts are common. In many vesicular basalts the vesicles are later filled with calcite, quartz, or some other mineral deposited by heated groundwater. Vesicles filled by secondary minerals are *amygdules* (Fig. 3.7).

FIGURE 3.4 Spectacular fountaining starts an eruption of a basaltic volcano in Iceland. Use of a telephoto lens foreshortens the field of view. The geologists who are measuring the height of the fountain (200 m) are many hundreds of meters away from the erupting lava.

FIGURE 3.6 Vesicular basalt. Note the grains of yellowish-green olivine. The specimen is 4.5 cm across

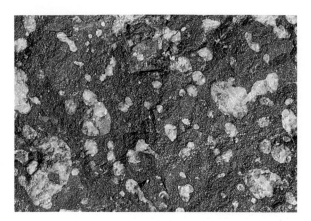

FIGURE 3.7 Amygdaloidal basalt formed when secondary minerals such as calcite and zeolites fill vesicles. The specimen is 4.5 cm across.

Explosive Eruptions

In viscous andesitic or rhyolitic magmas, gas bubbles can rise only very slowly. As such a magma moves upward toward the surface, the change in confining pressure causes dissolved gas to expand and escape explosively. Bubbles that form quickly in a huge mass of sticky rhyolitic magma can shatter into a froth of innumerable tiny glass-walled bubbles, producing a rock called *pumice.* Some pumice has such a low density that it will float in water. Beaches on mid-Pacific Islands, for example, are often littered with pumice that has floated in on currents from distant volcanic eruptions.

When little or no dissolved gas is present, a magma will be erupted as a lava flow regardless of composition. If dissolved gas is present, however, it must escape somehow, and the higher the viscosity, the greater the likelihood that the escaping gas will cause an explosive eruption.

Pyroclasts and Tephra

A fragment of rock ejected during a volcanic eruption is called a **pyroclast** (named from the Greek words *pyro,* meaning heat or fire, and *klastos,* meaning broken; hence, hot, broken fragments). Rocks formed from pyroclasts are **pyroclastic rocks.** Geologists also commonly refer to a deposit of pyroclasts as **tephra,** a Greek term for ash that was originally used by Aristotle and was revived in recent years by the Icelandic volcanologist Sigurdur Thorarinsson. Tephra is employed as a collective term for all airborne pyroclasts, including fragments of newly solidified magma as well as fragments of older broken rock. It includes individual pyroclasts that fall directly to the ground and those that move over the ground as part of a hot, moving flow. The terms used to describe tephra of different size—*bombs, lapilli,* and *ash*—are listed in Table 3.1 and illustrated in Figure 3.8.

The word ash is somewhat misleading because it means, strictly, the solid matter left after something flammable, such as wood, has burned. However, the fine tephra thrown out by volcanoes looks so much like true ash that it has become customary to use the word for this material also.

Eruption Columns and Tephra Falls

The largest and most violent explosive eruptions are associated with silica-rich magmas having a high dissolved gas content. As the rising magma approaches the surface, rapid decompression causes the dissolved gases to expand and produces the violent upward thrust of a dense mixture of hot gas and tephra. This hot, turbulent mixture rises rapidly in the cooler air above the vent to form an *eruption column* that may reach as high as 45 km in the atmosphere. The opening figure for this chapter shows a huge eruption column rising from Mount St. Helens. The rising, buoyant column is driven by heat energy released from hot, newly formed pyroclasts. At a height where the density of the material in the column equals that of the surrounding atmosphere, the column begins to spread laterally to form a mushroom-shaped cloud of the type so familiar in pictures of nuclear bomb explosions.

As an eruption cloud begins to drift with the upper atmospheric winds, the particles of debris fall out and eventually accumulate at the ground surface as tephra

TABLE 3.1 Names for Tephra and Pyroclastic Rock

Average Particle Diameter (mm)	Tephra (unconsolidated material)	Pyroclastic Rock (consolidated material)
>64 mm	Bombs	Agglomerate
2–64 mm	Lapilli	Lapilli tuff
<2 mm	Ash	Ash tuff

A.

B.

C.

FIGURE 3.8 Tephra. A. Large spindle-shaped bombs up to 50 cm in length cover the surface of a tephra cone on Haleakala volcano, Maui. B. Intermediate-sized tephra called lapilli cover the Kau Desert, Hawaii. The coin is about 1 cm in diameter. C. Volcanic ash, the smallest-sized tephra, blankets a farm in Oregon following the eruption of Mount St. Helens in 1980.

deposits (chapter 12). During exceptional explosive eruptions, tephra can be spread over distances of 1500 km or more. Some eruption columns reach such great heights that high-level winds are able to transport the fine debris and associated sulfur-rich gas completely around the world. Such atmospheric pollution, by blocking incoming solar energy, can lower average temperatures at the land surface by as much as 1°C for period of a year or longer and cause spectacular sunsets as the sun's rays are refracted by the airborne particles.

Pyroclastic Flows

A hot, highly mobile flow of tephra that rushes down the flank of a volcano during a major eruption is called a **pyroclastic flow.** These are among the most devastating and lethal forms of volcanic eruptions. Analysis of the worldwide geologic record of historic pyro-

clastic flows shows that they can travel 100 km or more from source vents and reach velocities of more than 700 km/h. One of the most destructive, on the Caribbean island of Martinique in 1902, rushed down the flanks of Mt. Peleé volcano and overwhelmed the city of St. Pierre, instantly killing some 29,000 people. Such a pyroclastic flow can be caused by the gravitational or explosive collapse of a mass of hot lava near the top of a volcano which produces a downrushing dense fluid mass of blocks, lapilli, ash, and hot gases. Geologists call the resulting poorly sorted deposit an *ignimbrite*. Pyroclastic flows can also be generated by the partial or continuous collapse of an eruption column. During the 1980 eruption of Mount St. Helens, for example, a number of hot (850°C) pyroclastic flows, likely caused by column collapse, travelled up to 8 km down the north side of the mountain and covered an area of about 15 km².

Lateral Blasts

The 1980 eruption of Mount St. Helens displayed many of the features of a typical large explosive eruption. Nevertheless, the magnitude of the event caught geologists by surprise. The events leading to this eruption are shown diagrammatically in Figure 3.9. As magma moved upward under the volcano, the north flank of the mountain began to bulge upward and outward. Finally, the slope became unstable, broke loose, and quickly slid toward the valley as a gigantic landslide of rock and glacier ice. The landslide exposed the mass of hot magma in the core of the volcano. With the lid of rock removed, dissolved gases underwent such rapid decompression that a mighty blast resulted that blew a mixture of pulverized rock and hot gases sideways as well as upward. The sideways blast, initially travelling at the speed of sound, roared across the landscape, killing David Johnson and others in the blast zone. Within the devastated area that extends as much as 30 km from the crater and covers some 600 km², trees in the formerly dense forest were blasted to the ground and covered with hot debris. Although Mount St. Helens provides the best-documented recent example of a lateral blast, a closely similar 1956 eruption of Bezmianny volcano in Kamchatka produced a devastating directed blast, a high eruption column, and associated pyroclastic flows.

VOLCANOES

Shield Volcanoes

The kind of volcano that is easiest to visualize is one built up of successive flows of very fluid lava. Such lavas are capable of flowing great distances down gentle slopes, and of forming thin sheets of nearly uniform thickness. Eventually the pile of lava builds up a **shield volcano,** which is a broad, roughly dome-shaped formation with an average surface slope of only a few degrees (Fig. 3.10).

The slope of a shield volcano is slight near the summit because the magma is hot and very fluid; it will readily run down a very slight slope. The further the lava flows down the flank, the cooler and more viscous it becomes and the steeper a slope must be in order for it to flow. Observe that a steepening of the slope of Mauna Kea can be seen in Figure 3.10. Slopes on young, growing shield volcanoes, such as Kilauea in Hawaii, typically range from less than 5° near the summit, to 10° on the flanks.

Shield volcanoes are characteristically formed by the eruption of basaltic lava—the proportions of ash

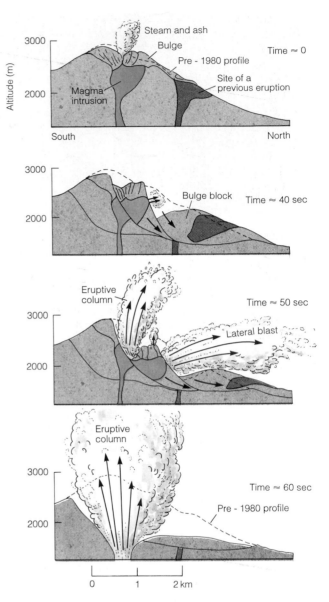

FIGURE 3.9   Sequence of events leading to the eruption of Mount St. Helens in May 1980. A. Earthquakes and then puffs of steam indicate that magma is rising; a small crater forms, and the north face of the mountain bulges alarmingly. B. On the morning of May 18, an earthquake shakes the mountain and the bulge breaks loose and slides downward. This reduced the pressure on the magma and initiated the lateral blast that killed David Johnson. C. The violence of the eruption causes a second block to slide downward, exposing more of the magma and initiating an eruption column. D. The eruption increases in intensity. The eruption column carries volcanic ash as high as 19 km into the atmosphere.

FIGURE 3.10 Mauna Kea, a 4200 m-high shield volcano on Hawaii, as seen from Mauna Loa. Note the gentle slopes formed by highly fluid basaltic lava. The view is almost directly north. A pahoehoe flow is in the foreground on the northeast flank of Mauna Loa.

and other tephra are small. Hawaii, Tahiti, Samoa, the Galapagos, and many other oceanic islands are the upper portions of large shield volcanoes.

Tephra Cones

Rhyolitic and andesitic volcanoes tend to eject large quantities of tephra. As the debris showers down, a **tephra cone** builds around the vent (Fig. 3.11). The slope of the cone is determined by the size of the tephra. Fine ash will stand at a slope angle of 30° to 35°, while lapilli generally stand at an angle of about 25°. The gradual decrease in the volume of fallout material away from the vent leads to more gentle slopes near the base of the cone.

Stratovolcanoes

Large, long-lived volcanoes, particularly those of andesitic composition, emit a combination of lava flows and tephra. **Stratovolcanoes** are defined as volcanoes that emit both tephra and viscous lava and build up steep conical mounds. The volume of tephra may equal or exceed the volume of the lava.

The slopes of stratovolcanoes, which may be thousands of meters high, are steep like those of tephra cones. Near the summit of a stratovolcano, the slope is about 30°, like that near the summit of a pyroclastic cone. Toward the base, the slopes of stratovolcanoes flatten to about 6° to 10°. The steep slopes near the summit of a stratovolcano are due in part to the short,

viscous lava flows that are erupted, and in part to the tephra. The lava flows are a major distinguishing factor between tephra cones and stratovolcanoes. As a stratovolcano develops, lava flows act as a cap to slow down erosion of the loose tephra, and thus the volcano becomes much larger than a typical tephra cone.

The beautiful, steep-sided cones of stratovolcanoes are among Earth's most picturesque sights (Fig. 3.12). The snow-capped peak of Mount Fuji in Japan has inspired poets and writers for centuries. Mount Rainier and Mount Baker in Washington and Mount Hood in Oregon are majestic examples in North America.

Craters, Calderas, and Other Volcanic Features

Certain features give volcanic terrains a unique character. Fractures may split the flanks of a large shield volcano or stratovolcano so that flank eruptions of lava and/or tephra can occur. Small cones of spattered lava or tephra may then develop above the fractures, peppering the slope of the main volcano like so many small pimples (Fig. 3.13A).

Gases that bubble up from magma far below the surface may emerge from either a central volcanic vent or from small, satellitic vents. The emitted gases tend to be mostly water vapor, but there may also be a certain amount of evil-smelling sulfurous gas present

FIGURE 3.15   Crater Lake, Oregon, occupies a caldera 8 km in diameter that crowns the summit of a once lofty stratovolcano, posthumously called Mount Mazama. Wizard Island, a small tephra cone, formed after the collapse that created the caldera.

FIGURE 3.16   The Pinnacles, Crater Lake National Park. Striking erosional forms developed in the thick tephra blanket left by the eruption of Mount Mazama 6600 years ago.

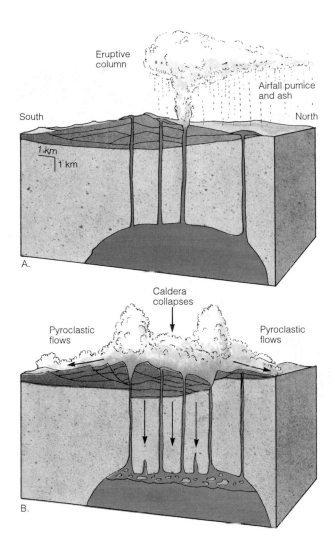

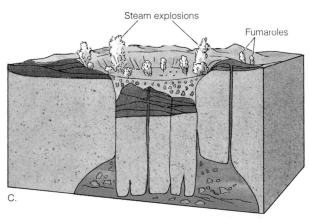

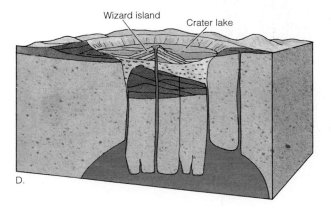

occur along these fractures, thus creating roughly circular rings of small cones. Crater Lake, Oregon, occupies a circular caldera 8 km in diameter, formed after a great tephra eruption about 6600 years ago (Fig. 3.15). The volcano that erupted has been posthumously called Mount Mazama. The tephra can still be seen at many places in Crater Lake National Park and over a vast area of the northwestern United States and adjacent parts of Canada (Fig. 3.16). After the outpouring of about 75 km³ of tephra, what remained of the roof collapsed into the partly empty magma chamber (Fig. 3.17). Yellowstone National Park contains several overlapping giant calderas that formed following gigantic tephra eruptions 2.0, 1.2, and 0.6 million years ago.

Resurgent Domes

A volcano does not necessarily cease activity following the formation of a caldera. Magma starts reentering the chamber, and in the process causes the uplifting of the collapsed floor of a caldera to form a structural dome. Such a feature is called a *resurgent dome*. Subsequently, small tephra cones and lava flows build up in the interior of the caldera. Wizard Island, in Crater Lake, is a cone having such an origin.

Lava Domes

When lava is extruded following a major volcanic eruption it tends to have very little dissolved gas left in it. If the lava is sticky and viscous it squeezes out to form a *lava dome*. A growing lava dome more than 200 m high now sits at the center of the crater of Mount St. Helens (Fig. 3.18).

Fissure Eruptions

Some lava reaches the surface via elongate fissures through the crust. Extrusion of lava along an extended fracture is called a *fissure eruption* (Fig. 3.19). Such eruptions, which are often very dramatic, are characteristically associated with basaltic magma, and the lavas that emerge from a fissure eruption on land tend to spread widely and to create flat lava plains called

FIGURE 3.17 Sequence of events that formed Crater Lake following the eruption of Mount Mazama 6600 years ago. A. Eruptive column of tephra rises from the flank of Mount Mazama. B. The eruption reaches a climax. Dense clouds of ash fill the air, the hot pyroclastic flows sweep down the mountain side. It was at this stage that the deposits shown in Figure 3.16 were formed. C. The top of Mount Mazama collapses into the partly empty magma chamber, forming a caldera 10 km in diameter. D. During a final phase of eruption, Wizard Island formed. The water-filled caldera is Crater Lake, shown in Figure 3.15.

85

FIGURE 3.18 A lava dome in the crater of Mount St. Helens, Washington, in May 1982. The plume rising above the dome is steam.

plateau basalts. A fissure eruption at Laki in Iceland in 1783 occurred along a fracture 32 km long. Lava flowed 64 km outward from one side of the fracture and nearly 48 km outward from the other side. Altogether it covered an area of 588 km². The volume of the lava extruded was 12 km³, making this the largest lava flow in historic times. It was also one of the most deadly. The flow destroyed homes and food supplies, killed stock, and covered fields. Famine followed and 9336 people died. There is good evidence to prove that larger eruptions have occurred in prehistoric times. The Roza flow, a great sheet of basaltic lava in eastern Washington State, can be traced over 22,000 km² and shown to have a volume of 650 km³.

FIGURE 3.19 Aerial view of a fissure eruption, Mauna Loa, Hawaii, in 1984. Basaltic lava is erupting from a series of parallel fissures. Note the tephra cone (upper left) formed during an earlier eruption.

FIGURE 3.20 Tubular-shaped pillows of basalt photographed in the central rift of the east Pacific Rise.

Pillow Basalts

The most extensive volcanic system on the Earth lies beneath the sea. There the fissures that split the center of the midocean ridges are channelways for the rising basaltic magma that forms new oceanic crust at the spreading margins of plates of lithosphere.

Seawater cools the basaltic magma so rapidly that a very distinctive lava form is observed. Close to a submarine volcanic fissure, where the lava temperature is highest, thin lava flows with rapidly quenched, glassy surfaces form. The flows build up piles of basalt in which each sheet may only be 20 cm or so thick. Farther away from a vent, where the temperature of the flowing lava has decreased, pillow structure develops. The term **pillow basalt** describes a structure characterized by discontinuous, pillow-shaped masses of basalt, ranging in size from a few centimeters to a meter or more in greatest dimension (Fig. 3.20).

Pillow structure forms when the surface of the basaltic lava is chilled quickly. The brittle, chilled surface cracks, making an opening for the still molten magma inside to ooze out like a strip of toothpaste. In turn, the newly oozed strip chills, its surface cracks, and the process continues. The end result is a pile of lava pillows that resemble a jumbled pile of sandbags. Most of the lavas on the oceanic crust are pillow basalts.

Volcanic Hazards

Volcanic eruptions are not rare events. Every year about 50 volcanoes erupt somewhere on the Earth. Eruptions of basaltic shield volcanoes are usually not dangerous. But tephra eruptions from andesitic or rhyolitic stratovolcanoes, such as Mount St. Helens and Krakatau, can be disastrous. Millions of people live on or close to stratovolcanoes and eruptions present five kinds of hazards.

1. Hot, rapidly moving pyroclastic flows and laterally directed blasts may overwhelm people before they can run away. The tragedies of Mt. Pelée in 1902 and Mount St. Helens in 1980 are examples.

2. Tephra and hot poisonous gases may bury people or suffocate them. Such a tragedy occurred in A.D. 79 when Mount Vesuvius, the supposedly dormant volcano in southern Italy, burst to life. Hot, poisonous gases killed people in the nearby Roman city of Pompeii, and then tephra buried them (Fig. 3.21).

3. Tephra can be dangerous long after an eruption has ceased. Rain or meltwater from snow can loosen tephra piled on a steep volcanic slope and start a deadly mudflow sweeping down the mountainside. In 1985, following a small and otherwise nondangerous eruption of the Columbian volcano Nevado del Ruíz, massive mudflows formed when glaciers at the summit melted, moved swiftly down the mountain, and killed 20,000 people (Chapter 8).

4. Violent undersea eruptions can cause giant sea waves called *tsunami*. Set off by the eruption of

FIGURE 3.21 Evidence of an ancient disaster. Casts of bodies of five citizens of Pompeii, Italy, who were killed during the eruption of Mount Vesuvius in 79 A.D. Death was caused by poisonous gases, then the bodies were buried by lapilli. Over the centuries the bodies decayed away but the body shapes were imprinted in the tephra blanket. When excavators discovered the imprints they carefully recorded them with plaster casts.

TABLE 3.2   Volcanic Disasters Since A.D. 1800 in Which a Thousand or More People Lost Their Lives

| Volcano | Country | Year | Primary Cause of Fatalities | | | |
			Pyroclastic Eruption	Mudflow	Tsunami	Famine
Mayon	Philippines	1814	1,200			
Tambora	Indonesia	1815	12,000			80,000
Galunggung	Indonesia	1822	1,500	4,000		
Mayon	Philippines	1825		1,500		
Awu	Indonesia	1826		3,000		
Cotopaxi	Ecuador	1877		1,000		
Krakatau	Indonesia	1883			36,417	
Awu	Indonesia	1892		1,532		
Soufriere	St. Vincent	1902	1,565			
Mt. Peleé	Martinique	1902	29,000			
Santa Maria	Guatemala	1902	6,000			
Taal	Philippines	1911	1,332			
Kelud	Indonesia	1919		5,110		
Merapi	Indonesia	1930	1,300			
Lamington	Papua-New Guinea	1951	2,942			
Agung	Indonesia	1963	1,900			
El Chichón	Mexico	1982	1,700			
Nevado del Ruíz	Colombia	1985	25,000			

Source: From a Report by the Task Group for the International Decade of Natural Disaster Reduction, published in *Bull. Volcano. Soc. Japan,* Series 2, Vol. 35, #1, 1990, p. 80–95, 1990.

Krakatau, tsunamis killed more than 36,000 coast dwellers on Java and other Indonesian islands.

5. A tephra eruption may wreak such havoc on agricultural land and livestock that people die from famine.

Since A.D. 1800 there have been 18 volcanic eruptions in which a thousand or more people died (Table 3.2). It is certain that other violent and dangerous eruptions will occur in the future. To some extent, volcanic hazards can be anticipated provided experts can gather data before, during, and after eruptions. The experts can then advise civil authorities when to implement hazard warnings and when to move endangered populations to areas of lower risk.

IGNEOUS ROCK

Igneous rock forms by the cooling and solidification of magma. **Extrusive igneous rocks** are formed by solidification of lava; **intrusive igneous rocks,** by contrast, are formed when magma solidifies within the crust or mantle. A great deal of information about how magma forms and moves below the Earth's surface can be derived from studies of intrusive igneous rocks. Before discussing the origin of magma, there-

fore, it is instructive to briefly examine how igneous rocks are classified, and then to discuss how intrusive igneous rocks occur in the crust.

Both extrusive and intrusive igneous rocks are classified and named on the basis of rock texture and the mineral assemblage.

Texture

The most obvious textural feature of an igneous rock is the size of the mineral grains of which it is comprised. Intrusive igneous rocks tend to be coarse grained because magma that solidifies in the crust or mantle cools slowly and has sufficient time to form large mineral grains. Figure 3.22A is an example of coarse-grained igneous rock.

Extrusive igneous rocks, by contrast, solidify on the Earth's surface. Lava cools rapidly and extrusive igneous rocks are fine grained or even glassy. Figure 3.22B is an example of a fine-grained igneous rock.

One special texture involves a distinctive mixture of large and small grains. Rock of such a texture is called a **porphyry,** meaning an igneous rock consisting of coarse mineral grains scattered through a mixture of fine mineral grains, as shown in Figure 3.22C. The isolated large grains in porphyry are called *phenocrysts,* and they form in the same way mineral

A.

B.

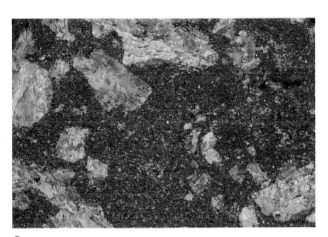

C.

FIGURE 3.22 Differing textures of A. basalt, B. gabbro, and C. basalt porphyry. Each of the rocks has the same composition. Basalt is fine grained because it cooled very rapidly. Gabbro is coarse grained because it cooled slowly. Basalt porphyry contains phenocrysts of plagioclase (white) set in a matrix that is so fine the individual mineral grains can barely be resolved with a microscope. The coarse phenocrysts formed during slow cooling at depth; the matrix formed when partly crystallized magma was suddenly extruded as a lava and rapidly cooled. In each case, the field of view is 7 cm across.

grains in coarse-grained igneous rocks do—by slow cooling of magma in the crust or mantle. The fine grained groundmass that encloses phenocrysts provides evidence that a partly solidified magma moved quickly upward. In the new setting, the magma cooled rapidly and as a result the later mineral grains, which form the groundmass, are all tiny. Many extrusive igneous rocks are porphyries.

Lava may sometimes cool and solidify so rapidly that its atoms do not have time to organize themselves into minerals. Glass, a mineraloid, forms instead. Extrusive igneous rocks that are largely or wholly glassy are called **obsidian.** Such rocks display a distinctive fracture pattern on a broken surface. The fracture pattern consists of a series of smooth, curved surfaces like a conch shell (Fig. 3.23).

An intrusive igneous rock that contains unusually large mineral grains is called a **pegmatite.** The term is used for rocks in which average grain diameters are 2 cm or larger. Sometimes individual mineral grains in pegmatites can be huge. Single grains of muscovite and feldspar several meters across have been reported.

FIGURE 3.23 Obsidian from the Jemez Mountains, New Mexico. The specimen is glass of a rhyolitic composition. The curved ridges are typical of the fracture pattern in glass broken by a sharp blow. The specimen is 10 cm across.

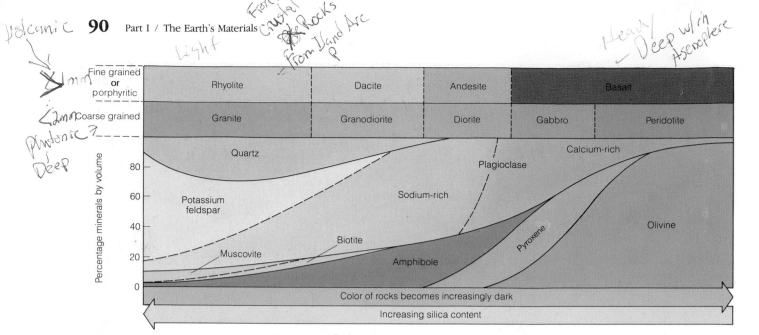

FIGURE 3.24 The proportions of minerals in the common igneous rocks. Boundaries between rock types are not abrupt but gradational, as suggested by the broken lines. To determine the composition range for any rock type, project the broken lines vertically downward, then estimate the percentages of the minerals by means of the numbers at the edge of the diagram.

Mineral Assemblage

When magma of a given composition solidifies, the mineral assemblage that forms is the same for intrusive and extrusive igneous rocks. The differences are textural. Once the texture of an igneous rock has been determined, therefore, specimens are named on the basis of mineral assemblage. To see how this is done, it is convenient to employ the diagram shown in Figure 3.24. All common igneous rocks are composed of one or more of these six minerals or mineral groups: quartz, feldspar (both potassium feldspar and plagioclase), mica (both muscovite and biotite), amphibole, pyroxene, and olivine. The vertical axis in Figure 3.24 records mineral percentages. When the percentage of each mineral in a rock specimen has been estimated, the correct place in Figure 3.24 is located and the corresponding rock name determined. The fine- or coarse-grained texture name is selected, whichever is appropriate. For example, suppose you have a rock specimen that is 30 percent olivine, 30 percent pyroxene, and 40 percent plagioclase. To name this specimen you estimate the point on Figure 3.24 where the curve separating olivine from pyroxene cuts a horizontal line drawn through 30 percent. Draw a vertical line through the point. You will see that this vertical line cuts the curve between pyroxene and plagioclase at a point close to 60 percent. Olivine (30%) plus pyroxene (30%) equal 60 percent. Plagioclase ac-

counts for the remaining 40 percent. If your rock specimen is coarse grained it is called gabbro, if fine grained, basalt.

When a rock has a porphyritic texture, we use the name determined by the mineral assemblage as an adjective and the term for the texture of the groundmass for the noun. For example, if the groundmass is fine grained, we would call it a rhyolite porphyry, say, or a dacite porphyry; if the groundmass is coarse grained, we would call it a granite porphyry or a granodionite porphyry. It should always be kept in mind when using Figure 3.24 that nature does not draw sharp boundaries between igneous rocks; every conceivable gradation in texture and composition can be found.

Whether a rock is light or dark in color is a useful piece of information. Quartz, feldspar, and muscovite are light; biotite, amphibole, pyroxene, and olivine are iron-bearing minerals and so are dark. Note that rocks to the left in Figure 3.24 are light colored and those to the right are increasingly dark. The boundary line between diorite and gabbro is drawn where the dark-colored minerals reach 50 percent of the total and exceed the light-colored feldspars. The boundary is carried through between the fine-grained equivalents of diorite and gabbro—andesite and basalt, respectively—on the basis of color.

FIGURE 3.25 Three coarse-grained igneous rocks. Compare their mineral assemblages by using the chart in Figure 3.24. Note the change in color from granite (left), which is light colored because it is rich in feldspar and quartz, through diorite (center), to gabbro (right), which is quartz free and rich in pyroxene and olivine and therefore darker in color. Each specimen is 7 cm across.

Varieties of Intrusive Igneous Rock

Granite and Granodiorite

Feldspar and quartz are the chief minerals in granite and granodiorite (Fig. 3.25). A mica, either muscovite or biotite, is usually present, and many granites contain scattered grains of hornblende.

The name granite is applied only to quartz-bearing rocks in which potassium feldspar is predominant. The name granodiorite applies to similar rocks in which plagioclase is the chief feldspar. Without special equipment the differences in feldspars are not always easily recognized, and in a general study the term granitic is extended to this whole group of rocks.

Diorite

The chief mineral in diorite is plagioclase. Quartz and mica are usually absent, but either or both amphibole and pyroxene are invariably present (Fig. 3.25). Diorite is a common igneous rock but not so common as granite and granodiorite.

Gabbro and Peridotite

Dark-colored diorite grades into gabbro (Fig. 3.25), in which the dark-colored minerals, pyroxene and olivine may exceed 50 percent. A coarse-grained igneous rock, related in origin to gabbro, that contains 90 percent or more olivine, is called a peridotite.

Varieties of Extrusive Igneous Rock

Rhyolite and Dacite

A fine-grained porphyritic rock with phenocrysts of quartz is either a rhyolite or a dacite. The quartz indicates the close chemical kinship to granite and granodiorite. The difference between rhyolites and dacites rests, as it does with granites and granodiorites, on the feldspars. In rhyolites potassium feldspar is more abundant than plagioclase; in dacites the reverse is true. It is rarely possible to identify very fine-grained feldspars without special microscopes, so dacites are very difficult to distinguish from rhyolites. Many geologists, when in doubt, simply call both rocks rhyolites, or even *rhyodacites*.

The colors of rhyolites and dacites are always pale—ranging from nearly white to shades of gray, yellow, red, or purple. Many obsidians have the same compositions as rhyolites or dacites. Such obsidians may appear dark, even black, and seem to contradict the rule that igneous rocks with a high silica content are light colored. But rhyolitic obsidian chipped to a thin edge appears white, even colorless. The dark appearance results from a small amount of dark mineral matter distributed evenly in the glass.

Andesite

A porphyritic igneous rock similar in appearance to a dacite but lacking quartz phenocrysts is an andesite. Named for the Andes, the major mountain system of western South America, andesite is equivalent in composition to a diorite and is a common rock around the world. Common colors are shades of gray, purple, and even dark green.

Basalt

The dominant rock of the oceanic crust is basalt, a fine-grained igneous rock, sometimes porphyritic, that is always dark gray or black. Compositionally equivalent to gabbro, basalt is the most common kind of extrusive igneous rock. Phenocrysts, when present in a basalt porphyry, can be either plagioclase, pyroxene, or olivine.

Varieties of Pyroclastic Rocks

There is an old saying that tephra is igneous on the way up but sedimentary on the way down. As a result, pyroclastic rocks are transitional between igneous and sedimentary rocks and the names of pyroclastic rock types reflect this fact. As we will see in chapter 4, the names of one class of sedimentary rocks are determined by the sizes of the sediment fragments. So, too, with pyroclastic rocks: They are called **agglomerates** when tephra are bomb sized, or **tuffs** when the tephra are either lapilli or ash (Fig. 3.26). As seen in Table 3.1, the appropriate rock names are *lapilli tuff,* and *ash tuff.* The igneous origin of a pyroclastic rock is indicated by the name for the mineral assemblage. For example, we would refer to a rock of appropriate mineral assemblage as a dioritic lapilli tuff if coarse grained, or andesitic lapilli tuff if fine grained.

Conversion of tephra to pyroclastic rock can happen in two ways. The first, and most common, way is through the addition of a cementing agent such as quartz or calcite introduced by groundwater. Figure 3.26A is an example of a rhyolitic lapilli tuff formed by cementation. The second way tephra is transformed to pyroclastic rock is through welding of hot, glassy, ash particles. When ash is very hot and plastic the individual particles can fuse together to form a glassy pyroclastic rock. Such a rock is called **welded tuff** (Fig. 3.26A).

A.

B.

FIGURE 3.26 Two kinds of tuff. A. Lapilli tuff, formed by cementation of lapilli and ash, from Clark County, Nevada. B. Welded tuff from the Jemez Mountains, New Mexico. The dark patches are glassy fragments of pumice flattened during welding. Note the fragments of other rocks in the specimen. Both samples are 4 cm across.

PLUTONS

Beneath every volcano there lies a complex of chambers and channelways through which magma reaches the surface. Naturally we cannot study the magmatic channels of an active volcano, but we can look at ancient channelways that have been laid bare by erosion, as seen in Figure 3.27. What we find is that these ancient channelways are filled by intrusive igneous rock because they are the underground sites where magma solidified.

All bodies of intrusive igneous rock, regardless of

shape or size, are called **plutons** after Pluto, the Greek god of the underworld. The magma that forms a pluton did not originate where we now find the pluton. Rather, the magma was intruded upward into the surrounding rock from the place where it formed.

Minor Plutons

Plutons are given special names depending on their shapes and sizes. The common plutons are illustrated in Figure 3.28 and each is briefly discussed below.

Dikes

The most obvious and familiar evidence of past igneous activity is a **dike,** a tabular, parallel-sided sheet of igneous rock that cuts across the layering of the rock it intrudes (Fig. 3.29A). Several dikes can be seen in the photograph of Shiprock, New Mexico, shown in Figure 3.27. A dike forms when magma squeezes into a fracture. The erupting fissure in Figure 3.19, for example, is probably now filled by a dike of gabbro, formed as a result of the cooling of the basaltic magma that filled the fissure at the end of the eruption.

A.

B.

FIGURE 3.27 Shiprock, New Mexico. A. The conical tephra cone that once surrounded this volcanic neck has been removed by erosion. B. Diagram of the way the original volcano may have appeared prior to erosion.

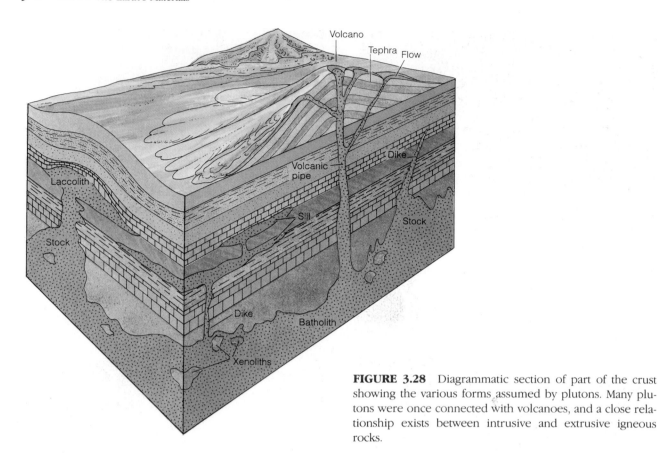

FIGURE 3.28   Diagrammatic section of part of the crust showing the various forms assumed by plutons. Many plutons were once connected with volcanoes, and a close relationship exists between intrusive and extrusive igneous rocks.

Sills

Tabular, parallel-sided bodies of intrusive igneous rock that are parallel to the layering of the rocks they intrude are called **sills** (Fig. 3.29B). Commonly, dikes and sills occur together as part of a network of plutons, as shown in Figure 3.28. Both dikes and sills can be very large. For example, the Great Dike in Zimbabwe is a mass of gabbro nearly 500 km long and about 8 km wide, with parallel, vertical walls. The Great Dike fills what must once have been a huge fracture in the crust. An example of a large and well-known sill can be seen in the cliffs of the Palisades that line the Hudson River opposite New York City. The Palisades Sill reaches a thickness of about 300 m, and like the Great Dike is gabbro. The sill was intruded between layers of ancient sedimentary rock about 200 million years ago. The sill is visible today because tectonic forces raised that portion of the crust upward and erosion then removed the covering sedimentary rocks.

Laccoliths

A variation of a sill is a *laccolith,* an igneous body intruded parallel to the layering of the rocks it intrudes, and above which the layers of the intruded rocks have been bent upward to form a dome.

Volcanic Pipes and Necks

A *volcanic pipe* is the approximately cylindrical conduit of igneous rock forming the feeder pipe immediately below a volcanic vent. A pipe that has been stripped of its surrounding rock by erosion is called a *volcanic neck*. Figure 3.27 shows a famous example of a volcanic neck, together with associated dikes, at Shiprock, New Mexico.

Major Plutons

Batholiths

A **batholith** is the largest kind of pluton. It is an intrusive igneous body of irregular shape that cuts across the layering of the rock it intrudes. Most batholiths are composite masses that comprise a number of separate intrusive bodies of slightly differing composition. The differences reflect variations in composition of the magma from which the batholith formed. Some batholiths exceed 1000 km in length and 250 km in width—the largest in North America is the Coast Range Batholith of British Columbia and northern Washington, which has a length of about 1500 km (Fig. 3.30).

It is not apparent in Figure 3.28 what the bottom

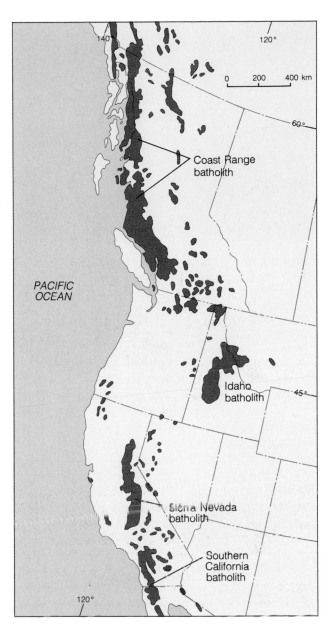

FIGURE 3.29 Sheet-like plutons. A. Dike of gabbro cutting across horizontally layered shales in Hance Rapid, Grand Canyon National Park, Arizona. B. Sill of gabbro (dark brown) intruded parallel to layering of sedimentary rocks above and below, in Big Bend National Park, Texas.

FIGURE 3.30 The Idaho, Sierra Nevada, and Southern California batholiths, largest in the United States, are dwarfed by the Coast Range Batholith in Southern Alaska, British Columbia, and Washington. Each of these giant batholiths formed from magma generated by the partial melting of continental crust, and each intrudes metamorphosed rocks.

of a batholith looks like. Where it is possible to see them, the walls of batholiths tend to be nearly vertical. This observation led to a commonly held perception that batholiths extend downward to great depths— possibly even to the base of the crust. Geophysical measurement and studies of very deeply eroded bodies of igneous rock suggest that this perception is incorrect. Batholiths seem to be only 20 to 30 km thick,

which is rather thin compared to their great widths and lateral extents.

Batholiths do not form by magma squeezing into fractures because there are no fractures large enough to account for batholiths. Despite their huge sizes, however, batholiths do move upward. Even though intruded rocks can be pushed upward by the slowly rising magma, some other process must also operate.

FIGURE 3.31 Xenoliths of metamorphic rock dislodged during intrusion of the Petersburg Granite. The outcrop is in the bed of the James River, Richmond, Virginia. The size of the xenoliths can be judged from the hammer.

The rising magma apparently dislodges fragments of the overlying rock by a process known as *stoping*. Dislodged blocks are more dense than the rising magma and will therefore sink. As they sink, those fragments may react with and be partly dissolved by the magma.

Not all stoped fragments dissolve. Instead they may sink all the way and reach the floor of the magma chamber. Any fragment of rock still enclosed in a magmatic body when it solidifies is known as a *xenolith* (Fig. 3.31) (from the Greek words *xenos*, stranger, and *lithos*, stone).

Stocks

Like batholiths, **stocks** are irregularly shaped intrusives no larger than 10 km in maximum dimension. As is apparent from Figure 3.28, a stock may merely be a companion body to a batholith or even the top of a partly eroded batholith.

THE ORIGIN OF MAGMA

We come now to the most difficult but also one of the most interesting questions concerning magmas and volcanoes—how and where do magmas form, and why are there three major kinds of magmas—basaltic, andesitic, and rhyolitic? Many clues to the question of origin can be learned from the distribution of the kinds of volcanoes from which the three different magma types erupt. A summary of present thinking about the distribution of the kinds of volcanoes is presented in Figure 3.32 and in the following discussion.

Distribution of Volcanoes

It has long been known that volcanoes that erupt rhyolitic magma are found abundantly on the continental crust. A few andesitic volcanoes in the ocean also

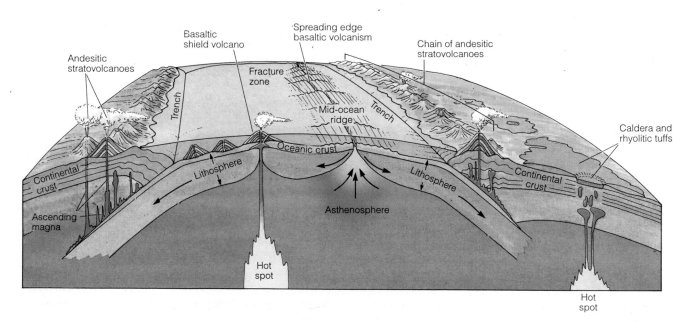

FIGURE 3.32 Diagram illustrating the locations of the major kinds of volcanoes in a plate-tectonic setting.

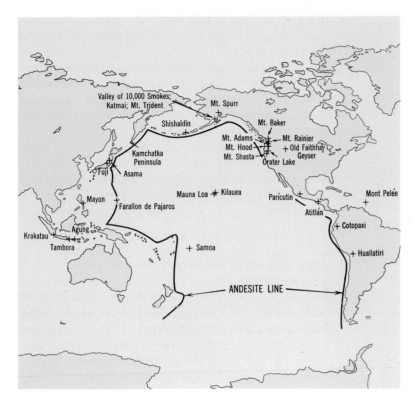

FIGURE 3.33 The Andesite line surrounds the Pacific Ocean basin and separates areas within the basins, where andesite magma is not found, from areas where it is common. Volcanoes that are inside the Pacific basin, such as Mauna Loa, erupt basaltic magma but not andesitic magma. Those outside the line, such as Mount Shasta and Mount Fuji, may erupt basaltic magma too, but they also erupt andesitic magma

erupt some rhyolitic magma. These are important observations because they suggest that the processes that form rhyolitic magma must be restricted to the continental crust, plus the crust beneath andesitic volcanoes, but do not occur in the oceanic crust. Nor, presumably, do the processes that form rhyolitic magma occur in the mantle, because if they did, the magma would be expected to rise to the surface regardless of the kind of crust above.

Volcanoes that erupt andesitic magma are found on both the oceanic crust and the continental crust. This suggests that andesitic magma forms in the mantle and rises up regardless of the overlying crust. An additional piece of information comes from the geographic distribution of andesitic volcanoes around the Pacific Ocean, shown in Figure 3.33. A line surrounds the Pacific separating andesitic volcanoes from those that erupt only basaltic lava. The *Andesite Line* is exactly parallel to the plate subduction margins shown in Figure 1.9. As we shall see later in this chapter, andesitic magma is probably formed as a result of the melting of old oceanic crust that has been subducted back into the mantle.

Volcanoes that erupt basaltic magma also occur on both the oceanic and the continental crust. The source of basaltic magma, therefore, must also be the mantle. The geographic distribution of basaltic volcanoes does not coincide with a specific crustal feature, however, which suggests that basaltic magma must be formed by melting of the mantle itself regardless of the kind of overlying crust.

Two observations concerning basaltic volcanoes do suggest something about the origin of basaltic magma, however. First, everywhere along the midocean ridges volcanoes erupt basaltic magma. Midocean ridges coincide with plate spreading margins, so we must consider the possibility that plate motion and the generation of spreading-margin magma might somehow be connected. The second observation concerns large basaltic volcanoes that are not located along midocean ridges. An example is found in the volcanoes of Hawaii, which sit on oceanic crust in the middle of the Pacific Plate, far from any plate edges. The Hawaiian volcanoes that are active today, Mauna Loa, Kilauea, and Loihi (a submarine volcano), are just the youngest members of a chain of mostly extinct volcanoes. The

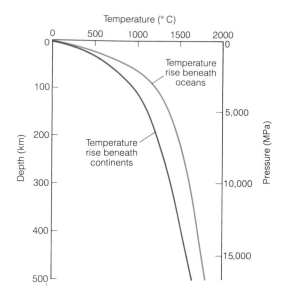

FIGURE 3.34   Geothermal gradients beneath the oceanic crust and continental crust. The graph is drawn so that depth and pressure on the vertical axis is increased downward, as they do in the Earth.

volcanic rocks in the Hawaiian chain are progressively older to the northwest.

The Hawaiian volcanic chain is believed to have formed as the Pacific Plate moved slowly northwest across a midocean hot spot above which frequent and voluminous eruptions built a succession of volcanoes. The hot spot is thought to have been building basaltic volcanoes on the moving plate for at least 70 million years. This must mean that somewhere deep in the mantle there is a long-lived source of basaltic magma. The exact causes of hot spot magmas are still conjectural, but as we shall see in later chapters, both spreading-margin magmas and hot-spot magmas are believed to play important roles in plate tectonics and are caused by convection in the mantle.

How Rock Melts

The making of a magma requires that temperatures be very high, so high, in fact, that the idea of a rock melting is a difficult one for many people to accept. So high, too, that one of the important turning points in the history of geology was the demonstration by a Scot, James Hall, almost 200 years ago, that common rocks can be melted and that such melts have the same properties as magma erupted from volcanoes.

Geothermal Gradient

No longer is there a question as to whether or not a rock can melt; rather, the kinds of questions we ask

now are, what kind of rock melted, did all of the rock melt or only part of it, how does temperature increase with depth, and how far below the Earth's surface did melting occur? Answers to such questions are not straightforward, and in order to approach them, we must consider (1) the **geothermal gradient** (which is the rate at which temperature increases as we go deeper and deeper into the Earth), and (2) the way pressure increases with depth (Fig. 3.34).

The geothermal gradient in and beneath the continental crust differs from that in and beneath the oceanic crust. This is so because the rocks in the two crusts differ significantly in their capacities to serve as thermal blankets to the mass of hot mantle rocks below.

The Effects of Pressure and Water on Melting

As can be seen in Figure 3.34, temperatures beneath both the oceanic and continental crusts rise to about 1000°C at rather shallow depths (less than 100 km down, which is in the upper part of the mantle). Measurements made on lava prove that magma is fluid at 1000°C, so an immediate question is, "Why isn't the Earth's mantle entirely molten?" The answer is that pressure influences melting temperatures. As the pressure rises, the temperature at which a compound melts also rises. For example, albite ($NaAlSi_3O_8$) melts at 1104°C at the Earth's surface, where the pressure is 0.1 MPa, but at a depth of 100 km, where the pressure is 35,000 times greater, the melting temperature is 1440°C (Fig. 3.35A). Therefore, whether a particular rock melts and forms a magma at a specified depth in the Earth depends on both the temperature and the pressure at that depth.

The effect of pressure on melting is straightforward provided a mineral is dry. When water or water vapor is present, however, a complication enters. At any given pressure, a wet mineral will melt at a lower temperature than a dry mineral of the same composition. The effect is the same as that of salt and ice. Salt can melt the ice on an icy road. This happens because a mixture of ice and salt melts to a salt solution at a lower temperature than pure ice melts to water. In the same way a mineral and water mixture melts to magma at a lower temperature than the melting temperature of the pure mineral. Furthermore, as the pressure rises, the effect of water on the melting temperature increases. This is so because the higher the pressure, the greater the amount of water that will dissolve in the melt. Therefore, increasing pressure decreases still further the temperature at which a wet mineral starts to melt (exactly the opposite of what happens with a dry mineral), as Figure 3.35B shows.

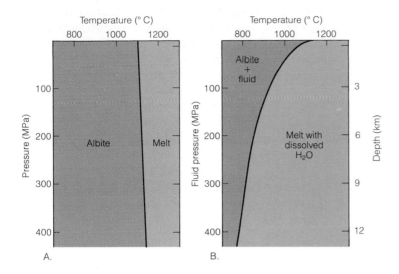

FIGURE 3.35 Influence of pressure on the melting temperature of albite ($NaAlSi_3O_8$). A. Dry-melting curve. Increasing pressure raises the melting temperature. B. Wet-melting curve. H_2O dissolves in the melt and decreases the melting temperature.

Partial Melting

Because it is a composite of several minerals, a rock does not melt at one specific temperature as a single mineral does. Instead, as shown in Figure 3.36, a rock melts over a temperature interval. The mineral with the lowest melting temperature starts to melt first. The melt will dissolve a little of the still unmelted minerals, but even so the melt and the mixture of unmelted minerals will differ in composition. Consider, for example, a rock composed of three minerals: pyroxene, a plagioclase with a composition halfway between albite and anorthite, and olivine. When melting starts, the pyroxene and the albite component of the plagioclase form a liquid, leaving an unmelted mixture of olivine and anorthite-rich plagioclase.

Suppose now that the magma from a partially melted rock is squeezed out and separated from the residue of unmelted minerals. Both the magma and the residual rock will have compositions that differ from each other and from the starting composition of the parent rock. The process of forming magmas with differing compositions through the incomplete melting of rocks is known as **magmatic differentiation by partial melting.**

The Origin of Basaltic Magma

The dominant minerals found in basalt are olivine, pyroxene, and plagioclase. Each of them is anhydrous. This fact suggests that basaltic magma is probably either a dry or a water-poor magma. Indeed, all evidence from observations of basaltic lava during eruption suggests that the water content of basaltic magma rarely exceeds 0.2 percent. It must be concluded, therefore, that basaltic magma originates by some sort of dry, partial-melting process in the mantle.

Much debate has centered on the question of the exact chemical composition of the mantle, but it appears that the upper portion contains rocks rich in olivine and garnet called *garnet peridotites*. Laboratory experiments on the dry, partial-melting properties of garnet peridotite show that at pressures and temperatures reached in the asthenosphere (100 km deep), a 10 to 15 percent partial melt will yield a magma of basaltic composition. While this leaves unanswered, for the moment, the question of a heat source, and why basaltic magma should develop in some parts of

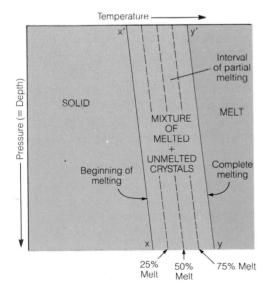

FIGURE 3.36 Dry melting of rock containing several kinds of minerals. The pressure effects are similar to those shown in Figure 3.35A. Line x-x' marks the onset of melting, curve y-y' the completion of melting. Between the two lines is a region in which melt and a mixture of unmelted crystals coexist.

the asthenosphere but not others, we can nevertheless confidently conclude that basaltic magma forms by dry partial melting of rocks in the upper mantle. We will return to the questions of a heat source and why melting occurs in chapter 16 when convection in the mantle is discussed.

The Origin of Andesitic Magma

The chemical composition of andesitic magma is close to the average composition of the continental crust. Igneous rocks formed from andesitic magma are commonly found in the continental crust. From these two facts it might be supposed that andesitic magma forms by the complete melting of a portion of the continental crust. Some andesitic magma may indeed be generated in this way, but andesitic magma is also extruded from volcanoes that are far from the continental crust. In those cases, the magma must be developed either from the mantle or from the oceanic crust. Laboratory experiments provide a possible answer.

In the laboratory, partial melting of wet oceanic crust under suitably high pressure yields a magma of andesitic composition. An interesting hypothesis suggests how this might happen in nature. When a moving plate of lithosphere plunges back into the asthenosphere, it carries with it a capping of basaltic oceanic crust saturated by seawater. The plate heats up and eventually the wet crust starts to melt. Wet partial melting that starts at a pressure equivalent to a depth of 80 km produces a melt having the composition of andesitic magma.

There are a number of details concerning the melting of wet oceanic crust that remain to be deciphered, but two pieces of evidence are very supportive of the idea that much of the andesitic magma forms in this manner. The first concerns active volcanoes and the andesite line (Fig. 3.33). The Andesite Line corresponds closely with plate subduction margins. The second piece of evidence comes from the distribution of andesitic volcanoes with respect to the subduction zone. On the upper surface of a plate of lithosphere, a subduction zone is marked by the presence of a deep-sea trench. Beyond the subduction zone, the lithosphere sinks into the asthenosphere, carrying with it its capping of wet oceanic crust. A horizontal distance of about 250 km, equivalent to a depth of about 80 km, marks the edge of a curved belt of volcanoes. The situation is nicely demonstrated by the andesitic stratovolcanoes of Japan, as shown in Figure 3.37.

The Origin of Rhyolitic Magma

Two observations suggest a continental origin for rhyolitic magma:

1. Volcanoes that extrude rhyolitic magma are confined to the continental crust or to regions of andesitic volcanism.

2. Volcanoes that extrude rhyolitic magma give off a great deal of water vapor, and intrusive igneous rocks formed from rhyolitic magma (granite) contain significant quantities of OH-bearing minerals

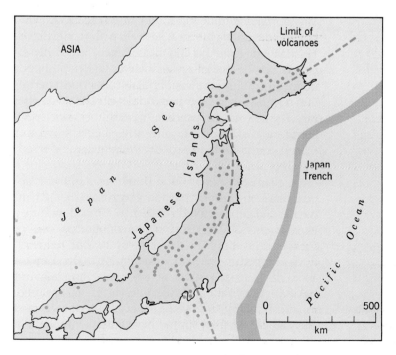

FIGURE 3.37 Relations between ocean trenches and arcs of volcanoes erupting andesitic magma. Arc-shaped Japanese islands are parallel to the Japan Trench. Andesitic volcanoes active during the last million years are also confined behind arcuate boundaries.

such as mica and amphibole. The OH must come from water dissolved in the magma.

These two points suggest that the generation of rhyolitic magma probably involves some sort of wet partial melting of rock with the composition of andesite; that is, with the average composition of the continental crust. Laboratory experiments bear out this suggestion. When, in the laboratory, water-bearing rocks having the average composition of the continental crust have partially melted, the composition of the magma that forms is rhyolitic. The source of heat for such melting to occur in the crust comes, apparently, from the mantle, as suggested in Figure 3.32.

Once a rhyolitic magma has formed, it starts to rise. However, the magma rises slowly because it has a high SiO_2 content (70%, as we learned earlier) and is therefore very viscous. As it slowly rises, the pressure on the magma decreases. As discussed earlier, the effectiveness of water in reducing the melting temperature is diminished by reduced pressure. Unless there is some way to heat it, therefore, a rising magma formed by wet partial melting will tend to solidify and form an intrusive igneous rock underground. But a rising magma traverses cool rock, and there is no source of heat to cause a temperature increase. As a result, the depth–temperature path of a rising body of rhyolitic magma brings it closer and closer to its solidification temperature. Therefore, most rhyolitic magma solidifies underground and forms *granitic batholiths*, rather than reaching the surface and forming lava or tephra.

SOLIDIFICATION OF MAGMA

Literally hundreds of different kinds of igneous rock can be found. Most are rare, but the fact that they exist suggests an important point: A magma of a given composition can crystallize into many different kinds of igneous rock.

Solidifying magma forms several different minerals, and those minerals start to crystallize from the cooling magma at different temperatures. The process is just the opposite of partial melting, and the temperature interval across which solidification occurs is simply the reverse of the melting interval discussed previously and illustrated in Figure 3.36. If at any stage during crystallization the melt becomes separated from the crystals, a magma with a brand-new composition results, while the crystals left behind form an igneous rock with a composition that is quite different from the composition of the magma.

There are a number of ways by which crystal–melt separations can occur. For example, compression can squeeze melt out of a crystal–melt mixture. Another mechanism involves the sinking of the dense, early crystallized minerals to the bottom of a magma chamber thereby forming a solid mineral layer covered by melt. However a separation occurs, the compositional changes it causes are called **magmatic differentiation by fractional crystallization.**

Bowen's Reaction Series

It was a Canadian-born scientist, N.L. Bowen, who first recognized the importance of magmatic differentiation by fractional crystallization. Because basaltic magma is far more common than either rhyolitic or andesitic magma, he suggested that basaltic magma may be primary and the other magmas may be derived from it by magmatic differentiation. At least in theory, Bowen argued, a single magma could crystallize into both basalt and rhyolite because of fractional crystallization. Such extreme differentiation rarely happens, we now know, but fractional crystallization is nevertheless an exceedingly important phenomenon in producing a wide range of rock compositions.

Bowen knew that plagioclases crystallized from basaltic magma are usually calcium rich (anorthitic), while those formed from rhyolitic magma are commonly sodium rich (albite). Andesitic magma, he observed, tends to crystallize plagioclases of intermediate composition. Bowen also knew that plagioclase grains in many igneous rocks have concentric zones of differing compositions such that the innermost core (and therefore earliest formed), is anorthitic, and that successive layers are increasingly albite rich (Fig. 3.38A).

Bowen's experiments provided a common explanation for these observations. He discovered that even though the composition of the first plagioclase that crystallized from a basaltic magma was anorthitic, the composition changed toward albite as crystallization proceeded and the ratio of crystals to melt increased. This means that all the plagioclase crystals, even the ones formed earliest, must continually change their compositions as the magma cools. A chemical balance between crystals and melt, if maintained, is referred to as chemical equilibrium.

As explained in chapter 2, the plagioclases involve a coupled ionic substitution in which $Ca^{2+} + Al^{3+}$ in the crystal are replaced by $Na^{1+} + Si^{4+}$. Bowen referred to such a continuous change of mineral composition in a cooling magma as a *continuous reaction series,* by which he meant that even though the composition changed continuously, the crystal structure remained unchanged. The speed at which the change occurs is controlled by the rates at which the four ions can diffuse through the plagioclase structure. To maintain equilibrium, some Ca^{2+} and Al^{3+} must dif-

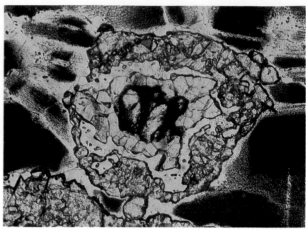

A.

B.

FIGURE 3.38   Textures illustrating Bowen's reaction series. A. Zoned plagioclase crystal in andesite, proof of continuous reaction, photographed in polarized light to enhance the zoning. Bands near center are anorthite rich, progressing to albite rich at the rim. The crystal is 2 mm long. B. A grain of olivine in a gabbro surrounded by reaction rims of pyroxene and amphibole demonstrates discontinuous reaction. The diameter of the outer rim is 3 mm.

fuse out of the early formed plagioclase and be replaced by Na^{1+} and Si^{4+} that diffuse into the crystal structure from the melt. Such diffusion is exceedingly slow. Equilibrium is rarely attained because the cooling rates of magmas are faster than the diffusion rates. As a result, zoned crystals of plagioclase are formed. The anorthite-rich inner zones are out of chemical equilibrium with the albite-rich outer zones and with the residual magma.

Bowen pointed out that the existence of zoned crystals has important implications. When anorthite-rich cores are present, the residual melt is necessarily richer in albite (and thus in sodium and silica) than it would have been if equilibrium had been maintained. According to Bowen, the anorthite-rich cores are an example of magmatic differentiation by fractional crystallization. If, in a partially crystallized magma containing zoned crystals, the melt were somehow squeezed out of the crystal mush, the result would be an albite-rich magma, and the residue would be an anorthite-rich rock.

Bowen identified several sequences of reactions besides the continuous reaction series of the feldspars. One of the earliest minerals to form in a cooling basaltic magma is olivine. Olivine contains about 40 percent SiO_2 by weight, while a basaltic magma contains 50 percent SiO_2. Thus crystallization of olivine will leave the residual liquid a little richer in silica. Eventually the solid olivine reacts with silica in the melt to form a more silica-rich mineral, pyroxene (Fig. 3.38B). The pyroxene in turn can react to form amphibole, and then the amphibole can react to form an even more siliceous mineral, biotite. Such a series of

reactions, where early formed minerals form entirely new compounds through reaction with the remaining liquid, is called a *discontinuous reaction series*.

If a core of olivine is shielded from further reactions by a rim of pyroxene, the remaining liquid will be more silica rich than it would be if equilibrium were maintained and all the olivine were converted to pyroxene. Bowen reasoned that if partial reactions occurred in both continuous and discontinuous reaction series, differentiation by fractional crystallization in a basaltic magma could, under extreme circumstances, even produce a residual magma with a rhyolitic composition (Fig. 3.39).

The answer to the question that Bowen posed—whether large volumes of rhyolitic magma can form from basaltic magma by fractional crystallization—is negative. The main evidence against Bowen's idea is (1) crystallization of a magma is complete long before a residual magma is siliceous enough to have a rhyolitic composition, (2) calculations show that only a tiny percentage of the volume of a basaltic magma could ever be differentiated to rhyolitic magma, and (3) rhyolitic magma forms in the continental crust. If rhyolitic magma formed by fractional crystallization of basaltic magma, we would surely expect to find some rhyolite in the oceanic crust, for it is there that basaltic magma is most common.

Although the principal manner by which rhyolitic magma must form is through partial melting of the continental crust, the importance of Bowen's reaction series has been demonstrated many times. Careful study of almost any igneous rock reveals evidence that fractional crystallization played a role in its formation.

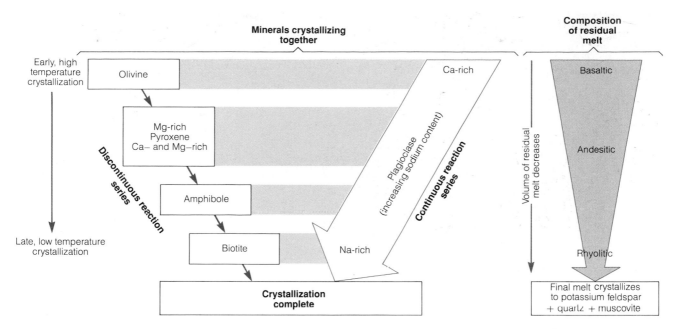

FIGURE 3.39 Bowen's reaction series. The earliest minerals that crystallize from a magma of basaltic composition are olivine and calcium-rich plagioclase (anorthite). As cooling and crystallization proceed, olivine (upper left) reacts with the remaining liquid to form pyroxene. Pyroxene, in turn, reacts to form amphibole, and amphibole forms biotite. The early plagioclase that cocrystallizes with olivine is calcium rich, but as cooling proceeds, the plagioclase reacts with the residual melt and continually changes its composition, becoming more sodium rich. The composition of the residual melt in contact with the crystallized minerals becomes increasingly silica rich, and eventually the final small fraction of melt has the composition of a rhyolitic magma.

Magmatic Mineral Deposits

The processes of partial melting and fractional crystallization in magmas sometimes lead to the formation of large and potentially valuable mineral deposits. Because magma is involved in the formation process, such deposits are called *magmatic mineral deposits*.

When a magma undergoes differentiation by fractional crystallization, the residual melt becomes progressively enriched in any chemical element that is not removed by the early crystallizing minerals. Separation and crystallization of the residual melt produces an igneous rock that contains the concentrated elements.

An important example of this kind of concentration process is provided by pegmatites, especially those formed as a result of the crystallization of rhyolitic magma. Pegmatites form by magmatic differentiation during the formation of granitic stocks and batholiths. Commonly, the pegmatites contain significant enrichments of rare elements such as beryllium, tantalum, niobium, uranium, and lithium, which, if sufficiently rich, can be mined.

Another form of magmatic differentiation by fractional crystallization occurs when early formed dense minerals sink and accumulate on the floor of a magma chamber. The process is called *crystal settling*, and in some cases the segregated minerals make desirable ores. Most of the world's chromium ores were formed in this manner by accumulation of the mineral chromite ($FeCr_2O_4$) (Fig. 3.40). The largest known chromite deposits are in South Africa, Zimbabwe, and the U.S.S.R. Similarly, vast deposits of ilmenite ($FeTiO_3$), a source of titanium, were formed by magmatic differentiation. Large deposits occur in the Adirondack Mountains.

A form of concentration similar to crystal settling occurs when, for reasons not clearly understood, certain magmas separate, as oil and water do, into two immiscible liquids. One, a sulfide liquid that is rich in copper and nickel, sinks to the floor of the magma chamber because it is denser. After cooling and crystallization, the resulting igneous rock has a copper or nickel ore at the base. The world's greatest known concentration of nickel ore, at Sudbury, Ontario, is believed to have formed in this fashion. Other great nickel deposits in Canada, Australia, and Zimbabwe formed in the same manner.

FIGURE 3.40 Layers of chromite (black) and plagioclase (white) formed by crystal settling during the crystallization of the Bushveld Igneous Complex. The location of this unusually fine outcrop is the Dwars River, South Africa.

VOLCANOES AND GEOTHERMAL ENERGY SOURCES

Iceland is an unusual place. It lies even farther north than the northernmost part of Hudson Bay, and, as the name suggests, it's a very cold place. But Iceland straddles the Mid-Atlantic Ridge, and so it is also a place of active volcanism. The people who live in Iceland have found many clever ways to harness volcanic heat in order to combat the cold climate. Using water warmed by hot volcanic rocks, they heat their houses, grow tomatoes in hot houses even though the temperature outside may be below freezing, and swim year-round in naturally heated pools. Icelanders also generate most of the electricity they need by using volcanically produced steam.

Everywhere the Earth's temperature increases with depth. In theory at least, everyone, not just Icelanders, should be able to drill water circulation holes deep enough so the Earth's internal heat could be tapped. In practice it turns out to be very difficult to do so. In order to be used efficiently, geothermal steam should be 200°C or hotter. In most parts of the world holes must be 5 to 7 km deep in order to reach rock that is 200°C. For the practical development of *geothermal energy,* as the Earth's store of heat is called, we need to be able to reach rock temperatures of 200°C or higher within 3 kilometers of the surface. So far, at least, it is also necessary for there to be a natural reservoir of underground water that is heated by the hot rocks, as shown in Figure B3.1. The places on the Earth where these conditions are met are places of recent volcanic activity. Most of the world's volcanic and magmatic activity is close to plate margins, and it is here, in places such as New Zealand, the Philippines, Japan, Italy, Iceland, and the western United States, that geothermal power is being used.

Unfortunately, the total amount of energy that can be recovered from natural geothermal fields

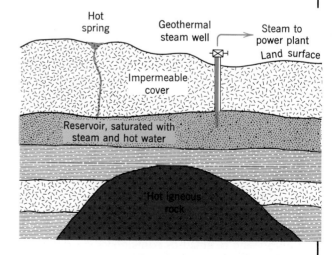

FIGURE B3.1 A typical geothermal steam reservoir. Water in a permeable aquifer, such as a tuff, is heated by magma or hot igneous rock. As steam and hot water are withdrawn through the well, cold water flows into the reservoir through the aquifer.

is not very large. Today geothermal power is of local importance only. It is an interesting question whether the situation may change. Many volcanoes, like Mount St. Helens, are too active and too dangerous to be considered as geothermal energy sources. Some places, like Yellowstone National Park, could produce a large amount of energy, but if we drilled and pumped out the steam and hot water reservoirs beneath Yellowstone, the famous hot springs and geysers would soon be dry.

Interesting geothermal experiments are now being conducted in New Mexico. The goal is to create our own geothermal fields. In the Jemez Mountains on the edge of an extinct (but still hot) volcano scientists have drilled two holes deep into the hot rock, as shown in Figure B3.2. Then they shattered the hot rock with explosives to create an artificial reservoir and pumped water through to produce steam. The first tests have been only partly successful. A major difficulty was that water did not flow uniformly through the hot rock but instead followed narrow flow paths, and the rocks that lined them soon cooled down. Further tests are planned, not only in the Jemez Mountains, but in France, England, and other countries also.

Hot dry rocks are much more abundant than geothermal steam fields. If the water flow troubles encountered in the Jemez Mountains experiments can be overcome, geothermal energy may someday play a major role in meeting the energy needs of society.

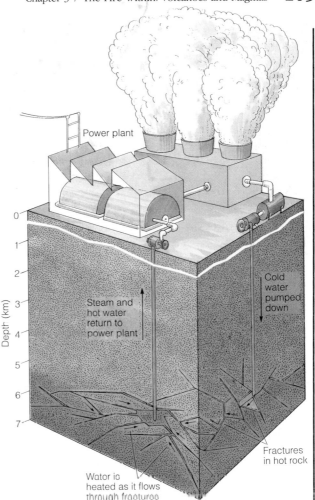

FIGURE B3.2 Geothermal energy from hot dry rocks. Cold water is pumped through fractures at the bottom of a 6-km-deep well, becomes heated, then flows back up to geothermal power plant, where heat energy is converted to electricity.

SUMMARY

1. Three kinds of magma predominate: basaltic, andesitic, and rhyolitic.

2. The principal controls on the physical properties of magma are temperature, SiO_2 content, and dissolved-gas content. High temperature, low SiO_2 content, and low dissolved-gas content result in fluid magma (basaltic). Lower temperature, a lot of dissolved gas, and high SiO_2 contents result in viscous magma (andesitic and rhyolitic magma).

3. Basaltic magma forms by dry partial melting of

rock in the mantle. Andesitic magma forms during subduction by wet partial melting of basalt in oceanic crust. Rhyolitic magma forms by wet partial melting of rock in the continental crust.

4. The sizes and shapes of volcanoes depend on the kind of material erupted, viscosity of the lava, and explosiveness of the eruptions.

5. Viscous magmas rich in SiO_2 erupt a lot of tephra and build steep sided tephra cones or stratovolcanoes.

6. Low-viscosity magmas low in SiO_2 tend to be

erupted as fluid lavas that build gently sloping shield volcanoes or lava plateaus.

7. Magma forms by the complete or partial melting of rock.

8. Igneous rock forms by the solidification and crystallization of magma.

9. Igneous rock may be intrusive (meaning it formed within the crust) or extrusive (meaning it formed on the surface). The texture and grain size of igneous rock indicate how and where the rock cooled.

10. Igneous rocks rich in quartz and feldspar, such as granite, granodiorite, and rhyolite, are characteristically found in the continental crust. Basalt, which is rich in pyroxene and olivine, is derived from magma formed in the mantle and is common in the oceanic crust.

11. All bodies of intrusive igneous rock are called plutons. Special names are given to plutons based on shape and size.

12. Modern volcanic activity is concentrated along plate margins. Andesitic volcanoes are found at subduction margins and basaltic volcanoes are concentrated along spreading margins. Rhyolitic volcanoes occur at collision margins, in continental rift valleys, and in places where rising basaltic magma causes the continental crust to melt.

13. Volcanism that occurs in the middle of an oceanic plate is apparently due to hot spots that are long-lived sources of basaltic magma deep in the mantle. As oceanic plates move over hot spots, long chains of basaltic volcanoes are formed.

14. Processes that separate remaining melt from already formed crystals in a cooling magma lead to the formation of a wide diversity of igneous rocks.

15. Magmatic mineral deposits, which form as a result of magmatic differentiation, are the world's major sources of nickel, chromium, vanadium, platinum, beryllium, and a number of other important industrial metals.

IMPORTANT TERMS TO REMEMBER

batholith (p. 94)

caldera (p. 82)
crater (p. 82)

dike (p. 93)

extrusive igneous rock (p. 88)

geothermal gradient (p. 98)

intrusive igneous rocks (p. 88)

lava (p. 74)

magma (p. 74)
magmatic differentiation by fractional crystallization (p. 101)
magmatic differentiation by partial melting (p. 99)

partial melting (p. 99)
pillow basalt (p. 87)
plateau basalt (p. 86)
pluton (p. 93)
porphyry (p. 88)
pyroclast (p. 78)

pyroclastic flow (p. 79)
pyroclastic rocks (p. 78)

shield volcano (p. 80)
sill (p. 94)
stock (p. 96)
stratovolcano (p. 81)

tephra (p. 78)
tephra cone (p. 81)

viscosity (p. 75)
volcano (p. 74)

IMPORTANT ROCK NAMES TO REMEMBER

agglomerate (p. 92)
andesite (p. 74)

basalt (p. 74)

dacite (p. 91)
diorite (p. 91)

gabbro (p. 91)
granite (p. 91)
granodiorite (p. 91)

obsidian (p. 89)

pegmatite (p. 89)
peridotite (p. 91)

rhyolite (p. 74)

tuff (p. 92)

welded tuff (p. 92)

QUESTIONS FOR REVIEW

1. What controls the grain size and texture in igneous rocks? Would you expect lava flows to be basalt or gabbro?

2. What are the distinguishing features of pyroclastic rocks? How might you tell the difference between a rhyolite that flowed as a lava and a rhyolitic tuff?

3. Do porphyries always contain phenocrysts? What minerals might you find as phenocrysts in a rhyolite porphyry? A basalt porphyry?

4. What is the major difference between the mineral assemblage of a diorite and a granodiorite? Between granite and granodiorite? Between a gabbro and a peridotite?

5. How is pumice formed and of what is it made? Can you suggest why pumice will float on water?

6. How does a lapilli tuff differ from a welded tuff? Could both rocks form as a result of eruptions from the same volcano?

7. Is the major oxide component of magma SiO_2, MgO, or Al_2O_3? Briefly describe the effect of the SiO_2-content on the fluidity of magma. What effect does temperature have on viscosity?

8. What does the term *partial melting* mean and what role does it play in formation of basaltic magma? Where in the Earth does basaltic magma form?

9. How does the magma forming process of dry partial melting differ from wet partial melting? Can you suggest an example of a magma type formed by each kind of melting?

10. What is the origin of andesitic magma such as that erupted by Mount St. Helens? With what kind of volcanoes are andesitic eruptions associated? What is the distribution of andesitic volcanoes with respect to today's tectonic plates?

11. How might it be possible for fractional crystallization to produce more than one kind of igneous rock from a single magma? Comment on the role of fractional crystallization in the formation of mineral deposits.

12. Comment on the importance of Bowen's Reaction Series for understanding differentiation processes in cooling magmas. How does a continuous reaction series differ from a discontinuous series?

13. Why does a shield volcano like Mauna Loa in Hawaii have a gentle surface slope, while a stratovolcano such as Mount Fuji in Japan has steep sides?

14. How does a lava dome such as the one that now lies in the crater of Mount St. Helens form? How, if at all, does a lava dome differ from a resurgent dome?

15. Describe the distinctive kind of basaltic lava flows formed by submarine fissure eruptions. How do such flows form?

16. How does a dike like the Great Dike of Zimbabwe differ from a sill such as the Palisades Sill?

17. How big can batholiths be, and by what mechanism are granitic magmas believed to move upward in the crust?

18. Briefly describe the plate tectonic settings of andesitic and basaltic volcanoes. Where does Kilauea Volcano in Hawaii sit with respect to the Pacific Plate? Where does Mount St. Helens sit with respect to a plate boundary?

Marine geologists collect fossiliferous samples of a deep-sea core to reconstruct past oceanic conditions.

CHAPTER 4

Sediments and Sedimentary Rocks

T he ultimate resting place of most of the sediment eroded from the land is in the world's oceans. Covering more than half the surface area of the Earth, the deep ocean floor was a terra incognita for most of human history and virtually inaccessible to scientists. In the last three decades things have changed dramatically as international teams of scientists aboard specially outfitted drilling ships have recovered long cores of sediment and rock from the deepest parts of the ocean. The result is a record of recent Earth history that rivals that on the continents and often is far more complete.

When this era of ocean exploration began, we knew little about how the Earth works. A key to understanding the dynamics of the Earth's mobile lithosphere has come from studies of the sediments and rocks beneath the deep ocean that ultimately led to the theory of plate tectonics. What many consider to be the most important and exciting modern breakthroughs in geology have thus resulted from exploration of the remote world of the ocean floor.

Like any "big science" project, the program of deep ocean drilling has produced important technological and scientific spinoffs. Knowledge of the character and distribution of deep sea sediments helps us to evaluate the feasibility of disposing of dangerous radioactive wastes beneath the sea floor. Information gained from ocean sediments has been used to reconstruct past climatic patterns on the Earth that are important for anticipating future long-term climatic trends. Experience in drilling through ocean sediments in deep water also has obvious applications to the search for offshore petroleum resources in sedimentary rocks of the continental shelves. So well known has the geology of the ocean basins now become, that the seafloor can no longer be considered terra incognita. Geology has become a global science.

SEDIMENTS AND SEDIMENTATION

Like a perpetually restless housekeeper, nature is ceaselessly sweeping regolith off the solid rock beneath it, carrying the sweepings away, and depositing them as sediment in river valleys, lakes, and innumerable other places. We can see sediment being transported by trickles of water after a rainfall and by every wind that carries dust. The mud on a lake bottom, the sand on a beach, even the dust on a windowsill is sediment. Because erosion and deposition of rock particles take place almost continuously, we find sediment nearly everywhere.

Stratification and Bedding

Sedimentary **stratification** results from the arrangement of sedimentary particles in distinct layers (Fig. 4.1). Each sedimentary **stratum** (plural = **strata**) is a distinct layer of sediment that accumulated at the Earth's surface. While layering is an obvious feature of most sediments and sedimentary rocks, it is seen also in the products of volcanic eruptions (lava flows and tephra deposits) and in many metamorphic rocks. A close look at sedimentary strata shows that they differ from one another because of differences in some characteristic of the particles or in the way in which they are arranged. For example, the average diameter of particles in one stratum may differ from the average diameter in another stratum.

The layered arrangement of strata in a body of sediment or sedimentary rock is referred to as **bedding**. Each **bed** within a succession of strata can be distinguished from adjacent beds by differences in thickness or character. The top or bottom surface of a bed is a **bedding plane.**

Clastic Sediment

A close look at sand and gravel beside a stream shows that pebbles and sand grains are simply bits of rock and minerals. A magnifying glass discloses that the finer sedimentary particles are also derived from broken-up rock, although generally these particles have undergone some chemical change; feldspars, for example, have been partly altered to clay. The loose, fragmental debris produced by the mechanical breakdown of older rocks is called **detritus** (from the Latin for "worn down") or **detrital sediment**. Such sediment also is referred to as **clastic sediment** (from the Greek word *klastos*, meaning "broken") (Fig. 4.2). Any individual particle of clastic sediment is a *clast*.

FIGURE 4.1   Multicolored layered sedimentary rocks in Capitol Reef National Park, Utah.

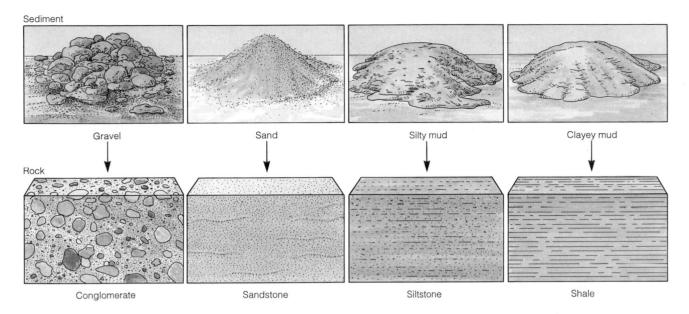

FIGURE 4.2 Principle kinds of clastic sediment and sedimentary rocks formed from them.

Particles of clastic sediment may range in size from the largest boulders down to submicroscopic clay particles. This range of particle size is the primary basis for classifying clastic sediments and clastic sedimentary rocks. In Table 4.1 we can see that clastic sediment can be divided into four main classes, which from coarsest to finest are gravel, sand, silt, and clay. Gravel is further subdivided on the basis of dominant clast size into boulder gravel, cobble gravel, and pebble gravel. If, for example, a gravel consists predominantly of clasts having a diameter between 64 and 256 mm, we would call it a cobble gravel.

Production of Clastic Sediment

Igneous rocks solidify from lavas erupted at high temperatures or from magmas within the crust where temperatures and pressures are far higher than at the Earth's surface. When exposed at the land surface, igneous rocks decompose and break down mechanically as they are subjected to new physical and chemical conditions. The processes involved are collectively referred to as weathering (chapter 7). The resulting detritus, consisting of mineral grains and rock fragments, forms a major component of clastic sediment. Clastic particles are also produced by the breakdown

TABLE 4.1 Definition of Clastic Particles, Together With the Sediments and Sedimentary Rocks Formed From Them

Name of Particle	Range Limits of Diameter (mm)[a]	Names of Loose Sediment	Name of Consolidated Rock
Boulder	More than 256	Boulder gravel	Boulder conglomerate[c]
Cobble	64 to 256	Cobble gravel	Cobble conglomerate[c]
Pebble	2 to 64	Pebble gravel	Pebble conglomerate[c]
Sand	1/16 to 2	Sand	Sandstone
Silt	1/256 to 1/16	Silt	Siltstone
Clay[b]	Less than 1/256	Clay	Mudstone and shale

Source: C. K. Wentworth, 1922, A scale of grade and class terms for clastic sediments: *Journal of Geology,* v. 30, p. 377–392.

[a]Note that size limits of sediment classes are powers of 2, just as are memory limits in microcomputers (for example, 2K, 64K, 256K, 512K).

[b]Clay, used in the context of this table, refers to particle size. The term should not be confused with clay minerals, which are definite mineral species.

[c]If the clasts are angular, the rock is called a *breccia* rather than a conglomerate.

of metamorphic and sedimentary rocks. They also may consist of the accumulated skeletal parts or shells of dead organisms.

Transport and Deposition of Clastic Sediment

Clastic sediment is transported in many ways. It may slide or roll down a hillside under the pull of gravity or be carried by a glacier, by the wind, or by flowing water. In each case, when transport ceases, the sediment is deposited in a fashion characteristic of the transporting mechanism. Deposition occurs because of a drop in energy. Sediment transported by sliding, rolling, or flowing downhill is deposited as the angle of slope decreases and the moving sediment loses momentum. Sediment is deposited by a glacier if the forward motion of the ice becomes very slow or ceases or the glacier margin begins to retreat. In the case of sediment moved by gravity down a hillslide or deposited directly by a glacier, the resulting deposit generally is a random mixture of particles of many sizes. Sedimentary particles transported by wind or water are deposited when the flowing water or moving air slows to a speed at which particles can no longer be carried. In a general way, the size of the grains in sediment moved by wind or water is related to the speed of the transporting agent: the faster the speed, the larger are the particles that can be moved.

Mineral Composition of Clastic Sediment

Most coarse clastic sediments consist of mineral grains and rock fragments, with those least susceptible to chemical and physical breakdown predominating. Many sands, for example, have a high content of quartz and/or potassium feldspar which are the common rock-forming minerals most resistant to weathering. As detrital rock fragments and mineral grains are transported, they are subjected to continuous chemical and physical breakdown. After several cycles of erosion and deposition, the result can be a sediment that consists almost entirely of quartz, the most resistant of the rock-forming minerals (Fig. 4.3).

Some Conspicuous Features of Clastic Sediments

The size of particles, the way they are packed together, and other distinctive features permit us to distinguish different types of sediment, as well as sedimentary rocks. Such characteristics also help us to infer the environment in which a sediment was deposited.

Sorting. **Sorting** is a measure of the range of particle size of sediments. A sediment having a wide range of particle size is said to be poorly sorted; if the range is small, the sediment is said to be well sorted (Figs.

4.3 and 4.4). In a clastic sediment, changes of grain size typically result from energy fluctuations in whatever agent is responsible for the deposit. Such energy changes, usually small, are not the exception but the rule.

Sorting of particles also can be related to differences in their *specific gravity* (ratio of the weight of a given volume of material to the weight of an equal volume of water). Particles of unusually heavy minerals (e.g., gold, platinum, and magnetite) are deposited quickly on streambeds or on beaches, whereas lighter particles are carried onward.

Most of the particles transported by water or wind, however, are common rock-forming minerals, such as quartz and feldspar, that have similar specific gravities. Therefore, such particles typically are sorted not according to specific gravity but according to *size* (Fig. 4.3).

Long-continued movement of particles by turbulent

FIGURE 4.3   Well-rounded grains of quartz sand from the St. Peter Sandstone of Wisconsin have been sorted by size and polished by constant shifting and abrasion in surf along an ancient shoreline.

water or air results in gradual destruction of the weaker particles, leaving behind the particles that can better survive in the turbulent environment. Very commonly the survivor is quartz because it is hard and lacks cleavage. In this case sorting is based on *durability*.

Particle Shape. Mechanically weathered particles broken from bedrock tend to be angular because breakage typically occurs along grain boundaries, fractures, and surfaces separating rock layers. Nearly all such weathered particles become smooth and rounded as they are transported by water or air and are abraded by other rock fragments. *Roundness*, as measured by the sharpness of a particle's edges, is not the same as *sphericity*, which is a measure of how closely particle shape approaches that of a sphere (Fig. 4.4). A flat particle bounded by cleavage or fracture surfaces may have well-rounded edges, but it also may have a low degree of sphericity. In general, the greater the distance of travel, the greater is the degree of rounding.

Rhythmic Layering. Some sediments display a distinctive alternation of parallel layers having different properties. Such alternation suggests that some naturally occurring rhythm has influenced sedimentation. A pair of such sedimentary layers deposited over the cycle of a single year is termed a **varve** (Swedish for cycle). Varves are most commonly seen in deposits of high-latitude or high-altitude lakes, where there is a strong contrast in seasonal conditions. In spring, as a cover of winter ice melts away, the inflow of sediment-laden water increases and coarse sediment is then deposited throughout the summer. With the onset of colder conditions in the autumn, streamflow decreases and ice forms over the lake surface. During winter, very fine sediment that has remained suspended in the water column slowly settles to form a thinner, darker layer above the coarse lighter-colored summer layer. Varved lake sediments are common in Scandinavia and New England where they formed beyond the retreating margins of Ice Age glaciers (Fig. 4.5) and they are also seen in some ancient sedimentary rocks.

FIGURE 4.4 Clastic sediments range from very poorly sorted to very well sorted depending on the extent to which the constituent grains are of equal size. Particles also range from angular to rounded, depending on the degree to which sharp edges have been worn off, and they have a high or low degree of sphericity depending on how closely they approach a spherical shape.

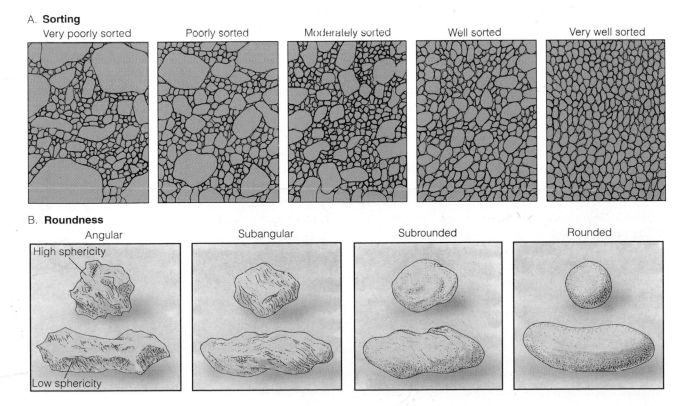

A. **Sorting**

| Very poorly sorted | Poorly sorted | Moderately sorted | Well sorted | Very well sorted |

B. **Roundness**

| Angular | Subangular | Subrounded | Rounded |

High sphericity

Low sphericity

FIGURE 4.5 Varves deposited in a glacial-age lake in southern Connecticut. Each pair of layers in a sequence of varves represents an annual deposit. Light-colored silty layers were deposited in summer, and the dark-colored clayey layers accumulated in winter. Each layer is well sorted.

Cross Bedding. **Cross bedding** refers to beds that are inclined with respect to a thicker stratum within which they occur (Fig. 4.6). Cross beds consist of particles coarser than silt and are the work of turbulent flow in streams, wind, or ocean waves. As they move along, the particles tend to collect in ridges, mounds, or heaps in the form of ripples, waves, or dunes that migrate slowly forward in the direction of the current. Particles accumulate on the downcurrent slope of the pile to produce beds having inclinations as great as 30° to 35°. The direction in which cross bedding is inclined tells the direction in which the related current of water or air was flowing at the time of deposition.

Graded Bedding. The way particles are arranged within a layer provides important information about the conditions of sedimentation. If a mixture of small solid particles having different diameters and about the same specific gravity is placed in a glass of water, shaken vigorously, and then allowed to stand, the particles will settle out and form a deposit on the bottom of the glass. The largest particles settle first, followed by successively smaller ones. The finest may stay in suspension for hours or days before they finally settle out at the top of the deposit. In the resulting **graded bed**, the particles are sorted more or less according to size, and grade upward from coarser to finer (Fig. 4.7). A graded bed also can form from a sediment-laden current. As the current slows down, the heaviest and largest particles settle out first, followed by lighter and smaller ones.

Nonsorted Sediment. In some sedimentary rocks the particles are a mixture of different sizes arranged chaotically, without obvious order. Such sediments are created, for example, by rockfalls, slow movement of debris down hillslopes, slumping of loose deposits on the seafloor, mudflows, and deposition of debris from glaciers or floating ice. Some nonsorted sediments are given specific names (for example, *till* is a nonsorted sediment of glacial origin, while the corresponding rock is a *tillite*; Fig. 4.8).

FIGURE 4.6 Ancient cross-stratified sand dunes that have been converted to sedimentary rock that crops out near Kanab, Utah. Cross strata are inclined to the right, in the direction toward which the ancient prevailing winds were blowing.

Chemical Sediment

Some sediment contains no clastic particles, yet the material composing the sediment has been transported. In such sediment, the components were dissolved, transported in solution, and precipitated chemically. Sediment formed by precipitation of minerals from solution in water is **chemical sediment**, and it forms in two principal ways: One way is through biochemical reactions resulting from the activities of plants and animals in the water. For example, tiny plants living in seawater can decrease the acidity of the surrounding water and thereby cause calcium carbonate to precipitate.

Chemical sediment also forms as a result of inorganic reactions in the water. When the water of a hot spring cools, it may precipitate opal or calcite. We can witness a similar effect when chemical sediment is deposited along the inside of a hot-water pipe, thereby constricting the flow of water and leading to expensive

FIGURE 4.8 The Gowganda Tillite, which crops out over an area of thousands of km² in eastern Canada, provides evidence of extensive glaciation during the Middle Precambrian (ca. 2500–2150 million years ago). The nonsorted tillite consists of boulders and cobbles of igneous and metamorphic rocks in dark, fine-grained matrix.

home repairs. Another common example is simple evaporation of seawater or lake water; as the water evaporates, dissolved matter is concentrated and salts begin to precipitate out as chemical sediment.

FIGURE 4.7 A mudflow deposit that originated at Mount St. Helens volcano is vertically sorted. Conspicuous cobbles and pebbles in the lower part grade upward into finer pebbles and sand near the top. This graded bed resulted from a prehistoric eruption that sent a flood of muddy debris down a nearby valley.

Biogenic Sediment

Many bodies of sediment contain **fossils**, the remains of plants and animals that died and were incorporated and preserved as the sediment accumulated. Sometimes the form of an original plant or animal is preserved (Fig. 4.9), but more commonly the remains are broken and scattered. A sediment composed mainly of fossil remains, and therefore produced directly by the physiological activities of organisms, is called **biogenic** sediment. The solid parts generally end up as fragments, or clasts. If the sediment is composed largely of such fragments we call it *bioclastic* sediment. Often, however, the remains are transformed into organic compounds that either are locally concentrated or widely dispersed in the enclosing sediment.

Calcareous and Siliceous Biogenic Sediment

One important kind of biogenic sediment is formed of calcium carbonate and is widespread in the oceans. While calcium carbonate can be precipitated chemically from seawater, most carbonate sediments result from biogenic activity, mainly in warm surface ocean waters. There, carbonate-secreting organisms precipitate calcite or aragonite in building their hard parts.

FIGURE 4.9 The fossil remains of a kauri pine lie next to the skeleton of a fossil fish on a bedding plane of a 175-million-year-old mudstone in Australia.

In doing so, the calcium and bicarbonate ions in the water are combined to form solid calcium carbonate:

$$Ca^{2+} + (HCO_3)^{1-} \rightarrow CaCO_3 + H^{1+}$$

| calcium ion | bicarbonate ion | calcium carbonate | hydrogen ion |

When floating microscopic marine organisms die, their remains settle and accumulate on the seafloor to form a muddy sediment called **deep-sea ooze** (Fig. 4.10). If such a sediment consists mainly of carbonate remains, it is *calcareous ooze*. Corals, algae, and other colonial organisms growing on the seafloor in shallow tropical waters also precipitate calcium carbonate, thereby creating extensive carbonate reefs.

Siliceous ooze resembles calcareous ooze, but is composed mainly of the siliceous remains of tiny floating protozoa (radiolarians) and algae (diatoms). The hard parts in this ooze are often delicate and lacelike (Fig. 4.10). Some kinds of diatoms also live in lakes where their remains accumulate as an important component of lake sediment.

Growth and Burial of Plant Matter

Nearly all living organisms derive their energy from the Sun. The chief energy-trapping mechanism is *photosynthesis*, a process by which plants use the Sun's energy to combine water and carbon dioxide to make carbohydrates (organic compounds containing C, O, and H) and oxygen. Animals that consume plants, therefore, are secondary consumers of trapped solar energy. When plants or animals die and decay, atmospheric oxygen combines with carbon and hydrogen in the organic compounds to form H_2O and CO_2 once again.

Whenever organic matter is buried, a small portion of it escapes decay. In this way some of the solar energy becomes stored in sediments and ultimately in sedimentary rocks. The total amount of trapped organic matter is far less than 1 percent of the organic matter formed by growing plants and animals. Nevertheless, during the past 600 million years, the amount of trapped organic matter has grown to be very large.

Fossil Fuels

Local large concentrations of organic substances in sediments and sedimentary rocks, in the form of coal,

FIGURE 4.10 Skeletons of calcareous foraminifera (smooth globular objects), siliceous radiolaria (delicate meshed objects), and siliceous rod-shaped sponge spicules from a deep-sea ooze, photographed by scanning electron microscope. The fossils are from a sediment core collected in the western Indian Ocean.

oil, and natural gas (collectively called the **fossil fuels**), provide most of the energy that runs our modern civilization. The nature and occurrence of each of these basic fuel resources depends on the kind of sediment, the kind of organic matter trapped, and the changes that have occurred during the long geological ages since the organic matter was buried.

Peat

On land, plants such as trees, bushes, and grasses contribute most of the trapped organic matter. These plants contain carbohydrates, but they are also rich in *resins, waxes*, and *lignins*, which tend to remain solid. In water-saturated environments, such as bogs or swamps, plant remains accumulate to form **peat**, an unconsolidated deposit of plant remains having a carbon content of about 60 percent (Fig. 4.11). Peat is the initial stage in the development of the combustible sedimentary rock we call coal, the most abundant of the fossil fuels.

Petroleum

In the ocean, microscopic phytoplankton (tiny floating plants) and bacteria (simple, single-celled organisms) are the principal sources of organic matter trapped in sediment. Most of the organic matter is trapped in clay that is slowly converted to shale. During this conversion, organic compounds are transformed into oil and natural gas. These two products are the main forms of **petroleum**, defined as gaseous, liquid, and semisolid naturally occurring substances that consist chiefly of **hydrocarbons** (chemical compounds of carbon and hydrogen).

Two kinds of evidence support the hypothesis that petroleum is a product of the decomposition of organic matter: (1) oil possesses optical properties known only in hydrocarbons derived from organic matter, and (2) oil contains nitrogen and certain compounds believed to originate only in living matter.

Oil is nearly always found in marine sedimentary rocks. Sampling on the continental shelves and along the base of the continental slopes has shown that fine muds beneath the seafloor contain up to 8 percent organic matter. Geologists therefore conclude that oil originated primarily as organic matter deposited with marine sediment.

A long and complex chain of chemical reactions apparently is involved in the conversion of organic matter to crude petroleum and natural gas. In addition, chemical changes may occur in oil and gas even after they have accumulated. This explains, for example, why chemical differences exist between the oil in one body of petroleum and that in another.

LITHIFICATION AND DIAGENESIS

Lithification (from the verb to *lithify*, meaning turn to stone) is the process whereby a newly deposited, unconsolidated sediment is slowly converted to sedimentary rock. During lithification, a number of changes commonly occur. Geologists refer collectively to all the chemical, physical, and biological changes that affect sediment after its initial deposition and during and after lithification as **diagenesis**.

The first and simplest diagenetic change is **compaction**, which occurs as the weight of an accumulating sediment forces the grains together. As the *pore space* (the space between grains) is reduced, water is forced out of the sediment. Substances dissolved in circulating pore water precipitate and cement the grains together, a process called **cementation**. Calcium carbonate is one of the most common cements

FIGURE 4.11 A peat cutter harvests dark organic-rich peat from a bog in western Ireland. The peat has formed in a cool moist climate that favors preservation of organic matter in wet environments. When dried, the peat provides fuel for heat and cooking.

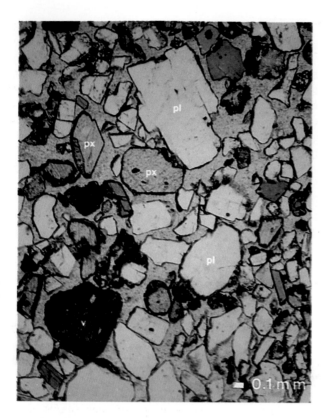

FIGURE 4.12 Sand grains in a thin section of a sandstone from central Washington are bonded together by calcite cement. Light-colored grains are plagioclase (pl), brownish grains are pyroxene (px), and the large dark grain is a fragment of volcanic rock (v).

(Fig. 4.12), but silica, a particularly hard cement, may also bond grains together.

As sediments accumulate, less-stable minerals may recrystallize to more stable forms. The process of **recrystallization** is especially common in porous reef limestone. Over time, the mineral aragonite, which forms the skeletal structure of living corals, recrystallizes to its polymorph calcite.

Important chemical alterations also affect sediments. In the presence of oxygen (an *oxidizing environment*), organic remains are quickly converted to carbon dioxide and water. If oxygen is lacking (a *reducing environment*), the organic matter does not completely decay but instead may be slowly transformed to solid carbon.

The formation of peat, and its slow transformation to coal, provides an example of diagenesis under reducing conditions. Geologists recognized long ago that the plants responsible for deposits of peat and coal must have lived in ancient swamps because (1) a complete physical and chemical gradation exists from peat to coal, and peat now accumulates mainly in swamps; and (2) only under swamp conditions is the conversion of plant matter to coal chemically probable. On dry land and in running water, oxygen is abundant. Under these conditions dead plant matter gradually rots away. Under stagnant or nearly stagnant swamp water, however, oxygen is used up and not replenished. Any plant matter lying in swamp water is attacked by anaerobic bacteria, which partly decompose it by splitting off some of the oxygen and hydrogen. These two elements escape, combined in various gases, and the carbon gradually becomes concentrated in the residue. Although they work to destroy the plant matter, the bacteria themselves are destroyed before they can finish the job, because the poisonous acidic compounds they liberate from the dead plants kill them. In this manner, plant matter is gradually converted to peat. This could not happen in a stream because the flowing water would bring in new oxygen to decompose the plants and would also dilute the poisons and permit the bacteria to complete their destructive work.

COMMON SEDIMENTARY ROCKS

Sedimentary rocks, like the sediments from which they are derived, fall into three categories: clastic, chemical, and biogenic.

Clastic Sedimentary Rocks

If a sedimentary rock is made up of mineral particles derived from the weathering and erosion of igneous rock, how can we tell it is sedimentary and not igneous? In addition to such obvious clues as sedimentary layering, rock texture also provides evidence that tells us whether a rock is sedimentary or igneous.

The mineral grains in igneous rock are irregular and interlocked, but the mineral grains and rock fragments in sedimentary rock commonly are rounded and show signs of the abrasion they received during transport (Fig. 4.12). Also, clastic sedimentary rock contains cement holding the particles together, whereas igneous rock consists of interlocking crystals. Fossils are another important feature for distinguishing between the two classes of rock. No organism can survive the high temperatures at which igneous rocks form, and so the presence of ancient shells or similar evidence of past life is an important clue to sedimentary origin.

Clastic sedimentary rocks are classified on the basis of predominant particle size, just as sediments are. The four basic classes are conglomerate, sandstone, siltstone, and shale, which are the rock equivalents of gravel, sand, silt, and clay (Fig. 4.2).

A **conglomerate** is a lithified gravel. Because the clasts in gravels can vary greatly in size, we often use a modifying term to indicate the predominant size of the particles in a conglomerate, just as we do with the corresponding sediment (e.g., a boulder conglomerate, a cobble conglomerate, or a pebble conglomerate). The large clasts in a conglomerate generally are pieces of preexisting rock, whereas fine particles between the clasts (collectively known as the *matrix*), consist mainly of mineral fragments. In a conglomerate, the clasts are rounded in shape; if the clasts are equally large but angular, the rock is a **breccia**. One kind of breccia is produced by explosive volcanic activity and is called a *volcanic breccia*.

Sandstone consists mainly of grains of sand, although coarser or finer particles may be present (e.g., a pebbly sandstone or a silty sandstone). Different types of sandstone are recognized on the basis of composition. Quartz, being very resistant to weathering, is a common mineral in sandstones. A quartz sandstone consists predominantly of quartz grains. If feldspars are a major component, the sediment is called an *arkose* or *arkosic sandstone*. A dark-colored sandstone containing quartz, feldspar, and a large amount of tiny rock fragments (*lithic* particles) is a *lithic sandstone* (also called a *greywacke*).

Siltstone is composed mainly of silt-size mineral fragments, predominantly quartz and feldspar. Clastic rocks of still finer grain size include *mudstone*, which breaks down into blocky fragments, and **shale**, which cleaves into sheetlike fragments when it weathers. In shales, the clay-size particles are so small that composition generally must be determined by X-ray methods. The most common components of shale are quartz, feldspar, calcite, and clay minerals.

Chemical Sedimentary Rocks

Chemical sedimentary rocks result from lithification of organic or inorganic chemical precipitates. Most of these rocks contain only one important mineral, which forms the basis for classification.

Common Rock Types

Among the most common chemical rocks are *rock salt* (halite = NaCl) and *gypsum*, both of which are formed by evaporation of seawater or lake water and have economic value (see below) (Fig. 4.13A).

Chert is mainly a hard, extremely dense sedimentary rock composed entirely of amorphous fine-grained silica. Its typical splintery to conchoical frac-

FIGURE 4.13 A. High evaporation rates in the desert basin of Searles Lake in eastern California have led to precipitation of salts that are interbedded with organic muds. The most common salts found in cores drilled through the lake sediments are halite (NaCl) and trona ($Na_2CO_3 \cdot NaHCO_3 \cdot 2H_2O$).

B. Irregular layers of black bedded chert occur in a thick section of fossiliferous light-gray limestone in the central Brooks Range of Alaska. The chert was used by prehistoric inhabitants of these mountains to make knives, points, and other tools.

ture made it useful to primitive people for tool production. It occurs as extensive continuous layers (bedded chert) or as nodules in carbonate rocks (Fig. 4.13B). Although most chert is probably an organic or inorganic precipitate, some contains the remains of siliceous organisms.

Mineral Deposits in Chemical Sedimentary Rocks

Sedimentary rocks provide employment for most of the world's geologists, for in these rocks are trapped vast accumulations of the sedimentary minerals, petroleum, and coal that help drive our modern civilization. For as long as society requires energy and mineral resources, there will be a need for geologists who have learned how these earthly treasures form and where to find them.

Iron Deposits.   Modern civilization is more dependent on iron than on any other industrial metal. Some of the world's most important concentrations of iron are found in ancient sedimentary rocks that are billions of years old (e.g., in Brazil, Canada, the U.S.S.R, South Africa, and Australia). These deposits are among the most unusual kinds of chemical sedimentary rocks known (Fig. 4.14). They consist of sediment wholly chemical, or possibly biochemical, in origin and are free of detritus, although they are commonly interbedded with clastic sedimentary rocks. Every aspect of the bedded iron deposits indicates chemical precipitation, but the cause of precipitation remains a problem. Possibly the precipitation can be explained by the chemis-

try of the ancient seawater. Very likely, the most important difference between modern seawater and the seawater in which the iron deposits accumulated was the oxygen content of the water. If the amount of oxygen in the surface waters was very low, large amounts of dissolved Fe^{2+} could have been transported and precipitated. This would further imply that the atmosphere then contained much less oxygen than it does today.

Phosphorus Deposits.   Sedimentary deposits of phosphorus minerals, which are a primary source of fertilizer, form through the precipitation of apatite $[Ca_5(PO_4)_3(OH,F)]$ from seawater. The surface waters of the ocean are depleted in phosphorus because fish and other marine animals extract phosphorus to make bone, scales, and other body parts. When the animals die and sink to the seafloor, their bodies slowly decay and release phosphorus to the deep ocean water. If such phosphorus-rich waters are brought to the surface by upwelling currents, precipitation of apatite can occur. Phosphorus-rich sediments are forming today off the western coasts of Africa and South America, but the process was much more common in the past, especially when shallow seas inundated broad areas of the continents.

Evaporite Deposits.   Economically important concentrations of sedimentary minerals can result when lake or ocean water evaporates to form a deposit called an **evaporite** (Fig. 4.15). Examples of salts that precipitate from lake waters are sodium carbonate (Na_2CO_3), sodium sulfate (Na_2SO_4), and borax ($Na_2B_4O_7 \cdot 10 H_2O$). These and other salts have many uses, including the production of paper, soap, detergents, antiseptics, and chemicals for tanning and dyeing. The most important salts that precipitate from seawater are gypsum ($CaSO_4 \cdot 2H_2O$), halite (NaCl), and carnallite ($KCl \cdot MgCl_2 \cdot 6H_2O$). Much of the common table salt we use daily, the gypsum used for plaster and construction materials, and the potassium used in fertilizers are recovered from marine evaporites.

Biogenic Sedimentary Rocks

Biogenic rocks result from the lithification of biogenic sediment or of sediment having a high organic component.

Limestone and Dolostone

Limestone is the most important of the biogenic rocks and accounts for a major proportion of the carbon dioxide stored in the Earth's crust. If this CO_2 were somehow released, it would significantly change the

FIGURE 4.14   Iron-rich sediments of the Brockman Iron Formation in the Hamersley Range of Western Australia. The white layers are largely chert, while the darker bluish and reddish layers consist mainly of iron-rich silicate, oxide, and carbonate minerals.

FIGURE 4.15 Evaporite salts encrust the surface of a desert playa on the floor of Death Valley, California. A shallow lake forms during rainy periods. As the water then evaporates and the playa lake dries up, salts crystallize out of the brine, and polygonal fractures form as the drying sediment contracts.

composition of the atmosphere and cause the surface temperature of the planet to heat up dramatically (chapter 19).

Limestone is formed chiefly of the mineral calcite (Fig. 4.13B). When calcite is replaced by dolomite, the resulting rock is *dolostone*. Limestone and dolostone are not easily classified because they can be either clastic or chemical in character. Bioclastic limestones consist of lithified shells or fragments of marine organisms that have carbonate hard parts; one coarse-grained type composed of shelly debris is called *coquina*. Other limestones consist of cemented reef organisms (*reef limestone*), the compacted carbonate shells of minute floating organisms (*chalk*), and accumulations of tiny, round, calcareous accretionary bodies (ooliths) that are 0.5 to 1 mm in diameter (*oolitic limestone*). Lime muds, composed of chemically precipitated calcite, can form fine-grained limestones such as *lithographic limestone,* an exceedingly dense, very fine-grained rock formerly used in lithography for engraving and the production of color plates.

Diatomite

Siliceous ooze consisting of radiolarian or diatom remains can become lithified to form a relatively soft,

light-colored rock called **diatomite** (Fig. 4.13). Although some diatomites formed in lakes, most are of marine origin. The rock is useful commercially as an abrasive and also, because of its very fine grain size and degree of porosity, as a filter.

Coal

Coal occurs in strata (miners call them *seams*) along with other sedimentary rocks, mainly shale and sandstone. Most coal seams are 0.5 to 3 m thick, although some reach more than 30 m (Fig. 4.16), and they tend to occur in groups. A look through a magnifying glass at a piece of coal reveals the shapes of bits of fossil wood, bark, leaves, roots, and other parts of land plants, chemically altered but still identifiable. Accordingly, **coal** can be defined as a black, combustible sedimentary rock, more than 50 percent of which consists of decomposed plant matter.

As peat is buried beneath more plant matter and accumulating sand, silt, or clay, both temperature and pressure rise. The increased temperature and pressure bring about a series of continuing changes. The peat is compressed, water is squeezed out, and the gaseous organic compounds, such as methane (CH_4), escape, leaving an increased proportion of carbon. The peat is thereby converted into *lignite* and eventually into *bituminous coal*, both of which are sedimentary rocks (Fig. 4.17).

Although peat can form even under subarctic conditions, the luxuriant plant growth needed to form thick and extensive coal seams develops most readily in a tropical or subtropical climate. This implies either that the global climate was warmer when the plant matter of coal accumulated, or else that the wet,

FIGURE 4.16 On average, about 300,000 tons of coal per day are extracted from a seam 20 to 30 m thick in this strip mine at Wyodak, Wyoming.

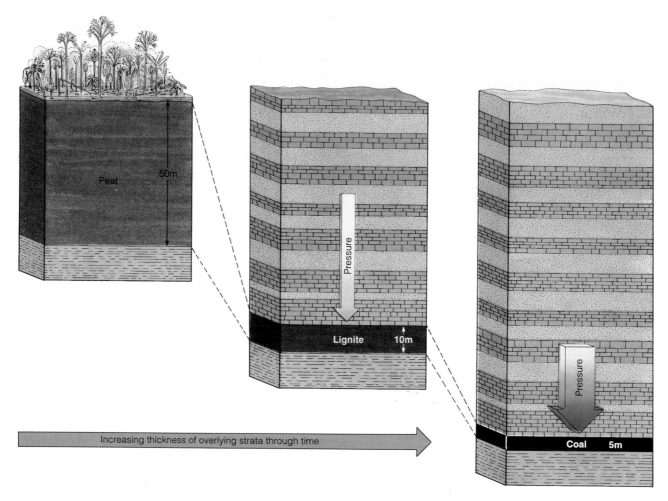

FIGURE 4.17 Plant matter is converted to coal by decomposition and increasing pressure as overlying sediments increase in thickness. By the time a layer of peat 50 m thick is converted to bituminous coal, its thickness has been reduced by 90 percent. In the process, the proportion of carbon has increased from 60 to 80 percent.

swampy environments in which most of the world's coal seams formed existed in the tropics, within about 20° of the equator. Probably both conditions were involved. Coal deposits that must have formed in warm low-latitude environments and now lie in frigid polar lands (e.g., northern Alaska and Antarctica) provide some of the most compelling evidence we have for the slow drift of continents over great distances.

Oil Shales

Although shale is a clastic sedimentary rock, some shales have an unusually high content of organic matter and are of economic value. The organic oils and fats contained in dead organisms that are buried in marine or lake muds may be converted to hydrocarbon residues. While these residues may ultimately form petroleum, in some shales burial temperatures never reach the levels required to break down the organic molecules completely. Instead, an alteration process occurs in which waxlike substances are

formed. If such a rock (called *oil shale*) is heated, the solid organic matter is converted to liquid and gaseous hydrocarbons similar to those in petroleum. Extensive oil shales in Colorado, Utah, and Wyoming can produce as much as 240 liters of oil per ton of rock (Fig. 4.18).

Environmental Clues in Sedimentary Rocks

We have seen already that the size, shape, and arrangement of particles in sediments, as well as the geometry of sedimentary strata, provide us with evidence about the geological environment in which sediments accumulate. These and other clues, several of which are mentioned below, enable us to demonstrate the existence of ancient oceans, coasts, lakes, streams, glaciers, swamps, and all the other places where sediments accumulate.

Features on Bedding Planes

Wavelike irregularities formed by currents moving across a sediment, together with cracks, grooves, and other minor depressions, can be preserved on the bedding plane of sandstone or siltstone. Such features, collectively called *sole marks*, are useful in reconstructing past current directions and bottom conditions.

Bodies of sand that are being moved by wind, streams, or coastal waves are often rippled, and such ripples may be preserved in sandstones and siltstones as *ripple marks* (Fig. 4.19).

Some mudstones and siltstones contain layers that are cut by polygonal markings. By comparing them with similar features in modern sediments, we infer that these are *mud cracks*, caused by shrinkage and cracking of wet mud as its surface dries (Fig. 4.20). Mud cracks imply former tidal flats, exposed streambeds, desert lake floors, and similar environments.

Footprints and *trails* of animals are often found with ripple marks and mud cracks (Fig. 4.21). Even *raindrop impressions* made during brief, intense showers may be preserved in strata. All provide evidence of moist surface conditions at the time of formation.

Fossils

Fossils provide significant clues about former environments. Some animals and plants are restricted to warm, moist climates, whereas others are associated only with cold, dry climates. By using the climatic ranges of modern plants and animals as guides and invoking the Principle of Uniformitarianism, we can infer the general character of the climate in which similar ancestral forms lived. For example, plant fossils can provide estimates of past precipitation and temperature for sites on land, while fossils of tiny floating organisms can tell us about former surface temperatures and salinity conditions in the oceans. Fossils are also the chief basis for telling the relative age of strata and are very important in reconstructing the past 600 million years of Earth history (chapter 6).

Color

The color of fresh sedimentary rock is determined by the colors of the minerals, rock fragments, and organic matter that compose it. Iron sulphides and organic detritus, buried with sediment, are responsible for most of the dark colors in sedimentary rocks and imply deposition in a reducing environment. Reddish and brownish colors result mainly from the presence of iron oxides, occurring either as powdery coatings

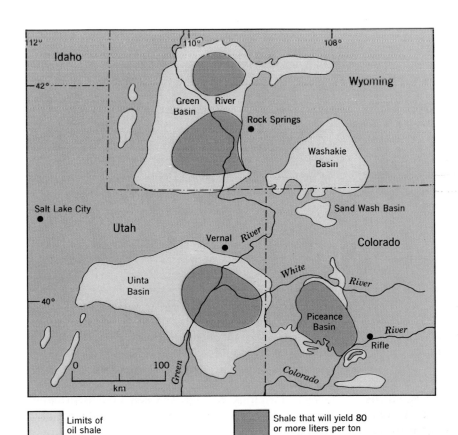

FIGURE 4.18 Vast areas of Colorado, Wyoming, and Utah are underlain by the Green River Oil Shale. The extensive deposits of oil shale formed as organic-rich sediment that accumulated in ancient freshwater lakes and was buried, compacted, and cemented. If heated, the solid organic matter in the shale is converted to hydrocarbons similar to those in petroleum.

on mineral grains or as very fine particles mixed with clay, and point to oxidizing conditions.

The weathered surface of a sedimentary rock may have a different color than a fresh, unweathered surface. For example, a sandstone that is pale gray on freshly broken surfaces may have a surface coating of iron oxide in weathered outcrop that gives it a yellowish-brown color. In this case, the surface color is derived from chemical breakdown of iron-bearing minerals in the rock.

FIGURE 4.19 Modern ripples and ancient ripple marks. A. Ripples forming in shallow water near the shore of Ocracoke Island, North Carolina. B. Ripple marks on the bedding surface of an ancient quartzite bed in the Baraboo region of south-central Wisconsin.

A.

B.

SEDIMENTARY FACIES AND DEPOSITIONAL ENVIRONMENTS

In much the same way that history books record the changing patterns and progress of civilization, sedimentary rocks record the history of our planet. Layers of sediment, like pages in a book, can reveal the changing environmental patterns of the Earth's surface and the progress of life over more than 3 billion years. If we can learn to read them, the sediments allow us to journey back through the ages and visualize how the world has evolved during its long, dynamic history.

Sedimentary Facies

If we examine a sequence of exposed sedimentary rocks, we will likely see differences as we move up from one layer to the next. These differences reflect changes over time in depositional conditions at a particular place. If any single layer in the sequence is traced away from the initial outcrop, it may change laterally. Most sedimentary strata change character laterally as a result of changes in the conditions under which the sediments accumulated.

A diversity of environments would be encountered if we could travel across the edge of a continent and into the adjacent ocean basin (Fig. 4.22). Distinctive sediments and associated organisms serve to identify each. For each environment, we can make a list of the distinctive physical, chemical, and biological characteristics that permit us to distinguish sediment accumulating there from sediment being deposited in another environment. The change in sediment character that takes place as we move from one depositional environment to another is referred to as a change of **facies** (pronounced *fay-sees*). A *sedimentary facies*, therefore, can be thought of as any sediment that can be distinguished from another, contemporary sediment which accumulated in a different depositional environment. A facies may be characterized, for example, by distinctive grain size, grain shape, stratification, color, chemical composition, depositional structures, or fossils. Adjacent facies can merge into one another either gradually or abruptly (Fig. 4.23). Coarse gravel and sand of a beach may pass very gradually offshore into finer sand, silt, and clay on the floor of the sea or a lake. Coarse, bouldery glacial sediment, on the other hand, may end abruptly against stream sediments at the margin of a glacier.

By determining the distinctive characteristics of different bodies of sediment or sedimentary rock, studying the relationship of different facies, and using these

A.   B.

FIGURE 4.20 Modern and ancient mud cracks. A. Mud cracks formed the surface of a dry lake floor. B. Ancient mud cracks preserved on the surface of a mudstone bed exposed at Ausable Chasm, New York.

FIGURE 4.21 Tracks of a three-toed dinosaur are exposed on the surface of a sandstone bed in the Painted Desert near Cameron, Arizona. All the tracks in the picture belong to a single species.

characteristics to identify original depositional settings, we can reconstruct a picture of the varied environments of a region during past geologic intervals.

Common Sediments of Nonmarine Environments

Sediment derived from the mechanical and chemical breakdown of rocks is moved inexorably toward the sea. En route it is moved by water, ice, wind, or gravity. The sediment may be temporarily stored and then reworked repeatedly by one or several of these agencies before reaching its final resting place, where it is slowly converted to sedimentary rock.

Stream Sediments

Streams constitute the principal agency for transporting sediment across the land (Figs. 4.22 and 4.23). Their deposits are seen nearly everywhere. Stream-deposited sediment differs from place to place depending on the type of stream, the energy available

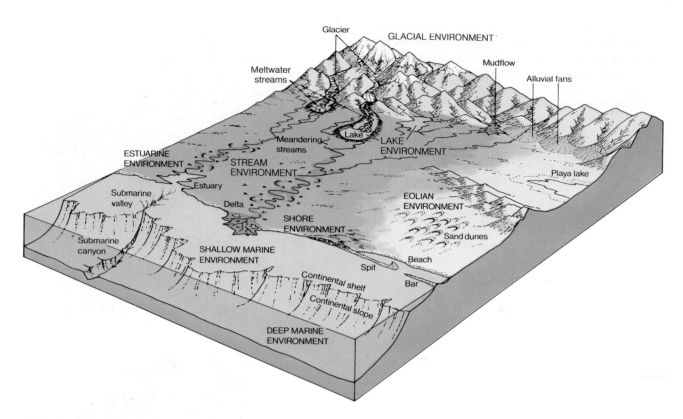

FIGURE 4.22 Various depositional environments are seen while traveling from the crest of a mountain range across the edge of a continent to the adjacent margin of a nearby ocean basin.

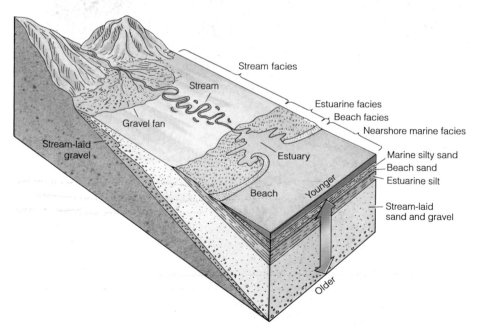

FIGURE 4.23 A section from a mountain valley across a gravelly plain and estuary to the adjacent ocean shows a variety of depositional environments in which distinctive facies are deposited. In this example, facies adjoin or merge into one another on the land surface, whereas in a vertical section offshore, they lie one above another. The boundaries between adjacent facies dip seaward, indicating that the boundaries have shifted progressively in the landward direction through time. This implies that a rise of sea level relative to the land has taken place.

for doing work, and the nature of the sedimentary load. A typical large, smoothly flowing stream may deposit well-sorted layers of coarse and fine particles as it swings back and forth across its valley. Silt and clay are deposited on the valley floor adjacent to the stream during spring floods, and organic sediments accumulate in abandoned sections of the stream channel that no longer carry flowing water. By contrast, a stream flowing from the front of a mountain glacier may divide into an intricate system of interconnected channels that change in size and direction as the volume of water fluctuates and as the stream copes with an abundance of detritus. The resulting sediments consist largely of coarse-grained stream-channel deposits. A large mountain stream flowing down a steep valley can transport an abundant load of sediment. When the stream reaches the mountain front, and is no longer constrained by valley walls, it is free to shift laterally back and forth across the more gentle terrain as its load is dropped. The result is a fan-shaped deposit in which the sediments may range from coarse, poorly sorted gravels to well-sorted, cross-stratified sands.

Lake Sediments

Sediments deposited in a lake (Fig. 4.22) accumulate chiefly on the lakeshore and on the lake floor. Lakeshore deposits of generally well-sorted sand and gravel form beaches, bars, and spits across the mouths of bays. The sediment load of a stream entering a lake will be dropped as the velocity and transporting ability of the stream suddenly decrease. The resulting deposit, built outward into the lake, is a *delta* (Fig. 4.22; chapter 9). Inclined, generally well-sorted layers on the front of a delta pass downward and outward into thinner, finer, and evenly laminated layers on the lake floor (Fig. 4.5).

Glacial Sediments

Sedimentary debris eroded and transported by a glacier is either deposited along the glacier's base or is released at the glacier margin as melting occurs. The sediment is then subjected to further reworking by running water. Debris deposited directly from ice commonly forms a random mixture of particles that range in size from clay to boulders and consist of all the rock types over which the ice has passed (Fig. 4.8). Such sediment characteristically is neither sorted nor stratified, in contrast to most other nonmarine sediments. Stones in glacially deposited sediment often are angular, and sand grains typically show distinctive fractures due to crushing and abrasion (Fig. 4.24A)

A.

B.

FIGURE 4.24 Surface features of sand grains, seen on enlarged pictures taken with a scanning electron microscope, aid in differentiating among transporting agencies. A. The surface of a quartz grain (0.1 mm diameter) that has been crushed and abraded during transport at the bed of a Swiss glacier displays distinctive concoidal fractures. B. The surface of a wind-transported quartz grain (0.5 mm diameter) from south-central Libya has a distinctive pitted appearance, thought to result from the long-continued action of dew that slowly dissolves mineral substance.

Eolian Sediments

Both wind activity and the geologic results of it are referred to as eolian, after Aeolus, the Greek god of wind. Sediment carried by the wind tends to be finer than that moved by other erosional agents. Sand grains are easily moved where strong winds blow and where vegetation is too discontinuous to stabilize the land surface, as along seacoasts and in deserts (Fig. 4.22). In such places, the sand may pile up to form dunes composed of well-sorted sand grains and with bedding inclined in the downwind direction (the direction toward which the air is flowing). Individual grains may have a frosted appearance (Fig. 4.24B), thought to result from long-continued action of dew which dissolves mineral substance on a grain, thereby creating a finely pitted surface texture. Using these characteristics, geologists can rather easily identify ancient dune sands in the rock record (Fig. 4.6).

Powdery dust picked up and moved by the wind is deposited as a blanket of sediment across the landscape (chapter 12). Such sediment is thickest and coarsest near its source and becomes progressively thinner and finer with increasing distance downwind. Although common as a sediment in many parts of the world, wind-blown silt is virtually unknown as sedimentary rock, probably because it is easily eroded from the landscape and therefore is unlikely to be widely preserved.

Sediments of the Continental Shelves

The world's rivers continuously transport detritus to the edges of the sea where it can accumulate near the mouths of streams, be moved laterally along the coast by currents, or be carried seaward to accumulate on the continental shelves, sometimes to great thicknesses. In part spurred by the search for large undersea reservoirs of oil and gas, geologists have learned a great deal about the sediments accumulating on the shelves.

Estuarine Sediments
Much of the load transported by a large river may be trapped in an **estuary**, a semi-enclosed body of coastal water within which seawater is diluted with freshwater (Figs. 4.22 and 4.23). Coarse sediment tends to settle close to land while fine sediment is carried in the seaward direction. Tiny individual particles of clay carried in suspension settle very slowly to the seafloor. However, when freshwater meets seawater, the clay particles tend to aggregate into clumps which, because of their greater mass, settle more rapidly to the bottom. If the rate of sedimentation is high, and the land is slowly subsiding, a thick body of estuarine sediment can form.

Deltaic Sediments
Marine deltas are built outward in the sea at places where streams reach the shore and deposit their sediment load (Fig. 4.22). Large deltas are complex deposits consisting of coarse stream-channel sediments, fine sediments deposited between channels, and still finer sediments deposited on the seafloor (chapter 9).

Beach Sediments
Quartz, the most durable of common minerals in continental rocks, is a typical component of beach sands. However, not all ocean beaches are sandy. Any beach consists of the coarsest rock particles contributed to it by erosion of adjacent sea cliffs, or carried to it by rivers or by currents moving along the shore. Beach sediments tend to be better sorted than stream sediments of comparable coarseness and typically display cross stratification. Dragged back and forth by the surf and turned over and over, particles of beach sediment become rounded by abrasion. Although beach gravel and gravelly alluvium may be similar in appearance, the pebbles and cobbles on many beaches acquire a distinctive flattened shape.

Offshore Sediments
Freshwater flowing through an estuary or past a river mouth may continue seaward across the submerged continental shelf as a distinct layer overlying denser, salty marine water. Some fine-grained sediment, carried in suspension, thereby reaches the outer shelf. The sediment then either settles slowly to the seafloor or is ingested by floating organisms that excrete it as small pellets which fall to the bottom. On the continental shelf of eastern North America up to 14 km of fine sediment has accumulated over the last 70 to 100 million years. To build the whole pile, an average of less than a millimeter of sediment need have been deposited each year.

Most coarse marine sediment is deposited within 5 to 6 km of the land after being dispersed by currents that flow parallel to the shore. Coarse sediment is also found as far offshore as the seaward limits of the shelves. Its observed patchy distribution is mostly the result of changing sea level. At times when the sea fell below its present level, the shoreline migrated seaward across the shelves, exposing new land. Bodies of coarse sediment deposited near shore or on the land at such times were submerged as sea level again rose across the shelves. As much as 70 percent of the sediment cover on the continental shelves is probably a relict of such past conditions.

Only about 10 percent of the sediment reaching the continental shelves remains in suspension long enough to arrive in the deep sea, so it is clear that the great bulk of the Earth's sedimentary strata is shelf strata whose sediment originated on the continents. The shelves, in effect, conserve continental crust which is continuously recycled within the continental realm.

FIGURE 4.25 Carbonate sediments consisting of fine skeletal debris and inorganic precipitates accumulate in the warm, shallow marine waters of a broad, flat carbonate shelf surrounding the numerous islands of Bermuda.

Carbonate Shelves

Carbonate sediments of biogenic origin accumulate on the continental shelves wherever the influx of land-derived sediment is minimal and the climate and sea-surface temperature are warm enough to promote the abundant growth of carbonate-secreting organisms. Carbonate sediments accumulate mainly on broad, flat carbonate shelves that border a continent or rise as platforms off the seafloor (Fig. 4.25). Most of the sediment consists of sand-sized skeletal debris, together with inorganic precipitates that form fine carbonate muds. Coarser debris is found mainly near coral and algal reefs or in areas of turbulence and strong currents.

Marine Evaporite Basins

Ocean water occupying a basin with restricted circulation that lies in a region of very warm climate will evaporate. This leads to the precipitation of soluble substances and the accumulation of marine evaporite deposits. Such deposits are widespread. In North America, for example, marine evaporite strata underlie as much as 30 percent of the entire land area.

The Mediterranean Sea is an example of an evaporite basin (Fig. 4.26). Were it not for continuous inflow of Atlantic water at its western end, the Mediterranean would gradually decrease in volume because of evaporation. It is estimated that if the Mediterranean were deprived of new water, evaporation would cause this landlocked sea to dry up completely in about 1000 years. In the process, a layer of salt about 70 m thick would be precipitated. Far thicker evaporites that underlie the Mediterranean Basin are regarded as evi-

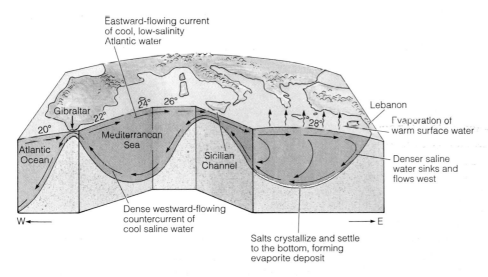

FIGURE 4.26 In the Mediterranean Sea, the inflow of fresh water from rivers in Europe and Africa is too small to offset the water evaporated from the sea surface in the hot eastern part of the basin. An inflowing current of Atlantic Ocean water therefore balances the loss. As the relatively cool ocean water (summer temperature 20°C) enters the basin and moves eastward it warms (to 28°C) and evaporates. The salinity of the remaining water rises, increasing its density and setting up a westward-flowing current of cool saline water at depth. Salts crystallizing from the salty water settle to the bottom to form evaporite-rich sediments.

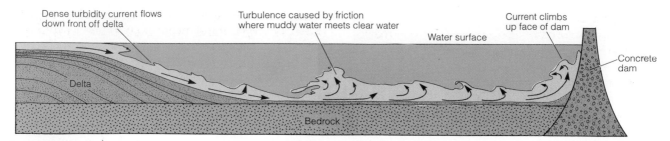

FIGURE 4.27   A turbidity current generated by a surge of sediment-laden water enters the quiet water of a reservoir behind a dam. Moving rapidly down the face of a delta, the current passes along the lake floor and climbs up the face of the dam before subsiding. As the sediment settles to the bottom, it forms a graded tubidite layer.

dence of former periods when a high evaporation rate, together with a continuous inflow of Atlantic water to supply the necessary salt, allowed evaporites 2 to 3 km thick to accumulate.

Sediments of the Continental Slope and Rise

Along most of their length, the shallowly submerged continental shelves pass abruptly into continental slopes that descend to depths of several kilometers. Sediments that reach the shelf edge are poised for further transport down the slope and onto the adjacent continental rise.

FIGURE 4.28   Deep-sea turbidite beds that have been tilted, uplifted, and exposed in a wave-eroded bench along the coast of the Olympic Peninsula, Washington.

Turbidity Currents and Turbidites

Thick bodies of coarse sediment of continental origin lie at the foot of the continental slope at depths as great as 5 km. The origin of these coarse accumulations at great depths in the oceans was difficult to explain until marine geologists demonstrated that the sediments could be deposited by **turbidity currents**. These are gravity-driven currents consisting of dilute mixtures of sediment and water having a density greater than the surrounding water. Such currents have been produced in the laboratory in water-filled tanks into which a dense mixture of water, silt, and clay has been introduced. They have also been documented moving across the floors of lakes and reservoirs (Fig. 4.27). In the oceans, turbidity currents have been set off by earthquakes, landslides, and major coastal storms. Off the mouths of rivers, they can be set in motion by large floods.

The accumulated evidence leads us to believe that turbidity currents are very effective geologic agents on continental slopes, where they can reach velocities greater than those of the swiftest streams on land. Some achieve a velocity of more than 90 km/h and transport up to 3 kg/m^3 of sediment, spreading it as far as 1000 km from the source.

A turbidity current typically deposits a graded layer of sediment called a **turbidite** (Fig. 4.28). Such a graded layer is the result of rapid, continuous loss of energy in the moving current. As a rapidly flowing turbidity current slows down, successively finer sediment is deposited.

At any site on the continental rise or an adjacent abyssal plain, a turbidite is deposited very infrequently, perhaps only once every few thousand years. In these places, far-distant from the sediment source, the deposits are mainly thin layers a few millimeters to 30 cm thick. Although deposition is infrequent, over millions of years turbidites can slowly accumulate to form vast deposits beyond the continental realm.

Deep-Sea Fans

Some large submarine canyons on the continental slopes are aligned with the mouths of major rivers, like the Amazon, Congo, Ganges, and Indus. At the base of many such canyons is a huge **deep-sea fan**, a fan-shaped body of sediment that spreads downward and outward to the deep seafloor (Fig. 4.29). The sediments, which are derived mostly from the land, include fragments of land plants, as well as fossils of shallow- and deep-water marine organisms. Also present are many graded layers that are interpreted as turbidites.

Deep-sea fans are a major exception to the generalization that the final deposition of land-derived sediment in the ocean is largely confined to the continental shelves. When shelves are exposed at times of lowered sea level and rivers extend across them nearly to the continental slope, the stage is set for the rapid building of deep-sea fans.

Sediment Drifts

Discrete bodies of sediment up to hundreds of kilometers long, tens of kilometers wide, and two kilometers high have been discovered along the continental margins bordering the North Atlantic Ocean. Called *sediment drifts*, these huge deposits are associated with deep ocean currents that flow near the base of the continental margins. Rippled sands along the axis of a sediment drift, where current velocities are high, grade laterally into muds deposited where currents decrease in velocity. Factors that lead to the formation of sediment drifts appear to include a deep current system in a stable location, a prevailing current velocity (<15 cm/s) that permits deposition, and a supply of sediment. At typical accumulation rates of tens of meters per million years, the formation of sediment drifts requires many millions of years.

Sediments of the Deep Sea

Analyses of samples from sediment cores allow geologists to sort out the various sources of seafloor sediment. The study of great numbers of samples indicates clearly that all the sediments are mixtures; while a large portion is produced by biologic activity in the surface waters, some sediment is transported over great distances from continental interiors and eventually reaches the deep sea.

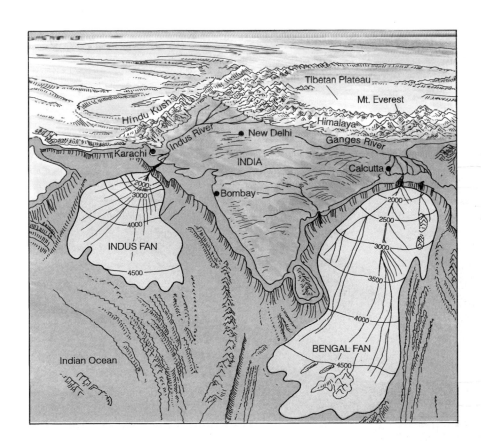

FIGURE 4.29 The Indus and Ganges–Bramaphutra rivers have built the vast Indus and Bengal deep-sea fans on the seafloor adjoining the Indian subcontinent. Most of the sediment in the fans was transported from the high Himalaya in the north as the mountain system was uplifted over tens of millions of years.

Deep-Sea Oozes

Calcareous ooze occurs over wide areas of ocean floor at low to middle latitudes where warm sea-surface temperatures favor the growth of carbonate-secreting organisms in the surface waters. However, calcareous ooze is not found in these same warm latitudes where the water is deeper than about 4 km. Cold deep-ocean waters are under high pressure and contain more dissolved carbon dioxide than shallower waters. As a result, they can easily dissolve any carbonate particles that settle to their level.

Other parts of the deep ocean floor are mantled with siliceous ooze, most notably in the equatorial Pacific and Indian oceans, and in a belt encircling the Antarctic region where siliceous organisms predominate. These are areas where surface waters have a high biological productivity, in part related to the rise of deep ocean water rich in nutrients.

Land-Derived Sediment

In addition to biogenic oozes, deep-sea strata include sediments carried to the oceans by rivers, eroded from coasts by wave action, transported by wind (fine desert dust and volcanic ash), and released from floating ice. Over much of the North Pacific and part of the central South Pacific the ocean floor lies at depths of more than 4 km and lacks calcareous ooze. In these regions, the seafloor is covered by oxidized reddish or brownish clay composed mainly of extremely fine clay minerals, quartz grains, and micas that originated on land.

SEDIMENTATION AND TECTONICS

The energy that drives sedimentation comes ultimately from the two major energy sources of the Earth: internal heat and the Sun. Internal heat energy drives the motions of the highly mobile lithosphere, including uplift of the land, and is the ultimate source of energy that drives plate tectonics. Sediment produced by weathering and erosion of rocks in uplifted regions is transferred downslope toward sea level under the pull of gravity, and eventually reaches the ocean basins. Streams, glaciers, and ocean waves and currents, the major movers of sediment, are all part of the hydrologic cycle, which derives its energy from solar radiation.

Sedimentation rates are high adjacent to regions of active tectonic uplift, far lower on relatively stable continental interiors, and still lower in regions of the deep sea that are far from terrestrial sources of sediment. In most tectonically active regions, uplift rates exceed erosion rates and high mountain ranges persist as prominent features of the landscape. The magnitude of uplift can be impressive. Marine sedimentary rocks that crop out at the top of Mount Everest demonstrate at least 9 km of uplift since the sediments were deposited in a shallow sea about 100 million years ago. Here in the Himalaya, as in other major mountain ranges of the world, we see clear evidence that ancient sediments deposited on an ocean floor have been converted to rock, added to a continent, and uplifted to high altitudes by tectonic forces.

Exceptionally thick accumulations of strata are related to specific plate-tectonic settings. For example, wherever continents are split apart by a spreading center, thick sedimentary wedges may slowly accumulate along the new continental margins as streams transport sediment to the growing ocean basin; the Atlantic margin of North America is an example of this process (Fig. 4.30A). A large portion of the thick pile of strata beneath the continental shelf consists of shallow marine sedimentary rocks, which means that this sedimentary wedge must slowly subside as accumulation takes place.

In zones where continents collide, coarse stream sediments shed from a rising mountain system can accumulate to great thicknesses in adjacent sedimentary basins (Fig. 4.30B). The Himalaya–Hindu Kush mountain arc of south-central Asia is a dramatic example. Thick conglomerates and sandstones along the southern margin of the chain record the rise of these ranges. Finer-grained detritus has been transported by major streams to the Indian Ocean, where extensive submarine fans have accumulated since the uplift began (Fig. 4.27).

Along active subduction zones near continental margins, as along the western margin of South America, sediments are shed into deep ocean trenches, where they may accumulate to great thicknesses (Fig. 4.30C). Because volcanoes are commonly associated with such tectonic belts, the accumulating clastic sediments contain a high proportion of volcanic detritus. As the moving plates slowly converge, the sediments are crushed against the continent and become part of it. In this way, sediments are cycled from continent to ocean and back to continent, where continuing uplift causes the process to begin anew.

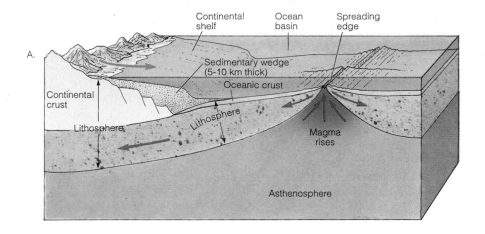

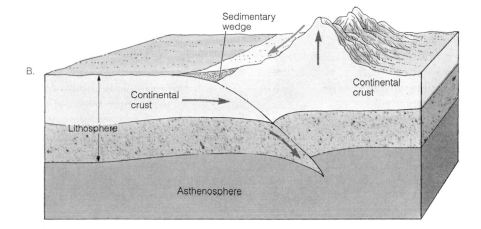

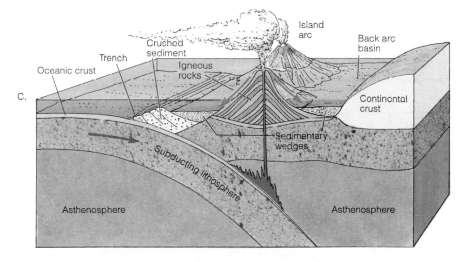

FIGURE 4.30 Examples of thick sedimentary accumulations of sediment in different plate-tectonic settings. A. A thick wedge of sediment slowly accumulates along a new continental margin that has developed as a continent split along a spreading ridge and the resulting fragments moved apart. B. In a zone of continental collision, sediment is shed from a growing mountain system to form a thick sedimentary wedge. C. Sediment shed into a deep-sea trench from an adjacent continent, bordered by an arc of active volcanoes, forms a wedge that is compressed and crushed as the ocean plate is subducted.

GEOLOGY AND THE POLITICS OF MIDDLE EAST OIL

Petroleum is the premier fossil fuel of twentieth-century civilization. We use it to fuel our cars, trucks, ships, and airplanes, to run generators that produce electricity, and to provide heat for our homes. Industrial societies are so dependent on petroleum that it constitutes a key component of their economies. In the best of all worlds, each nation would have its own reserve of petroleum suitable to meet its present and future needs. Unfortunately, however, geopolitical boundaries and geologic resources seldom coincide. A few of the major industrial countries, like Canada and the U.S.S.R., have adequate supplies of petroleum for present needs. The United States can supply most, but not all, of the petroleum it uses. Other countries, like Japan, France, and Germany, have either no petroleum or miniscule supplies and must import the oil and gas they require. The reason for this disparity lies in the unequal distribution of petroleum, which is a function both of sedimentary geology and tectonic history.

Sedimentary rocks that contain large quantities of oil are concentrated in a few relatively restricted geologic regions. Their distribution is a function of a long and often complex history of sediment accumulation and tectonic events related to moving lithospheric plates. Countries like the United States and the Soviet Union cover such large geographic areas that the laws of geologic chance make it probable that they will contain at least some large petroleum deposits. For smaller countries, however, the chance is almost like a roll of the dice. Many small countries con-

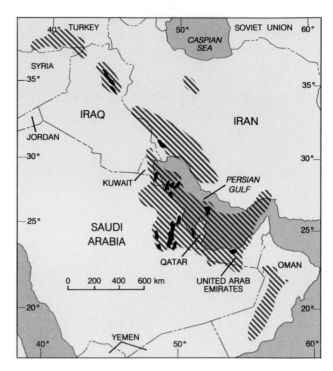

FIGURE B4.1 Strata containing oil and gas underlie large areas of the Middle East. Areas outlined enclose the more than 400 separate oil fields that have been discovered so far. The outlines of the largest individual fields are highlighted.

tain little or no petroleum and thus are losers, while others are big winners.

The largest known reserve of petroleum happens to lie in a limited geographic region centered around the Persian Gulf (Fig. B4.1). Here the sedimentary and tectonic laws of chance

SUMMARY

1. Sediment is transported by streams, glaciers, wind, slope processes, and ocean currents. After deposition, it experiences compaction and cementation as it is transformed into sedimentary rock.

2. Stratification results from the arrangement of sedimentary particles in layers. Each bed in a succession of strata is distinguished by its distinctive thickness or character.

3. Clastic sediment consists of fragmental rock debris resulting from weathering, together with the

broken remains of organisms. Chemical sediment forms where substances carried in solution are precipitated.

4. Particles of sediment become rounded and sorted during transport by water and air, but not during transport by glaciers.

5. Various arrangements of the particles in strata are seen in parallel strata and cross strata, graded bedding, and nonsorted layers.

6. Oil and gas probably originated as organic matter deposited in seafloor sediment and then de-

have converged to produce huge underground reservoirs of fossil fuel in a region otherwise blessed with few natural resources. A few relatively small countries surrounding the gulf control more than half of the world's known petroleum resources. As a result, their economic and political significance has skyrocketed in the decades since oil was discovered in this region (Fig. B4.2).

Always a politically sensitive region, the Middle East lies at the center of the world political stage. International conflicts, the legacy of centuries of intertribal and religious disputes, now include petroleum as an added complicating factor, for control of oil resources is associated with both economic and political power. While this oil-rich region may remain politically unstable well into the future, eventually its importance will decline as continued exploitation causes the nonrenewable petroleum resources to diminish and the world turns to alternative sources of power.

FIGURE B4.2 Geologic exploration teams working in the harsh desert environment of the Persian Gulf region have discovered vast reservoirs of petroleum that comprise more than half the world's known supplies. These discoveries quickly brought the oil-rich Middle East countries to economic and political prominence.

composed chemically. Later these fluids moved into geologic traps to form pools. Oil shales are a major potential source of fossil fuel, but they must be heated to release oil and gas.

7. Coal originated as plant matter in ancient swamps, and is both abundant and widely distributed.

8. During diagenesis, compaction, cementation, and recrystallization act to transform sediment into sedimentary rock.

9. Clastic sedimentary rocks, like sediments, are classified mainly on the basis of predominant particle size. Conglomerate, sandstone, siltstone, and shale are the rock equivalents of gravel, sand, silt, and clay.

10. Limestone is a widespread biogenic rock that forms primarily in warm marine environments and stores carbon dioxide in the Earth's crust.

11. An extensive body of strata may possess several facies, each determined by a different depositional environment. The boundaries between facies may be abrupt or gradational.

12. Most sedimentary strata are built of continental detritus that is transported to the submerged continental margins. Some detritus is trapped in basins on land where it is deposited by nonmarine processes. A small percentage reaches the deep sea.

13. Depositional environments of nonmarine and shallow-marine sediments can be inferred from such properties as texture, degree of sorting and rounding, character of stratification, and types of contained fossils.

14. Coarse land-derived sediment reaching the continental margins is deposited close to shore where it is reworked by longshore currents. Finer sediment is deposited on the continental shelves and slopes and in the deep sea. Extensive areas of the shelves are covered by relict sediments deposited at times of lower sea level.

15. Carbonate shelves are found in low latitudes where warm waters promote growth of carbonate-secreting organisms.

16. Evaporite deposits accumulate in restricted marine basins where evaporation is high and continuous inflow provides a supply of saline water.

17. By depositing turbidites, turbidity currents have built large deep-sea fans at the base of the continental slope.

18. The chief kinds of sediment on the deep seafloor are brownish or reddish clay, calcareous ooze, and siliceous ooze. Their distribution is largely related to surface water temperature, water depth, and surface productivity.

19. Sediment is constantly being recycled, moving from continent to ocean and back to continent.

IMPORTANT TERMS TO REMEMBER

bed (p. 110)
bedding (p. 110)
bedding plane (p. 110)
biogenic sediment (p. 115)

cementation (p. 117)
chemical sediment (p. 115)
clastic sediment (p. 110)
compaction (p. 117)
cross bedding (p. 114)

deep-sea fan (p. 131)
deep-sea ooze (p. 116)
detrital sediment (p. 110)
detritus (p. 110)
diagenesis (p. 117)

eolian (p. 127)
estuary (p. 128)
evaporite (p. 120)

facies (p. 124)
fossil (p. 115)
fossil fuels (p. 117)

graded bed (p. 114)

hydrocarbons (p. 117)

lithification (p. 117)

peat (p. 117)
petroleum (p. 117)

recrystallization (p. 118)

sorting (p. 112)
stratification (p. 110)
stratum (pl = strata) (p. 110)

turbidite (p. 130)
turbidity current (p. 130)

varve (p. 113)

IMPORTANT ROCK NAMES TO REMEMBER

breccia (p. 119)

coal (p. 121)
conglomerate (p. 119)

diatomite (p. 121)

limestone (p. 121)

sandstone (p. 119)
shale (p. 119)
siltstone (p. 119)

QUESTIONS FOR REVIEW

1. On what basis are sediments and sedimentary rocks classified?

2. What obvious clues can be used to tell a clastic sedimentary rock from an igneous rock?

3. What chemical reactions can lead to precipitation of chemical sediments?

4. What features in a sediment or sedimentary rock are responsible for stratification?

5. Describe the processes involved in the conversion of sediment to sedimentary rock.

6. Explain why petroleum is thought to originate in marine sediments.

7. Describe cross stratification and graded layers, and explain what they tell us about conditions of deposition.

8. What clues regarding depositional environment are provided by features on sedimentary bedding surfaces?

9. How would you tell alluvium from sediment deposited by wind or in a lake?

10. If estuaries are generally shallow bodies of water, what hypothesis can you suggest to explain the occurrence of thick estuarine accumulations in the sedimentary record?

11. What explanation can be given for the occurrence of widespread relict sediments on the continental shelves?

12. If the Mediterranean Sea, which is an enclosed marine basin, were to evaporate completely, less than 100 m of evaporite sediments would be deposited. How, then, might *continuous* evaporite deposits under the Mediterranean that are more than 2 km thick have formed?

13. What distinctive features of the sediments in deep-sea fans provide clues about the way in which the sediments were transported?

14. What factors explain the distribution of calcareous and siliceous oozes on the ocean floors?

15. Explain where and why you would expect to find exceptionally thick accumulations of sedimentary rock.

Elevated temperatures and pressures cause sedimentary strata to be metamorphosed. As a result of a continental collision, strata that were once horizontal have been contorted and changed in texture and mineral assemblage. This example is in Shining Rock Wilderness, Blue Ridge Mountains, North Carolina.

CHAPTER 5

New Rocks from Old: Metamorphism and Metamorphic Rocks

The term **metamorphism** is used to describe all changes in mineral assemblage and rock texture that take place in rocks within the Earth's crust as a result of changes in temperature and pressure.

Metamorphic rocks are of particular interest because the changes happen in the solid state. As tectonic plates move and crustal fragments collide, rocks are squeezed, stretched, bent, heated, and changed in complex ways. But even if a rock has been altered two or more times, vestiges of its earlier forms are usually preserved because the changes occurred in the solid state. Solids, unlike liquids and gases, tend to retain a memory of the events that changed them. In many ways, therefore, metamorphic rocks are the most complex, but also the most interesting of the rock families. In them is preserved the story of all the things that have happened to the crust. Deciphering the record is an exceptional challenge for geologists. For example, when tectonic plates collide, distinctive kinds of metamorphic rocks form along the plate edges. Geologists are attempting to determine where the boundaries of ancient continents once were by studying metamorphic rocks. Geologists are also trying to use evidence derived from metamorphic rocks to determine how long plate tectonics has been active on the Earth. So far the evidence suggests that plate tectonics has been operating for at least 2 billion years!

THE LIMITS OF METAMORPHISM

Before we discuss metamorphism in detail, it is important to define the limits of the process. Metamorphism describes changes in mineral assemblage and texture in sedimentary and igneous rocks subjected to temperatures above 200°C and pressures in excess of about 300 Mpa (the pressure caused by a few thousand meters of overlying rock). Metamorphism does not refer to changes caused by weathering (chapter 1) or by diagenesis (chapter 4), because both weathering and diagenesis take place at temperatures below 200°C and pressures below 300 Mpa.

There is, of course, an upper limit to metamorphism, because at sufficiently high temperatures rock will melt. Remember then, metamorphism refers only to changes in solid rock, not to changes caused by melting. Changes due to melting involve igneous phenomena, as we discussed in chapter 3.

Because at least a small amount of H_2O is present in most rocks, the upper limit of metamorphism in the crust is determined by the onset of wet partial melting, as discussed in chapter 3 and shown in Figure 5.1. The H_2O present controls the temperature at which wet partial melting commences and the amount of magma that can form from a metamorphic rock. The upper limit of metamorphism is therefore a temperature range that depends on the amount of H_2O present. As shown in Figure 5.1, the upper limits of metamorphism overlap with the region of tempera-ture and pressure where magmatism commences. When a tiny amount of H_2O is present, only a small amount of melting occurs and the melt stays trapped as small pockets in the metamorphic host. In many places it is possible to find evidence that melting started in H_2O-rich layers of metamorphic rock even though adjacent, drier layers do not show any sign of melting. Composite volumes of rock containing an igneous component formed by a small amount of melting plus a metamorphic portion, are called **migmatites.** When large volumes of magma develop by partial melting they will rise upward and intrude the metamorphic rock above. Eventually, a large volume of rising magma formed by wet partial melting will solidify as an intrusive igneous rock. As a result we observe that batholiths of granitic rock and large volumes of metamorphic rock tend to be closely associated. The geological setting of this igneous–metamorphic rock association is along subduction and collision margins of tectonic plates.

Low-grade metamorphism refers to metamorphic processes occurring at temperatures from about 200° to 320°C, and at relatively low pressures (Fig. 5.1). **High-grade metamorphism** refers to metamorphic processes at high temperature (above about 550°C) and high pressure.

CONTROLLING FACTORS IN METAMORPHISM

In a simplistic way you can think of metamorphism as cooking. When you cook, what you get to eat depends on what you start with and on the cooking conditions. So too with rocks; the end product is controlled by the initial composition of the rock and by the metamorphic (or cooking) conditions. The chemical composition of a rock undergoing metamorphism plays a controlling role in the new mineral assemblage; so do changes in temperature and pressure. The controls of temperature and pressure on metamorphism are not entirely straightforward, however, because they are strongly influenced by such things as the presence or absence of fluids, how long a rock is subjected to high pressure or high temperature, and whether it is simply compressed or is twisted and broken during metamorphism.

Chemical Reactivity Induced by Fluids

The innumerable open spaces between the grains in a sedimentary rock and the tiny fractures in many ig-

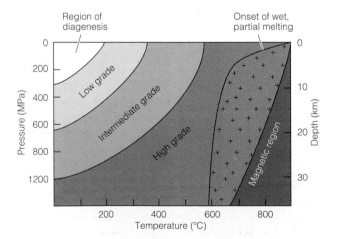

FIGURE 5.1 Regions of temperature and pressure (equivalent to depth) under which metamorphism occurs in the crust. At the low temperatures and pressures, changes are due to diagenesis (as discussed in chapter 4). At the highest temperatures, magmatism and metamorphism overlap (marked with crosses) depending on the amount of H_2O present, as explained in the text.

neous rocks, are called pores, and all pores are filled by a watery fluid. The fluid is never pure water, for it always has dissolved within it small amounts of gases and salts plus traces of all the mineral constituents that are present in the enclosing rock; and at high temperature it is more likely to be a vapor than a liquid. Nevertheless, the **intergranular fluid,** for that is its best designation, plays a vital role in metamorphism.

When the temperature and pressure of a rock undergoing metamorphism change, so does the composition of the intergranular fluid. Some of the dissolved constituents move from the fluid to the new minerals growing in the metamorphic rock. Other constituents move in the other direction, from the minerals to the fluid. In this way the intergranular fluid serves as a transporting medium that speeds up chemical reactions in much the same way that water in a stew pot speeds up the cooking of a tough piece of meat.

When intergranular fluids are either absent or else present only in trace amounts, metamorphic reactions are very slow. When a dry rock is heated, few changes occur because the growth of new minerals means that atoms must move by diffusing through the solid minerals. Diffusion through solids is an exceedingly slow process. If somehow an intergranular fluid is introduced, perhaps because pores are created as a result of the rock being crushed, diffusion of the atoms from one place to another can take place through the intergranular fluid. This is a vastly faster process, and as a result, new minerals grow rapidly and the metamorphic effects are pronounced.

As pressure increases due to burial of a rock, and metamorphism proceeds, the amount of pore space decreases and the intergranular fluid is slowly driven out of the rock. As the temperature of the rock increases, hydrous minerals recrystallize to anhydrous minerals and in the process water is released. The released water joins the intergranular fluid and is also slowly driven out of the metamorphic rock. For this reason, rock subjected to high-grade metamorphism contains fewer hydrous minerals, and less intergranular fluid, than does low-grade metamorphic rock.

The metamorphic changes that occur while temperatures and pressures are rising, and while abundant intergranular fluid is present, are termed **prograde metamorphic effects.** Those that occur as temperature and pressure are declining, and after much of the intergranular fluid has been expelled, are called **retrograde metamorphic effects.** Not surprisingly, because of the lack of fluid, retrograde metamorphic effects happen less rapidly, and are less pronounced, than prograde effects. Indeed, it is only because retrograde reactions happen so slowly that we see high-grade metamorphic rocks at the Earth's surface. If retrograde reactions were rapid, all metamorphic rocks would react back to clays and other low-grade minerals.

Pressure and Temperature

When a mixture of flour, salt, sugar, yeast, and water is baked in an oven, the high temperature causes a series of chemical reactions—new compounds grow and the final result is a loaf of bread. When rocks are heated, new minerals grow, and the final result is a metamorphic rock. In the case of the rocks, the source of heat is the Earth's internal heat. Rock can be heated simply by burial or by a nearby igneous intrusion. But burial and intrusion can also cause a change in pressure. Therefore, whatever the cause of the heating, metamorphism can rarely be considered to be entirely due to the rise in temperature. The effects due to changing temperature and pressure must be considered together.

The combined influence of temperature and pressure on the melting properties of rocks and minerals was discussed in chapter 3 (see Fig. 3.35). Metamorphic transformations of mineral assemblages are also controlled by the dual effects of temperature and pressure. By measuring the effects of temperature and pressure on assemblage transformations in the laboratory, we can delineate the ranges of pressure and temperature conditions under which metamorphism occurs in the crust. We will return to the important issue of the controls exerted by temperature and pressure on mineral assemblages later in the chapter.

Differential Stress
When talking about deformation in a solid and development of new textures, we often use the term *stress* instead of pressure. We do so because stress has the connotation of direction. Rocks are solids and solids can be squeezed more strongly in one direction than another; that is, stress in a solid, unlike a liquid, can be different in different directions. The textures in many metamorphic rocks record **differential stress** (meaning not equal in all directions) during metamorphism. Most igneous rocks, by contrast, have textures formed in **uniform stress** (meaning equal in all directions), because igneous rocks crystallize from liquids.

The most visible effect of metamorphism in a differential stress field involves the texture of silicate minerals, such as micas and chlorites, that contain polymerized $(Si_4O_{10})^{4-}$ sheets. Compare Figure 5.2A and B. Figure 5.2A is a granite that has a typical texture of randomly oriented mineral grains that grew in a uniform stress field. Figure 5.2B, on the other hand, is a high-grade metamorphic rock containing the same minerals that formed in a differential stress field. Note that in Figure 5.2B all the mica grains are parallel, giving the rock a distinctive texture. Another example of a metamorphism in a differential stress field is shown in Figure 5.3. A conglomerate which originally consisted of uniformly rounded pebbles has been squeezed so that the pebbles are now flattened discs.

Note that it is the texture, not the mineral assemblage of a metamorphic rock that is controlled by differential versus uniform stress. For this reason geologists often use the terms stress and pressure interchangeably where mineral assemblages and metamorphic grades are concerned.

Time

Chemical reactions involve energy, and two compounds will react only if a lower state of energy is reached in the process. At the lowest possible energy state, reaction ceases, and we say a state of equilibrium has been reached. A certain amount of time is needed for any chemical reaction to reach equilibrium. Some reactions, such as the burning of methane gas (CH_4) to yield carbon dioxide and water, happen so rapidly

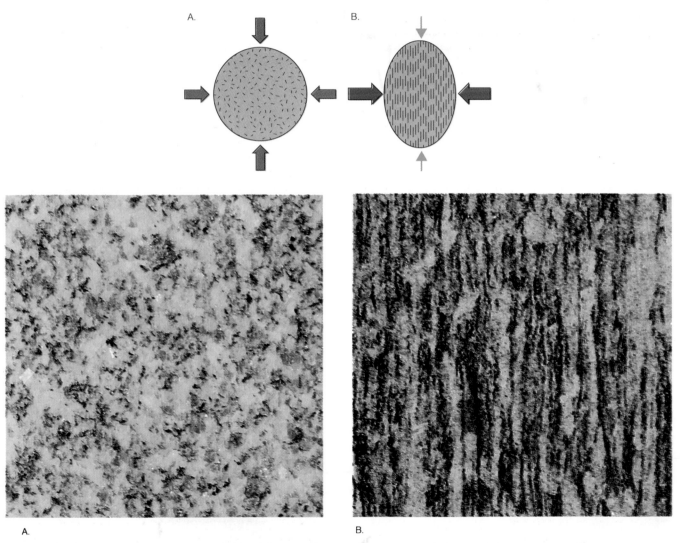

FIGURE 5.2 Comparison of textures developed in rocks of the same composition under uniform and differential stress. A. Granite, consisting of quartz, feldspar, and biotite that crystallized in a uniform stress field. Note that biotite grains are randomly oriented. B. High-grade metamorphic rock, also consisting of quartz, feldspar, and biotite, that crystallized in a differential stress field. Biotite grains are parallel, giving the rock a distinct foliation.

FIGURE 5.3 Deformation of a conglomerate during metamorphism. Sandstone pebbles, originally round, have been flattened.

that they create explosions. At the other end of the scale are reactions that require millions of years to proceed to completion.

Many of the chemical reactions that occur in rocks undergoing metamorphism are of the latter kind. No reliable ways have yet been developed to determine exactly how long a given metamorphic rock has remained at a given temperature and pressure. However, it can be readily demonstrated in the laboratory that high temperature, high pressure, and long reaction times produce large mineral grains. Thus, it is possible to draw the interesting general conclusion that coarse-grained rocks are the products of long-sustained metamorphic conditions (possibly over millions of years) at high temperatures and pressures, while fine-grained rocks are products of lower temperatures, lower pressures, and shorter reaction times.

METAMORPHIC RESPONSES TO CHANGES IN TEMPERATURE AND PRESSURE

Textural Responses

Most metamorphic rocks form in a differential stress field and as a result they develop conspicuous, directional textures. As metamorphism proceeds and the sheet-structure minerals, such as mica and chlorite, start to grow, the minerals are oriented so that the sheets are perpendicular to the direction of maximum

stress as shown in Figure 5.2B. The new, parallel flakes of mica produce a planar texture called **foliation** (Fig. 5.4), named from the Latin word, *folium*, meaning leaf. Foliated rocks tend to split into thin, leaflike flakes. Foliation may be pronounced or subtle, but when present it provides strong evidence of metamorphism.

Slaty Cleavage

During the earliest stages of low-grade metamorphism, stress tends to be caused by the weight of the overlying rock. The new sheet-structure minerals, and therefore the foliation, tend to be parallel to the bedding planes of the sedimentary rock being metamorphosed. But with deeper burial, or when compression from a plate collision deforms the flat sedimentary layers into folds, the sheet-structure minerals and the foliation are no longer parallel to the bedding planes (Fig. 5.5). Low-grade metamorphic rocks tend to be so fine grained that the new mineral grains can be seen only with the microscope; the foliation is then called **slaty cleavage,** which is defined as the property by which a low-grade metamorphic rock breaks into platelike fragments along planes (Fig. 5.6).

Schistosity

At intermediate and high grades of metamorphism grain sizes increase and individual mineral grains can be seen with the naked eye. Foliation in coarse-grained metamorphic rocks is called schistosity and is not necessarily planar, **Schistosity** is derived from *schistos*, a Latin work meaning cleaves easily, and refer

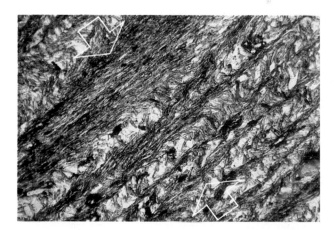

FIGURE 5.4 Microscopic thin section of a metamorphic rock showing pronounced foliation due to the parallel arrangement of mica grains. The pale-colored grains are mainly quartz. Note that the foliation is enhanced because mica grains and quartz grains have become segregated into separate layers. Direction of maximum stress is indicated by arrows. The sample is 1 cm wide.

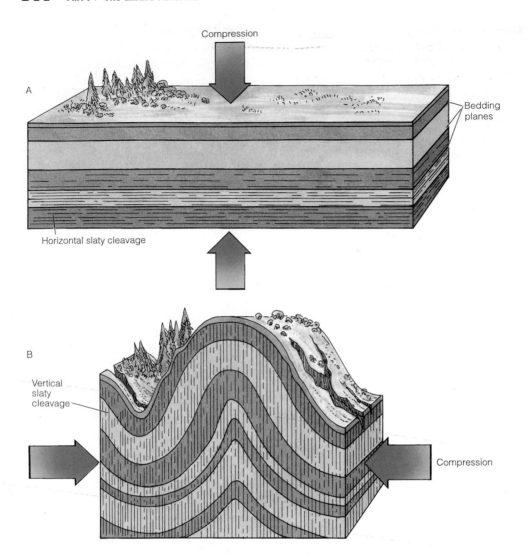

FIGURE 5.5 Slaty cleavage developed in low-grade metamorphic rocks. A. Maximum stress is perpendicular to bedding. Slaty cleavage is therefore parallel to bedding. B. Strata are deformed and folded. Maximum stress is indicated by arrows. Slaty cleavage cuts across bedding.

ring to the parallel arrangement of coarse grains of the sheet-structure minerals. Schistosity differs from slaty cleavage mainly in grain size. Intermediate and high-grade metamorphic rocks tend to break along wavy, or slightly distorted, surfaces, reflecting the presence and orientation of grains of quartz, feldspar, and other minerals.

Assemblage Responses

Metamorphism produces new mineral assemblages as well as new textures. As temperature and pressure rise, one new mineral assemblage follows another. For any given rock composition, each assemblage is characteristic of a given range of temperature and pressure. A few of these minerals are found rarely (or not at all) in igneous and sedimentary rocks. Their presence in a rock is usually evidence enough that the rock has been metamorphosed. Examples of these metamorphic minerals are chlorite, serpentine, epidote, talc, and the three polymorphs of Al_2SiO_5: kyanite, sillimanite, and andalusite. An illustration of the way mineral assemblages change with grade of metamorphism as a shale is metamorphosed is given in Figure 5.7.

KINDS OF METAMORPHIC ROCK

The naming of metamorphic rocks is based partly on texture, partly on mineral assemblage. The most

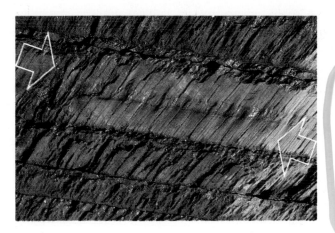

FIGURE 5.6 Slaty cleavage cuts across nearly horizontal bedding. The cleavage is developed in the Martinsburg Formation, which was subjected to low-grade metamorphism. Arrows indicate the direction of maximum stress. The sample is 70 cm across and is near Palmerton, Pennsyvania.

widely used names are those applied to metamorphic derivatives of shales, sandstones, limestones, and basalts. This is so because shales, sandstones, and limestones are the most abundant sedimentary rock types, while basalt is by far the most abundant igneous rock.

Metamorphism of Shale and Mudstone

Slate

The low-grade metamorphic product of either shale or mudstone is **slate.** The minerals usually present in both shale and mudstone include quartz, clays of various kinds, calcite, and possibly feldspar. Under conditions of low-grade metamorphism, muscovite and/or chlorite crystallize. Although the rock may still look like a shale or mudstone, the tiny new mineral grains produce slaty cleavage. The presence of slaty cleavage is clear proof that a rock has gone from being a sedimentary rock to a metamorphic rock.

Phyllite

Continued metamorphism of a slate to intermediate grade produces both larger grains of mica and a changing mineral assemblage; the rock develops a pronounced foliation and is called **phyllite** (from the Greek, *phyllon,* a leaf). In a slate it is not possible to see the new grains of mica with the unaided eye, but in a phyllite they are just large enough to be visible.

Schist and Gneiss

Still further metamorphism beyond that which produces a phyllite leads to a coarse-grained rock with

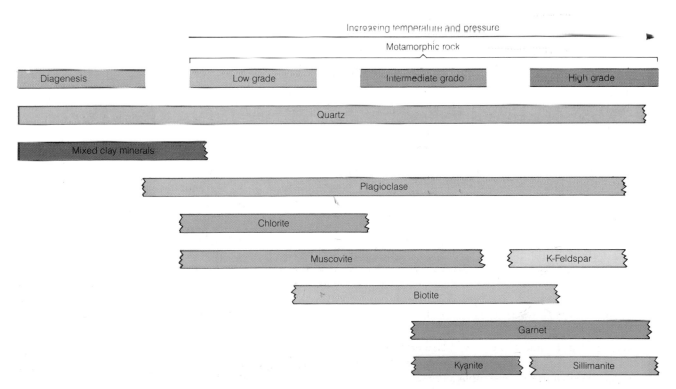

FIGURE 5.7 Changing mineral assemblages as a shale is metamorphosed from low to high grade. Kyanite and sillimanite are polymorphs (Al_2SiO_5) that are found only in metamorphic rocks.

pronounced schistosity, called **schist.** The most obvious differences between slate, phyllite, and schist are in grain size, but, as can be seen in Figure 5.8A, increasing grain size is only one of a number of changes. At the high grades of metamorphism characteristic of schists, minerals may start to segregate into separate bands. A high-grade rock with coarse grains and pronounced foliation, but with layers of micaceous minerals segregated from layers of minerals such as quartz and feldspar, is called a **gneiss** (pronounced nice, from a word in the German *gneisto,* meaning to sparkle) (Fig. 5.9).

The names *slate* and *phyllite* describe textures and are commonly used without adding mineral names as adjectives. The names of the coarse-grained rocks, schists and gneiss, are also derived from textures, but in these cases mineral names are commonly added as adjectives; for example, we refer to a quartz–plagioclase–biotite–garnet gneiss. The difference arises because minerals in coarse-grained rocks are large enough to be seen and readily identified.

Metamorphism of Basalt

Greenschist

The main minerals in basalt are olivine, pyroxene, and plagioclase, each of which is anhydrous. When a basalt is subjected to metamorphism under conditions where H_2O can enter the rock and form hydrous minerals, distinctive mineral assemblages develop (Fig. 5.8B). At low grades of metamorphism, mineral assemblages such as chlorite + plagioclase + epidote + calcite form. The resulting rock is equivalent in metamorphic grade to a slate but has a very different appearance. It has pronounced foliation as a phyllite does, but it also has a very distinctive green color because of its chlorite content; it is termed *greenschist.*

Amphibolite and Granulite

When a greenschist is subjected to an intermediate grade of metamorphism, chlorite is replaced by amphibole; the resulting rock is generally coarse grained, and is called an **amphibolite** (Fig. 5.10). Foliation is present in amphibolites but is not pronounced because micas and chlorite are usually absent. At highest grade metamorphism, amphibole is replaced by pyroxene and an indistinctly foliated rock called a *granulite* develops.

Metamorphism of Limestone and Sandstone

Marble and *quartzite* are, respectively, the metamorphic derivatives of limestone and sandstone. Neither

A.

Intensity of metamorphism			
	Not metamorphosed	Low grade	Intermediate grade	High grade
Rock name	Shale ⟶	Slate ⟶	Phyllite	Schist / Gneiss
Foliation	None	Subtle, slaty cleavage	Distinct; schistosity apparent	Conspicuous; schistosity and compositional layering
Size of mica grains	Microscopic	Microscopic	Just visible with hand lens	Large and obvious
Typical mineral assemblage	Quartz, clays, calcite	Quartz, chlorite, muscovite, plagioclase	Quartz, biotite, garnet, kyanite, plagioclase	Quartz, biotite, garnet, sillimanite, plagioclase

B.

Intensity of metamorphism			
	Not metamorphosed	Low grade	Intermediate grade	High grade
Rock name	Basalt $\xrightarrow{+H_2O}$	Greenschist ⟶	Amphibolite ⟶	Granulite
Foliation	None	Distinct schistosity	Indistinct;, when present due to parallel grains of amphibole	Indistinct because of absence of micas
Size of mica grains	Visible with hand lens	Visible with hand lens	Obvious by eye	Large and obvious
Typical mineral assemblage	Olivine, pyroxene, plagioclase	Chlorite, epidote, plagioclase, calcite	Amphibole, plagioclase, epidote, quartz	Pyroxene, plagioclase, garnet

FIGURE 5.8 Progressive metamorphism of shale and basalt. Mineral assemblage and foliation change as a result of increasing temperature and differential stress. A. Shale. B. Basalt.

FIGURE 5.9   A coarse-grained gneiss. Minerals present are feldspar and quartz (both light colored), and biotite (dark). The parallel orientation of biotite grains proves that the rock is metamorphic in origin. The specimen is 8 cm across.

Figure 5.10   Amphibolite resulting from metamorphism of a pillow basalt. Compare this with Figure 3.20. The pillow structure was compressed during metamorphism but can be discerned by the borders of pale-yellow epidote formed from the original glassy rims of the pillows. The outcrop is in Namibia.

limestone nor quartz sandstone (when pure) contain the necessary ingredients to form sheet- or chain-structure minerals. As a result marble and quartzite commonly lack foliation.

Marble

Marble consists of a coarsely crystalline, interlocking network of calcite grains. During recrystallization of a limestone, the bedding planes, fossils, and other features of sedimentary rocks are largely obliterated. The end result, as shown in Figure 5.11A, is an even-grained rock with a distinctive, somewhat sugary texture. Pure marble is snow white in color and consists entirely of pure grains of calcite. Such marbles are favored for marble gravestones and statues in cemeteries, perhaps because white is considered to be a symbol of purity. Many marbles contain impurities such as organic matter, pyrite, limonite, and small quantities of silicate minerals, that impart various colors.

Quartzite

Quartzite is derived from sandstone by the filling in of the space between the original grains with silica, and by recrystallization of the entire mass (Fig. 5.11B). Sometimes, the ghostlike outlines of the original sedimentary grains can still be seen, even though recrystallization may have rearranged the original grain structure completely.

KINDS OF METAMORPHISM

The processes that result from changing temperature and stress, and that cause the metamorphic changes observed in rocks, can be grouped under the terms *mechanical deformation* and *chemical recrystallization*. Mechanical deformation includes grinding, crushing, and the development of foliation. The deformed conglomerate shown in Figure 5.3 is an example of mechanical deformation. Chemical recrystallization includes all the changes in mineral composition, in growth of new minerals, and the losses of H_2O and CO_2 that occur as rock is heated. Different kinds of metamorphism reflect the different levels of importance of the two processes.

Cataclastic Metamorphism

Mechanical deformation of a rock can sometimes occur without any accompanying chemical recrystallization, but such deformation is rare and usually localized. For example, when a coarse-grained rock such as granite is subjected to intense, differential stresses, individual mineral grains may be shattered and pulverized as shown in Figure 5.12. This sort of deformation occurs in brittle rocks and is called **cataclastic metamorphism.** As cataclastic metamorphism proceeds, grain and rock fragments become elongated and a foliation develops. The sequence of photographs shown in Figure 5.13 illustrates the progressive change that can be produced in a granite by cataclastic metamorphism.

Contact Metamorphism

Contact metamorphism occurs adjacent to bodies of hot magma that are intruded into cool rocks of the

A.

B.

FIGURE 5.11 Textures of nonfoliated metamorphic rocks seen in thin sections and viewed in polarized light. Notice the interlocking grain structure produced by recrystallization during metamorphism. Each specimen is 2 cm across. A. Marble, composed entirely of calcite. All vestiges of sedimentary structure have disappeared. B. Quartzite. Arrows point to faint traces of the original rounded quartz grains in some of the grains.

crust. Such metamorphism involves chemical recrystallization and happens in response to a pronounced increase in temperature. Mechanical deformation is minor or absent because the stress around a mass of magma tends to be homogeneous. Rock adjacent to the intrusion becomes heated and metamorphosed, developing a well-defined shell, or **metamorphic aureole,** of altered rock (Fig. 5.14).

The width of an aureole of contact metamorphosed rock depends on the size of the intrusive body and on the amount of H_2O in the rock being metamorphosed. With a small intrusion, such as a dike or sill a few meters thick, and in the absence of a fluid, the aureole may be only a few centimeters wide. The metamor-

phic rock is a hard, fine-grained rock composed of an interlocking mass of uniformly sized mineral grains, and is called a *hornfels*. A large intrusion contains much more heat energy than a small one and it may also give off a lot of H_2O vapor. When an intrusion is a kilometer or more in diameter, the aureole may reach a hundred meters or more in width and the metamorphic rocks are likely to be coarse grained.

Within a large metamorphic aureole through which fluid has passed, several different, and roughly concentric, zones of mineral assemblages can usually be identified, as shown in Figure 5.14. Each zone is characteristic of a certain temperature range. Immediately adjacent to the intrusion, where temperatures are high, we find anhydrous minerals such as garnet and pyroxene. Beyond them are found hydrous minerals such as epidote and amphibole, and beyond them micas and chlorites. The exact assemblage of minerals in each zone depends, of course, on the chemical composition of the intruded rock and on that of the invading fluid, as well as on temperature and pressure.

Burial Metamorphism

Sediments, together with interlayered pyroclastics, may attain temperatures of 300°C or more when buried deeply in a sedimentary basin. Abundant pore water is present in buried sediment, and this water speeds up chemical recrystallization and helps new minerals to grow. But water-filled sediment is weak

FIGURE 5.12 Cataclastic metamorphism of a granitic dike intruded into a biotite-rich gneiss. Brittle granite, rich in feldspar (pink) and quartz (light gray), was fragmented during deformation. The gneiss did not fragment because it was ductile and tended to flow plastically. The field of view is about 50 cm across.

A.   B.

FIGURE 5.13 Development of foliation in a granite by cataclastic metamorphism. From Groothoek, South Africa. A. Undeformed granite consisting of quartz, feldspar, and biotite. The dark patch in the center of the field of view is a xenolith of amphibolite. Foliation is not present in the granite. B. The original granitic texture has been completely changed, and the granite has been transformed to a gneiss with a distinct foliation. Amphibolite xenoliths have been flattened and elongated.

and acts more like a liquid than a solid. The stress during burial metamorphism tends, therefore, to be homogeneous. As a result, there is little mechanical deformation involved in burial metamorphism and the texture of the metamorphic rock that results looks like that of an essentially unaltered sedimentary rock, even though the mineral assemblage is completely different.

The family of minerals that particularly characterize the conditions of burial metamorphism are *zeolites,* a group of silicate minerals with fully polymerized crystal structures containing the same chemical elements as feldspars but also containing water.

Burial metamorphism, which is the first stage of metamorphism following diagenesis, is usually observed in deep sedimentary basins, such as trenches

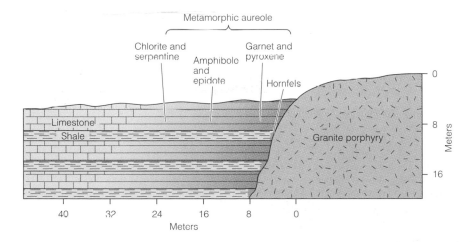

FIGURE 5.14 Contact metamorphism around an intrusion of granite porphyry near Breckenridge, Colorado. Limestone composition was changed by fluids released by the cooling granite and as a result a metamorphic aureole up to 20 m wide developed. New minerals form a series of concentric shells in the aureole, each with a distinct mineral assemblage. Shale between limestone layers was impervious to fluids and except for a narrow band of hornfels immediately adjacent to the granite was not affected.

on the margins of tectonic plates. As temperatures and pressures increase, burial metamorphism grades into regional metamorphism.

Regional Metamorphism

The most common metamorphic rocks of the continental crust occur through areas of tens of thousands of square kilometers, and the process that forms them is called **regional metamorphism.** Unlike contact or burial metamorphism, regional metamorphism involves differential stress and a considerable amount of mechanical deformation in addition to chemical recrystallization. As a result, regionally metamorphosed rocks tend to be distinctly foliated.

Slate, phyllite, schist, and gneiss are the most common varieties of regionally metamorphosed rocks, and they are usually found in mountain ranges, or the eroded remnants of former mountain ranges. Mountain ranges that contain rocks formed by regional metamorphism form as a result of subduction or collision between fragments of continental crust. During a collision between continents, sedimentary rock along the margin of a continent is subjected to very intense differential stresses. The foliation that is so characteristic of slates, schists, and gneisses is a consequence of those intense stresses. Regional metamorphism is therefore a consequence of plate tectonics. Greenschists and amphibolites are also products of regional metamorphism. They tend to be found where segments of ancient oceanic crust of basaltic composition have been incorporated into the continental crust and later metamorphosed.

In order to see what happens during regional metamorphism, consider a segment of the crust that is subjected to a horizontal compressive stress. Rock in the crust becomes folded and buckled. The folding and buckling cause the crust to become locally thickened, as shown in Figure 5.15. The bottom of the thickened mass is pushed deeper into mantle where temperatures are higher. As a result, the rocks near the bottom of the thickened pile are subjected to both elevated stress and higher temperature. New minerals start to grow. However, rocks are poor conductors of heat; so the heating-up process can be very slow. The temperatures reached depend both on depth and on how long a rock is buried in the thickened pile. If the folding and thickening is very slow, heating of the pile keeps pace with the temperature of adjacent parts of the crust and mantle (i.e., a normal continental geothermal gradient is maintained). However, if burial is very fast, as it is with sediment dragged down in a subduction zone, the pile has insufficient time to heat up, and conditions of high pressure but rather low temperature prevail. Depending on the rate of burial, therefore, the same starting rock can yield two quite different metamorphic rocks because different pressures and temperatures are reached.

Metamorphic Zones

The first geologists to make a systematic study of a regionally metamorphosed terrain did so in the Scottish Highlands. They observed that rocks having the same overall chemical composition (that of a shale) could be subdivided into a sequence of zones, each zone having a distinctive mineral assemblage. Each assemblage was characterized by the appearance of new minerals. They selected characteristic *index minerals,* which, proceeding from low-grade rocks to high-grade rocks, marked the appearance of each new mineral assemblage. Their index minerals were, in order of appearance, chlorite, biotite, garnet, staurolite, kyanite, and sillimanite. By plotting on maps the places where each of the index minerals first appeared in rocks having the chemical composition of shale, the workers in the Scottish Highlands defined a series of isograds. An **isograd** is a line on a map connecting

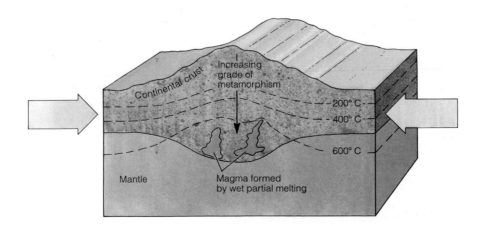

FIGURE 5.15 Compression causes the crust to deform and thicken. In the lower part of the thickened mass, regional metamorphism occurs.

points of first occurrence of a given mineral in metamorphic rocks. The concept of isograds is now widely used to study metamorphic rocks of all kinds; it is just as applicable to burial and contact metamorphism as it is to regional metamorphism.

The regions on a map between isograds are known as **metamorphic zones.** We speak of the chlorite zone, the biotite zone, and so forth, and it is the zones that are commonly depicted on maps showing the relationships between metamorphic rocks (Fig. 5.16).

METAMORPHIC FACIES

Careful study of metamorphic rocks around the world has demonstrated that the chemical compositions of most rocks are little changed by metamorphism. The main changes that do occur are the addition or loss of volatiles such as H_2O and CO_2, but the principal constituents of rocks, such as SiO_2, Al_2O_3, and CaO, remain fixed. The changes brought about during metamorphism, then, are changes in the mineral assemblage, not changes in the overall chemical composition of the rocks. The conclusion to be drawn from this observation is that the mineral assemblages in the metamorphic derivatives of common sedimentary and igneous rocks must be determined by the temperatures and stress to which the rocks are subjected during metamorphism. Based on this conclusion, the famous Finnish geologist Pennti Eskola proposed, in 1915, the concept of **metamorphic facies.** The concept is that for a given rock composition, the assemblages of minerals that reach equilibrium during metamorphism within a specific range of physical conditions belong to the same metamorphic facies. Eskola drew his conclusions from studies of metamorphosed basalts that were interlayered with rocks of entirely different composition.

Another analogy with cooking is appropriate; think of a large roast of beef. When it is carved, one sees that the center is rare, the outside well done, and in between is a region of medium-rare meat. The differences occur because the temperature was not uniform throughout. The center, or "rare-meat" facies, is a low-temperature facies; the outside, or "well-done" facies, is a high-temperature one. The composition of beef varies little, if at all, from roast to roast, so the facies depends not on composition, but on the temperature. So too with rocks, although in any rock of a given composition the stress field as well as the temperature determines the mineral assemblage.

Metamorphic facies were originally described in terms of recurring mineral assemblages; for each assemblage there was assumed to be a specific set of temperature and stress conditions. The realization that temperature, stress, and rock composition each play a role in determining the mineral assemblage provided the link geologists needed to determine metamorphic conditions to be determined through laboratory experiments, and eventually to prove Eskola's suggestion. The concept is now applied to a very wide range of temperatures and stresses. The principal metamorphic facies, together with geothermal gradients to be expected under three different geological conditions, are shown in Figure 5.17.

Because Eskola was studying metamorphosed basalts when he proposed the metamorphic facies concept, most of the names he gave to metamorphic facies reflect the mineral assemblages developed in rocks of basaltic composition. It is important to remember, however, as shown in Table 5.1, that mineral assemblages are just as much a result of original rock composition as they are of the temperature and stress of metamorphism. When comparing mineral assemblages of rocks subjected to different grades of metamorphism, therefore, one must be certain that they have the same overall chemical composition.

METASOMATISM

The metamorphic processes we have discussed so far involve essentially fixed compositions and relatively small amounts of fluid. The amount of fluid is small because the pore volume in rocks undergoing metamorphism tends to be small, and because the release of H_2O and CO_2 from minerals involved in metamorphic reactions happens slowly rather than all at once. Geologists use the jargon expression "a small water–rock ratio" to describe the metamorphic environment. What they mean by the expression is that the weight ratio of fluid (mainly water) to rock is about 1:10 or smaller. This is enough fluid to serve as a metamorphic juice, but not enough to dissolve a lot of the rock and so change the rock composition noticeably.

There are, however, a few circumstances where large water–rock ratios occur. An open rock fracture through which a lot of fluid flows is one example. Under such circumstances a water–rock ratio can be 10:1 or even 100:1, and the rocks adjoining the fracture can be drastically altered by addition of new ions, removal of material in solution, or both. The term **metasomatism** (from *meta*, change, and *soma*, derived from the Latin word for juice) is applied to the process whereby rocks have their chemical compositions distinctively altered by the addition or removal of ions in solution.

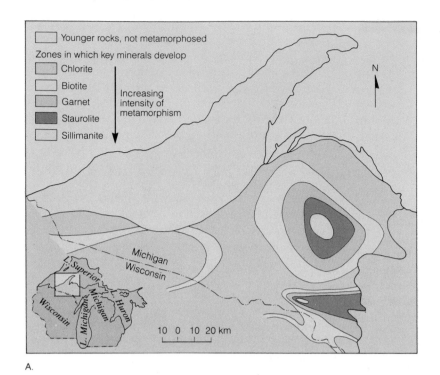

A.

B.

FIGURE 5.16 Metamorphic zones resulting from regional metamorphism. A. Michigan. B. Scottish Highlands.

Metasomatism is commonly associated with contact metamorphism, especially where the rocks being metamorphosed are limestones, as in the example shown in Figure 5.14. Metasomatic fluids released by a cooling magma pass outward, through the volume of rock undergoing contact metamorphism. Because the fluids may carry constituents such as silica, iron, and magnesium in solution, the composition of the limestone close to the cooling magma can be drastically changed even though the limestone distant from the magma, beyond reach of invading fluids remains unchanged. Figure 5.18 is a photograph of a contact metamorphic rock that was originally a limestone. Without addition of new material, the limestone would have become a marble, but through metasomatism it was changed to an assemblage of garnet, a green pyroxene called diopside, and calcite.

Hydrothermal Solutions and Mineral Deposits

Most of the fluids that cause metasomatism are rich in H_2O and tend to be hot—250°C or higher. Such fluids are called **hydrothermal solutions** (*hydro* being the Greek word for water, thermal derived from *therme*, the Greek word for heat).

Hydrothermal solutions form veins by depositing dissolved constituents such as quartz or calcite in cracks they flow through. As shown in Figure 5.19, hydrothermal solutions can also produce distinctive

Chapter 5 / New Rocks from Old: Metamorphism and Metamorphic Rocks **153**

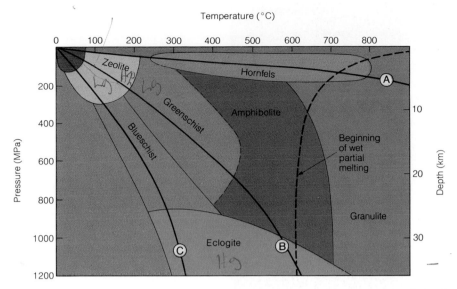

FIGURE 5.17 Metamorphic facies plotted with respect to temperature and depth. —equlibrium
Curve A is a typical thermal gradient around an intrusive igneous rock that is causing
contact metamorphism. Curve B is a normal continental geothermal gradient. Curve C
is the geothermal gradient developed in a subduction zone. Note that the zeolite facies
overlaps in temperature and pressure with the conditions of digenesis.

changes in the rocks they flow through. By alteration of existing rocks and by deposition of veins, hydrothermal solutions play an important role in the formation of many valuable mineral deposits. Probably more mineral deposits have been formed by deposition from hydrothermal solutions than in any other way. However, despite the importance of such deposits, the origins of the solutions are often difficult to decipher. Whatever the origin of a given hydrothermal solution, deposition of any minerals carried in solution occurs deep underground where it cannot be seen; by the time a deposit is finally uncovered by erosion, the hydrothermal solution that formed it is no longer present. Nevertheless, clues have been

TABLE 5.1 Characteristic Minerals of Differing Metamorphic Facies for Selected Rocks[a]

Facies Name	Precursor Rock Type	
	Basalt	**Shale**
Granulite	Pyroxene, Plagioclase, Garnet	Biotite, K-Feldspar, Quartz, Andalusite
Amphibolite	Amphibole, Plagioclase, Garnet, Quartz	Garnet, Biotite, Muscovite, Sillimanite, Quartz
Epidote–Amphibolite	Amphibole, Epidote, Plagioclase, Garnet, Quartz	Garnet, Chlorite, Muscovite, Biotite, Quartz
Greenschist	Chlorite, Amphibole, Plagioclase, Epidote	Chlorite, Muscovite, Plagioclase, Quartz
Blueschist	Blue-Amphibole, Chlorite, Ca-rich Silicates	Blue-Amphibole, Chlorite, Quartz, Muscovite, Lawsonite
Eclogite	Pyroxene (variety Jadeite), Garnet, Kyanite	Not observed
Hornfels	Pyroxene, Plagioclase	Andalusite,[b] Biotite, K-Feldspar, Quartz
Zeolite	Calcite, Chlorite, Zeolite (variety Laumontite)	Zeolites, Pyrophyllite, Na-Mica

[a]For temperature and pressure conditions of each facies, refer to Figure 5.17.
[b]*Andalusite is a polymorph* of kyanite and sillimanite.

FIGURE 5.18   Metasomatically altered marble. The white mineral is calcite, brown is garnet, green is a pyroxene, and purple is fluorite. Materials needed to form the garnet, pyroxene, and fluorite were added to the marble by metasomatic fluids. The sample is 8 cm wide and comes from King Island, Australia.

found so that many details of the process of deposition are now understood. Even so, a great deal of research is still to be done.

Composition of the Solutions

The principal ingredient of hydrothermal solutions is water. The water is never pure and always contains dissolved salts such as sodium chloride, potassium chloride, calcium sulfate, and calcium chloride. The amounts vary, but most solutions range from about the saltiness of seawater (3.5 percent dissolved solids by weight) to about ten times the saltiness of seawater. A hydrothermal solution is therefore a brine, and brines, unlike pure water, are capable of dissolving minute amounts of seemingly insoluble minerals such as gold, chalcopyrite, galena, and sphalerite.

Origins of the Solutions

Hydrothermal solutions have many sources. One way they form is by the cooling and crystallization of magma formed by wet partial melting (chapter 3). Most of the water that causes the wet partial melting is released when such a magma solidifies. Instead of being pure water, however, it carries in solution both the most soluble constituents in the magmas, such as NaCl, and chemical elements such as gold, silver, copper, lead, zinc, mercury, and molybdenum, which do not readily enter quartz, feldspar, and other common minerals by ionic substitution.

High temperatures increase the effectiveness of brines in forming hydrothermal minerals deposits. Volcanism and high temperatures go together. It is not surprising, therefore, that many mineral deposits are associated with hot volcanic rocks that were invaded by deep circulating water that started as rainwater or seawater. Nor is it surprising that a great many mineral deposits are found in the upper portions of volcanic

FIGURE 5.19   Dark-colored metasomatic alteration produced in a calcareous siltstone by hydrothermal solutions. Hot solutions flowed through joint-controlled openings introducing iron and other chemical elements and allowing a mineral assemblage of amphiboles, pyroxenes, and garnets to grow.

piles, where they were deposited when upward-moving hydrothermal solutions cooled and precipitated the ore minerals. Figure 5.20 is a simplified diagram of hydrothermal solutions being generated by seawater penetrating oceanic crust, being heated, and rising upward by convection. Heated seawater reacts with the rocks it is in contact with, causing changes in the chemical and mineral compositions. As the minerals are transformed, trace metals such as copper and zinc, present by ionic substitution, are released and become concentrated in the hot seawater.

Hydrothermal solutions formed beneath the sea can become so hot that they rise rapidly through fractures and form jetlike eruptions of hot, hydrothermal solutions into cold seawater. Figure 5.21A is a photograph of such a submarine hot spring at 21°N latitude on the East Pacific Rise, the plate spreading edge that runs through the Pacific Ocean. Many such eruptions have been observed along midocean ridges. When a jetting hydrothermal solution cools, it deposits its minerals in a massive blanket around the erupting vent. Figure 5.21B is an example of an assemblage of valuable ore minerals deposited by the jetting hot spring shown in Figure 5.21A.

PLATE TECTONICS, METAMORPHISM, AND METASOMATISM

One of the triumphs of the revolution in geology brought about by plate tectonics theory is that it pro-

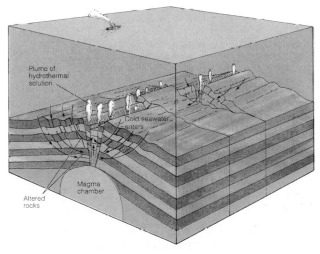

FIGURE 5.20 Seawater penetrates volcanic rocks on the seafloor. Heated by a magma chamber, seawater becomes a hydrothermal solution, produces metasomatic alterations, and rises at a midocean ridge as a hydrothermal plume.

A.

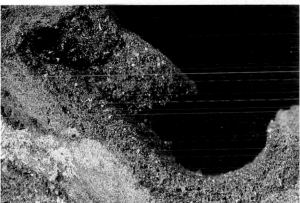

B.

FIGURE 5.21 Hydrothermal solution forming an ore deposit on the seafloor. A. A so-called "black smoker" photographed at a depth of 2500 m below sea level on the East Pacific Rise at 21°N latitude. The "smoker" has a temperature of 320°C. The rising hydrothermal solution is actually clear; the black color is due to fine particles of iron sulfide and other minerals precipitated from solution as the plume is cooled through contact with cold seawater. The chimneylike structure is composed of pyrite, chalcopyrite and other ore minerals deposited by the hydrothermal solution. B. A sample of the blanket of sulfide minerals that forms around the seafloor vents at 21°N. This specimen is part of a system of chimneys. Lining the chimney is pyrite, and scattered through the pyrite are grains of chalcopyrite and sphalerite. Surrounding the massive pyrite is a mixture of gypsum, pyrite, and chalcopyrite. The specimen is 28 cm across.

vides, for the first time, an explanation for the distribution of metamorphic zones in regionally metamorphosed rocks. Regional metamorphism occurs at the subduction boundary of a plate, as shown in Figure 5.22.

Burial metamorphism is believed to occur in the lower portions of the thick piles of sediment that accumulate on the continental shelf and continental slope. Such metamorphism is known to be happening today in the great pile of sediments accumulated in the Gulf of Mexico, off the mouth of the Mississippi River.

The temperatures and pressures characteristic of *blueschist* and *eclogite facies* metamorphism are reached when crustal rocks are dragged down by a rapidly subducting plate. Under such conditions pressure increases more rapidly than temperature, and as a result rock is subjected to high pressure and relatively low temperature. Rocks subjected to blueschist and eclogite facies metamorphism are widespread in the Coast Ranges of California. Blueschist metamorphism is probably happening today along the subducting margin of the Pacific Plate, where it plunges under the coast of Alaska and the Aleutian Islands.

The metamorphic conditions characteristic of *greenschist* and *amphibolite facies* metamorphism occur where crust is either thickened by continental collision or heated by rising magma. Continental collision is the most common setting for regional metamorphism, and such rocks can be observed throughout the Appalachians and the Alps. Such metamorphism is no doubt occurring today beneath the Himalaya, where the continental crust is thickened by collision, and beneath the Andes, where it is both thickened and also heated by rising magma. If the crust is sufficiently thick, rocks subjected to amphibolite facies or higher grade metamorphism can reach temperatures at which wet partial melting commences, and metamorphism passes into migmatism and eventually to magmatism.

Metasomatism and the generation of hydrothermal solutions can also be linked to plate tectonics because metasomatism is closely related to regional metamorphism and magmatic activity. A striking example can be seen in the distribution of chalcopyrite-rich copper deposits in North and South America. As shown in Figure 5.23, a pronounced belt of deposits formed in, or associated with, old stratovolcanoes lies along the western margin of the Americas. The magmas that produced the stratovolcanoes were formed by wet partial melting of subducted oceanic crust. The magmas also served as the heat sources for the hydrothermal solutions that altered the mineral assemblages adjacent to the conduits through which the solutions flowed, and in which the solutions formed the ore deposits. In no small measure, the abundance of mineral resources we enjoy on the Earth is due to the combined effects of magmatic, metamorphic, and metasomatic processes, all of which occur because of plate tectonics.

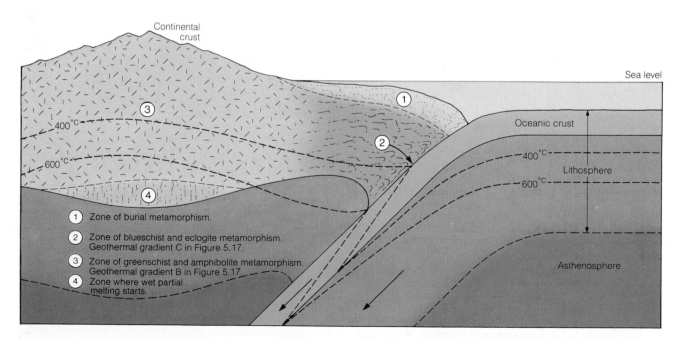

FIGURE 5.22. Diagram of a convergent plate boundary showing the different regions of metamorphism. Dashed lines indicate temperature contours.

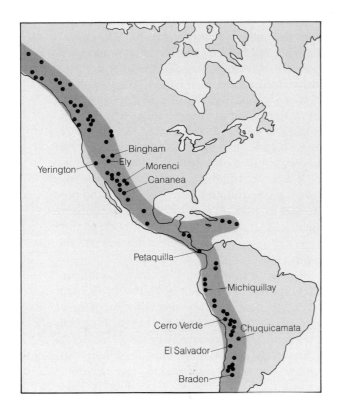

FIGURE 5.23 Chalcopyrite-rich copper deposits formed by hydrothermal solutions create a well-defined belt parallel to the subduction edges of the South and North American plates.

PRESSURE–TEMPERATURE–TIME PATHS FOR METAMORPHIC ROCKS

Rock subjected to metamorphism develops a distinctive mineral assemblage characteristic of the grade of metamorphism reached. In order to reach the temperature and stress at which the highest grade of metamorphism occurs in a given rock, that rock must first have been subjected to lower temperatures and stresses, and thus previously must have contained other mineral assemblages. Prograde metamorphism is a dynamic process, and mineral assemblages replace one another in succession.

One of the triumphs of modern scientific instrumentation has been the development of techniques by which tiny mineral fragments can be examined and analyzed. In many metamorphic rocks microscopic relicts of those earlier mineral assemblages remain (Fig. B5.1); by analyzing these relicts, scientists can decipher the way stress (here called pressure, P, for simplicity), and temperature (T) changed with time (t) during metamorphism. In the language of the scientists who carry out such research, it is possible to determine the *P-T-t* path of a metamorphic rock. Surprising as it may seem, metamor-

FIGURE B5.1 A grain of garnet in a garnet–biotite gneiss. The small inclusions are relics from an earlier mineral assemblage formed under $P-T$ conditions different from those that formed the garnet–biotite gneiss. The garnet grain is 2 mm across.

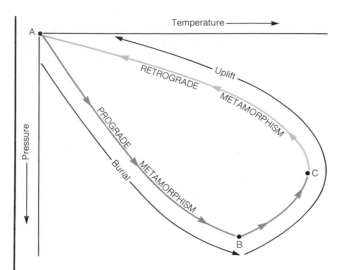

FIGURE B5.2   *P—T* path for a body of rock undergoing metamorphism along a subduction zone.

phic rocks are rarely subjected simultaneously to the highest pressures and highest temperatures. This can only mean that the *P-T-t* paths of rocks undergoing metamorphism are only approximated by the geotherms shown in Figure 5.17. The actual *P-T-t* paths must be more complicated as shown in Figure B5.2.

Consider the case of rock subjected to metamorphism along a plate subduction boundary. The rate of burial, and therefore the rate of pressure increase can be so rapid that thermal equilibrium cannot be maintained. The maximum pressure is reached at point *B* in Figure B5.2, before the maximum temperature, which is only reached at point *C*. When the pressure is released a little, and uplift starts, heating due to radioactive elements will still be in progress, and thus the highest temperature and the highest-grade mineral assemblages will be produced at point *C*. Uplift, like burial, tends to be rapid, and so the *P-T-t* path for a body of rock subjected to retrograde metamorphism will follow the curve *C* → *A*.

By a further advance of modern scientific instrumentation, it is sometimes possible to obtain radiometric ages for the metamorphic mineral assemblages (chapter 6). Knowing the *P-T-t* path of metamorphism and two or more time points on the path, we can calculate *P-T-t* rates and thereby compare the subduction rate of former tectonic plates with present subduction rates. They turn out to be very similar. Bodies of metamorphic rock are turning out to be sensitive monitors of large-scale tectonic processes.

SUMMARY

1. Metamorphism involves changes in mineral assemblage and rock texture and occurs in the solid state as a result of changes in temperature and pressure.

2. Mechanical deformation and chemical recrystallization are the processes that affect rock during metamorphism.

3. The presence of intergranular fluid greatly speeds up metamorphic reactions.

4. Foliation, as expressed by directional textures such as slaty cleavage and schistosity, arises from parallel growth of minerals formed during metamorphism.

5. Cataclastic metamorphism involves mechanical deformation but little or no chemical recrystallization.

6. Heat given off by bodies of intrusive igneous rock causes contact metamorphism and creates contact metamorphic aureoles. Contact meta-

morphism involves chemical recrystallization but very little mechanical deformation.

7. Regional metamorphism, which involves both mechanical deformation and chemical recrystallization, is the result of plate tectonics. Regionally metamorphosed rocks are produced along subduction and collision edges of plates.

8. Rocks of the same chemical composition that are subjected to identical metamorphic environments react to form the same mineral assemblages. Each mineral assemblage defines a metamorphic facies, and each facies forms under a specific range of stress and temperature.

9. Metasomatism involves the changes in rock composition that occur when material in solution is added to the rock, or taken away, as the result of fluids flowing through a rock.

10. Hydrothermal solutions are naturally formed hot brines that are capable of dissolving, transporting, and precipitating minerals.

11. Metamorphism can be explained by plate tectonics. Burial metamorphism occurs within the thick piles of sediment at the foot of continental slopes; regional metamorphism is found in regions of subduction and continental collision.

12. Metasomatism due to hydrothermal solutions is linked to plate tectonics because the solutions tend to form in, or be associated with, stratovolcanoes. Stratovolcanoes are formed above subduction zones.

IMPORTANT TERMS TO REMEMBER

amphibolite facies (p. 146)

burial metamorphism (p. 148)

cataclastic metamorphism (p. 147)

contact metamorphism (p. 147)

differential stress (p. 141)

foliation (p. 143)

gneiss (p. 146)

high-grade metamorphism (p. 140)
hydrothermal solution (p. 152)

intergranular fluid (p. 141)
isograd (p. 150)

low-grade metamorphism (p. 140)

metamorphic aureole (p. 148)
metamorphic facies (p. 151)
metamorphic zones (p. 151)
metamorphism (p. 139)
metasomatism (p. 151)
migmatite (p. 140)

phyllite (p. 145)
prograde metamorphic effects (p. 141)

regional metamorphism (p. 150)
retrograde metamorphic effects (p. 141)

schist (p. 146)
schistosity (p. 143)
slate (p. 145)
slaty cleavage (p. 143)

uniform stress (p. 141)

QUESTIONS FOR REVIEW

1. Briefly discuss the factors that control metamorphism.

2. How and why does slaty cleavage form?

3. What is schistosity? How does it differ from slaty cleavage?

4. What is the difference between a schist and a gneiss?

5. How does a quartzite differ from a sandstone?

6. What is a metamorphic aureole?

7. What is regional metamorphism?

8. What is the geological setting of regional metamorphism? Can you name two places in the world where regional metamorphism is probably happening today?

9. What is burial metamorphism? Suggest some place on the Earth where it is probably happening today.

10. Geologists have found that the concept of metamorphic zones is very helpful in making geological maps. Suggest a reason why.

11. What is the metamorphic facies concept and how does it help in the study of metamorphic rocks?

12. Under what conditions of pressure and temperature does blueschist facies metamorphism occur? What is the geologic environment where such temperatures and pressures are found? Suggest some place on the Earth where blueschist metamorphism is probably happening today.

13. Name three minerals that are found only in metamorphic rocks.

14. What are the two main volatiles that are added to, or lost from, rocks undergoing metamorphism? What roles do volatiles play in metamorphism?

15. What is cataclastic metamorphism? How would you distinguish it from contact metamorphism of an impure limestone?

16. Discuss the importance of metasomatism and describe how metasomatic processes can form valuable mineral deposits.

Excavations seeking evidence of paleoindian settlements, Katmai National Park, Alaska. The age of strata containing artifacts can be determined from the layers of volcanic ash (white), and by radiocarbon dates of organic matter in soil horizons.

CHAPTER 6

Geologic Time

G eologists deal with two kinds of time, relative and absolute. Relative time is the order in which a sequence of past events occurred. Absolute time is the time in years, when a specific event happened.

James Hutton was the first scientist to understand the profound significance of relative time in geology. Hutton had no way to measure absolute time where the Earth's history is concerned, but, as discussed in the Introduction, the evidence at Siccar Point in Scotland (Fig. I.8), proved to Hutton that a sequence of past geological events is preserved in the rock record.

Charles Lyell (1797–1875), a Scot like Hutton, was responsible above all others for demonstrating how Hutton's discovery can be used to determine the relative timing of all geologic events. Lyell realized that the slowness of processes like erosion implied that relative geologic time must equate to truly vast amounts of absolute time. However, Lyell could only speculate about absolute geologic time because, like Hutton, he lacked an absolute time clock.

Lyell never did know the duration of geologic time because radioactivity, which is an accurate way to measure absolute time, was not discovered until 1896. Radioactivity is a natural clock that ticks continuously and leaves a record of its ticking in rocks. The radioactivity clock records that the Earth is 4.6 billion years old. Geologic time is vastly longer than Lyell or any of his associates ever imagined.

Comprehension of a time span as great as 4.6 billion years is very difficult. We inevitably judge history in terms of human life spans and a life span is a mere instant in geologic time. One way to grasp the magnitude of geologic time is through a device used cleverly by Don L. Eicher in his book *Geologic Time* (1976).

Compress . . . the entire 4.6 billion years of geologic time into a single year. On that scale, the oldest rocks we know date from about mid-March. Living things first appeared in the sea in May. Land plants and animals emerged in late November and the widespread swamps that formed the Pennsylvanian coal deposits flourished for about four days in early December. Dinosaurs became dominant in mid-December, but disappeared on the 26th at about the time the Rocky Mountains were first uplifted. Manlike creatures appeared sometime during the evening of December 31st, and the most recent continental ice sheets began to recede from the Great Lakes area and from northern Europe about 1 minute and 15 seconds before midnight on the 31st. Rome ruled the Western world for 5 seconds, from 11:59:45 to 11:59:50. Columbus discovered America 3 seconds before midnight, and the science of geology was born with the writings of James Hutton just slightly more than one second before the end of our eventful year of years.

STRATIGRAPHY

The historical information that geologists have to work with is largely in the form of layered rocks that crop out at the Earth's surface or that can be penetrated by drilling. If we examine the rocks exposed in the upper walls of the Grand Canyon (Fig. 6.1), where the Colorado River has cut nearly 2 km into the Earth's crust, we can see many nearly horizontal layers. These strata formed one atop the other as sediment accumulated on the floor of a shallow sea. Such rocks contain important clues about past environments at and near the Earth's surface. Their sequence and relative ages provide the basis for reconstructing much of Earth history.

The study of strata is called **stratigraphy.** Knowledge of stratigraphic principles and of the relative ages of rock sequences make it possible to work out many of the fundamental principles of physical geology. Two straightforward and simple, but nevertheless very powerful, laws govern stratigraphy: the law of original horizontality and the principle of stratigraphic superposition.

Original Horizontality

Most sedimentary rocks are laid down in a lake or beneath the sea, generally in relatively shallow waters. Under such conditions, each new layer is laid down horizontally over older ones. This observation is consistent with the **law of original horizontality** which states that water-laid sediments are deposited in strata that are horizontal or nearly horizontal, and parallel or nearly parallel to the Earth's surface. From this generalization we can infer that rock layers now inclined, or even buckled and folded, must have been disturbed since the time when they were deposited in a horizontal position.

Stratigraphic Superposition and the Relative Ages of Strata

Toward the end of a cold winter, it is often possible to see layers of old snow that are compact and perhaps also dirty, overlain by fresh, looser, clean snow deposited during the latest snowstorm. Here are layers, or strata, that were deposited in sequence, one above the other. The simple principle involved here also applies to layers of sediment and sedimentary rock. Known as **the principle of stratigraphic superposition**, it states that in any sequence of sedimentary strata, the order in which the strata were deposited is from bottom to top. The relative ages of any two strata can therefore be determined according to whether one of the layers lies above or below the other. Figures 6.1 and 6.2 offer examples of the principle of stratigraphic superposition. The horizontal strata at the top of the Grand Canyon are younger than the horizontal strata below.

The principle of stratigraphic superposition must be employed with a certain amount of care. Observe in Figure 6.3 that tilting and buckling of strata, such as happens during continental collisions, can sometimes be so severe that overturning can bring older strata to overlie younger strata. Observations made on overturned strata would obviously lead to incorrect conclusions concerning relative timing of deposition unless the overturning were recognized. Evidence such as ripple marks, graded beds, and cross-stratified

FIGURE 6.1   The Grand Canyon of the Colorado River. Flat-lying strata, nearly 2000 m thick and accumulated over 300 million years, were laid down on a basement of tilted and tectonically deformed igneous and metamorphic rocks.

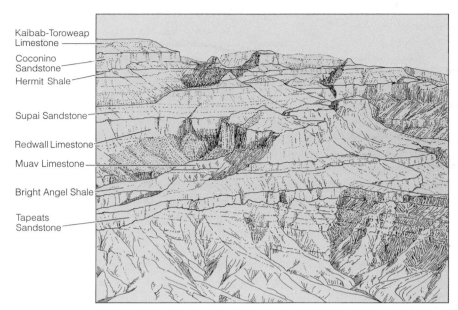

Kaibab-Toroweap Limestone

Coconino Sandstone

Hermit Shale

Supai Sandstone

Redwall Limestone

Muav Limestone

Bright Angel Shale

Tapeats Sandstone

FIGURE 6.2 Sequence of horizontal sedimentary strata laid down on the tilted and deformed basement rocks of the Grand Canyon, from the Tapeats sandstone (oldest) to the Kaibab-Toroweap limestone (youngest) illustrates both the principle of stratigraphic superposition and the law of original horizontality.

beds, as discussed in chapter 4, can be used to determine whether strata are correctly oriented or overturned.

Breaks in the Stratigraphic Record

Lyell and other geologists of the nineteenth century speculated that it might be possible to determine absolute ages by using the stratigraphic record. If one measures the rate of sedimentation in the sea they argued, and if one determines the thickness of all strata, it should be possible to calculate how long it has taken for all the sediments in the stratigraphic record to accumulate. Two assumptions must be correct for the method to work. First it must be assumed that the rate of sedimentation has been constant throughout geologic time. Second, it must be assumed that all strata are *conformable,* meaning they have been deposited layer after layer without interruption. In other words, there must not be any gaps in the stratigraphic record due to erosion or nondeposition.

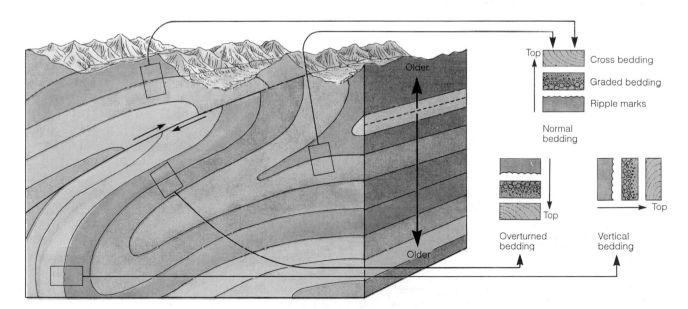

Top — Cross bedding
— Graded bedding
— Ripple marks

Normal bedding

Older

Older

Top

Top

Overturned bedding

Vertical bedding

FIGURE 6.3 Sketch illustrating how some sedimentary features can indicate whether strata are normal (right side up), vertical, or overturned as a consequence of tectonic deformation.

Estimates of hundreds of millions of years for geologic time were made, but they are wrong because both assumptions are false. The first assumption is false because it can be observed today that sedimentation rates vary widely from place to place on the Earth, and there is a lot of evidence to suggest that rates have varied even more widely throughout geologic history. The second and even more important assumption is false because sedimentation can be disrupted periodically by major environmental changes, such as sea level changes and tectonic activity, that lead to intervals of erosion or nondeposition. There are numerous such breaks in the sedimentation record, with no way of knowing how much time the breaks represent. Because erosion destroys some of the record, the part that is preserved is necessarily incomplete and marked by discontinuities where intervals of geologic time, some brief and others very long, are not represented by deposits.

An **unconformity** is a substantial break or gap in a stratigraphic sequence. It records a change in either environmental conditions that caused deposition to cease for a considerable time, or erosion that resulted in loss of part of an earlier-formed depositional record, or a combination of both.

Kinds of Unconformities

There are three important kinds of unconformities found in sedimentary rocks (Fig. 6.4). The most obvious is the **angular unconformity,** which is marked by angular discontinuity between older and younger strata. It is labeled (2) in Figure 6.4. An angular unconformity implies that the older strata were deformed and then truncated by erosion before the younger layers were deposited across them. The outcrop at Siccar Point (Fig. I.8) observed by James Hutton is obviously an angular unconformity. The second kind of unconformity is called a **disconformity;** it is an irregular surface of erosion between parallel strata. The surface numbered (3) in Figure 6.4 is a disconformity. A disconformity implies a cessation of sedimentation, plus erosion, but no tilting. Disconformities can be hard to recognize because the strata above and below are parallel. The way disconformities are usually discovered is through the recognition that fossils of very different ages are present in adjacent strata.

The third kind of unconformity, labeled (1) in Figure 6.4, is a **nonconformity,** where strata overlie igneous or metamorphic rocks.

The three types of unconformity can be seen in the Grand Canyon, as shown in Figure 6.5. At the base of the sedimentary sections is a nonconformity, and some distance above it is an obvious angular unconformity; still higher are three disconformities. Some of the same unconformities can be seen in Figures 6.1 and 6.2; see if you can pick them out.

A study of unconformities brings out the close relationship between tectonics, erosion, and sedimentation. All of the Earth's land surface is a potential surface of unconformity. Some of today's surface will be destroyed by erosion, but some will be covered by sediment and preserved as a record of the present landscape. For example, the Swiss Alps, which were elevated by plate tectonic movements, are being rapidly eroded away. Meanwhile, the eroded material is being carried away by streams and deposited in the Mediterranean Sea. The Mediterranean seafloor was once dry land, but tectonic forces depressed it, just as tectonic forces elevated the Alps. A surface of unconformity separates the young, river-transported sediments and the older rocks of the seafloor on which the sediments are being piled. In a sense, accumulation in one place compensates for destruction in another. The many surfaces of unconformity exposed in rocks of the Earth's crust are evidence that former land surfaces were uplifted by the tectonic forces and exposed to erosion. Preservation of a surface of erosion occurs when later tectonic forces depress the surface so it, in turn, becomes a site of deposition of sediment. Unconformities testify that interactions between the internal and external processes have been going on throughout the Earth's long history.

STRATIGRAPHIC CLASSIFICATION

Every rock stratum can tell us something about the physical and biological character of a part of the Earth at some time in the geologic past. Anyone counting strata would quickly realize that the rock record is like a vast library consisting of thousands upon thousands of volumes. Like a library too, the rock record employs a complex cataloging system. Only the most important cataloging terms used in stratigraphic classification are introduced here.

Three related concepts are employed in stratigraphic classification. The first two are based on the rock units, while the third is somewhat abstract in that it concerns intervals of geologic time.

The first unit of stratigraphic classification uses any distinctive rock unit that differs from the strata above and below. An example of a *rock-stratigraphic unit* is the Navajo Sandstone in Zion National Park, Utah seen in Figure 6.6. This striking sandstone is very different from the strata immediately above and below and can be easily mapped in the field.

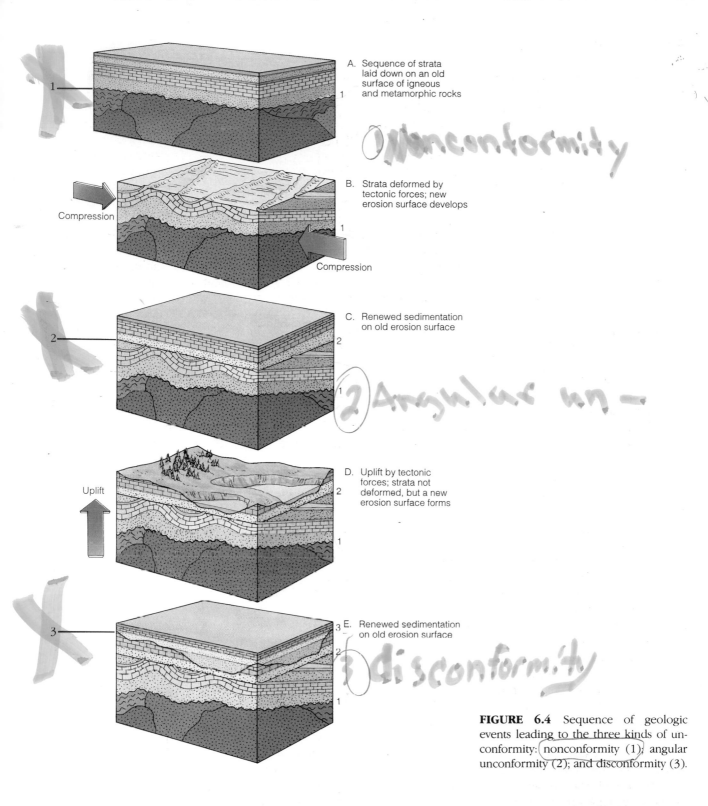

A. Sequence of strata laid down on an old surface of igneous and metamorphic rocks

① nonconformity

B. Strata deformed by tectonic forces; new erosion surface develops

Compression

Compression

C. Renewed sedimentation on old erosion surface

② Angular un —

D. Uplift by tectonic forces; strata not deformed, but a new erosion surface forms

Uplift

E. Renewed sedimentation on old erosion surface

③ disconformity

FIGURE 6.4 Sequence of geologic events leading to the three kinds of unconformity: nonconformity (1); angular unconformity (2); and disconformity (3).

The basic unit of rock stratigraphy is the **formation,** which is a group of similar strata that are sufficiently different from adjacent groups of strata so that on the basis of physical properties they constitute a distinctive, recognizable unit that can be used for geologic mapping over a wide area. The Navajo Sandstone is a formation (each formation is given a name: in North America it typically is the name of a geographic locality near which the unit is best exposed).

The second unit used in stratigraphic classification is one representing all the rocks that formed during a specific interval of geologic time. Each of the boundaries of a *time-stratigraphic unit,* upper and lower, is everywhere the same age.

A formation is defined only on the basis of its material characteristics, and its upper and lower boundaries lie where a recognizable change in physical properties occur. As illustrated in Figure 6.7, the ages

165

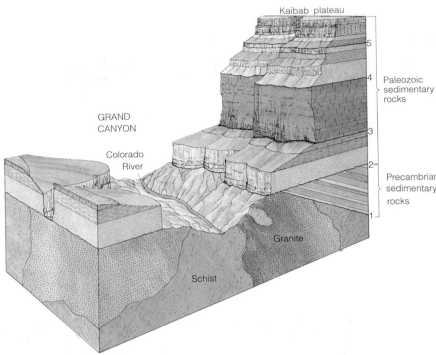

FIGURE 6.5 Geologic section through rocks exposed in the Grand Canyon. The lowest unconformity (1), separating tilted sedimentary strata from older crystalline rocks, is a non-conformity. An angular unconformity (2) separates the tilted strata from horizontally layered strata above, while three disconformities (3, 4, and 5) are seen still higher in the section. Unconformities 2, 3, 4, and 5 are visible in Figure 6.1.

FIGURE 6.6 Spectacular view of the Navajo Sandstone, Zion National Park, Utah. The Navajo Sandstone is an example of a rock-stratigraphic unit.

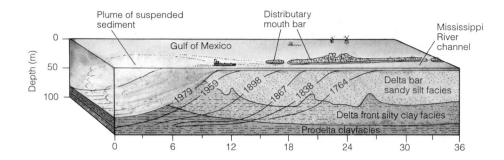

FIGURE 6.7 Section through the Southwest Pass lobe of the Mississippi River delta, which is expanding outward into the Gulf of Mexico. Dated time lines show successive positions of the delta front between 1764 and 1979. The boundaries separating three sedimentary facies intersect each time line. Should this deposit be preserved and eventually converted to sedimentary rock, the resulting formations, corresponding to the three facies, will each be younger in the direction of delta growth.

of the boundaries of a formation can differ from place to place. A time-stratigraphic unit, in contrast, may include more than one rock type, and its upper and lower boundaries may not necessarily coincide with a formational boundary, but as demonstrated in Figure 6.7, each boundary is everywhere the same age.

The primary time-stratigraphic unit is a **system** which is chosen to represent a time interval sufficiently great so that such units can be used all over the world. Names for systems arose from nineteenth-century studies of strata, mainly in Europe, and often derive from geographic localities. Most systems are now known to encompass absolute time intervals of tens of millions of years.

The final units used in stratigraphic classification are the intervals of geologic time during which strata in time-stratigraphic units accumulated. Units of geologic time are nonmaterial, while time-stratigraphic units are material. The primary unit of geologic time is a geologic *period;* it is the time during which a geologic system accumulated.

Because the geologic record is incomplete and punctuated by numerous unconformities, a given area may not contain a complete depositional record for the corresponding geologic time interval (Fig. 6.4). However, many of the gaps can be bridged by piecing together sequences of strata from different geographic areas, thereby providing us with a more complete picture of Earth history. The piecing together involves correlation.

CORRELATION OF ROCK UNITS

William Smith was an English land surveyor who was active around the beginning of the nineteenth century.

His profession gave him an ideal opportunity to observe not only the landscape but also the rocks that underlie it. While surveying for the construction of new canals in western England, he observed many sedimentary strata and soon realized that they lay, as he put it, "like slices of bread and butter" in a definite, unvarying sequence. He became familiar with the physical characteristics of each layer, with the fossils each contained, and with the sequence of the layers. By looking at a specimen of sedimentary rock collected from anywhere in southern England, he could name the layer from which it had come and, of course, the position of the layer in the sequence.

Smith's discovery was not considered by him to reflect any particular scientific principle; it was purely practical. Nevertheless, it opened the door to the correlation of sedimentary strata over increasingly wide areas. **Correlation** means the determination of equivalence in time-stratigraphic age of the succession of strata found in two or more different areas. Smith correlated strata on the dual basis of physical similarity and fossil content initially over distances of several kilometers, and later over tens of kilometers. By means of fossils alone, it ultimately became possible to correlate through hundreds and then thousands of kilometers.

Correlation involves two main tasks. One is to determine the relative ages one to another, of units exposed in local sections within an area being studied. Then the ages of the units relative to a standard scale of geologic time must be found. To accomplish these goals, a geologist employs various physical and biological criteria; one is not necessarily more dependable or precise than the others.

Continuous exposures are not common, and so we are often faced with correlating between widely spaced outcrops. A stratum or even a formation may

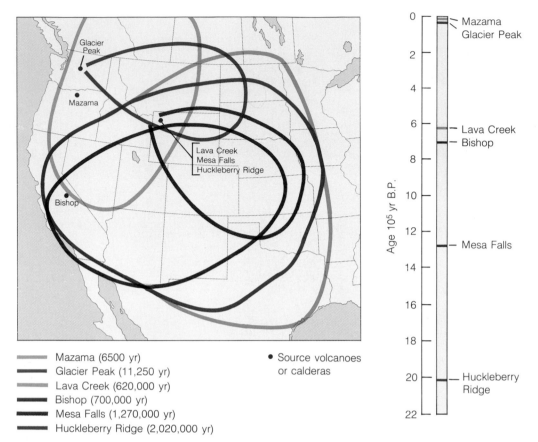

Mazama (6500 yr)
Glacier Peak (11,250 yr)
Lava Creek (620,000 yr)
Bishop (700,000 yr)
Mesa Falls (1,270,000 yr)
Huckleberry Ridge (2,020,000 yr)

• Source volcanoes
  or calderas

FIGURE 6.8 Map showing distribution of widespread tephra layers in western North America which erupted during the last 2 million years. Each tephra layer is a key bed.

be deformed and eroded so that only parts of it remain. The physical matching of the remnants generally involves the use of rock characteristics such as grain size, color, and sedimentary structures that permit the unit to be distinguished from others. In some instances quite different formations look almost identical, so correlation must be done with care. The Navajo Sandstone in Zion National Park shown in Figure 6.6, for example, is very similar to the Coconino Sandstone in the Grand Canyon (Fig. 6.1), 120 km to the south.

FIGURE 6.9 *Modocia typicalis,* a trilobite that lived about 530 million years ago, during the middle of the Cambrian Period, is a distinctive index fossil. This specimen came from the Marjum Shale in the House Range, Utah.

However, correlation of the two formations based on physical criteria would be incorrect, because careful mapping reveals that the Navajo Sandstone sits much higher in the stratigraphic section than the Coconino Sandstone.

A thin and generally widespread sedimentary bed with characteristics so distinctive that it can be easily recognized but not confused with any other bed is called a *key bed*. Such beds are exceedingly useful in correlating major rock sections. In areas of volcanic activity, ash layers can serve as distinctive key beds for purposes of regional correlation as shown in Figure 6.8.

We call a fossil that can be used to identify and date the stratum in which it is found, an *index fossil*. To be most useful, an index fossil should have common occurrence, a wide geographic distribution, and a very restricted age range. The best examples are swimming or floating organisms that underwent rapid evolution and quickly became widely distributed (Fig. 6.9). If a distinctive index fossil is recognizable at an outcrop, a rapid and reliable means of correlation is available (Fig. 6.10). While some genera and species permit long-range correlation of rocks in different geographic areas or even on different continents, more often close dating and correlation involves using as-

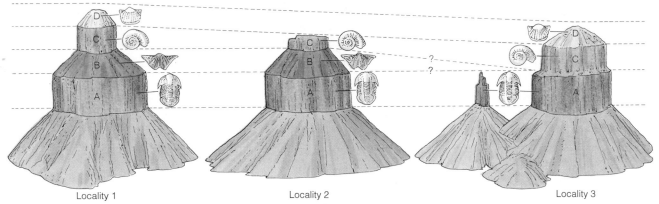

FIGURE 6.10 Correlation of strata exposed at three localities, many kilometers apart, on the basis of similarity of the fossils they contain. The fossils show that at Locality 3 stratum B is missing because C directly overlies A. Either B was never deposited there, or it was deposited and later removed by erosion before the deposition of layer C.

semblages of fossils of as many different types as possible.

THE GEOLOGIC COLUMN AND THE GEOLOGIC TIME SCALE

One of the great successes of the nineteenth-century geologists was the demonstration, through stratigraphic correlation, that time-stratigraphic sequences are the same on all continents. Through worldwide correlation those nineteenth-century geologists assembled a **geologic column,** which is a composite columnar section containing in chronological order the succession of known strata, fitted together on the basis of their fossils or other evidence of relative age. This worldwide standard is still being added to and refined as more rock units are described and mapped.

Standard names have evolved for the subdivisions of the geologic time units corresponding to the rock units of the geologic column. The units of the geologic time scale, which, like the geologic column, can be used worldwide, are eons, eras, periods and epochs as shown in Figure 6.11.

Eons

An **eon** is the largest interval into which geologic time is divided, and there are four of them. The term *Hadean* (Greek for beneath the Earth) is given to the oldest eon. This is the earliest part of the Earth's history, an interval for which no rock record is known. However, rocks of this age are present on other plan-

ets whose earliest crustal rocks have been little modified since they accumulated. The **Archean** (Greek word for ancient) eon follows the Hadean. Archean rocks are the oldest rocks we know of on the Earth, and they contain microscopic life-forms of bacterial character. The **Proterozoic** (earlier life) eon follows the Archean. Proterozoic rocks include evidence of multicelled organisms that lacked preservable hard parts. Understandably, the record of the ancient Archean and Proterozoic is not as well known as the record of younger rocks because many of these ancient rocks have been intensely deformed, metamorphosed, and eroded. The **Phanerozoic** (visible life) is the most recent of the four eons. Phanerozoic rocks often contain plentiful evidence of past life in the form of well-preserved hard parts. Most examples of fossils that we see displayed in museums or illustrated in books are from the Phanerozoic eon.

Eras

Eons are divided into eras. Geologic **eras** encompass major spans of time that also are defined on the basis of the life-forms found in the corresponding rocks. No formal eras are yet widely recognized for Archean and Proterozoic rocks, but the Phanerozoic Eon is divided into the **Paleozoic** (old life), **Mesozoic** (middle life), and **Cenozoic** (recent life) eras, each name reflecting the relative stage of development of the life of these intervals (Fig. 6.11). Paleozoic forms of life progress from marine invertebrates to fishes, amphibians, and reptiles. Early land plants also appeared, expanded, and evolved. The Mesozoic Era saw the rise of the dinosaurs, which became the dominant vertebrates on land. Toward the end of that era, mammals

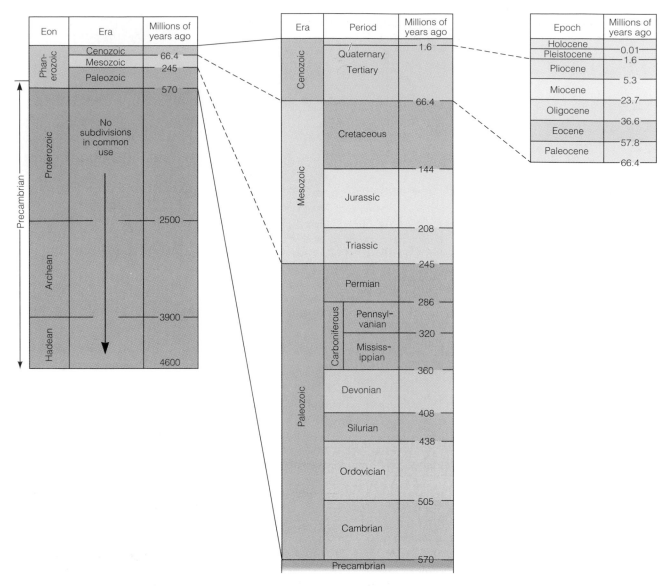

FIGURE 6.11 The geologic time scale. Absolute ages obtained from radiometric dates. Note that the Pennsylvanian and Mississippian Periods are equivalent to the Carboniferous Period of Europe. The time boundary between the Archean and Hadean is uncertain as no rocks of the Hadean Eon are known on the Earth. Hadean rocks are known to exist on other planets in the solar system.

first appeared and later dominated the Cenozoic Era. The Mesozoic Era also witnessed the evolution of flowering plants, while during the Cenozoic Era grasses appeared and became an important food for grazing mammals.

Periods
The eras of the Phanerozoic eon are divided into **periods.** The geologic periods have a haphazard nomenclature. They were defined over an interval of nearly 100 years on the basis of strata that crop out in Britain, Germany, Switzerland, Russia, and the United States.

The names are partly geographic in origin, but in some cases they are based on characteristics of the strata in the place of original study (Table 6.1).

The lowermost period of the Paleozoic Era, the Cambrian Period, is the time when animals with hard shells first appear in the geologic record. Prior to the Cambrian Period all animals were soft bodied and the fossil evidence they left is very sparse. Rocks formed during the Archean and Proterozoic Eons cannot be readily differentiated on the basis of the fossils they contain, so geologists often refer to the entire time preceding the Cambrian as simply Precambrian.

TABLE 6.1   Origin of Names for Periods of the Paleozoic, Mesozoic and Cenozoic Eras, and the Epochs of the Quaternary and Tertiary Periods

Era	Period	Epoch	Origin of Name
Cenozoic	Quaternary[a]	Holocene	Greek for wholly recent
		Pleistocene	Greek for most recent
	Tertiary[a]	Pliocene	Greek for more recent
		Miocene	Greek for less recent
		Oligocene	Greek for slightly recent
		Eocene	Greek for dawn of the recent
		Paleocene	Greek for early dawn of the recent
Mesozoic	Cretaceous	↑	Latin for chalk, after chalk cliffs of southern England and France
	Jurassic	Epoch	Jura Mountains, Switzerland and France
	Triassic	Names	Threefold divison of rocks in Germany
Paleozoic	Permian	Used	Province of Perm, Russia
	Pennsylvanian	Only	State of Pennsylvania
	Mississippian	By	Mississippi River
	Devonian	Specialists	Devonshire, county of Southwest England
	Silurian	↓	Silures, ancient Celtic tribe of Wales
	Ordovician		Ordovices, ancient Celtic tribe of Wales
	Cambrian		Cambria, Roman name for Wales

[a]Derived from eighteenth- and nineteenth-century geologic time scale that separated crustal rocks into a fourfold division of Primary, Secondary, Tertiary, and Quaternary, based largely on relative degree of lithification and deformation.

Epochs

The **epochs** of the Tertiary Period were defined in a piecemeal fashion. Studies of marine strata in sedimentary basins of France and Italy led Charles Lyell to subdivide the rocks into groupings based on the percentage of their fossils that are represented by still living species (Fig. 6.11). Each of the various periods of the Paleozoic and Mesozoic eras are also subdivided into epochs, the names of which are primarily geographic in origin. They are used mainly by specialists concerned with detailed studies of these strata and their contained fossils.

The names of the geologic time scale constitute the standard time language of geologists the world over. Through their use one can begin to comprehend numerous details of Earth history that have led to the discovery of many of the important principles of physical geology discussed in this book.

MEASURING ABSOLUTE GEOLOGIC TIME

The scientists who worked out the geologic column and time scale were challenged by the question of absolute time. They knew the relative time order in which the different systems had formed but they also wished to know whether the sediments in each system had accumulated during the same length of time. They sought answers to questions such as these: "How much time elapsed between the end of the Cambrian Period and the beginning of the Permian Period?" "How long was the Tertiary Period?" The question of absolute time is as important as the geologic column and time scale. Absolute ages must be determined in order to answer such challenging and important questions as the age of the Earth, the age of the ocean, how fast mountain ranges rise, and how long humans have inhabited the Earth.

Indirect Attempts to Measure Absolute Geologic Time

During the nineteenth century many attempts were made to develop a scale of years for the geologic time scale by indirect methods. One widely used method was discussed earlier in this chapter. It consisted of estimates of the time during which the rock cycle has been at work based on rates of sedimentation and the thickness of sedimentary rocks. Estimates for the age of the earth, which was presumed to be the same as the duration of the rock cycle, ranged from 3 million to 1.5 billion years!

A clever suggestion for estimating the age of the ocean concerned the saltiness of seawater. Because sea salts come from the erosion of common rocks and reach the sea dissolved in river water, why not measure the salts in modern river water and calculate the time needed to transport all the salts now in the sea? The first person to suggest that sea salt might be used to date the ocean was Edmund Halley in 1715. There is no record that Halley (for whom the comet is named) ever made the necessary measurements. John Joly finally made the necessary measurements and calculations in 1889. His answer for the ocean's age, 90 million years, does not represent the age of the ocean at all. The composition of the ocean is essentially constant because all the salts in the sea, like all other chemical constituents, are cyclic. Salts are added both by erosion and by submarine volcanism, but salts are also removed from solution. Some are removed as evaporite minerals (chapter 4), while others are removed by reaction with hot volcanic rocks on the seafloor.

Perhaps the most interesting estimates were those made by Lord Kelvin, a physicist, who attempted to calculate the time the Earth has been a solid body. The Earth started as a very hot object, he argued. Once it had cooled sufficiently to form a solid outer crust, it could continue to cool only by the conduction of heat through solid rock. By measuring the thermal properties of rock and estimating the present temperature of the Earth's interior, he calculated the time for the Earth to cool to its present state. Kelvin's logic was faultless and his mathematical calculations correct. However, his estimate of 100 million years for the maximum age of the earth is incorrect because he made an incorrect assumption. Kelvin assumed that no additional heat has been added since the Earth was formed. When Kelvin made his calculations radioactivity was not known. Radioactivity continuously supplies

heat to the Earth's interior, so that instead of cooling rapidly, the Earth's interior is cooling so slowly it has a nearly constant temperature over periods as long as hundreds of millions of years.

What was needed in order to resolve the dilemma of absolute geologic time was a way to measure geologic time by some process that runs continuously, that is not reversible, that is not influenced by other processes and other cycles, and that leaves a continuous record without gaps in it. In 1896, the discovery of radioactivity provided the needed method.

Radioactivity and the Measurement of Absolute Time

Natural Radioactivity

We learned in chapter 2 that the atomic number of a given element—that is the number of protons in the atomic nucleus of the element—is constant and characteristic of that element. However, an atomic nucleus also contains neutrons, and the number of neutrons can vary without changing the number of protons. For example, all carbon atoms contain 6 protons, but the protons can be joined by 6, 7, or 8 neutrons. Atoms of an element that contain different numbers of neutrons are called **isotopes** (Fig. 6.12). An isotope is identified by its **mass number,** which is the sum of the neutrons plus protons. Carbon, therefore, has three isotopes with mass numbers 12, 13, and 14 respectively which are written ^{12}C, ^{13}C, and ^{14}C. All common chemical elements are mixtures of several isotopes.

Most of the isotopes of the chemical elements found in the Earth are stable and not subject to change. However, a few, such as ^{14}C, are radioactive because of an instability in the nucleus. The instability arises because there are limits within which the mass numbers of the isotopes of any element can vary. The nucleus of a radioactive isotope will transform spontaneously to a nucleus of either a more stable isotope of

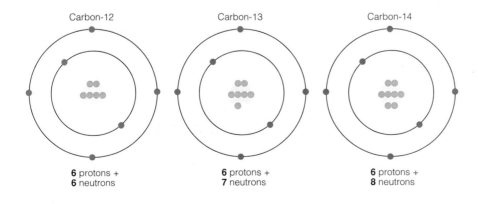

Carbon-12   Carbon-13   Carbon-14

6 protons +   6 protons +   6 protons +
6 neutrons   7 neutrons   8 neutrons

FIGURE 6.12 The three isotopes of carbon. Note that in each case there are 6 protons in the nucleus and 6 electrons in the energy-level shells. The differences lie in the numbers of neutrons in the nucleus.

the same chemical element or an isotope of a different chemical element. The rate of transformation is different for each isotope. Even though the process is one of transformation—from an unstable nucleus to a more stable one—it has become common practice to call the process radioactive decay. An atomic nucleus undergoing radioactive decay is said to be a **parent;** the product arising from radioactive decay is called a **daughter.** ^{14}C decays to ^{14}N and ^{238}U decays to ^{206}Pb; ^{14}C and ^{238}U are parents, ^{14}N and ^{206}Pb daughters.

Radioactive decay is a phenomenon of the atomic nucleus; electrons that orbit the nucleus are not involved. To understand radioactivity we need to expand the previous discussion of the nucleus in chapter 2. A proton in the nucleus (written p) can combine with an electron (written β and pronounced beta) to make a neutron (written n). Thus,

$$p + \beta \rightarrow n.$$

Similarly, a neutron can be transformed to a proton by emitting an electron, so that

$$n \rightarrow p + \beta.$$

Radioactive decay can happen in three ways (see Fig. 6.13):

1. By emission of a β particle, in which case the mass number is unchanged but the atomic number increases by one (Fig. 6.13A).

2. By capture of a β particle, which reduces the atomic number by one but leaves the mass number unchanged (Fig. 6.13B).

3. Emission from the nucleus of a heavy atomic particle consisting of two neutrons and two protons ($2p + 2n$), called an α (alpha) particle (Fig. 6.13C). Loss of an α particle reduces the mass number by four and the atomic number by two. Each of the radioactive decay systems can also be accompanied by the emission of electromagnetic radiation, called γ rays (gamma rays). γ rays do not affect either the atomic number or the mass number.

Rates of Decay

Many of the radioactive isotopes that were once in the Earth have decayed away and are no longer present.

FIGURE 6.13 The three ways radioactive isotopes decay. Note that in each case the atomic number (number of protons) of the daughter differs from the atomic number of the parent.

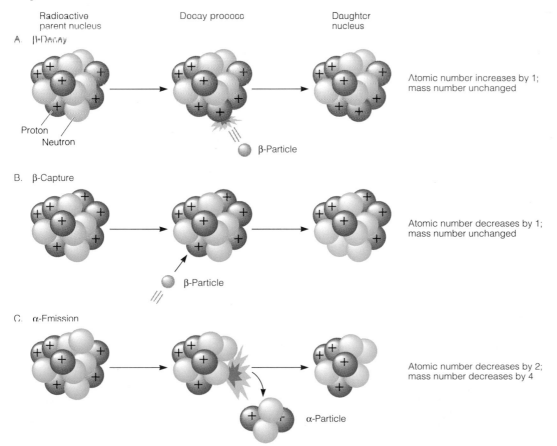

Radioactive parent nucleus   Decay process   Daughter nucleus

A. β-Decay

Proton
Neutron

β-Particle

Atomic number increases by 1; mass number unchanged

B. β-Capture

β-Particle

Atomic number decreases by 1; mass number unchanged

C. α-Emission

α-Particle

Atomic number decreases by 2; mass number decreases by 4

This is so because their rates of spontaneous decay are fast. A few radioactive isotopes that transform very slowly are still present, however. Careful study of radioactive isotopes in the laboratory has shown that decay rates are unaffected by changes in the chemical and physical environment. Thus, the decay rate of a given isotope is the same in the mantle, or a magma, as it is in a sedimentary rock. This is a particularly important point because it leads to the conclusion that rates of radioactive decay are not influenced by geologic processes.

Each radioactive element decays according to a distinct and measurable timetable, but all decay timetables follow the same basic law that is depicted in Figure 6.14. The law, stated in words, is that the proportion of parent atoms that decay during each unit of time is always the same. The number of decaying parent atoms continuously decreases while the number of daughter atoms continuously increases.

The rate of radioactive decay is measured by the **half-life** which is the time needed for the number of parent atoms to be reduced by one-half. For example, if the half-life of a radioactive isotope is 1 hour and we started an experiment with 1000 atoms, only 500 parent atoms would remain at the end of an hour and 500 daughter atoms would have formed. At the end of a second hour there would be 250 parent and 750 daughter atoms, and after hour 3, 125 parents and 875 daughters.

In the graphic illustration of radioactive decay in Figure 6.14 the time units marked are half-lives. Of course, the time units are of equal length, but at the end of each unit the number of parent atoms, and therefore the radioactivity of the sample, has decreased by exactly one-half. Figure 6.14 also shows that the growth of daughter atoms just matches the decline of parent atoms. When the number of remaining parent atoms (N_p) is added to the number of daughter atoms (N_d), the result is N_o, the number of parent atoms that a mineral sample started with. That fact is the key to the use of radioactivity as a means of measuring geologic time and determining ages.

Potassium-Argon ($^{40}K/^{40}Ar$) Dating

We have selected one of the naturally radioactive isotopes, potassium-40 (^{40}K), to illustrate how minerals can be dated. Potassium has three natural isotopes: ^{39}K, ^{40}K, and ^{41}K. Only one, ^{40}K, is radioactive and its half-life is 1.3 billion years. The decay of ^{40}K is interesting because two different decay schemes occur.

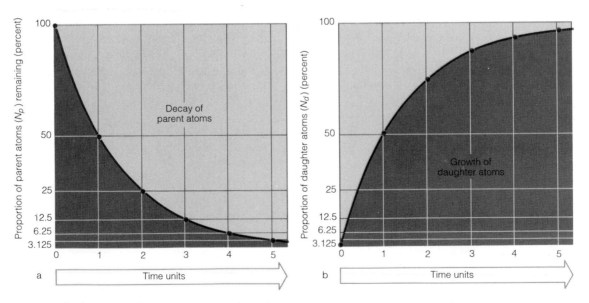

FIGURE 6.14   Curves illustrating the basic law of radioactivity. A. At time zero, a sample consists of 100 percent radioactive parent atoms. During each time unit, half the atoms remaining decay to daughter atoms. B. At time zero, no daughter atoms are present. After one time unit corresponding to a half-life of the parent atoms, 50 percent of the sample has been converted to daughter atoms. After two time units, 75 percent of the sample is daughter atoms, 25 percent parent atoms. After three time units, the percentages are 87.5 and 12.5, respectively. Note that at any given instant N_p, the number of parent atoms remaining, plus N_d, the number of daughter atoms, equals N_o, the number of parent atoms at time zero.

Twelve percent of the ^{40}K atoms decay by capture to ^{40}Ar, an isotope of the gas argon. The remaining 88 percent of the ^{40}K atoms change by decay to ^{40}Ca. The appropriate equations are:

$$^{40}K + \beta \rightarrow {}^{40}Ar$$

and

$$^{40}K \rightarrow {}^{40}Ca + \beta$$

It is important to know that the fraction of ^{40}K atoms decaying to ^{40}Ar is always 12 percent; the percentage is not affected by changes in physical or chemical conditions.

When a potassium-bearing mineral crystallizes from a magma, or grows within a metamorphic rock, it includes some ^{40}K in its crystal structure. As soon as the mineral is formed, ^{40}Ar and ^{40}Ca daughter atoms start accumulating in the mineral, because they are trapped, like the parent ^{40}K atoms, in the crystal structure. Because the ratio of ^{40}Ar to ^{40}Ca daughter atoms is always the same, it is only necessary to measure either ^{40}Ar or ^{40}Ca daughter atoms in order to know how many ^{40}K atoms have decayed. It is more accurate to use ^{40}Ar because argon is an element that has unusual atomic properties.

There is another reason, too, why ^{40}Ar is measured rather than ^{40}Ca. The electron energy-level shells of argon are filled, so atoms of argon do not readily form chemical bonds. Thus, ^{40}Ar atoms are not chemically bound in minerals, they are trapped by the crystal lattice but do not form bonds. The trapping is only effective at low temperatures, so at high temperatures argon rapidly diffuses out of a mineral. At lower temperatures argon cannot diffuse out and stays trapped. When the argon content of a mineral is measured, therefore, what is determined is the ^{40}Ar accumulated during the time since a mineral started trapping and retaining argon. Even if some ^{40}Ar is present in a magma when a mineral crystallizes, therefore, a potassium-bearing mineral will not retain any initial argon because magmatic temperatures are far above trapping temperatures. All the ^{40}Ar atoms in a potassium-bearing mineral in an extrusive igneous rock such as a rhyolite or andesite must have come from decay of ^{40}K and must have accumulated since the temperature fell below the trapping temperature. Because extrusive igneous rocks cool very quickly, the trapping time and eruption time are essentially the same.

All that has to be done to determine the absolute time of eruption of an extrusive igneous rock is to select a potassium-bearing mineral, measure the amount of parent ^{40}K that remains, and the amount of trapped ^{40}Ar. The half-life of ^{40}K being known, it is a straightforward matter to calculate the radiometric age—the length of time a mineral has contained in its built-in radioactivity clock.

Calculation of a $^{40}K/^{40}Ar$ Age

Let's calculate an example. Chemical analysis of a potassium feldspar sample from a pyroclastic rock shows that for every 20,000 parent atoms of ^{40}K present there are 1200 atoms of ^{40}Ar. We know that the ratio $^{40}Ar/^{40}Ca$ is constant, so 1200 atoms of ^{40}Ar means that 8800 atoms of ^{40}Ca are also present. This in turn means that N_d, the number of daughter atoms, is 1200 + 8800, or 10,000.

The equation for radioactive decay (i.e., the equation for the curve in Fig. 6.14), is:

$$\frac{N_p}{N_o} = (1 - \lambda)^y$$

where N_p is the number of parent atoms now, in our example 20,000.

N_o is the number of parent atoms when the mineral formed. Because each parent atom that decays produces only one daughter atom, $N_o = N_p + N_d = 20,000 + 10,000 = 30,000$

λ is the decay constant, which is the fraction of parent atoms that decays per unit time.

y is the number of time units

We can simplify the calculation in two ways. First, we select the unit of time to be equal to the half-life, which means $\lambda - 0.5$. This in turn makes y equal to the number of half-lives since the mineral formed. The second simplification is to put the equation in a logarithmic form:

$$\text{thus,} \frac{N_p}{N_o} = (1 - 0.5)^y = 0.5^y$$

or, in logarithmic form,

$$\log \frac{N_p}{N_o} = -0.3y$$

or

$$\log N_o - \log N_p = 0.3\,y$$

Thus

$$\log 30,000 - \log 20,000 = 0.3\,y$$

or

$$y = \frac{4.477 - 4.301}{0.3} = 0.587$$

The feldspar therefore formed 0.587 half-lives ago.

The half life of ^{40}K has been measured in the laboratory to be 1300 million years. Thus, the age of the mineral is 1300 x 0.587 = 760 million years.

Flexibility of ^{40}K/^{40}Ar Dating

Dating by ^{40}K is not limited to minerals that contain potassium as a major element. Even minerals that contain small amounts of potassium will serve the purpose. Thus, hornblende, a calcium–iron–magnesium silicate, can be used for ^{40}K/^{40}Ar dating because it generally contains a small quantity of potassium present by ionic substitution.

^{40}K/^{40}Ar dating is most successfully applied to volcanic and pyroclastic rocks because their solidification and cooling are rapid. As a result, they have solidification ages that are essentially coincident with their trapping ages. Because argon analyses can be performed with great accuracy, and because contamination by initial argon at the time of crystallization is generally not a problem, the method can be used for volcanic rocks as young as 50,000 years. For this reason, ^{40}K/^{40}Ar dating has proved very useful in studies of archaeology as well as geology.

Other Radiometric Dating Methods

Many naturally radioactive isotopes can be used for radiometric dating, but six predominate in geologic studies. These are the two radioactive isotopes of uranium plus the single radioactive isotopes of thorium, potassium, rubidium, and carbon. These isotopes occur widely in different minerals and rock types, and they have a very wide range of half-lives, so that many geologic materials can be dated radiometrically (Table 6.2).

Radiocarbon Dating

Among the radiometric dating methods listed in Table 6.2, the one based on ^{14}C (also known as radiocarbon) is unique for two reasons. The first is that the half-life of ^{14}C is short by comparison with the half-lives of ^{40}K, ^{87}Rb, and the isotopes of thorium and uranium. The second reason is that the number of daughter atoms cannot be measured.

Radiocarbon is continuously created in the atmosphere through bombardment of nitrogen-14 (^{14}N) by neutrons created by cosmic radiation (Fig. 6.15). ^{14}C, with a half-life of 5730 years, decays back to ^{14}N by β

TABLE 6.2   Some of the Principal Isotopes Used in Radiometric Dating

Isotopes			Half-Life of Parent (years)	Effective Dating Range (years)	Minerals and Other Materials That Can Be Dated
Parent	Decay System	Daughter			
Uranium-238	α + β decay	Lead-206	4.5 billion	10 million–4.6 billion	Zircon and Uraninite
Uranium-235	α + β decay	Lead-207	710 million	10 million–4.6 billion	Zircon and Uraninite
Thorium-232	α + β decay	Lead-208	14 billion	10 million–4.6 billion	Zircon and Uraninite
Potassium-40	β capture β decay	Argon-40 Calcium-40	1.3 billion	50,000–4.6 billion	Muscovite Biotite Hornblende Whole volcanic rock
Rubidium-87	β decay	Strontium-87	47 billion	10 million–4.6 billion	Muscovite Biotite Potassium feldspar Whole metamorphic or igneous rock
Carbon-14	β decay	Nitrogen-14	5,730 ± 30	100–70,000	Wood, charcoal, peat, grain, and other plant material Bone, tissue, and other animal material Cloth Shell Stalactites Groundwater Ocean water Glacier ice

Magnatizm
young rocks
+ oceans

A. ^{14}C created by neutron capture

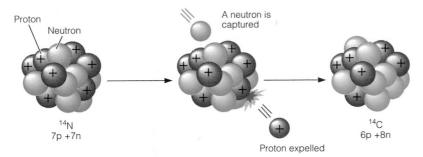

B. ^{14}C decays to ^{14}N by β-decay

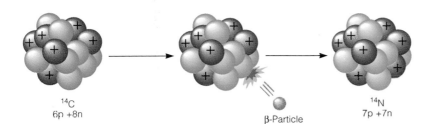

FIGURE 6.15 Production and decay of radiocarbon. A. The nucleus of a ^{14}N atom captures a neutron that was created by cosmic ray bombardment of the atmosphere and expels a proton, thereby changing ^{14}N to ^{14}C. B. The nucleus of a ^{14}C atom is radioactive. By β-decay, ^{14}C reverts back to ^{14}N.

decay. The ^{14}C mixes with ^{12}C and ^{13}C and diffuses rapidly through the atmosphere, hydrosphere, and biosphere. Because the rates of mixing and exchange are rapid compared with the half-life, the proportion of ^{14}C is nearly constant throughout the atmosphere. As long as the production rate remains constant, the radioactivity of natural carbon remains constant because rate of production balances the rate of decay.

While an organism is alive it will continuously take in carbon from the atmosphere, and so will contain the balanced proportion of ^{14}C. However, at death the balance is upset, because replenishment by life processes such as feeding, breathing, and photosynthesis ceases. The ^{14}C in dead tissues continuously decreases by radioactive decay. The analysis for the radiocarbon date of a sample involves only a determination of the radioactivity level of the ^{14}C it contains. This is done by measuring the particles emitted as a result of radioactive decay. The daughter isotope, ^{14}N, cannot be measured successfully because it leaks away and because of atmospheric contamination.

Because of its application to organisms (by dating fossil wood, charcoal, peat, bone, and shell material) and its short half-life, radiocarbon has proved to be enormously valuable in establishing dates for prehistoric human remains and for recently extinct animals. In this way it is of extreme importance in archaeology. It is also of great value in dating the most recent part of geologic history, particularly the latest glacial age.

For example, the dates of many samples of wood taken from trees overrun by the advance of the latest great ice sheet and buried in the rock debris thus deposited show that the ice reached its greatest extent in the Ohio–Indiana–Illinois region about 18,000 to 21,000 years ago. It is even possible to date young ice, such as that in the Greenland ice sheet, directly. As the ice forms, bubbles of air are trapped in it. The carbon dioxide in the air bubbles can be liberated in the laboratory and dated, providing an age for the time of ice formation.

Similarly, radiocarbon dates afford the means for determining rates of geologic processes, such as: the rate of advance of the last ice sheet across Ohio; the rate of rise of the sea against the land while glaciers melted throughout the world; average rates of circulation of water in the deep ocean; the rates of local uplift of the crust that raised beaches above sea level; and even the frequency of volcanism.

Absolute Time and the Geologic Time Scale

Through the various methods of radiometric dating, geologists have determined the dates of solidification of many bodies of igneous rock. Many such bodies have identifiable positions in the geologic column, and because of this it becomes possible to date, approximately, a number of the sedimentary layers in the column.

The standard units of the geologic column consist of sedimentary strata containing characteristic fossils, but the typical rocks from which radiometric dates (other than radiocarbon dates) are determined are igneous rocks. It is necessary, therefore, to be sure of the relative time relations between an igneous body that is datable and a sedimentary layer whose fossils closely indicate its position in the column.

Figure 6.16 shows how ages of sedimentary strata are approximated from the ages of igneous bodies. A sequence of sedimentary strata containing fossils of known ages are separated by an unconformity and two disconformities. Intrusive stock *A* cuts strata 1 and 2 but is truncated by the disconformity at the top of stratum 2. Thus, *A* must be younger than strata 1 and

2 but older than stratum 3, which was laid down on the erosion surface at the top of stratum 2 and contains weathered fragments of *A* among the sedimentary particles. Similarly, the combination of dikes and sills that make up the intrusive igneous complex *B* are truncated by the disconformity at the top of stratum 3, and they must be younger than stratum 3 but older than stratum 4. Lava flow *C* above the disconformity at the top of stratum 3 must also be younger than stratum 3 and younger than the dike–sill complex *B*. Lava flow *C* must be older than stratum 4, however, because it is covered by stratum 4, and lava flow *D* must be even younger because it overlies stratum 4.

From the radiometric dates of the igneous bodies and the relative ages of the geologic relations shown in Figure 6.16, we can draw the following inferences about the ages of the sedimentary strata.

FIGURE 6.16   Idealized section illustrating the application of radiometric dating to the geologic column. For method, see the text discussion.

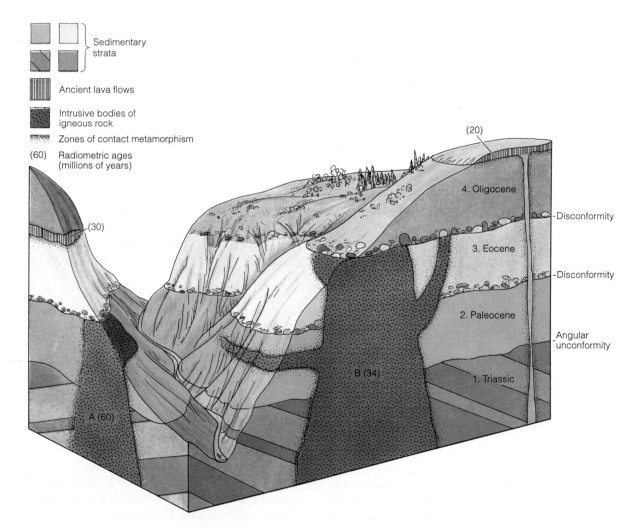

Stratum	Age (millions of years)	Interpretation
4	<34 (younger than *B*) <30 (younger than *C*) >20 (older than *D*)	Age lies between 20 and 30 million years
3	<60 (younger than *A*) >34 (older than *B*) >30 (older than *C*)	Age lies between 34 and 60 million years
2 1	>60 (older than *A*) >60 (older than *A*)	Age of both is greater than 60 million years

Through a combination of geologic relations and radiometric dating methods, twentieth-century scientists have been able to fit a scale of absolute time to the geologic column worked out in the nineteenth century. The scale is being continuously refined, and so the numbers given in Figure 6.11 should be considered the best available now. Further work will make them more accurate.

It is a great tribute to the work of geologists during the nineteenth century that the geologic column they established by the ordering of strata into relative ages has been fully confirmed by radiometric dating. It is interesting too, to see just how wrong Lord Kelvin was in his estimate of 100 million years for the age of the Earth.

The earliest record, as indicated in Figure 6.11, comes from the great assemblage of metamorphic and igneous rocks formed during the Precambrian that all lack fossils with hard parts. Of the many radiometric dates obtained from them, the youngest are around 600 million years, the oldest about 3.9 billion years. The Precambrian unit of the geologic column, then, existed during a *minimum* time equal to 3.9 billion minus 600 million years, or 3.3 billion years—a span about five times as long as the entire Phanerozoic Eon.

Given that some Precambrian rocks are nearly 4 billion years old, the beginning of the Earth's history must be still farther back in time. The oldest radiometric dates, 4.1 billion years, have been obtained on individual mineral grains in clastic sedimentary rocks from Australia. Dates that are almost as old—3.9 billion years—have been obtained from granitic rocks from Canada. The existence of ancient granite proves that continental crust was present 3.9 billion years ago while the 4.1 billion year sedimentary grains prove that the rock cycle was operating then. Further confirmation of the ancient age of continental crust comes from another body of very ancient Precambrian rock, a 3.6 billion-year-old granite in South Africa. Although

itself an igneous rock, this ancient granite contains xenoliths of quartzite. At an earlier time, before it became enveloped by the granite magma, the quartzite must have been part of a layer of sandstone. Before that, it must have been part of a layer of loose sand. And even earlier still, an igneous rock must have been subjected to weathering and erosion to produce the grains of quartz sand. Clearly, therefore, the rock cycle must have been operating in its present manner well before the granite magma solidified. Hence, as far back as we can see through the geologic column, we find evidence of the rock cycle and, because we see ancient sediment that must have been transported by water, we know that when that sediment was deposited there must have been a hydrosphere.

The ancient rocks we have been speaking of are all Archean. How long did the Hadean Eon last and therefore how much older might our planet be? Strong evidence suggests that the Earth formed at the same time as the Moon, the other planets, and meteorites (small independent bodies that have "fallen" onto the Earth). Through various methods of radiometric dating and, in particular, the Rb/Sr and U/Pb systems, it has been possible to determine the ages of meteorites and of "Moon dust" (brought back by astronauts) as 4.6 billion years. By inference, the time of formation of the Earth, and indeed of all the other planets and meteorites in the solar system, is believed to be 4.6 billion years ago.

THE MAGNETIC POLARITY TIME SCALE

The Earth is like a gigantic magnet. It has an invisible magnetic field that permeates everything. Because of an unusual characteristic of the Earth's magnetism, it is possible to use the magnetic properties of rocks as a dating technique. If a small magnet is allowed to swing freely in the Earth's magnetic field, the magnet will become oriented so that the axis of the small magnet points to the Earth's magnetic north pole (Fig. 6.17). This is true for all places on the Earth. All free-swinging magnets will point to the north magnetic pole. Certain rocks became permanent magnets as a result of the way they form, and like free-swinging magnets they point to the north magnetic pole. That property leads them to be useful as dating tools in the following manner.

Magnetism in Rocks

Magnetite and certain other iron-bearing minerals can become permanently magnetized. This property arises because orbital electrons spinning around a nucleus

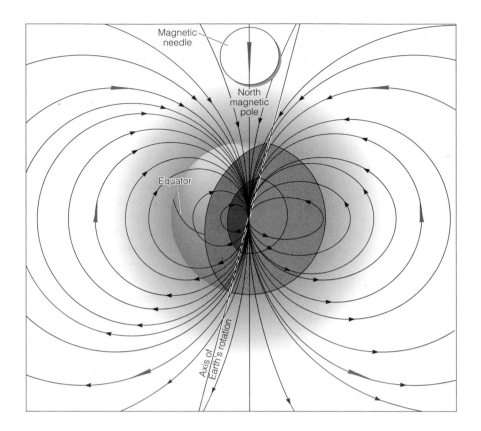

FIGURE 6.17 The Earth is surrounded by a magnetic field generated by the magnetism of the core. A small magnet, if allowed to swing freely, will always line up parallel to the magnetic field and point to the magnetic north pole.

create a tiny atomic magnetic field. In minerals that can become permanent magnets, the atomic magnets line up in parallel arrays and reinforce each other. In nonmagnetic minerals, the atomic magnets are oriented in random directions.

Above a temperature called the **Curie point,** the thermal agitation of atoms is such that permanent magnetism is impossible. The Curie point for magnetite is about 580°C. Above that temperature, the magnetic fields of all the iron atoms are randomly oriented and cancel each other out. Below the Curie point, the magnetic fields of adjacent iron atoms reinforce each other (Fig. 6.18). When an external magnetic field is present, all magnetic domains (regions) parallel to the external magnetic field become larger and expand at the expense of adjacent, nonparallel domains. Quickly, the parallel domains become predominant, and a permanent magnet is the result.

Consider what happens when lava cools. All the minerals crystallize at temperatures above 700°C—well above the Curie points of any magnetic minerals present. As the crystallized lava continues to cool, the temperature will drop below 580°C, the Curie point for magnetite. When the temperature drops below the Curie point all the magnetite grains in the rock be-

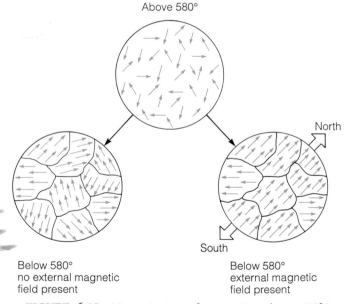

FIGURE 6.18 Magnetization of magnetite. Above 580°C (the Curie point), the thermal motion of atoms is so great that the magnetic poles of individual atoms, shown as arrows, point in random directions. Below 580°C, atoms in small domains influence one another and form tiny magnets. In the absence of an external field, the domains are randomly oriented. In the presence of a magnetic field (lower right), most pole directions tend to be parallel to that of the external field and magnetite becomes permanently magnetized. The example in the lower right is that for a magnetitie grain in a crystallizing lava.

come tiny permanent magnets with the same polarity as the Earth's field. Grains of magnetite locked in a lava cannot move and reorient themselves the way a freely swinging bar magnet can. So long as that lava lasts, therefore (until it is destroyed by weathering or metamorphism), it will carry a record of the Earth's magnetic field at the moment it passed through the Curie point.

Sedimentary rocks can also acquire weak but permanent magnetism through the orientation of magnetic grains during sedimentation. As clastic sedimentary grains settle through ocean or lake water, or even as dust particles settle through the air, any magnetite particles present will act as freely swinging magnets and orient themselves parallel to the magnetic lines of force caused by the Earth's magnetic field. Once locked into a sediment, the grains make the rock a weak permanent magnet.

The Polarity-Reversal Time Scale

From a study of magnetism in lavas, it was discovered early in the twentieth century that some rocks contain a record of reversed polarity. That is, when their magnetism was measured, some lavas indicated a south magnetic pole where the north magnetic pole is today, and vice versa (Fig. 6.19). Just why the Earth's poles reverse polarity is not yet understood, but the fact that they do provides a very useful dating technique. The ages of lavas can be accurately determined using radiometric dating techniques, especially the $^{40}K/^{40}Ar$ method. Through combined radiometric dating and magnetic polarity measurements in thick piles of lava extruded over several million years, it has been possible to determine when magnetic polarity reversals occurred (Fig. 6.20). A detailed record of all changes back to the Jurassic Period has now been assembled,

and still earlier reversals are the topic of ongoing research.

The polarity record for the past 20 million years is shown in Figure 6.20. Periods of predominantly normal polarity (as at present), or predominantly reversed polarity, are called **magnetic chrons.** The four most recent chrons have been named for scientists who made great contributions to studies of magnetism: Brunhes, Matuyama, Gauss, and Gilbert.

Use of magnetic reversals for geological dating differs from other dating methods. One magnetic reversal looks like any other in the rock record. When evidence of a magnetic reversal is found in a sequence of rocks, the problem is to know which of the many reversals it actually is. Additional information is needed. When a continuous record of reversals can be found, starting with the present, it is simply a matter of counting backward. This is the technique used in the dating of oceanic crust which is discussed in chapter 16.

Magnetism in sedimentary rocks has proved to be a very sensitive and important dating technique. When fossils are present, an approximate age can be given to sedimentary rocks. Knowing an approximate age, geologists can determine the exact age from the magnetic reversals. Sediment cores recovered from the seafloor can be dated very accurately using a combination of fossils and magnetic reversals. The measurements are so good that magnetic reversals can even provide an accurate way to measure rates of sedimentation in the world ocean.

In the later chapters of this book many examples of actual rates of geological processes are mentioned. Few if any examples would be possible without the absolute dates obtained through magnetic and radiometric dating. The ability to determine absolute dates has, more than any other contribution by geologists, changed the way we humans think about the world.

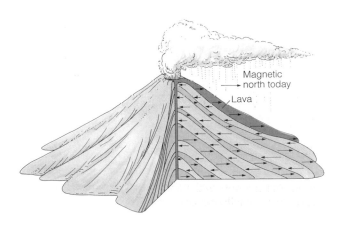

FIGURE 6.19 Lavas retain a record of the polarity of the Earth's magnetic field at the instant they cool through the Curie point. A pile of lava flows, like those in the volcanoes of the Hawaiian islands, may record several field reversals, each of which can be dated using the potassium–argon dating method.

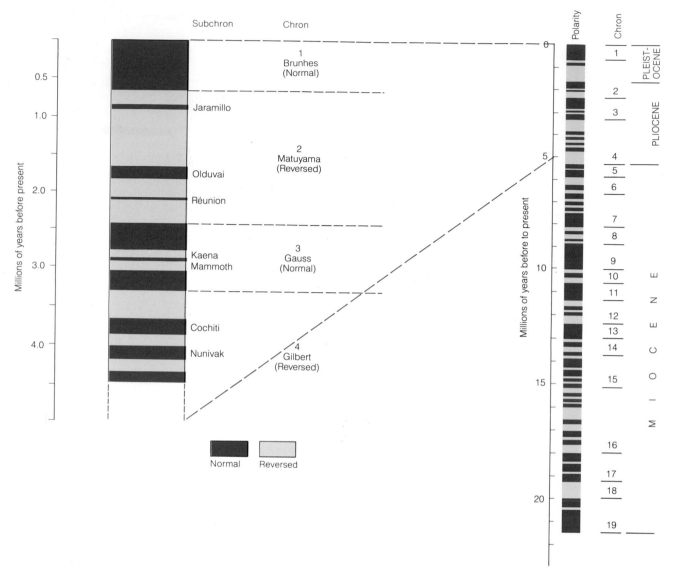

FIGURE 6.20 Polarity reversals back 20 million years. Period of normal polarity, as today, and periods of primarily reversed polarity are called magnetic chrons. Nineteen magnetic chrons have been identified to the beginning of the Miocene. Within each chron, one or more subchrons may occur. During a normal chron, subchrons are times of reversed polarity. Polarity reversals have been dated back to the mid-Jurassic, approximately 162 million years ago.

DATING PREHISTORIC EVENTS

The ability to attach dates to events in the geologic record has been an enormous boon for geologists. It is now possible to obtain quantitative answers to many questions that a few years ago could only be approached in a descriptive way. Two examples demonstrate how radiometric dating can be used.

During the recent ice age, ice caps and glaciers existed in many places that are ice-free today. The problem is to determine whether the age of the ice was everywhere the same. The age problem is especially severe on oceanic islands. On Mauna Kea, a 4200 m high shield volcano on Hawaii, an ice cap formed, then expanded and contracted several times. Mauna Kea is now dormant, but it was active when the ice cap existed. On the slopes of the volcano, flows of potassium-rich lavas are interlayered with sediment deposited by the ice as shown in Figure B6.1. The lavas have been dated by the $^{40}K/^{40}Ar$

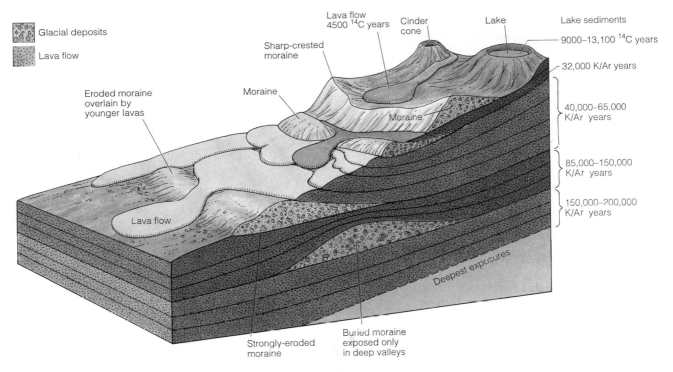

FIGURE B6.1 An example of radiometric dating. Glacial deposits on Mauna Kea Volcano, Hawaii, are interlayered with basaltic lava flows that have been dated by the ^{40}K/^{40}Ar method. See text for discussion.

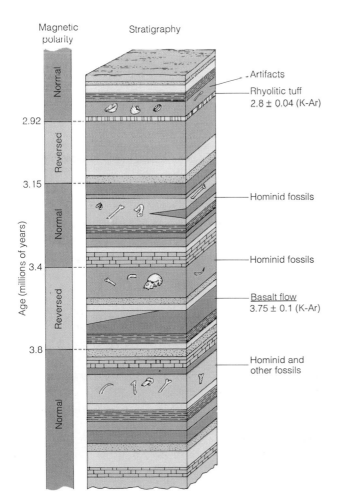

method; organic matter in sediment from a postglacial lake has been dated by the ^{14}C method.

The lavas consist of multiple flows, so ranges of ages were obtained from each set of flows. The ^{40}K/^{40}Ar ages given in Figure B6.1 are the youngest and oldest obtained for each set. The oldest glacial sediments are close to 150,000 years, which is the age of bracketing lavas; the second oldest sediment is bracketed by ages of about 85,000 and 65,000 years, and the youngest by a ^{40}K/^{40}Ar age of 32,000 years, and ^{14}C ages of 9000 to 13,000 years, the age of organic matter in lake sediment. The dates establish that glacial ages in Hawaii are essentially contemporaneous with glacial ages on North America.

A second dating example concerns the Haddar region of northern Ethiopia, one of the most productive places in the world for finding fossils of ancestral human beings (hominids). The sediments give good magnetic signals (Fig. B6.2).

FIGURE B6.2 Example of several dating techniques brought to bear on a geological problem.

The problem is to know where in the magnetic polarity time scale the Haddar reversals fall. $^{40}K/^{40}Ar$ dates were obtained on a tuff and a basalt flow. The two radiometric dates determine the magnetic reversal ages unambiguously and indicate that early hominids lived in the region until about 3.15 million years ago; that is, they lived there during the Pliocene Epoch.

SUMMARY

1. Strata provide a basis for reconstructing Earth history and past surface environments. Most strata were horizontal when deposited (law of original horizontality), and all strata accumulated in sequence from bottom to top (principle of stratigraphic superposition).

2. Stratigraphic superposition concerns relative time. The relative ages of two strata can be fixed according to whether one of the layers lies above or below the other.

3. Unconformities are physical breaks in a stratigraphic sequence marking a period of time when sedimentation ceased and erosion removed some of the previously laid strata. Angular unconformity results when rocks are disturbed by tectonic activity prior to deposition of overlying strata.

4. A formation is a fundamental rock unit for field mapping distinguished on the basis of its distinctive physical characteristics and named for a geographic locality.

5. Systems are rock sequences that accumulated during a specified time interval. Systems are the primary time-stratigraphic units used to construct the geologic column.

6. Geologic time units are based on time-stratigraphic units and represent the time intervals during which the corresponding systems accumulated.

7. Correlation of strata from place to place is based on physical and biological criteria that permit demonstration of time equivalence. Reliability of correlation is greatest if several criteria are used.

8. The geologic column is a composite section of all known strata, arranged on the basis of their contained fossils or other age criteria.

9. The geologic time scale is a hierarchy of time units established on the basis of corresponding time-stratigraphic units. Systems (time-stratigraphic units) and periods (geologic-time units) are based on type sections or type areas in Europe and North America. The geologic time scale constitutes the global standard to which geologists correlate local sequences of strata.

10. Decay of radioactive isotopes of various chemical elements is the basis of radiometric dating. The main radioactive isotopes and their daughters are $^{49}K/^{40}Ar$, $^{238}U/^{206}Pb$, $^{235}U/^{207}Pb$, $^{232}Th/^{208}Pb$, $^{87}Rb/^{87}Sr$, and $^{14}C/^{14}N$.

11. A sedimentary rock layer can be dated radiometrically when it is bracketed between two bodies of igneous rock to which a radiometric dating method can be applied.

12. Radiocarbon dating is only effective for relatively young organic materials (less than 70,000 years).

13. The age of the Earth, determined by uranium-lead dating, is 4.6 billion years.

14. Magnetism in rocks and the polarity-reversal time scale are useful for dating oceanic crust, lavas, and young sedimentary rocks.

IMPORTANT TERMS TO REMEMBER

angular unconformity (p. 164)
Archean Eon (p. 169)

Cenozoic Era (p. 169)
conformity (p. 163)
correlation (of strata) (p. 167)
Curie point (p. 180)

daughter (from radioactive decay) (p. 173)
disconformity (p. 164)

eon (p. 169)
epoch (p. 171)
era (p. 169)

formation (p. 165)

geologic column (p. 169)

Hadean Eon (p. 169)
half-life (p. 174)

isotope (p. 172)

magnetic chron (p. 181)
mass number (p. 172)
Mesozoic Era (p. 169)

nonconformity (p. 164)

original horizontality (law of)
(p. 162)

Paleozoic Era (p. 169)
parent (radioactive) (p. 173)
period (geologic) (p. 170)
Phanerozoic Eon (p. 169)
Proterozoic Eon (p. 169)

stratigraphy (p. 162)
stratigraphic superposition
(principle of) (p. 162)
system (p. 167)

unconformity (p. 164)

QUESTIONS FOR REVIEW

1. How do the law of original horizontality and the principle of stratigraphic superposition help geologists unravel the history of deformed belts of sedimentary rock? Is the history so determined known in absolute or relative time?

2. What geologic events are implied by an angular unconformity? By a disconformity?

3. How does a rock-stratigraphic unit differ from a time-stratigraphic unit?

4. How can strata be correlated from place to place? Can you identify an important geological advance that came about through correlation?

5. What is the geologic column? Is the column in North America the same as the column in Australia? How could you prove that your answer is correct?

6. How is a newly discovered formation placed in its correct position in the geologic column?

7. The geologic time scale is divided into four eons. Name them in the correct order, starting with the most ancient.

8. What are the three eras of the Phanerozoic Eon and what do their names mean?

9. What features make radioactivity an ideal way to measure geologic time? Would the radioactivity ages of a rock on the Moon and one on the Earth, formed at the same instant be the same? Why?

10. What radiometric dating method would be suitable for obtaining the age of (a) A rhyolite thought to about 100 million years old; (b) An Archean granite containing uraninite; and (c) Charcoal from an archaeological site thought to be about 10,000 years old.

11. How has ^{14}C been used to date the advance of the last great ice sheet in central North America?

12. What is the Curie point and why is it important for magnetic dating?

13. Polarity reversals of the Earth's magnetic field are recorded in rocks in two different ways. What are they?

14. How can magnetic polarity reversals be used to determine the timing of past geologic events?

15. Can you explain why the vast time span known as the Precambrian does not have a detailed geologic time scale of periods and epochs? Which dating schemes would you consider using in Precambrian igneous rocks?

16. The radiometric age for a sheetlike mass of igneous rock is 20 million years. The sheet of igneous rock is parallel to the layering of a sandstone below, a shale above. How could you tell if the sheet is a lava flow or a sill? What could you say about the age of the sandstone if the sheet is a sill? A lava flow?

17. Uranium has an atomic number of 92. When ^{238}U decays to lead it does so in a series of steps, emitting in the process 8 α-particles and 6 β-particles. What are the atomic and mass numbers of the daughter isotope of lead?

18. The half-life of ^{235}U decaying to ^{207}Pb is 710 million years. What is the age of a grain of zircon that contains $^{235}U/^{207}Pb$ atoms in the ratio 3:1? What would the age be if the ratio were 1:3?

PART II

Processes that Shape the Earth's Surface

Kaskawulsh glacier

Hawaiian coast

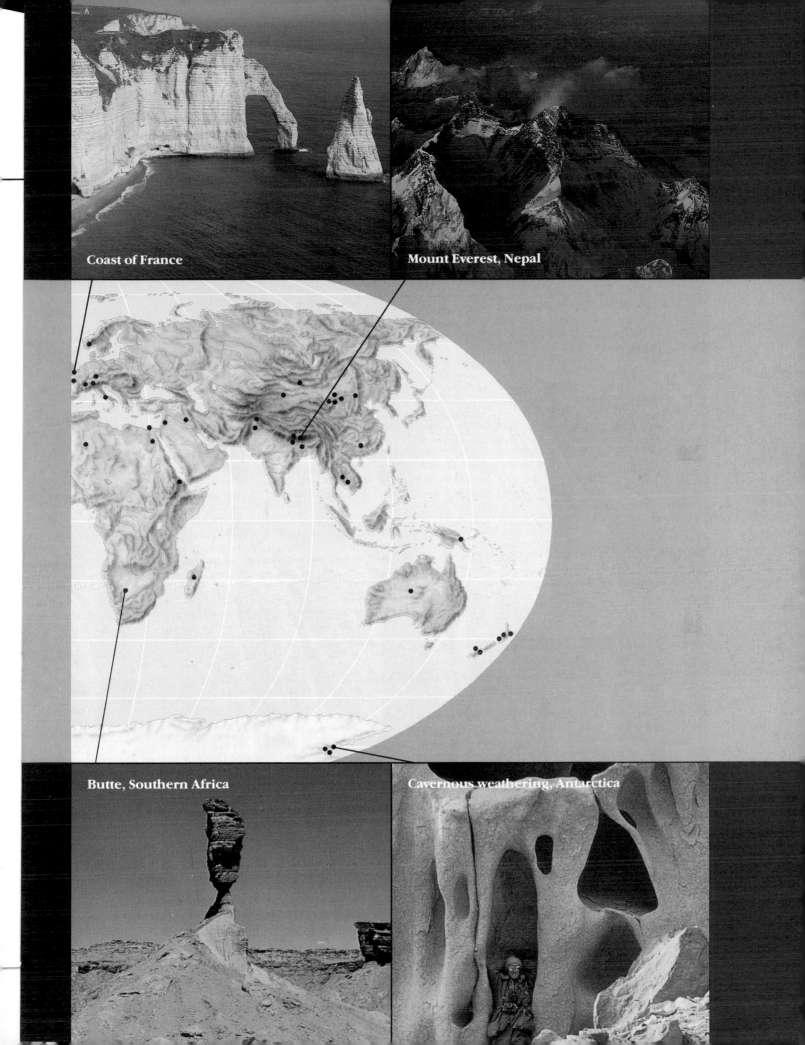

Coast of France

Mount Everest, Nepal

Butte, Southern Africa

Cavernous weathering, Antarctica

WEATHERING

Whether it occurs rapidly or slowly, mechanical and chemical alteration of rock takes place throughout the zone where the lithosphere, hydrosphere, biosphere, and atmosphere mix. The zone extends downward into the ground to whatever depth air and water penetrate. Within it, the rock constitutes a porous framework, full of fractures, cracks, and other openings, some of which are very small but all of which make the rock vulnerable. This open framework is continually being attacked, both chemically and physically, by water solutions. Given sufficient time, the result is conspicuous alteration of the rock.

FIGURE 7.1 A roadcut on the island of Hawaii exposes a weathering profile through a thick lava flow. Loose, earthy regolith grades downward into discolored, weathered rock that retains its organized appearance. Fresh, virtually unweathered rock is seen at the base of the exposure near the figure.

When exposed to the atmosphere, no rock (whether bedrock or structures built of stone) escapes the effects of **weathering,** the chemical alteration and mechanical breakdown of rock and sediment when exposed to air, moisture, and organic matter. The results of weathering are often seen in landslide scars and large excavations that expose bedrock. In Figure 7.1, loose, unorganized earthy regolith in which the texture of the bedrock is no longer apparent grades downward into rock that has been altered but still retains its organized appearance, and then into unaltered bedrock. It is evident from such exposures that alteration of the rock progresses from the surface downward.

A close look at the bedrock in Figure 7.1 would show that near the bottom of the exposure, the cleavage surfaces of feldspar grains are bright and reflective, while higher up such surfaces are lusterless and stained. Soft, earthy material near the top of the outcrop no longer resembles the former feldspar, which has largely decomposed. The changes that have occurred result mainly from **chemical weathering,** the decomposition of rocks and minerals as chemical reactions transform them into new chemical combinations that are stable at or near the Earth's surface.

Sometimes regolith consists of fragments identical to the adjacent bedrock. The mineral grains are fresh or only slightly altered. Piles of loose rock fragments are commonly found at the base of bedrock cliffs from which the debris quite obviously has been derived. When compared with the bedrock, the coarse rock fragments show little or no evidence of chemical weathering, implying that bedrock can be broken down not only chemically but also mechanically. Although we consider **mechanical weathering,** which is the disintegration (physical breakup) of rocks, as being distinct from chemical weathering, the two processes generally work hand in hand, and their effects are inseparably blended.

Mechanical Weathering

Mechanical weathering of rock is common in nature and takes place whenever crystals of ice or salt grow in rock fractures, whenever rock is heated by fire, and whenever the growth of plant roots disrupts rock.

Development of Joints

Rock masses buried deep beneath the ground surface are subjected to enormous confining pressures due to the weight of overlying rock. As erosion wears down the surface, the weight of overlying rock, and therefore the pressure, is reduced. The rock adjusts to this unloading by expanding upward. As it does so, singly. Most commonly they occur as a widespread set of parallel joints (Fig. 7.3). Intersecting joints strongly influence the way a rock breaks apart. Once formed, joints act as passageways by which rainwater can enter a rock and promote both mechanical and chemical weathering.

One class of joints is restricted to tabular bodies (having a flat shape, like a table top) of igneous rock—such as dikes, sills, lava flows, and welded tuffs—that cooled rapidly at or close to the land surface. When such a body of igneous rock cools, it contracts and may fracture into pieces, in much the same way that a very hot glass bottle contracts and shatters when plunged into cold water. Unlike shattered glass, cooling fractures in igneous rock tend to form regular patterns. For joints that split igneous rocks into long prisms or columns, the term *columnar joints* is used (Fig. 7.4).

Crystal Growth

Groundwater moving slowly through fractured rocks contains ions that may precipitate out of solution to form salts. The force exerted by salt crystals growing either within rock cavities or along grain boundaries can be enormous and results in rupturing or disaggregation of rocks. The effects of such mechanical weathering can often be seen in desert regions, where salt crystals grow as rising groundwater evaporates and its dissolved salts are precipitated.

FIGURE 7.2 Sheetlike jointing in massive granite forms stair-step surface on a mountainside in Yosemite National Park, California.

closely spaced fractures can develop that give the rock the appearance of a stacked deck of cards (Fig. 7.2). These are **joints,** defined as fractures in a rock along which no observable movement has occurred. Generally joints disappear below a depth of about 50 m, as the thickness and pressure of overlying rock become too great for joints to form. Rarely do joints occur

FIGURE 7.3 Three sets of joints, one horizontal and two vertical, intersect at nearly right angles to form a spectacular rocky vantage point, called the Pulpit, overlooking Lysefjord in southwestern Norway. A widening crack along a vertical joint suggests that the Pulpit eventually will collapse and plunge into the icy waters of the fjord far below.

FIGURE 7.4 Columnar jointing in igneous rock near San Miguel Regla, Mexico, offers a challenge to a rock climber jamming his way up a crack between two adjacent columns.

FIGURE 7.5 Granitic bedrock on the flank of Gondola Ridge in Antarctica has been so strongly weathered that it resembles Swiss cheese. Such cavernous weathering results from granular disintegration as salts crystallize in small cavities and along grain boundaries.

In the ice-free valleys of coastal Antarctica, bizarre cavernous landforms have been produced by the mechanical weathering of granite boulders and bedrock outcrops (Fig. 7.5). When salts crystallize out of solutions confined within the pores and fine cracks between mineral grains in the granite, the rock slowly disintegrates. The resulting debris is transported away by strong winds blowing off the nearby ice sheet, leav-

ing an unusual landscape in which the exposed rocks resemble Swiss cheese.

Frost Wedging

Wherever temperatures fluctuate about the freezing point for part of the year, water in the ground periodically freezes and thaws. When water freezes to form ice, its volume increases by about 9 percent. The high stresses resulting from this volume increase lead to disruption of rocks. As freezing occurs in the pore spaces of a rock, water is strongly attracted to the growing ice, thereby increasing the stresses against the rock. The process leads to a very effective type of mechanical weathering known as **frost wedging,** the formation of ice in a confined opening within rock, thereby causing the rock to be forced apart. Frost wedging is strong enough to force apart not only tiny particles but huge blocks, some weighing many tons (Fig. 7.6). Frost wedging probably is most effective at temperatures of −5° to −15°C. At higher temperatures ice pressures are too low to be very effective, and at lower temperatures the rate of ice growth drops because the water necessary for crack growth is less mobile.

Frost wedging is responsible for most of the rock debris seen on high mountain slopes. At lower altitudes this process is likely to be most effective wherever the number of yearly freeze–thaw cycles is high.

Effects of Heat

Some geologists have speculated that daily heating of rock in bright sunlight followed by cooling each night should cause mechanical breakdown of rocks because the common rock-forming minerals expand by different amounts when heated. Surface temperatures as high as 80°C have been measured on desert rocks, and daily temperature variations of more than 40° have been recorded on rock surfaces. Highest temperatures are achieved by dark-colored rocks, like basalt, and rocks that do not easily conduct heat inward. Nev-

FIGURE 7.6 An extensive block field covers the summit of Mount Whitney, the highest peak in California's Sierra Nevada. Frost wedging, the result of melting snow that percolates downward into cracks where it refreezes, has disrupted the granite bedrock, producing a vast litter of angular boulders.

FIGURE 7.7 A large forest fire, moving rapidly through a pine forest in Yellowstone National Park, Wyoming, has caused extensive spalling of a surface boulder. Fresh, light-colored spalls, which flaked off as heat caused differential expansion of the rock surface, litter the fire-blackened ground.

ertheless, despite a number of careful laboratory experiments, no one has yet demonstrated that such daily heating and cooling cycles have noticeable physical effects on rocks. These experiments, however, have been carried out only over relatively brief time intervals. Possibly, thermal fracturing occurs only after repeated, extreme natural temperature fluctuations over long periods of time.

Fire, on the other hand, can be very effective in disrupting rocks, as anyone knows who has witnessed a rock beside a campfire become overheated and shatter explosively. Because rock is a relatively poor conductor of heat, an intense fire heats only a thin outer shell, which expands and breaks away as a *spall*. The heat of forest fires and brush fires can lead to the spalling of rock flakes from exposed bedrock or boulders (Fig. 7.7). Studies of fire history in forested regions show that large natural fires, most started by lightning, may recur every several hundred years. Therefore, over long intervals of geologic time, fires may contribute significantly to the mechanical breakdown of surface rocks.

Plant Roots

Seeds germinate in cracks in rocks to produce plants that extend their roots farther into the cracks. As trees grow, their roots wedge apart adjoining blocks of bedrock. In much the same way, roots also disrupt stone pavements, sidewalks, garden walls, and even buildings (Fig. 7.8). Large trees swaying in the wind can cause cracks to widen, and, if blown over, can pry rock apart. Although it would be difficult to measure, the total amount of rock breakage done by plants must be very large. Much of it is obscured by chemical weathering, which takes advantage of the new openings as soon as they are created.

Chemical Weathering

Minerals in igneous and metamorphic rocks that formed at high temperatures and pressures are chemically unstable when exposed to the lower temperatures and pressures at the Earth's surface. Such minerals break down and their components form new, more stable minerals.

The principal agents of chemical weathering are water solutions that behave as weak acids. The effects of chemical weathering are most pronounced in regions where both precipitation and temperature are high, two conditions that promote chemical reactions.

FIGURE 7.8 Tree roots force apart the walls of ruined buildings of Ta Prohm at Angkor in the Cambodian jungle. In the same manner, roots growing in cracks in bedrock force the rock apart.

Chemical Weathering of Rock-Forming Minerals

As rainwater falls through the atmosphere, it dissolves small quantities of carbon dioxide, producing **carbonic acid.** As the slightly acid water moves downward and laterally through the soil, additional carbon dioxide is dissolved from decaying vegetation. The carbonic acid ionizes to form hydrogen ions and bicarbonate ions (Table 7.1, reaction 1). The hydrogen ions are so small they can enter a crystal and replace other ions, thereby changing the composition.

Hydrolysis.   The effectiveness of the H^{1+} ion in decomposing minerals is illustrated by the way potassium feldspar, a common rock-forming mineral, is decomposed by carbonic acid (Table 7.1, reaction 2). The H^{1+} ions enter the potassium feldspar and replace potassium ions, which then leave the crystal and pass into solution. Water combines with the remaining aluminum-silicate molecule to create the clay mineral **kaolinite.** The kaolinite was not present in the original rock and was created by the chemical reaction. Kaolinite, a common member of the group of very insoluble minerals that constitute clay, forms a substantial part of the regolith. Such a chemical reaction, in which the H^{1+} or OH^{1-} ions of water replace ions of a mineral, is called **hydrolysis.** It is one of the chief processes involved in the chemical breakdown of common rocks.

Leaching.   Another common process of chemical weathering is **leaching,** the continued removal, by water solutions, of soluble matter from bedrock or regolith. For example, when silica is released from rocks by chemical weathering, some of it remains in the clay-rich regolith and some is slowly taken into solution by water moving through the ground. Many of the potassium ions weathered from the rock also escape in solution. Soluble substances leached from rocks during weathering are present in all surface water and groundwater. Sometimes their concentrations are high enough to give the water an unpleasant taste.

Oxidation.   Iron is a normal constituent of many common rock-forming minerals, including biotite, augite, and hornblende. When any of these minerals is chemically weathered, iron is released and rapidly oxidized from Fe^{2+} to Fe^{3+} if oxygen is present. Typically this results in the growth of a yellowish mineral, **goethite** (Table 7.1), through a combination of oxidation and **hydration,** the incorporation of water into a crystal structure. Goethite may later be dehydrated (which means it loses water) to form **hematite,** a brick-red mineral (Table 7.1). The intensity of these

---

TABLE 7.1   Common Chemical Weathering Reactions

1. Production of carbonic acid by solution of carbon dioxide:

$$H_2O \ + \ CO_2 \ \rightleftharpoons \ H_2CO_3 \ \rightleftharpoons \ H^{1+} \ + \ HCO_3{}^{1-}$$

Water   Carbon   Carbonic   Hydrogen   Bicarbonate
         dioxide   acid   ion   ion

2. Hydrolysis of potassium feldspar:

$$4KAlSi_3O_8 \ + \ 4H^{1+} \ + \ 2H_2O \ \rightarrow \ 4K^{1+} \ + \ Al_4Si_4O_{10}(OH)_8 + 8SiO_2$$

Potassium   Hydrogen   Water   Potassium   Kaolinite   Silica
feldspar   ions   ions

3. Oxidation of iron (Fe^{2+}) oxide to form goethite:

$$4FeO + 2H_2O \ + \ O_2 \ \ \rightarrow \ \ 4FeO{\cdot}OH$$

Iron   Water   Oxygen   Goethite
oxide

4. Dehydration of goethite to form hematite:

$$2FeO{\cdot}OH \ \rightarrow \ Fe_2O_3 \ + \ H_2O$$

Goethite   Hematite   Water

5. Dissolution of carbonate minerals by carbonic acid:

$$CaCO_3 \ + \ H_2CO_3 \ \rightarrow \ Ca^{2+} \ + \ 2(HCO_3)^{1-}$$

Calcium   Carbonic   Calcium   Bicarbonate
carbonate   acid   ion   ions

colors in weathered rocks and soils can provide clues to how much time has elapsed since weathering began and to the degree or intensity of weathering.

Effects of Chemical Weathering on Common Rocks

The minerals and soluble ions that result when an igneous rock weathers chemically depend on the original mineral composition of the rock. Granite has a higher silica content than basalt, and a different mineralogical makeup. A typical granite contains quartz, which is relatively inactive chemically, potassium-bearing minerals such as muscovite and potassium feldspar, and minerals rich in iron and magnesium. When a granite decomposes by hydrolysis (Table 7.1, eq. 2), the feldspar, mica, and Fe/Mg minerals weather to clay minerals and soluble Na^{1+}, K^{1+}, and Mg^{2+} ions (Fig. 7.9). The quartz grains, being relatively inactive chemically, remain essentially unaltered. Because basalt lacks both quartz and potassium feldspar, quartz grains and K^{1+} ions are not among its chemical weathering products. Like granite, its feldspar and Fe/Mg minerals weather to clay minerals and soluble ions (Na^{1+}, Ca^{2+}, and Mg^{2+}), while iron-rich magnetite weathers to goethite.

Carbonate rocks, such as limestone, are chemically weathered in a different way. Limestone consists mainly of calcium carbonate, which is only slightly soluble in water. In the presence of carbonic acid, the calcium and bicarbonate ions dissolve (Table 7.1, eq. 5), leaving behind only the nearly insoluble impurities (chiefly clay and quartz) that are always present in small amounts in limestone. Therefore, as limestone weathers chemically, the residual regolith that develops from it consists mainly of clay and quartz.

Concentration of Stable Minerals

Not only quartz but a number of other minerals are extremely resistant to chemical attack at the Earth's surface. Minerals such as gold, platinum, and diamond persist in weathered regolith, are eroded, and become sediment. Because some of these minerals have higher specific gravities than common minerals such as quartz, they concentrate at the beds of streams or along ocean beaches (chapters 9 and 13). Some may end up sufficiently concentrated to form mineral deposits of economic value.

Weathering Rinds

If you crack open a cobble of weathered basalt, you usually will see a light-colored rind surrounding a darker core of unaltered rock (Fig. 7.10). The rind is composed of the solid products resulting from chemical weathering and is called a **weathering rind.** Weathering begins at the freshly exposed surface of a dark, unaltered cobble and proceeds slowly inward. Commonly this involves oxidation of iron-rich minerals to produce goethite (Table 7.1, eq. 3), which imparts a light brownish color to the developing rind. Such a rind forms on all but the most chemically stable

FIGURE 7.9 When a basalt weathers chemically, its silicate minerals and magnetite are converted to clay minerals, goethite, and soluble cations. The weathering products of a granite not only include clay minerals, geothite, and soluble cations, but grains of quartz that are resistant to chemical breakdown.

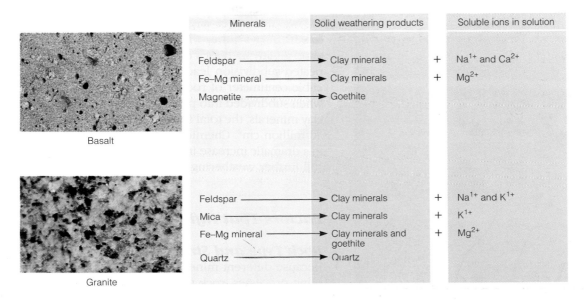

Minerals	Solid weathering products	Soluble ions in solution
Feldspar →	Clay minerals	+ Na^{1+} and Ca^{2+}
Fe–Mg mineral →	Clay minerals	+ Mg^{2+}
Magnetite →	Goethite	

Basalt

Minerals	Solid weathering products	Soluble ions in solution
Feldspar →	Clay minerals	+ Na^{1+} and K^{1+}
Mica →	Clay minerals	+ K^{1+}
Fe–Mg mineral →	Clay minerals and goethite	+ Mg^{2+}
Quartz →	Quartz	

Granite

composition. Quartz is so resistant to chemical breakdown that rocks rich in quartz are also resistant. In the Appalachian Mountains, resistant quartzite strata form ridges that stand prominently above valleys that are underlain by more erodible rocks containing less quartz. Granite also is resistant to weathering, for it consists of minerals (such as quartz, muscovite, and potassium feldspar) that are resistant to chemical breakdown (Fig. 7.9). Like quartzite, granite typically forms hilly or mountainous terrain.

The rate of weathering of a rock is influenced also by its texture and structure. Even a rock that consists entirely of quartz may break down rapidly if it contains closely spaced joints or other partings that make it susceptible to frost action.

Contrasts in local topography may be due to *differential weathering,* which means that weathering has occurred at different rates as a result of variations in the composition and structure of rocks (Fig. 7.14). In a sequence of alternating shale and quartz sandstone, for instance, the shale is likely to weather more easily, leaving the sandstone beds standing out in relief. If the beds are horizontal, the result is likely to be a

FIGURE 7.14 Differential weathering of Miocene marine strata at Nelson, New Zealand, has etched away erodible mudstone from between layers of siltstone, leaving siltstone standing as ridges above a wave-eroded platform.

stepped topography, with the sandstone forming abrupt cliffs between more gentle slopes of shale. If the bedding is inclined, the sandstone will stand as ridges separated by linear depressions underlain by shale.

Slope

A mineral grain loosened by weathering on a steep slope may be washed downhill by the next rain. On such a slope the solid products of weathering move quickly away, continually exposing fresh bedrock to renewed attack. As a result, weathered rock seldom extends far beneath the surface. On gentle slopes, however, weathering products are not easily washed away and in places may accumulate to depths of 50 m or more.

Climate

Moisture and heat promote chemical reactions. Not surprisingly, therefore, weathering is more intense and generally extends to greater depths in a warm, moist climate than in a cold, dry one (Fig. 7.15). In moist tropical lands, like Central America or southeast Asia, obvious effects of chemical weathering can be seen at depths of 100 m or more. By contrast, in cold, dry regions like northern Greenland and Antarctica, chemical weathering proceeds very slowly. Instead, the effects of mechanical weathering are generally obvious, for bedrock surfaces typically are littered with rubble dislodged by frost action.

Dramatic contrasts in weathering can be seen in the case of carbonate rocks that crop out in different climatic regions. Rocks such as limestone and marble, which consist almost entirely of soluble calcite, are highly susceptible to chemical weathering in a moist climate and commonly form low, gentle landscapes. In a dry climate, however, the same rocks form bold cliffs because, with scant rainfall and only patchy vegetation, little carbonic acid is present to dissolve carbonate minerals.

Burrowing Animals

Large and small burrowing animals (for example, rodents and ants) bring partly decayed rock particles to the land surface, where they are exposed more fully to chemical action. More than 100 years ago, Charles Darwin made careful observations in his English garden and calculated that every year earthworms bring particles to the surface at the rate of more than 2.5 kg/m^2 (25 tons per hectare). After a study in the Amazon River basin, geologist J. C. Branner wrote that the soil there "looks as if it had been literally turned inside out by the burrowing of ants and termites." Although burrowing animals do not break down rock

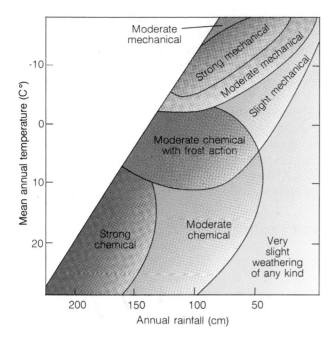

FIGURE 7.15 Climate plays a major role in controlling the type and effectiveness of weathering processes. Mechanical weathering is dominant where rainfall and temperature are both low. High temperature and precipitation favor chemical weathering. Places where some different weathering regimes predominate are shown on the map.

directly, the amount of disaggregated rock moved by them during many millions of years must be enormous. These examples illustrate once again the cumulative effect of small forces acting over vast intervals of geologic time.

Time
Studies of the decomposition of the stone in ancient buildings and monuments show that hundreds or even thousands of years are required for hard rock to decompose to depths of only a few millimeters. Granite and other hard bedrock surfaces in New England, Scandinavia, the Alps, and elsewhere still display polish and fine grooves made by ice-age glaciers before they disappeared about 10,000 years ago. In such regions, where cool climate and successive glaciations have significantly reduced the rate and effectiveness of chemical weathering, it takes many tens of thousands of years to create weathered regolith like that shown in Figure 7.1. However, in regions that were not repeatedly glaciated and have been continuously

exposed to weathering, the zone of weathering often extends to great depths. In some tropical areas, where weathering has continued for millions of years, mining operations have exposed bedrock that has been thoroughly decomposed to depths of 100 m or more.

The rates at which rocks weather have been estimated in several ways. First, experiments have been designed in which the length of the experiment provides time control and the processes were speeded up by increasing temperature and available water, and decreasing particle size. Second, studies have been made of the degree of weathering of ancient architectural structures of known age. Third, radiometric ages of rock or sediments that have been exposed to weathering for varying periods of time can provide estimates of average rates over geologically long intervals. The results of such investigations suggest that on geologic time scales, weathering rates that are initially rapid tend to decrease steadily with time as the weathering profile approaches a steady state, or equilibrium, condition (Fig. 7.16).

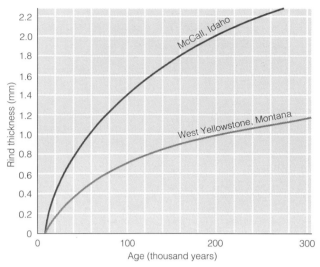

FIGURE 7.16 Graph showing change in weathering rates through time for two localities in the northwestern United States. The thickness of weathering rinds on basaltic stones is plotted as a function of their estimated or known age. Weathering rates are initially rapid during the first few tens of thousands of years (steep part of curves), but then decline steadily with increasing age; after several hundred thousand years, the rates are extremely low. The curve for McCall, Idaho, is steeper than the one for West Yellowstone. This indicates a higher rate of weathering, probably the result of a different climatic environment at the two sites.

SOILS

Soils are one of our most important natural resources. Soils support the plants that are the basic source of our nourishment and provide food for domesticated animals. Soils also help maintain the Earth's surface environment in a condition of relative equilibrium. They accomplish this by supporting vegetation, which, as it releases water to the atmosphere, is an integral part of the hydrologic cycle. In addition, soils store organic matter, thereby influencing the quantity of carbon that is cycled to the atmosphere as carbon dioxide, and they also trap pollutants.

Origin

The physical and chemical breakdown of solid rock by weathering processes is the initial step in the formation of soil. However, soil also contains organic matter mixed in with the mineral component. This organic fraction is essential in the usual definition of **soil:** the part of the regolith that can support rooted plants. Nevertheless, soils can develop even in the absence of vegetation, as seen in cold glacier-free areas fringing the Antarctic ice sheets.

The organic matter in soil is derived from the decay of dead plants and animals. Living plants are nour-ished by the nutrients released from decaying organisms, as well as by those released during weathering of mineral matter. Plants draw these nutrients upward, in water solution, through their roots. Therefore, through their life cycle, plants are directly involved in the manufacture of the fertilizer that will nourish future generations of plants. These activities are an integral part of a continuous cycling of nutrients between the regolith and biosphere. With its partly mineral, partly organic composition, soil forms an important bridge between the Earth's lithosphere and its teeming biosphere.

Soil Profile

As bedrock and regolith weather, soil gradually evolves. As a soil develops from the surface downward, an identifiable succession of subhorizontal weathered zones, called **soil horizons,** forms. Each horizon has distinctive physical, chemical, and biological characteristics. Although soil horizons may resemble a sequence of deposits, or layers, they are not strata. Instead, they represent physical and chemical changes to the regolith exposed at the land surface. Taken together, the soil horizons constitute a **soil profile,** which consists of the succession of soil horizons lying between the surface and the underlying **parent material** (the regolith from which a soil develops) (Fig. 7.17).

Soil profiles generally display two or more horizons. The uppermost horizon may be a surface accumulation of organic matter (*O horizon*) that overlies mineral soil. The organic matter is in various stages of decomposition.

An **A horizon** may either underlie an O horizon or lie directly beneath the surface. Typically the A horizon is dark grayish or blackish (at least near its top) because of the presence of *humus,* the decomposed residue of plant and animal tissues, which is mixed with mineral matter. The A horizon has lost some of its original substance through the downward transport of clay particles and, more important, through the chemical leaching of soluble minerals.

An *E horizon,* sometimes present beneath the A horizon, has a grayish or whitish color. The light color is mainly due to the lack of a darker oxide coating on light-colored mineral grains. E horizons commonly are found in acidic soils that develop beneath forests of evergreen trees.

The **B horizon** underlies the surface horizons and commonly is brownish or reddish. This horizon is enriched in clay and/or iron and aluminum hydroxides produced by the weathering of minerals within the horizon and also transported downward from

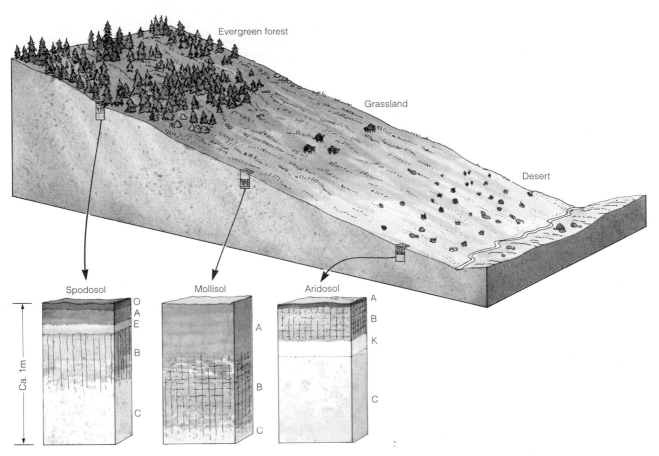

FIGURE 7.17 Examples of soil horizons developed under different climatic and vegetation conditions (See Table 7.2).

overlying A and E horizons. The B horizon often has a distinct structure that causes it to break into blocks or prisms. Where clay migration is an important process, each block or prism may be coated with clay. Although the B horizon generally is penetrated by plant roots, it contains less organic matter than the humus-rich A horizon.

A *K horizon,* present in some arid-zone soils beneath the B horizon, is densely impregnated with calcium carbonate which coats all mineral grains and constitutes up to 50 percent of the volume of the horizon.

The **C horizon** is the deepest horizon and constitutes parent material in various stages of weathering, but it lacks the distinctive properties of the A and B horizons. Oxidation of parent material in the C horizon generally imparts a light yellowish-brown color.

Soil Types

An astute observer traveling across the landscape will note that soils are not everywhere the same. Different soils result from the influence of six soil-forming factors: climate, vegetation cover, soil organisms, composition of parent material, topography, and time. The soil forming under prairie grassland differs from soil in a boreal forest or that of a tropical rainforest. The character of a soil may change abruptly as we move from basalt to limestone or from a gentle slope to a steep slope, and it also will change with the passage of time.

Soil scientists classify soils according to their physical and chemical properties in much the same way that geologists classify rocks. Such classification makes it easier to map, study, and understand how soils are distributed on the landscape and how their properties reflect the six soil-forming factors (Fig. 7.18). In the soil classification scheme now in standard use in the United States, soils are classified into ten orders, distinguished on the basis of easily recognizable characteristics (Table 7.2). An additional order—the Andisols—has been proposed for soils developed on volcanic ash.

TABLE 7.2   Orders of the Soil Classification System Used in the United States

Soil Order (Meaning of Name)	Main Characteristics
Alfisol (Pedalfer[a] Soil)	Thin A horizon over clay-rich B horizon, in places separated by light-gray E horizon. Typical of humid middle latitudes.
Aridosol (Arid Soil)	Thin A horizon above relatively thin B horizon and often with carbonate accumulation in K horizon. Typical of dry climates.
Entisol (Recent Soil)	Soil lacking well-developed horizons. Only a thin incipient A horizon may be present.
Histosol (Organic soil)	Peaty soil, rich in organic matter. Typical of cool, moist climates.
Inceptisol (Young soil)	Weakly developed soil, but with recognizable A horizon and incipient B horizon lacking clay or iron enrichment. Generally occurs under moist conditions.
Mollisol (Soft soil)	Grassland soil with thick dark A horizon, rich in organic matter. B horizon may be enriched in clay. E and K horizons may be present.
Oxisol (Oxide soil)	Relatively infertile soil with A horizon over oxidized and often thick B horizon.
Spodosol (Ashy soil)	Acidic soil marked by highly organic O and A horizons, an E horizon, and iron/aluminum-rich B horizon. Occurs in cool forest zones.
Ultisol (Ultimate soil)	Strongly weathered soil characterized by A and E horizons over highly weathered and clay-rich B horizon. Characteristic of tropical and subtropical climates.
Vertisol (Inverted soil)	Organic-rich soil having very high content of clays that shrink and expand as moisture varies seasonally.
Andisol (proposed) (Dark soil)	Soil developed on pyroclastic deposits and characterized by low bulk density and high content of amorphous minerals.

[a]Soils rich in iron and aluminum.

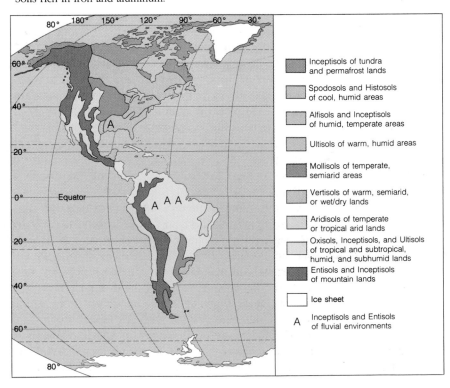

FIGURE 7.18   Map of the Western Hemisphere showing distribution of the major soil types.

Polar Soils

In cold, high-latitude deserts, like those found in Greenland, northern Canada, and Antarctica, soils generally are well drained and lack well-developed horizons (therefore classified as *Entisols*). Weakly oxidized parent material may underlie a layer of coarse frost-churned stones. In wetter high-latitude and high-altitude environments, matlike tundra vegetation overlies perennially frozen ground that prevents water from percolating downward. This leads to waterlogged soils that are rich in organic matter (*Histosols*). Only on well-drained sites do soils develop characteristic A and B horizons (*Inceptisols*). Because the cold climate retards chemical processes, well-drained soils generally do not develop the thick clay-enriched B horizon typical of highly developed temperate-latitude soils.

Temperate-Latitude Soils

Soils of the temperate zone vary largely in response to differences in climate and in the resulting vegetation cover. *Alifsols*, characteristic of deciduous woodlands, typically have a clay-rich B horizon beneath a light-gray E horizon. Acidic *Spodosols* developed in cool, moist evergreen forests have an organic-rich A horizon, an ashlike E horizon, and an iron-rich B horizon. Mountainous terrains, where a cool climate and steep, eroding slopes maintain low rates of soil development, frequently have minimally developed profiles (*Entisols*) or display weakly developed B horizons lacking clay enrichment (*Inceptisols*). Grasslands and prairies typically develop *Mollisols* having thick dark-colored, organic-rich A horizons. Soils formed in subtropical climates commonly display a strongly weathered B horizon (*Ultisols*).

Desert Soils

In dry climates, where lack of moisture reduces leaching, carbonates accumulate in the profile leading to the development of *Aridosols*. These strongly alkaline soils contrast with the more acid soils of humid regions that lack a carbonate-rich K horizon. An important part of the carbonate accumulation results from evaporation of water that rises in the ground, bringing dissolved salts from below. Recently, soil scientists have found that wind-blown dust also contributes to the accumulation of salts in arid-land soils. Over extensive arid regions of the southwestern United States, carbonates have in this way built up in the soil profile a solid, almost impervious layer of whitish calcium carbonate known as **caliche** (Fig. 7.19).

Tropical Soils

Soils forming where rainfall is high and the average temperature is very warm are characterized by ex-

FIGURE 7.19 A soil profile in semiarid central New Mexico includes whitish caliche forming a prominent K horizon between a yellowish-brown C horizon beneath and a reddish-brown B horizon above.

treme chemical alteration of the parent material. These soils (*Oxisols*) are intensely weathered and infertile because essential nutrients have been leached away. *Vertisols* of equatorial and tropical regions, where the climate alternates between a wet and a dry season, have a high clay content that causes the soil alternately to swell and shrink in response to seasonal wetting and drying. During the dry season, open cracks extend from the surface down into the soil profile.

Of the many minerals formed during chemical weathering, goethite and hematite are among the least soluble. In tropical regions where the climate is very wet and warm, rock-forming minerals are slowly leached away, leaving a soft, mottled, reddish-gray residue, rich in iron. Geologists generally refer to this product of deep weathering as **laterite.** One constituent of lateritic weathering is ferric hydroxide ($Fe(OH)_3$). As a result of climatic change or deforestation, the upper part of a laterite may dry out and become hardened because the ferric hydroxide dehydrates to goethite (Fe(OH)$_3$ → FeO·OH + H$_2$O). The resulting stonelike material, called *lateritic crust* or *ironstone*, is so hard that blocks of it can be used in construction (Fig. 7.20). The name laterite, which comes from the Latin word for brick (*latere*) recognizes this property.

Rate of Soil Formation

Although soil development is part of the complex process of chemical weathering, soil formation and weathering are not the same thing. Chemical weathering chiefly concerns the decomposition of bedrock, a process that takes a very long time. The time required to form a soil profile in regolith can be much shorter.

A soil profile can form rapidly in some environments. A study in the Glacier Bay area of southern Alaska shows us how soils develop when retreating glaciers leave unweathered parent material exposed at the land surface. Within a few years after a glacier

FIGURE 7.20   Reddish-brown blocks of laterite were used by the ancient Khymer people to construct a temple wall at Angkor Wat in the Cambodian jungle. The color and texture of the laterite contrasts with the smooth, grayish sandstone used to construct the adjacent ornately carved temple structure.

begins to recede, an A horizon develops on the newly exposed and revegetated landscape (Fig. 7.21). In this area, moderate temperatures and high rainfall promote rapid leaching of the parent material, which is carbonate-rich glacial sediment. As the plant cover becomes denser, carbonic and organic acids acidify the soil, and leaching becomes more effective. After about 50 years, a B horizon appears and the combined thickness of the A and B horizons reaches about 10 cm. Over the next 165 years, a mature forest develops on the landscape, the A and B horizons increase in thickness to 15 cm, and iron oxides accumulate in the developing B horizon. This entire succession of vegetation and soils is clearly visible on the deglaciated landscape in front of the retreating glaciers.

In less humid climates, soil forms more slowly, and it may take thousands of years for a detectable B horizon to appear. In the midcontinental United States, B horizons that have developed during the last 10,000 years contain little clay and lack structure, whereas those dating back about 100,000 years generally are rich in clay and have a distinctive structure. As an extreme example of the effect of wetness on rate of soil formation, the glacier-free cold deserts of Antarctica are so dry and cold that sediments more than a million years old have only weakly developed soils.

Paleosols

If a surface soil is buried, it becomes part of the geologic record. The soil is now a **paleosol**, defined as a soil that formed at the ground surface and subsequently was buried and preserved (Fig. 7.22). Its top is therefore an unconformity. Paleosols have been identified in rocks and sediments of many different ages, and are especially common in unconsolidated deposits of the Quaternary Period.

Distinctive and widespread paleosols have been used to subdivide, correlate, and date sedimentary sequences. They also can provide important clues regarding the nature of former landscapes, vegetation cover, and climate.

Soil Erosion

Global agricultural production increased dramatically during the first half of this century, when world population was still less than 2.5 billion people. With world population now approaching 5 billion, increasing competition for a finite amount of agricultural land frequently is causing disastrous problems of soil erosion (Fig. 7.23). In many Third World nations, farmers have been forced beyond traditional farm and grazing

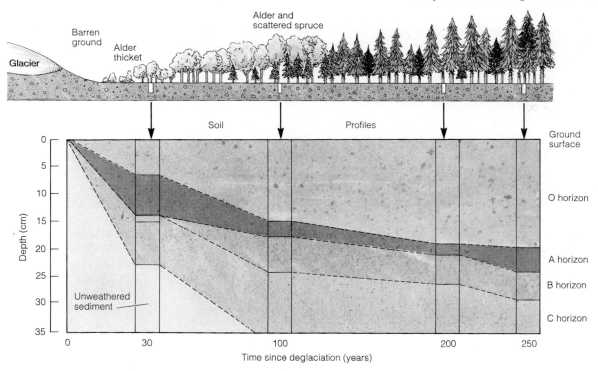

FIGURE 7.21 Progressive soil development in Glacier Bay, Alaska, over the past 250 years can be seen as changes in soil profiles from sites of different age. With time, organic litter forming an O horizon increases in thickness and A and B horizons develop. For the first 40 years, the A horizon directly overlies the C horizon. The B horizon then begins to develop and after another 60 years has reached a thickness of 7 cm. As vegetation changes from alder thicket toward a mature spruce-hemlock forest, organic litter continues to pile up and reaches a thickness of 20 cm after 150 years. The age of each study site was determined by counting the annual growth rings of the oldest surrounding trees. (Based on data from F. C. Ugolini)

lands onto steep, easily eroded slopes or into semiarid regions where periodic crop failure is a fact of life and plowed land is prone to severe wind erosion. At the same time, economic pressures in the more developed countries have increasingly led farmers to shift from ecologically favorable land-use practices to the planting of profitable row crops that often leave the land vulnerable to increased rates of erosion.

Although soil erosion results from natural changes in topography, climate, or vegetation cover, the results of human activities have overwhelmed natural systems in many parts of the modern world. Widespread felling of trees has led to accelerated rates of surface runoff and destabilization of soils due to loss of anchoring roots. Soils in the humid tropics, when

FIGURE 7.22 A thick reddish-brown paleosol (beside figure) developed on a pumice layer near Guatemala City, Guatemala, is overlain by another layer of pumice and pyroclastic flows that are separated by thinner brownish paleosols. A fault has displaced the layers about 2 m vertically.

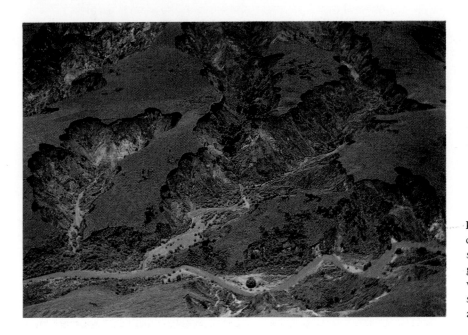

FIGURE 7.23 Hillsides in Madagascar that have been deforested for slash-and-burn farming are deeply gullied by erosion. Streams run red with irreplaceable topsoil that is destabilized as anchoring tree roots disappear.

stripped of their natural vegetation cover and cultivated, quickly lose their fertility (Fig. 7.24). So widespread are the effects of soil erosion and degradation that the problem has been described as "epidemic."

Indirect Effects of Soil Erosion

Much of the topsoil eroded from agricultural lands is transported down rivers and deposited along valley floors, in marine deltas, or in reservoirs behind large dams. The resulting impact on society, often unanticipated, can be significant. For example, the designers of a major dam and reservoir in Pakistan projected a life expectancy for the reservoir of at least a century. However, increased population pressure on the region above the dam has resulted in greatly increased soil erosion, leading to such a high rate of sediment production that the reservoir is now expected to be filled with eroded soil within 50 years, making it unusable.

Rates of Soil Loss

The upper layers of a soil contain most of the organic matter and nutrients that support crops. When the A and B horizons are eroded away, not only the fertility but the water-holding capacity of a soil diminishes. Because it generally takes between 80 and 400 years to form one centimeter of topsoil, for all practical purposes, soil erosion is tantamount to mining the soil: instead of remaining a continuously productive natural resource, it becomes an ever-decreasing and degraded resource. In the United States, the amount of farmland soil lost to erosion each year exceeds the amount of newly formed soil by more than 2 billion tons. It is estimated that farmers in the United States are now losing about 5 tons of soil for every ton of

FIGURE 7.24 Deforestation in Rondonia, Brazil, has led to accelerated surface runoff and created an eroded, devastated landscape from once-luxurious rain forest. Stripped of their natural vegetation cover and planted with crops, soils on this landscape quickly lose their natural fertility.

grain they produce. Soil loss in the Soviet Union is at least as rapid, while in India the soil erosion rate is estimated to be more than twice this high. Worldwide, the estimated loss exceeds 25 billion tons a year. Expressed another way, at the current rate of loss, the world's most productive soils are being depleted at the rate of 7 percent each decade. One recent estimate projected that as a result of excessive soil erosion and increasing population, only two-thirds as much topsoil will be available to support each person at the end of the century as was available in 1984.

Soil erosion is a worldwide problem of massive proportions, and it affects each and every one of us, as the following numbers make clear. A person eats about 750 kg of food a year, and there now are 5 billion people on the Earth. Food production leads to an annual loss of 25 billion tons of topsoil, or 5 tons (5000 kg) per person. Thus, for every kilogram of food we eat, the land loses 6.7 kg of soil.

Control of Soil Erosion

Although soil erosion and degradation are severely impacting many countries, effective control measures can substantially reduce these adverse trends.

In places where crops are grown, the surface is ordinarily bare during part of each year. On unvegetated, sloping fields, on pastures that are too closely grazed, and in areas planted with widely spaced crops such as corn, rates of erosion can be high (Fig. 7.25).

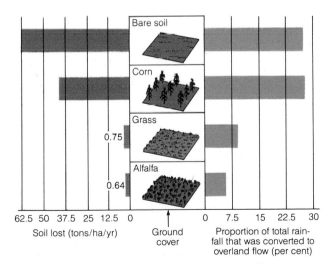

FIGURE 7.25 The effect of plant cover on rates of soil erosion measured over four years at Bethany, Missouri. The soil is silty, the slope is 4.5°, and the annual rainfall is 1000 mm. The measurements show that grass and alfalfa, with their continuous network of roots and stems, are nearly 300 times as effective as "row crops" such as corn in holding the soil in place. Erosional loss from bare soil in this area occurs at a rate of about 45 cm/100 years.

Wise farmers therefore reduce areas of bare soil to a minimum and prevent the grass cover on pastures from being weakened by overgrazing. If crops such as corn, tobacco, and cotton must be planted on a slope, strips of such crops are often alternated with strips of grass or similar plants that help resist soil erosion.

Another method of reducing soil loss involves crop rotation. A study in Missouri showed that land which lost 49.25 tons of soil per hectare when planted continually in corn lost only 6.75 tons per hectare when corn, wheat, and clover crops were rotated. In this case, the bare land exposed between rows of corn is far more susceptible to erosion than land planted with a more continuous cover of wheat or clover.

The most serious soil erosion problems occur on steep hillslopes. In Nigeria, for example, land planted with cassava (a staple food source) and having a gentle 1 percent slope lost an average of 3 tons of soil per hectare each year. On a 5 percent slope, however, the annual rate of soil loss increased to 87 tons per hectare. At this rate, a 15 cm thickness of topsoil would disappear in a single generation (about 20 years). On a 15 percent slope, the annual erosion rate increased to 221 tons per hectare, a rate that would remove all topsoil within a decade. Despite these grim statistics, steep slopes can be exploited through terracing.

In central China, where steep hillslopes are underlain by erodible deposits of wind-blown silt, terracing is a major factor in reducing soil loss (Fig. 7.26). In the arid regions of western China where strong winds and restricted vegetation cover can cause rapid soil loss, rows of trees planted as wind screens can substantially reduce erosion on the downwind side.

Soil Erosion and the World Economy

Because agriculture is the foundation of the world economy, progressive loss of soil signals a potential crisis that could undermine the economic stability of many societies. Halting the worldwide loss of soils is a formidable task that will require effective programs at the national and international level. Soils can be viewed as a renewable resource only over geologically long intervals of time. Over the lifetime of individuals, or even of nations, they must be considered nonrenewable resources that must be carefully utilized and preserved if a sustainable global economy is to be maintained.

MINERAL DEPOSITS FORMED BY WEATHERING

By removing soluble minerals and concentrating less-soluble ones, chemical weathering leads to economically valuable deposits.

FIGURE 7.26 A terraced hillside near Lanzhou, China, creates productive agricultural fields from steep hillslopes carved in deposits of windblown dust. Where unterraced, hillslopes are rapidly eroding and deeply gullied.

Lateritic Concentrations

Laterite is an example of enrichment by weathering. The most valuable laterite is a variety called **bauxite,** which is an aluminous laterite formed in tropical regions with gentle topography. As readily soluble mineral matter is leached out of the soil during weathering, chemically stable aluminum hydroxide minerals concentrate to form bauxite, the preferred source of aluminum.

Laterites developed on some plutonic igneous rocks are also enriched in nickel. Nickel-rich laterites are mined in such places as New Caledonia, Cuba, and Oregon.

Secondary Enrichment

When a preexisting sulfide mineral deposit is chemically weathered, the result can sometimes be a spectacular upgrading in the metal content. This process is called **secondary enrichment.** For example, iron-rich sedimentary rocks (chapter 4) may contain up to 25 percent iron by weight. Such rocks are too lean to be mined at a profit, but if chemical weathering removes silica in solution, a secondarily enriched deposit containing as much as 66 percent iron can result. Many of the world's great iron deposits have arisen through secondary enrichment. The same process is important in the formation of rich manganese deposits.

Secondary enrichment is also important in many hydrothermal deposits containing sulfide minerals. One example is the great copper deposits found in the arid southwestern United States and the Chilean Andes. The primary minerals are *pyrite* (FeS) and *chalcopyrite* ($CuFeS_2$). These minerals are oxidized by rainfall and the atmosphere to form a copper-bearing solution of sulfuric acid and a residue of limonite. The acidic solution percolates down and reacts with unoxidized pyrite and chalcopyrite by dissolving iron and precipitating copper to form *chalcocite* (Cu_2S). The chalcocite-rich blanket lies below the weathered residue and above the leaner primary mineral deposit below.

WEATHERING AND THE ROCK CYCLE

Weathering is an integral part of the global rock cycle, which is linked to plate tectonics. If fresh rocks were not continually brought to the surface by uplift and volcanism, erosion would lower the land and a deep weathering profile would eventually develop. Therefore, the processes associated with plate tectonics provide the grist for the weathering mill.

Silicate minerals that crystallize at high temperatures when igneous rocks form tend to weather most easily at the Earth's surface (Table 7.3). They include

TABLE 7.3 Order of Stability of Common Minerals Under Attack by Chemical Weathering

MOST STABLE (Least susceptible to chemical weathering)	Iron oxides and hydroxides Aluminum oxides and hydroxides Quartz Clay minerals Muscovite Potassium feldspar Biotite Sodium feldspar (albite-rich plagioclase) Amphibole Pyroxene
LEAST STABLE (Most susceptible to chemical weathering)	Calcium feldspar (anorthite-rich plagioclase) Olivine

olivine as well as calcium-rich feldspar, pyroxene, and amphibole. Biotite and sodium-rich feldspar crystallize at lower temperatures and are less easily weathered. Quartz, which crystallizes at a still lower temperature, is among the most resistant rock-forming minerals and degrades chemically only at an extremely slow rate. Nevertheless, given enough time, even quartz can be dissolved. Resistance to weathering is not solely related to temperature, however. It is due in part to the degree of silicate polymerization and in part to composition. Both these properties are controlled by the way magma crystallizes.

A glance at the mineral-stability sequence (Table 7.3) shows that the order in which the minerals are ranked is the reverse of that found in Bowen's reaction series (Fig. 3.39). This reversed order is not surprising, considering that the minerals least resistant to chemical weathering are those forming in crustal environments that differ most from those at the Earth's surface.

PLATE TECTONICS AND GLOBAL WEATHERING RATES

We can estimate the rate at which a tombstone in a graveyard is being weathered chemically by measuring how much surface decomposition has occurred since the date was chiseled on the stone. Measuring the average global rate of chemical weathering is not quite as easy. The best approach is to measure the amount of dissolved substances delivered by rivers to the oceans, for the dissolved matter in a stream results from chemical weathering of rocks and sediments in the stream's drainage basin. Although a multitude of streams enter the sea, the three largest contributors of dissolved substances are the Yangzte River which drains the high Tibetan Plateau of China, the Amazon River which drains the northern half of the Andes in South America, and the Ganges-Brahmaputra river system which drains the Himalaya in India. (Fig. B7.1). Collectively, these three streams deliver about 20 percent of the water and the dissolved matter entering the oceans. If we add all the other streams draining these three highland regions, we can conclude that chemical weathering and erosion in Tibet, the Andes, and the Himalaya must provide a substantial part of the total dissolved load reaching the world's oceans. In other words, a direct relationship apparently exists between the occurrence of high-altitude land masses and the global rate of chemical weathering. If true, then we might expect to see changes in weathering rates over geologic time, as mountain systems are uplifted and then worn away.

High rates of chemical weathering and high mountains are related for several reasons. First, high mountains are areas of rapid uplift. They occur where plates of lithosphere converge, causing mountain systems to form. Rapid uplift goes hand in hand with rapid mechanical breakdown and erosion of rock. This disintegration

FIGURE B7.1 The Indus River, which drains the high western Himalaya and Karakorum ranges, carries a high dissolved load to the Indian Ocean. This mountain region has one of the highest measured uplift rates in the world. Chemical and physical weathering rates are also high and lead to high rates of erosion.

exposes large quantities of rock and mineral debris to chemical weathering.

Second, the amount of dissolved matter in streams is greatest in areas where easily eroded sedimentary rocks are exposed. If we examine the geology of high mountains like the Himalaya, the Andes, and the Alps, we find that these areas of active, rapid uplift are dominated by sedimentary rocks of the type that form along continental margins. Lowland regions in continental interiors tend to be dominated by ancient, less erodible metamorphic and plutonic rocks.

Third, high mountains block moisture-bearing winds and generally receive large amounts of precipitation. This means high rates of stream runoff and high erosion rates. This effect is especially pronounced in southern Asia where the intense monsoon rainfall on the southern flank of the Himalaya leads to unstable slopes, intense erosion, and a high discharge of dissolved matter and suspended sediment to the Indian Ocean. Studies in South America have shown that more than three-quarters of the dissolved substances carried by the Amazon River to the sea originated in the Andean highlands.

Past weathering rates apparently have not always been as high as they are today. Evidence from ocean sediments points to an increase in the amount of dissolved matter reaching the oceans during the past 5 million years. This implies an increase in global weathering rates during this time. Furthermore, the major mountain systems of the world have not always been as high as we find them today. Evidence points to increased uplift rates in the Andes, Himalaya, Tibetan Plateau, and other tectonic regions like western North America and the Alps since the end of the Miocene Epoch (ca. 5.3 million years ago; Fig. 6.11). For example, sediments shed from the rising Himalaya coarsen upward, from silts of early Pliocene age to gravels of middle and late Pleistocene age. This change implies that streams draining the mountains gained increasing energy with time as mountain slopes and stream channels steepened.

The changes we can detect in the rates of mountain uplift parallel changes in weathering rates that we can infer from the sedimentary record. Mountain uplift rates are believed to be closely related to rates of sea-floor spreading. Thus, global rates of chemical weathering must be inextricably linked to plate tectonics.

SUMMARY

1. Weathering extends to whatever depth air and water penetrate the Earth's crust. Water solutions, which enter the bedrock along joints and other openings, attack the rock chemically and physically, causing breakdown and decay.

2. Mechanical and chemical weathering, although involving very different processes, generally work together.

3. Growth of crystals, especially ice and salt, along fractures and other openings in bedrock is a major process of mechanical weathering. Others include intense fires that cause spalling and fractur-

ing, the wedging action of plant roots, and the churning of rock debris by burrowing animals. Each of these processes can have large cumulative effects over time.

4. Chemical weathering involves the transformation of minerals formed at high temperatures and/or pressures into forms that are stable at the Earth's surface. The principle processes are hydrolysis, leaching, oxidation, hydration, and dissolution.

5. Subdivision of large blocks into smaller particles increases surface area and thereby accelerates chemical weathering.

6. The effectiveness of weathering depends on rock type and structure, surface slope, local climate, and the time over which weathering processes operate.

7. Because heat and moisture speed chemical reactions, chemical weathering is far more active in moist, warm climates than in dry, cold climates.

8. Soils consist of weathered regolith capable of supporting plants. Soils develop distinctive horizons the character of which is a function of climate, vegetation cover, soil organisms, parent material, topography, and time.

9. The A horizon of a soil is rich in organic matter and has lost soluble minerals through leaching. Clay accumulates in the B horizon, together with substances leached from the A horizon. Both overlie the C horizon, which is slightly weathered parent material.

10. The classification system used in the United States places soils in ten soil orders based on their physical characteristics.

11. Paleosols are buried soils that can provide clues about former landscapes, plant cover, and climate and are useful for subdividing, correlating, and dating the strata in which they are found.

12. Soil erosion and degradation are global problems that have been increasing as world population rises. Effective control measures include crop rotation, terracing, and tree planting, but halting widespread loss of soils is a formidable challenge.

13. Chemical weathering leads to concentration of economically valuable mineral deposits that are primary sources of aluminum, nickle, iron, manganese, and copper.

14. Weathering is an integral part of the rock cycle. The minerals least resistant to weathering are those which form in crustal environments most different from those at the Earth's surface.

15. Global rates of chemical weathering are linked to the presence or absence of high mountains, and thus to plate tectonics.

IMPORTANT TERMS TO REMEMBER

A horizon (p. 202)

B horizon (p. 202)
bauxite (p. 208)

C horizon (p. 202)
caliche (p. 203)
carbonic acid (p. 194)
chemical weathering (p. 190)

exfoliation (p. 196)

frost wedging (p. 192)

hydration (p. 194)
hydrolysis (p. 194)

joints (p. 191)

laterite (p. 203)
leaching (p. 194)

mechanical weathering (p. 190)

paleosol (p. 204)
parent material (p. 201)

secondary enrichment (p. 208)
soil (p. 201)
soil horizon (p. 201)
soil profile (p. 210)
spheroidal weathering (p. 196)

weathering (p. 190)
weathering rind (p. 195)

IMPORTANT MINERAL NAMES TO REMEMBER

goethite (p. 194)

hematite (p. 194)

kaolinite (p. 194)

QUESTIONS FOR REVIEW

1. Explain the difference between decomposition and disintegration of rocks?

2. In what way do joints affect rock weathering?

3. How does acid rain in industrialized regions contribute to weathering?

4. Why is frost wedging likely to be most effective at temperatures between about -5 and $-15°C$?

5. What causes flakes of rock to spall off exposed boulders during a forest fire?

7. Explain why minerals such as gold or platinum can become concentrated in sediments.

8. Why does the physical breakup of a rock increase the effectiveness of chemical weathering?

9. Explain how rock type, structure, slope, climate, and time can influence weathering.

10. How does chemically weathered regolith formed on limestone differ from that formed on an igneous rock, and why?

11. Describe the horizons in the profile of a well-developed soil in an evergreen forest; in an arid region.

12. If a buried soil is characterized by well-developed caliche in the top of the C horizon, what can you infer about the climate at the time the soil formed?

13. Why is the top of a paleosol an unconformity?

A huge volcanic mudflow overwhelmed the city of Armero, Colombia in 1985, killing at least 20,000 people.

CHAPTER 8

Mass-Wasting

The high Andes of South America include numerous active volcanoes and rugged peaks thrust up along converging lithospheric plates. The steep, unstable slopes rise above densely populated valleys, conditions spelling potential disaster in a landscape where major earthquakes and volcanic eruptions can bring sudden death and destruction.

In Colombia, the Andes culminate in a group of lofty active volcanoes lying west of Bogotá, the capital. Nevado del Ruíz (5400 m) has a history of volcanic activity extending back to at least 1595, when thunderous eruptions of tephra occurred and volcanic mudflows generated by melting ice and snow rushed down several valleys. In late 1984, the dormant volcano awakened and began belching clouds of steam and ash which continued through the autumn of 1985. People in the city of Armero, far downvalley from the volcano, grew alarmed. The local authorities seemed unconcerned and reassured them, even though recent geologic studies of the volcano had disclosed a history of repeated large volcanic mudflows. In early November, when the volcano showed signs of increasing activity, geologists warned that such mudflows could pose a danger for Armero in the event of an eruption. At 3 P.M. on the 13th, a technical emergency committee urged that Armero be evacuated, but the warning went unheeded. That night, as the local radio station played cheerful music and urged people to be calm, the volcano erupted. Torrents of water released from rapidly melting ice and snow near the summit sent huge waves of muddy debris surging downslope into surrounding valleys. The largest of several mudflows moved rapidly in the direction of Armero. Just after 11 P.M., as most of the population was sleeping soundly, a turbulent wall of mud came rushing out of a mountain canyon and inundated the city. At least 20,000 citizens of Armero perished, buried in a tomb of sulfurous volcanic mud. The geologists' prediction, based on a careful analysis of the geologic record, proved correct. Had their warning been heeded in time, the resulting human tragedy might have been avoided.

MASS MOVEMENT AND SLOPES

The landscapes we see about us may outwardly appear fixed and unchanging, but if we could make a time-lapse motion picture of almost any hillslope, the slope would seem almost alive and constantly changing. Much of the recorded motion would be the result of **mass-wasting**, the movement of regolith downslope, solely as a result of the pull of gravity, without the aid of a transporting medium, such as water, ice, or wind.

Because of gravity, a loose particle of regolith will always tend to move downslope. The beginning of the particle's journey downslope can be very slow or very fast, but in either case the movement is controlled primarily by gravity. Under natural conditions, a slope evolves toward an angle that allows the quantity of regolith reaching any point from upslope to be balanced by the quantity that is moving downslope. Such a slope is said to be in a balanced, or *steady-state*, condition.

Mass-wasting is not confined to the land. It occurs in lakes and evidence of it is seen over vast areas of the seafloor. As on land, mass-wasting in the oceans is controlled by gravity and takes place wherever slopes exist.

ROLE OF GRAVITY

A smooth, vegetated slope may appear outwardly stable and show little obvious evidence of geologic activity. Yet if we examine the regolith beneath the surface, it is likely we will find some rock particles derived from bedrock farther upslope. We can deduce, therefore, that the particles have moved downslope.

In any body of rock or rock debris located on a slope, two opposing forces determine whether the body will remain stationary or will move. The first of these, **shear stress**, is the force acting to cause movement of the body parallel to the slope. The primary factor influencing shear stress is the pull of gravity, and this pull is related to slope steepness. On a horizontal surface, gravity holds objects in place by pulling on them in a direction perpendicular to the surface (Fig. 8.1). On any slope, however, gravity can be resolved into two component forces. The *perpendicular component of gravity* (g_p in Fig. 8.1) acts at right angles to the slope and tends to hold objects in place. The *tangential component of gravity* (g_t in Fig. 8.1) acts along and down the slope, and it is this force that causes objects to move downhill. As a slope becomes steeper, the tangential component increases

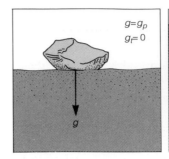
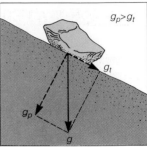

FIGURE 8.1 Effects of gravity on a rock lying on a hillslope. Gravity acts vertically and can be resolved into two components. One is perpendicular (g_p) and one parallel (g_t) to the surface.

relative to the perpendicular component, and the shear stress becomes larger.

The second force, **shear strength,** is the internal resistance of the body to such movement. Shear strength is governed by factors inherent in the body of rock or regolith. These factors include frictional resistance and cohesion between particles, and the binding action of plant roots.

As long as shear strength exceeds shear stress, the rock or debris will not move. However, as these two forces approach a balance, the likelihood of movement increases. Another way to express this relationship is as a ratio of shear strength to shear stress, a ratio known as the *safety factor (Fs)*:

$$Fs = \text{Shear Strength/Shear Stress}$$

When this factor is less than 1, failure of a slope is imminent. Steepening a slope by erosion, jolting it by an earthquake, or shaking it by blasting can cause an increase in shear stress. Shear strength might be reduced by weathering, by the decay of plant roots, or by saturation during a heavy rain.

The relationship between shear stress and slope angle (the steeper the slope, the greater the shear stress) means that conditions favoring mass movement tend to increase as slope angle increases. Steep slopes, of course, are most common in mountainous areas, so it is not surprising that mass-wasting is especially important in high mountains. Because the distribution and altitude of mountains is related to the interaction of lithospheric plates, mass-wasting on mountain slopes is a natural consequence of plate tectonics. It also is a basic part of the rock cycle: weathering, mass-wasting, and erosion constitute a continuum of interacting processes, the end result of which is the gradual breakdown of solid bedrock and the redistribution of its weathered components.

ROLE OF WATER

Water is almost always present within rocks and regolith near the Earth's surface. Although, by definition, water is not a transporting agent in mass-wasting, it nevertheless has a very important role to play. Unconsolidated sediments behave in different ways depending on whether they are dry or wet, as anyone knows who has constructed a sand castle at the beach. Dry sand is unstable and difficult or impossible to mold, but when some water is added, the sand gains strength and can be shaped into vertical castle walls. The water and sand grains are drawn together by a force called _capillary attraction_. The attraction results from surface tension, a property of liquids that causes the exposed surface to contract to the smallest possible area. This force tends to hold the wet sand together as a cohesive mass. However, the addition of too much water saturates the sand and turns it into a slurry that easily flows away, as the sand-castle builder sees with dismay when the rising tide on the beach destroys the elaborate work of an afternoon.

Moist or weakly cemented fine-grained sediments, such as fine silt and clay, may be so cohesive that they can stand in near-vertical cliffs, like the walls of a sand castle at the beach. However, if the silt or clay becomes saturated with water and the internal fluid pressure rises above a critical limit, this fine-grained sediment may also become unstable and begin to flow.

The movement of some large masses of rock has been attributed to the effects of increased water pressure in voids in the rock. If the voids along a nearly horizontal surface separating two rock masses are filled with water, and the water is under pressure, a buoying effect may result. In other words, the water pressure may be high enough to support the weight of the overlying rock mass, thereby reducing friction along the contact. The result can be sudden _failure,_ the collapse of a rock mass in which the stress has exceeded a critical limit. An analogous situation, in which water pressure buoys up a heavy object, can make driving in a heavy rainstorm extremely dangerous. When water is compressed beneath the wheels of a moving car, the increasing fluid pressure can cause the tires to "float" off the roadway. The driver quickly loses control of the vehicle, a condition known as hydroplaning.

An experiment illustrating this same principle was described by geologists M. King Hubbert and William Rubey in 1959. An empty beverage can is placed in an upright position on the wetted surface of a sheet of glass (Fig. 8.2a). If the glass is then slowly tilted, the can will not begin to slide until a critical angle (approximately 17°) is reached. This angle is characteristic for the substances used in this demonstration (metal and wet glass).

Next, the can is chilled in a freezer, and the experiment is repeated. With the can placed on the wetted glass with its open end upward, the angle at which sliding begins is seen to be the same. Finally, the cold can is placed with its open end downward on the wetted glass which is tilted so it has a 1° slope (Fig. 8.2b). The can will slide the length of the glass and then stop abruptly at the edge. The reason it now moves on such a gentle slope is that the conditions at the base of the can have changed. As the can warms, the air inside expands, increasing the pressure. The increased air pressure at the base buoys up the can, reducing the friction between metal and glass, and allows the can to glide down the gentle slope. The can stops at the edge of the glass as the pressure is

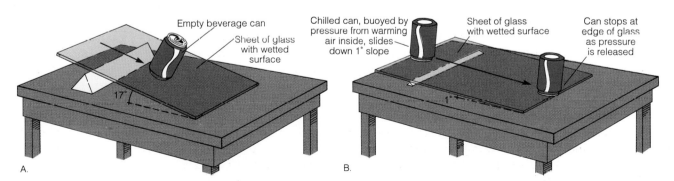

FIGURE 8.2 An experiment illustrating how compressed air reduces friction at the base of a mass resting on a slope. A. An empty beverage can placed on wet sheet of glass will begin to slide down the surface when the angle reaches about 17°. B. A chilled empty can, placed upside down on wet glass, will slide when the glass has a slope of only 1° because friction at the base is reduced as warming air in the can expands. Sliding ceases when the can reaches the edge of the glass and pressure at the base is suddenly released.

suddenly released. In an analogous way, high water pressure at the base of a large rock mass may promote downslope movement of the rock.

As seen from these examples, water can be instrumental in reducing shear strength and thereby promoting movement of rock and sediment downslope under the pull of gravity. It does so in two important ways: (1) by reducing the natural cohesiveness between grains, and (2) by reducing friction at the base of a rock mass through increased fluid pressure.

MASS-WASTING PROCESSES

Mass-wasting processes all share one characteristic: they take place on slopes. Any perceptible downslope movement of a mass of bedrock, regolith, or a mixture of the two is commonly referred to as a **landslide.** However, we can recognize many different kinds of slope movements, and because they often grade into one another, no simple and ideal classification exists. The composition and texture of the sediment involved, the amount of water and air mixed with the sediment, and the steepness of slope all influence the type and velocity of movement. In effect, a progressive transition exists from the flow of clear streamwater, to sediment-laden streamwater, to an array of mass-wasting processes ranging from those in which water promotes flow to those in which water plays no direct or significant role.

The approach we will take here is to separate mass-wasting processes into those involving (1) the sudden failure of a slope that results in the downslope transfer of relatively coherent masses of rock or rock debris by slumping, falling, or sliding, and (2) the downslope flow of mixtures of sediment, water, and air. In the latter group, which involves internal motion of flowing masses of debris, processes are distinguished on the basis of their velocity and the concentration of sediment in the flowing mixture. We also will examine some processes and deposits representative of mass-wasting in cold-climate regions and on the ocean floors. ⟶

Slope Failures

The constant pull of gravity makes all hillslopes and mountain cliffs susceptible to failure. When failure does occur, rock debris is transferred downslope and a stable slope condition is reestablished.

Slumps
A **slump** is a type of slope failure in which a downward and outward rotational movement of rock or regolith occurs along a curved concave-up surface (Fig. 8.3). The top of the displaced block usually is tilted backward, producing a reversed slope. Slumps may occur singly or in groups, and they can range in size from small displacements only a meter or two in dimension, to large slump complexes that cover hundreds or even thousands of square meters.

Slumps are one of the types of mass movement that

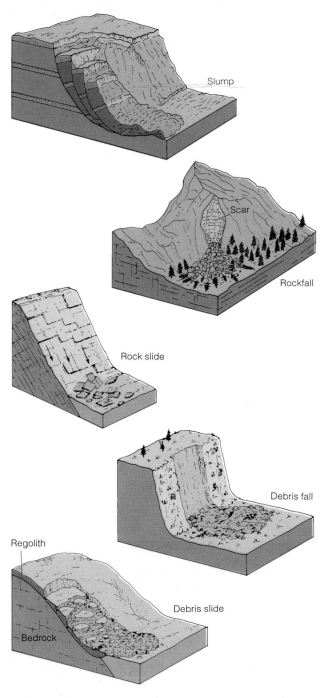

FIGURE 8.3 Examples of slope failures giving rise to slumps, falls, and slides.

FIGURE 8.4 A large slump in a high gravel terrace beside the Yakima River in central Washington has broken up a major highway and displaced it more than 100 m laterally into the river channel.

we are most likely to see, for many result from artificial modification of the landscape. They are numerous along roads and highways where bordering slopes have been oversteepened by construction activity (Fig. 8.4). We can also see them along river banks or seacoasts where currents or waves undercut the base of a slope.

Slumps frequently are associated with heavy rains or sudden shocks, such as earthquakes. Distinct episodes of slumping may be related to changing climatic conditions. Slumping may recur seasonally and be associated with seepage of water into the ground during the rainy season. In parts of the western United States, increased slumping during recent decades appears to be correlated with an overall rise in average rainfall. A similar increase about 5000 years ago apparently was related to a shift from a warm, dry climate in the middle Holocene to cooler and wetter conditions since.

Falls and Slides

Ask mountain climbers about the greatest dangers associated with their sport and it is likely they will place falling rock near the top of the list. **Rockfall,** the free falling of detached bodies of bedrock from a cliff or steep slope, is common in precipitous mountainous terrain, where rockfall debris forms conspicuous deposits at the base of steep slopes (Fig. 8.3). As a rock falls, its speed increases. Knowing the distance of fall (*H*), we can calculate the velocity (*v*) on impact as

$$v = \sqrt{2gH}$$

where *g* is the acceleration due to gravity. This formula tells us that a rock of a given size will be traveling at a much higher velocity if it falls free from high on a steep mountain face than from a low cliff.

A rockfall may involve the dislodgment and fall of a single rock fragment, or it may involve the sudden collapse of a huge mass of rock that plunges hundreds of meters, gathering speed until it breaks on impact into a vast number of smaller pieces. These pieces continue to bounce, roll, and slide downslope before friction and decreasing slope angle bring them to a halt.

When a mountain slope collapses, not only rock but overlying sediment and plants are generally involved. The resulting **debris fall** is similar to a rockfall, but it consists of a mixture of rock and weathered regolith, as well as vegetation (Fig. 8.3).

Slides, like falls, involve the rapid displacement of masses of rock or sediment. A **rockslide** is the sudden downslope movement of detached masses of bedrock (or of debris, in the case of a **debris slide**) along an inclined surface, such as a bedding plane (Fig. 8.3). Like falling rock and debris, rock slides and debris slides are common in high mountains where steep slopes abound. When large rockslides occur, the resulting deposit generally is a chaotic, jumbled mass of rock, with individual boulders sometimes measuring tens of meters across (Fig. 8.5).

Accumulations of angular rock fragments are a common sight at the base of steep cliffs. The rock debris commonly ranges in size from sand grains to large boulders. Such a body of debris sloping outward

FIGURE 8.5 A jumble of boulders, many the size of houses, forms the deposit of a massive landslide that descended a steeply sloping bedrock surface on a high Andean mountain in Argentina. Other large masses of jointed rock are poised at the top of this slope, ready to produce future rockslides.

FIGURE 8.6 A talus at the base of steep cliffs in the Brooks Range, northern Alaska. When most rockfall debris moves downslope via a gully, the resulting deposit at the base of the gully is a talus cone.

from the cliff that supplies it is a **talus** (Fig. 8.6). From cliff to talus, the movement is chiefly by falling, sliding, and rolling. The rock fragments come to rest at the steepest angle (measured from the horizontal) at which the debris remains stable (Fig. 8.7). This angle, referred to as the **angle of repose,** typically lies between 33° and 37°. Fine particles falling onto a talus tend to settle into open voids between coarser fragments. Large falling rocks have more momentum than

small particles and tend to move farther down the slope. Some may bound beyond the toe of the talus and form a scattered array of isolated boulders.

Sediment Flows

When a sufficiently large force is applied, any deformable material will begin to flow. In mass-wasting,

FIGURE 8.7 Coarse, angular limestone blocks stand at the angle of repose (about 30°) in a talus below steep cliffs in the central Brooks Range, Alaska.

the force is gravity and the material consists of dense mixtures of sediment and water (or sediment, water, and air). Mass-wasting processes that involve flow of such mixtures are called **sediment flows.**

Factors Controlling Flow

At above-freezing temperatures, the way a sediment flows depends on (1) the relative proportion of solids, water, and air, and (2) the physical and chemical properties of the sediment.

All streams carry at least some sediment, but if the sediment becomes so concentrated that the water no longer can transport it, a sediment-laden stream will change into a very fluid sediment flow. Because the water in a sediment flow is not a transporting medium, a sediment flow, by definition, is a mass-wasting process. The entrained water helps promote flow, but the pull of gravity remains the primary reason for movement.

In Figure 8.8, sediment flows are subdivided into two classes based on sediment concentration. A **slurry flow** is a moving mass of water-saturated sediment. A **granular flow** is a mixture of sediment, air, and water, but unlike a slurry flow it is not saturated with water; instead, the full weight of the flowing sediment is supported by grain-to-grain contact or collision between grains. Each of these two classes is further subdivided into several processes on the basis of flow velocity (e.g., creep is a very slow type of granular flow, measured in millimeters or centimeters per year, whereas a debris avalanche is measured in kilometers per hour). In this classification of sediment flows, the boundaries between processes are placed only approximately and depend on the grain-size distribution, sediment concentration, and other factors.

Slurry Flows

In slurry flows, the sediment mixture is often so dense that large boulders can be suspended in it. Boulders too large to remain in suspension may be rolled along by the flow. When flow ceases, fine and coarse particles remain mixed, resulting in a nonsorted sediment.

Solifluction. The very slow downslope movement of saturated soil and regolith is known as **solifluction.** As can be seen in Figure 8.8, this process lies at the lower end of the velocity scale for flowing sediment–water mixtures. Rates of movement are less than about 30 cm/yr, generally so slow as to be detectable only by measurements made over several seasons. The slow movement results in distinctive surface features, including lobes and sheets of debris that sometimes override one another (Figs. 8.9 and 8.10). Solifluction occurs on hillslopes in temperate and tropical latitudes, where sediment remains saturated with water for long intervals.

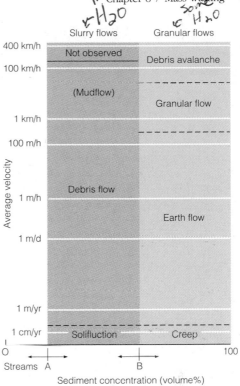

FIGURE 8.8 Classification of sediment flows on the basis of their average velocity and sediment concentration. The transition from a sediment-laden stream to a slurry flow occurs when the sediment concentration becomes so high that the stream no longer acts as a transporting agent; instead, gravity becomes the primary force causing the saturated sediment to flow. As the percentage of water decreases further, a transition from slurry flow to granular flow takes place. Now the sediment may contain water and/or air. The boundaries between muddy streams and slurry flows (A) and between slurry and granular flows (B) are not assigned sediment-concentration percentages because the position of the boundaries can shift to the left or right depending on the physical and compositional characteristics of the sediment + water + air mixture. Different types of slurry and granular flows are recognized on the basis of their mean velocity.

Debris Flows. A **debris flow** involves the downslope movement of unconsolidated regolith, the greater part being coarser than sand, at rates ranging from only about 1 m/yr to as much as 100 km/h (Fig. 8.8). In some cases a debris flow begins with a slump or debris slide, the lower part of which then continues to flow downslope (Figs. 8.9 and 8.11). A typical debris flow, once mobilized, moves along a stream channel and may then spread across the surface of an alluvial fan, where it consolidates to form a poorly sorted deposit.

Debris flow deposits commonly have a tonguelike front. They also have a very irregular surface, often with concentric ridges and depressions that resemble

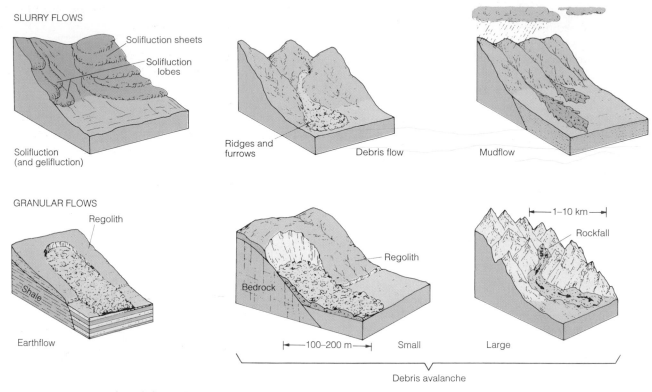

SLURRY FLOWS

Solifluction sheets

Solifluction lobes

Solifluction (and gelifluction)

Ridges and furrows

Debris flow

Mudflow

GRANULAR FLOWS

Regolith

Shale

Earthflow

Bedrock

Regolith

100–200 m

Small

1–10 km

Rockfall

Large

Debris avalanche

FIGURE 8.9   Examples of slurry flows and granular flows.

the surface of deposits left by mountain glaciers. Debris flows are frequently associated with intervals of extremely heavy rainfall that lead to oversaturation of the ground.

Mudflows.   A debris flow that has a water content sufficient to make it highly fluid is commonly called a **mudflow** (i.e., the term *mudflow* is a synonym for a rapidly moving debris flow). In Figure 8.8, the velocity range of mudflows lies at the upper range of debris flows (more than about 1 km/h). Most mudflows are highly mobile and tend to travel rapidly along valley floors (Fig. 8.9).

If you were to scoop up a handful of moving mudflow sediment, its consistency could range from that of freshly poured concrete to a souplike mixture only slightly denser than very muddy water. After heavy rain in a mountain canyon, a mudflow can start as a muddy stream that continues to pick up loose sediment until its front becomes a moving dam of mud and rubble, extending to each wall of the canyon and urged along by the force of the flowing water behind

FIGURE 8.10   A meter-thick solifluction lobe has slowly moved downslope and covers glacial deposits on the floor of the Orgière Valley in the Italian Alps.

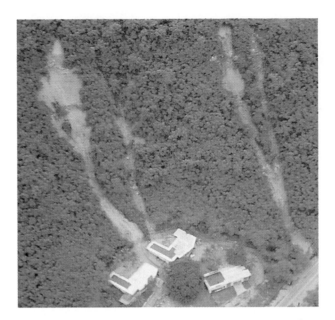

FIGURE 8.11 Debris slides that turned into debris flows on a steep vegetated mountainside in southern Puerto Rico have stripped away vegetation and inundated two houses at the base of the slope.

it (Fig. 8.12). On reaching open country at the mountain front, the moving dam collapses, floodwater pours around and over it, and mud mixed with boulders is spread as a wide, thin sheet. Sediment fans at the foot of mountain slopes in such arid regions as the central Andes, the Hindu Kush, and the eastern slope of the Sierra Nevada in California consist largely of superposed sheets of mudflow sediments interstratified with stream sediments (Fig. 8.13).

On active volcanoes in wet climates, layers of tephra and volcanic debris commonly cover the surface and are easily mobilized as mudflows. As we saw in the case of Armero, Colombia, highly fluid mudflows can travel great distances and at such high velocities that they constitute one of the major hazards associated with volcanic eruptions. Mudflow sediments in valleys surrounding many large explosive volcanoes in the Cascade Range contain a large percentage of the total volume of volcanic rocks erupted at each vent. A particularly large mudflow that originated on the slopes of Mount Rainier about 5700 years ago traveled at least 72 km. The sediment spread out beyond the mountain front as a broad lobe as much as 25 m thick. Its volume is estimated to be well over a billion cubic meters. Mount St. Helens, an unusually active volcano, has produced mudflows throughout much of its history. The most recent occurred during the huge eruption of May 1980 (Fig. 8.14).

Granular Flows

The sediment of granular flows may be largely dry, with air filling the pores, or it may be saturated with

water but have a range of grain sizes and shapes that allows water to escape easily.

Creep and Colluvium. Most of us have seen old fences, telephone poles, or gravestones leaning at an angle on hillslopes, or have seen evidence of the downslope displacement of fractured road surfaces (Fig. 8.15). All are common evidence of **creep,** an imperceptibly slow downslope movement of regolith. Steeply inclined rock strata are sometimes bent over in the downslope direction just below the ground surface, further evidence of creep.

A.

B.

C.

FIGURE 8.12 Passage of a muddy debris flow along a canyon bottom near Farmington, Utah, in June 1983. A. The bouldery front of a muddy debris flow advances from left to right along a stream channel in the wake of an earlier surge of muddy debris. B. The steep bouldery front, about 2 m high and advancing at 1.3 m/s, acts as a moving dam, holding back the flow of muddy sediment upstream. C. The main slurry, having a sediment concentration of about 80 percent and now moving at about 3 m/s, is viscous enough to carry cobbles and boulders in suspension.

FIGURE 8.13 A roadcut in northern Pakistan exposes angular mudflow deposits that have accumulated as a fan along the side of a steep mountain valley.

A number of factors contribute to creep (Table 8.1). Although creep occurs at a rate too slow to be seen, careful measurements of the downslope displacement of objects at the surface record the rates involved (Fig. 8.16). As might be expected, rates tend to be higher on steep slopes than on gentle slopes. Measurements in Colorado, for example, document a creep rate of 9.5 mm/yr on a slope of 39° but only 1.5 mm/yr on a 19° slope. Rates also tend to increase as soil moisture increases. However, in wet climates vegetation density

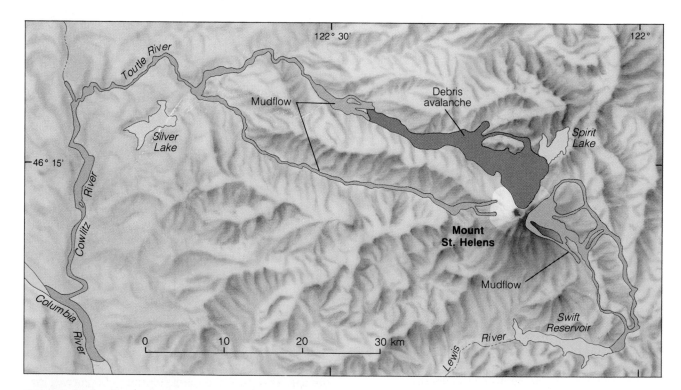

FIGURE 8.14 During the 1980 eruption of Mount St. Helens in Washington, volcanic mudflows were channeled down valleys west and east of the mountain. Some mudflows reached the Columbia River, having traveled more than 90 km. Flow velocities were as high as 40 m/s and averaged 7 m/s.

FIGURE 8.15 Effects of creep on surface features and on bedrock. Steeply inclined strata have been dragged over near the surface by creep, so they appear folded. Telephone poles and fence posts affected by creep are tilted, stone fences are deformed, roadbeds are locally displaced, and gravestones are tilted or fallen.

also increases and the roots of plants, which bind the soil together, tend to inhibit creep. The creep rate measured on one grassy hillside in England having a slope of 33° is only 0.02 mm/yr. Despite a slow rate of movement, creep affects all hillslopes covered with regolith, and its cumulative effect is therefore very great.

Loose, incoherent deposits on slopes that are moving mainly by creep are termed **colluvium.** The parti-

TABLE 8.1 Factors contributing to creep of regolith

Frost Heaving	Freezing and thawing, without necessarily saturating the regolith, causing lifting and subsidence of particles
Wetting and Drying	Causes expansion and contraction of clay minerals
Heating and Cooling Without Freezing	Causes volume changes in mineral particles
Growth and Decay of Plants	Causes wedging, moving particles downslope; cavities formed when roots decay are filled from upslope
Activities of Animals	Worms, insects, and other burrowing animals displace particles, as do animals trampling the surface
Dissolution	Mineral matter taken into solution creates voids in bedrock that tend to be filled from upslope
Activity of Snow	A seasonal snow cover tends to creep downslope and drag with it particles from the underlying ground surface

cles in colluvium tend to be angular and lack obvious sorting. These characteristics generally make it possible to distinguish colluvium from sediments deposited by flowing water or air which tend to consist of rounded particles, sorted and deposited in layers.

Earthflows. An **earthflow,** among the more common mass-wasting features on the landscape, is a granular flow having a velocity that falls in the range of about 1 m/day to several hundred m/h (Fig. 8.8). Earthflows may remain active for several days, months, or even years. Even after initial motion ceases, they may be highly susceptible to renewed movement. Like debris flows, earthflows are often made up of weak or weathered regolith, predominantly of silt or clay size, and occur on gentle to moderately steep slopes (2° to 35°). Earthflows occur where the ground is saturated, at least intermittently, and they frequently are associated with intervals of excessive rainfall.

An earthflow typically heads in a steep *scarp,* which is a cliff formed where slide material has moved away from undisturbed ground upslope (Fig. 8.17). An earthflow generally has a long and narrow tonguelike shape and a rounded, bulging front. Earthflows range in size from features several meters long and wide and less than a meter deep, to features as much as several hundred meters wide, more than 1 km long, and more than 10 m deep. Many exist as parts of large earthflow complexes. In a longitudinal profile from head (top) to toe (leading edge), an earthflow is concave upward near the head and convex upward near the toe. This profile implies thinning of the earthflow sediment in the upper part of the displaced mass and thickening in the lower part during movement. Although some internal deformation occurs in earthflows, mostly it is confined to or near the margins.

A.
B.

FIGURE 8.16 Colored targets placed in a straight line across a hillslope in Greenland (A) had moved differentially downslope by creep when photographed a year later (B). The maximum recorded movement along the slope averaged 12 cm/yr.

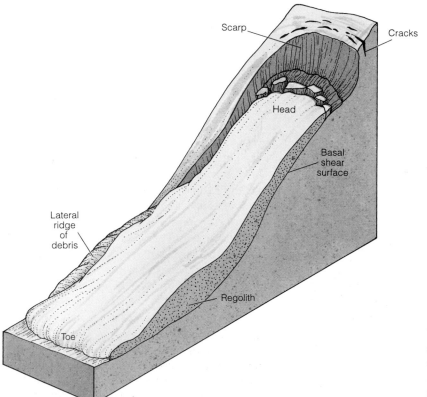

FIGURE 8.17 In this section through an idealized earthflow, regolith moves downslope across a basal shear surface, leaving a scarp at the head where the sediment separated from the slope above. A bulging toe protrudes beyond ridges of sediment that have piled up along the lower margins of the earthflow.

Field studies show that earthflows become mobilized when water pressure along shear surfaces rises. Pressures can reach such high values that movement will occur on slopes of only a few degrees.

A special type of earthflow occurs in wet, highly porous sediments consisting of clay- to sand-size particles. Such units will weaken if shaken suddenly, as by an earthquake. An abrupt shock increases shear stress and causes a buildup of water pressure in pore spaces, which decreases the shear strength. The result is rapid fluidization of the sediment and abrupt failure, a process known as **liquifaction.** Any structure built on such sediments or in their path may be quickly demolished (Fig. 8.18).

Grain Flows. If you have ever walked along the crest of a sand dune and stepped too close to the steep slope that faces away from the wind, your footstep likely started a cascade of sand flowing down the dune face. This example illustrates a type of mass-wasting called *grain flow* that involves movement of a dry or nearly dry granular sediment with air filling the pore spaces. Such grain flow occurs naturally when accumulating sand grains produce a slope that exceeds the angle of repose, leading to failure. During flow, moving grains collide frequently. Average velocities of the moving sediment typically fall between 0.1 and 35 m/s.

Debris Avalanches. A large, rapidly moving **debris avalanche** is a rare but spectacular event. This type of granular flow travels at high velocity (tens to hundreds of km/h) and can be extremely destructive (Fig. 8.9 and Table 8.2). A debris avalanche frequently involves huge masses of falling rock and debris that break up, pulverize on impact, and then continue to travel downslope, often for great distances.

Large rockfalls that give rise to debris avalanches have had the greatest human impact in populated mountain regions like the Alps and the Andes. In September 1717, a large mass of rock and ice fell onto Triolet Glacier from a mountain crest near Mont Blanc along the French–Italian border (Fig. 8.19). Pulveriz-

FIGURE 8.18 Chaotically tilted trees and houses in suburban Anchorage, Alaska show how violent shaking of the ground during the great 1964 earthquake caused sudden liquefaction of underlying clays and widespread slumping.

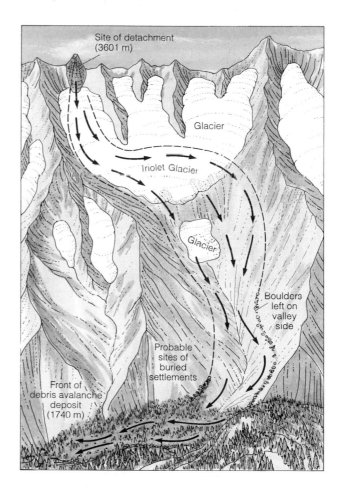

FIGURE 8.19 A large debris avalanche in upper Val d'Aosta, Italy, traveled 7 km from its source high on a mountain spur. Within only a few minutes the debris buried two settlements on the valley floor, killing all the inhabitants and livestock. The reconstructed trajectory of the debris avalanche (arrows), which occurred in 1717, is based on deposits left along the valley sides.

TABLE 8.2 Characteristics of some large debris avalances

Locality	Date	Volume (million m³)	Vertical Movement (m)	Horizontal Movement (km)	Calculated Velocity (km/h)
Huascaran, Peru	1971	10	4000	14.5	400
Sherman Glacier, Alaska	1964	30	600	5.0	185
Mr. Rainier, Washington	1963	11	1890	6.9	150
Madison, Wyoming	1959	30	400	1.6	175
Elm, Switzerland	1881	10	560	2.0	160
Triolet Glacier, Italy	1717	20	1860	7.2	≥125
Black Hawk, California	prehistoric	280	1220	8.0	120
Saidmarreh, Iran	prehistoric	2000	1650	14.5	340

ing on impact, the fragmented debris moved rapidly downvalley before its leading edge came to rest about 7 km from, and 1860 m lower than, the site of detachment. The estimated velocity of the mass on impact was close to 320 km/h. As the sheet of debris rushed downward to the floor of the main valley, its momentum carried it up the opposite valley wall to a height of at least 60 m. At this point, just as the debris overwhelmed two small mountain villages killing all the inhabitants and their livestock, its velocity is estimated to have been at least 125 km/h. From the estimated velocities and distance of travel, it is clear that the total travel time over the entire 7 km was between 2 and 4 minutes. Such rapid travel times mean that escape from large and destructive debris avalanches is seldom possible.

Because large debris avalanches are infrequent and extremely difficult to study while they are moving, few observational data are available about the process. Their extreme mobility has been attributed to the debris riding on a layer of compressed air. If true, debris avalanches behave somewhat like a commercial hovercraft that travels across land or water on air compressed by a large propeller. Alternatively, air trapped and compressed within the moving debris may reduce friction between particles and cause the mass to behave in a highly fluid manner.

The flanks of steep, unstable stratovolcanoes are especially susceptible to collapse that leads to the production of debris avalanches. The deposits of such debris avalanches can be difficult to recognize because of their huge dimensions. A broad valley that extends some 40 km north of towering Mount Shasta in northern California contains a complex of hills and mounds of volcanic rock that many geologists had interpreted as relics of numerous minor eruptions from isolated vents. Recognition of the close similarity between these deposits and those of a huge debris avalanche associated with the 1980 Mount St. Helens eruption prompted a reassessment. A new study led to the remarkable conclusion that the entire array of

features resulted from a similar, but far larger, collapse of a flank of the Shasta volcano about 300,000 years ago. In this case, the volume of rock involved was at least 26 km³—almost ten times the volume of the landslide on Mount St. Helens. Whereas the St. Helens deposits cover an area of 60 km², the Shasta debris avalanche overwhelmed at least 450 km² (Fig. 8.20).

Mass-Wasting in Cold Climates

Mass-wasting is especially active at high latitudes and high altitudes where average temperatures are very low. These are regions where much of the landscape is underlain by perennially frozen ground and frost action is an important geologic process.

Frost Heaving and Creep
When water freezes, it increases in volume. Ice forming in saturated regolith therefore pushes the ground surface up. This lifting of regolith by the freezing of contained water is called **frost heaving.**

Frost heaving strongly influences downslope creep of sediment in cold climates. When freezing occurs, the ground surface is lifted essentially at right angles to the slope. As the ground thaws, each particle tends to drop vertically, pulled downward by gravity. Thus, a particle's net motion during each freeze–thaw cycle is a very short distance downslope (Fig. 8.21). The net result of repeated episodes of freezing and thawing, during which a particle experiences a succession of up and down movements, is slow but progressive downslope creep.

Gelifluction
In cold regions underlain year-round by frozen ground, a thin surface layer thaws in summer and then refreezes in winter. During the summer, this thawed layer becomes saturated with meltwater and is very unstable, especially on hillsides. As gravity pulls the

FIGURE 8.20 A massive prehistoric debris avalanche from the northwest flank of Mount Shasta volcano in northern California left a chaotic deposit (hills in middle distance) that extends 34 km from the volcano and covers at least 450 km².

thawed sediment slowly downslope, distinctive lobes and sheets of debris are produced. Similar to solifluction in temperate and tropical climates, this process is known as **gelifluction.** Although measured rates of movement are low, generally less than 10 cm/yr, gelifluction is so widespread on high-latitude landscapes that it constitutes a highly important agent of mass transport. Hillslopes in arctic Alaska and Canada, for example, commonly are mantled by superimposed sheets or lobes of geliflucted regolith.

Rock Glaciers

A **rock glacier,** another characteristic feature of many cold, relatively dry mountain regions, is a tongue or lobe of ice-cemented rock debris that moves slowly downslope in a manner similar to glaciers (Fig. 8.22). Rock glaciers generally originate below steep cliffs, which provide a source of rock debris. Active rock glaciers may reach a thickness of 50 m or more and advance at rates of up to about 5 m/yr. They are especially common in high interior mountain ranges like

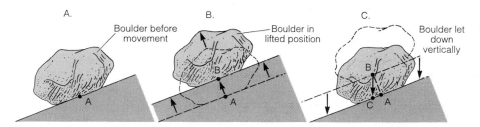

FIGURE 8.21 Stone moved downslope by alternate freezing and thawing of the ground. As freezing occurs (B), the stone is raised perpendicular to the ground surface, which also rises; when the ground thaws and settles (C), gravity pulls the stone down vertically, giving it a small but significant component of movement downslope. (A → C).

FIGURE 8.22 A jumbled mass of angular rock debris, supplied from a steep cliff in the Wrangell Mountains of southern Alaska, moves slowly downslope as a rock glacier.

the Swiss Alps, the Argentine Andes, and the Rocky Mountains.

Subaqueous Mass-Wasting

As geologists have extended their search for petroleum to the continental shelves and slopes, their explorations have shown that mass-wasting is an extremely common and widespread means of sediment transport on the sea floor. Mass-wasting also has been documented in lakes. As on land, the potential for gravity-induced movement of rock and sediment exists wherever there are subaqueous (underwater) slopes.

Submarine slope failures can give rise to turbidity currents (chapter 4), a type of subaqueous sediment flow, that travel down submarine canyons and deposit turbidites on the continental rise. In some regions subaqueous mass-wasting also accounts for a substantial portion of shelf deposits, especially in areas seaward of large rivers.

Marine Deltas
Major marine deltas commonly display surface features and sediments attributable to slope failures. In

such subaqueous environments, failure can occur even on slopes as low as 1°. The slope failures generally display three distinct zones: a source region where subsidence (sinking or downward settling) and slumping take place, a central channel where sediment is transported, and a zone where sediment is deposited, often in the form of overlapping lobes of debris. These and other features are common on the submarine slopes of the Mississippi delta, which is among the best-studied deltas in the world (Fig. 8.23). Slides and sediment flows are extremely active on the delta front. In places, more than 30 m of sediment has been deposited by sediment flows in the last 100 years.

Mass-Wasting in the Western North Atlantic
Extensive studies of the continental slope of eastern North America using a variety of modern techniques (deep ocean drilling, piston coring, side-scan sonar, echograms, submersible vessels) have shown that vast areas of the seafloor are disrupted by submarine slumps, slides, and flows (Fig. 8.24). Some large slide complexes cover areas of more than 40,000 km² and reach depths as great as 5400 m. Series of slumps heading in abrupt scarps up to 100 m high and 50 km long lie upslope from slide complexes. The slides generally have affected the uppermost 50 m of seafloor sediment. Cores taken within the slide masses show disturbed and contorted layering, overlain by silty turbidite beds. Buried deposits have been identified within slide complexes, pointing to multiple episodes of movement.

Localized slumping and small-scale slides mark the walls of submarine canyons. Some geologists postulate that large-scale mass-wasting may have started these canyons, which were then enlarged through continuous slumping on canyon walls and the repeated movement of turbidity currents down the canyons.

Submarine slope failures often occur when shear strength of the sediments is reduced by an abrupt increase in the pore pressure. One way this can happen is by a sudden earthquake shock. Several major turbidity currents and slumps off eastern North America were contemporaneous with strong earthquakes (e.g., 1929: Grand Banks, Newfoundland; 1886: Charleston, South Carolina; 1755: Cape Ann, Massachusetts).

The dating of numerous sediment cores from submarine slump masses and debris flows has shown that most of the mass-wasting events occurred during the last glacial age, when world sea level was at least 100 m lower than now, and during the subsequent rise of world sea level that resulted from the melting of

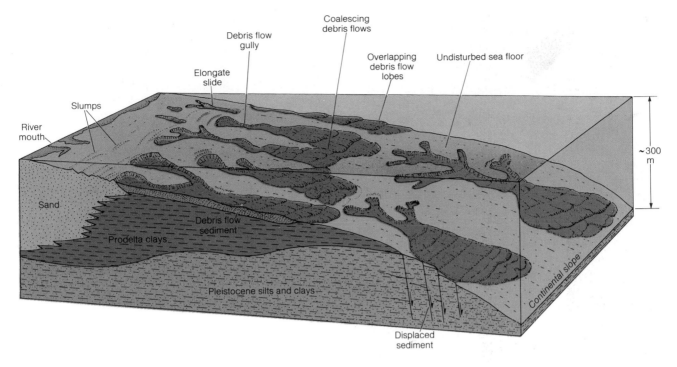

FIGURE 8.23 Block diagram showing various mass-wasting features on the submarine surface of the Mississippi Delta.

continental ice sheets. At such times, large quantities of sediment were dumped in the ocean by rivers crossing the emergent continental shelf and were then transported by mass-wasting into the deep sea. The cyclic rise and fall of sea level during numerous glacial cycles over the last several million years has slowly built a thick wedge of sediment along the base of the Atlantic continental slope.

Hawaiian Submarine Landslides

The largest volcanoes on the Earth rise 7000 m or more from the floor of the Pacific Ocean. Like their smaller counterparts on the continents, they are composed of huge piles of well-jointed lava flows and unstable rubble. Chaotic topography along the submerged lower margins of the Hawaiian volcanoes has been interpreted as evidence of repeated massive landslides on the volcano flanks. Irregular terrain on the seafloor adjacent to East Molokai suggests that much of the northern half of this volcano has slid downslope from a steep landslide scarp visible on submarine surveys. Anomalous coral-bearing gravels found to altitudes of 326 m on Lanai and nearby islands have recently been attributed to a giant wave that deposited the coral fragments high above sea level. The wave is believed to have resulted from a huge submarine landslide off the west coast of the island of Hawaii that traveled nearly 100 km from its source near sea level to depths of 4800 m. Based on

dating of the corals, the landslide occurred about 105,000 years ago.

TRIGGERING OF MASS-WASTING EVENTS

Mass-wasting events sometimes seem to occur at random, with no apparent reason. However, the largest, most disastrous, and most numerous events generally are related to some extraordinary activity or occurrence.

Shocks

Sudden shocks, such as an earthquake, may release so much energy that slope failures of many types and sizes are triggered simultaneously. In 1929 a major earthquake in northwestern South Island, New Zealand triggered at least 1850 landslides larger than 2500 m² within an area of 1200 km² near the quake's center. An estimated 210,000 m³ of debris was displaced, on average, in each 1 km² of land. Landslides were reported to be most numerous on well-bedded and well-jointed mudstones and fine sandstones. The

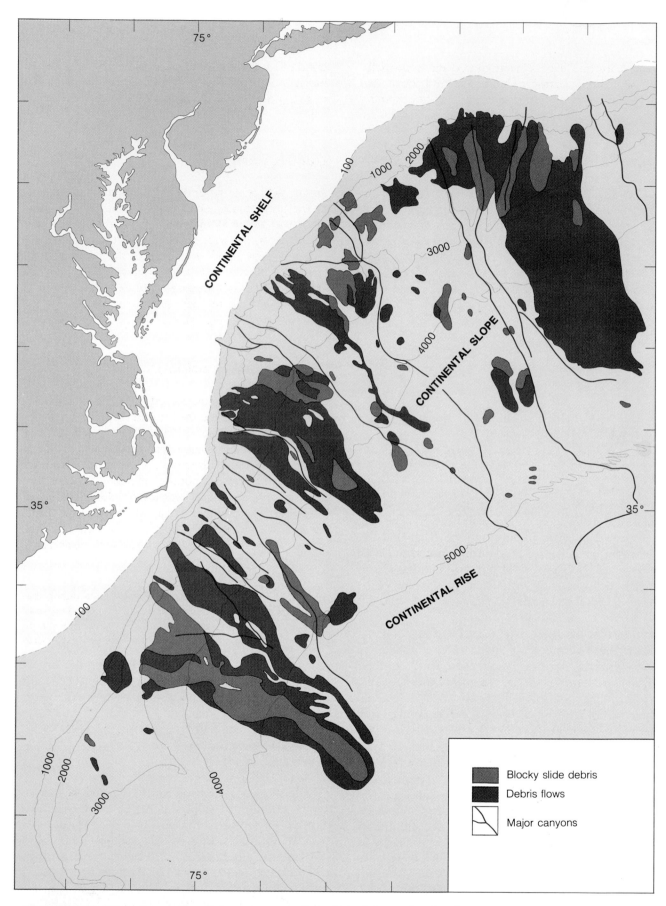

FIGURE 8.24 Map of a region off the east coast of the United States showing the distribution of large blocky landslide and debris-flow deposits on the continental slope and rise.

232

Alaska earthquake of 1964 triggered many rockfalls, one of which became a huge rock avalanche that swept across the surface of Sherman Glacier, burying it with up to several meters of coarse, angular debris.

Slope Modification

Landslides often result when natural slopes are modified by human activities. Slides often occur, for example, where roads have been cut into regolith or unstable rock, creating an artificial slope that exceeds the angle of repose (Fig. 8.25). Such landslides are especially common along the coastal cliffs of California where roads have been carved into deformed sedimentary rocks. Retaining walls may reduce the likeli-

hood of landslides in thick colluvium, but unless such barriers are very strong, persistent downslope creep of the colluvial debris may ultimately cause them to fail.

Undercutting

Slumps and other types of landslides can be triggered by the undercutting action of a stream along its bank or by surf action along a coast. Coastal landslides are often associated with major storms that direct their energy against rocky headlands or the bases of cliffs of unconsolidated sediments. Windward coasts of the Hawaiian Islands retreat as pounding surf removes jointed lava from the base of steep seacliffs, causing them to collapse (Fig. 8.26).

Exceptional Precipitation

Landslides are often associated with heavy or persistent rains that saturate the ground and make it unstable. Such was the case in 1925 when prolonged rains, coupled with melting snow, started a large debris flow in the Gros Ventre River basin of western Wyoming. The water saturated a porous sandstone that overlies impermeable shale and dips toward the valley floor. This saturated condition was an ideal trigger for slope failure. An estimated 37 million m^3 of rock, regolith, and organic debris moved rapidly downslope and cre-

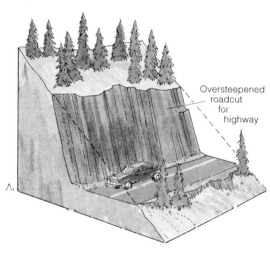

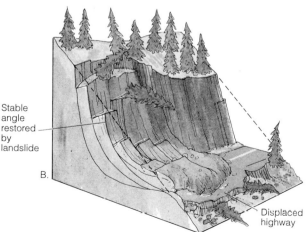

FIGURE 8.25 Modification of a slope in building a road leads to slope failure. A. In constructing a road, the natural angle of repose of the sediment is exceeded. B. The oversteepened slope fails, and a landslide buries the road. In the process, the natural angle of repose is reestablished.

FIGURE 8.26 Steep seacliffs of jointed basalt along the windward coast of Hawaii are undercut by pounding surf. When a cliff collapses, the resulting landslide debris is rapidly reworked by surf and currents, and the process begins anew.

ated a natural dam that ponded the river. Two years later the dam failed, causing a flood that led to several deaths. Today, more than 65 years after the debris flow began, the scar at the head of the slide is still quite obvious, as is the distinctive chaotic topography downslope.

Volcanic Eruptions

Volcanic eruptions are another means of initiating mass-wasting events. Large stratovolcanoes consist of inherently unstable accumulations of interstratified lava flows, rubble, and pyroclastic layers that form steep slopes. On high, ice-clad volcanoes, slopes may be further oversteepened by glacial erosion. During eruptions, slope failure is often widespread. Large volumes of water, released when summit glaciers and snowfields melt during eruption of hot lavas or pyroclastic debris, can combine with unconsolidated deposits to form mudflows that move rapidly down valleys.

Submarine Slope Failures

Factors leading to unstable conditions on continental slopes and delta fronts include (1) high internal water pressures resulting from rapid deposition of sediment and the inability of water trapped in accumulating sediment to escape; (2) generation of methane gas from organic matter deposited with sediments, which increases pressure in pores between grains; (3) local oversteepening of slopes, a result of high rates of sedimentation; (4) displacements along vertical fractures in seafloor sediments or rocks, which leads to oversteepening of slopes; and (5) shocks induced by earthquakes.

MASS-WASTING HAZARDS

As the human population increases and cities and roads expand across the landscape, the likelihood that mass-wasting processes will affect people increases. Landslides occur worldwide, and their impact, in terms of loss of lives and property, can be devastating (Table 8.3). In the United States alone, landslides in a typical year cause at least $1 billion in economic losses and 25 to 50 deaths, and the figures are rising. Although it may not always be possible to predict accurately the occurrence of significant mass-wasting events, a knowledge of the processes and their relationship to local geology can lead to intelligent plan-

TABLE 8.3 Fatalities resulting from some major landslides during this century[a]

Year	Location[b]	Fatalities
1916	Italy, Austria	10,000
1920	China[e]	200,000
1945	Japan[f]	1200
1949	USSR[e]	12,000–20,000
1954	Austria	200
1962	Peru	4000–5000
1963	Italy	2000
1970	Peru[e]	70,000
1985	Colombia[v]	23,000
1987	Ecuador[e]	1000

[a]*Source:* National Research Council (1987).
[b]Landslides related to earthquakes ([e]), floods ([f]), and volcanic eruptions ([v])

ning that will help reduce the loss of lives and property.

Assessments of Hazards

Assessments of potential hazards resulting from major mass-wasting events are based mainly on reconstructions of similar past events in order to evaluate their magnitude and frequency. From such information, it is possible to calculate how often an event of a certain magnitude is likely to recur.

Maps showing potential areas of impact of mass-wasting events are important tools for land-use planners. For example, large debris avalanches and small rockfalls are ever-present hazards in the northern Italian Alps (Fig. 8.27). Field studies have shown that large debris avalanches similar to that of 1717 (Fig. 8.19) have repeatedly blanketed valley floors with rocky debris during the last 3000 years. From this evidence, a map has been constructed showing areas that could be affected by future rockfalls having various trajectories and distribution patterns. A number of small communities are at hazard from small- to intermediate-size rock avalanches (traveling 3 to 5 km), while several large communities, including one village with a population of several thousand, could be affected by large debris avalanches (traveling up to 7 km) like some recorded in deposits on the valley floors.

Valleys in the Cascade Range of Washington and Oregon contain deposits of large mudflows that spread from high volcanoes repeatedly during the last 10,000 years. Based on the number and extent of such deposits, hazards maps have been prepared, like that shown for Mount Rainier and vicinity in Figure 8.28. This map shows that risk from mudflows is high within

FIGURE 8.27 A new apartment building at the base of a steep mountain slope in the Italian Alps was struck by a large boulder falling from the cliffs above. This relatively small rockfall, which occurred just one day before the new owners were to move in, demolished the bedroom and most of the living room.

FIGURE 8.28 Map of the southeastern Puget Lowland, Washington, showing areas of low, moderate, and high risk from mudflows and floods originating at Mount Rainier volcano. Also shown is the extent of the huge prehistoric Osceola mudflow that was associated with a summit eruption about 5700 years ago.

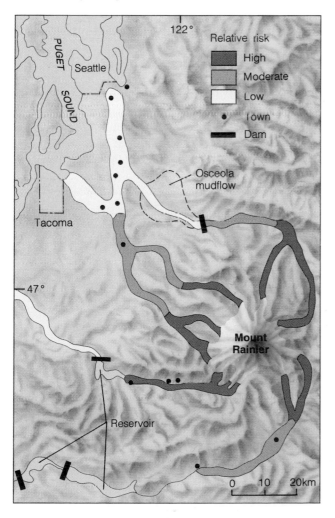

about 25 km of the volcano's slopes, and that risk exists even at distances of 100 km or more along densely populated valley floors. A similar map prepared prior to the 1980 eruptions of Mount St. Helens proved prophetic, for mudflows generated during that series of eruptions had distributions closely similar to those predicted on the basis of geologic studies of past events.

Mitigation of Hazards

The impact of mass-wasting processes on human environments often can be reduced or eliminated by careful advanced planning. Slopes subject to creep can be stabilized by draining or pumping water from saturated sediment, while oversteepened hillslopes can be prevented from slumping if they are regraded to angles equal to or less than the natural angle of repose. In some mountain valleys subject to mudflows from active volcanoes, water-filled reservoirs can be quickly emptied so that dams will pond potentially destructive mudflows before they reach population centers (Fig. 8.28). Although we generally have no way of anticipating or preventing large rockfalls and debris avalanches, eliminating or restricting human activities in possible impact zones offers the best means of mitigating such hazards.

HIMALAYAN LANDSLIDES AND PLATE TECTONICS

If we were to plot on a map the location of the world's major historic and prehistoric landslides, we will find that most tend to cluster along belts that lie close to the boundaries between converging lithospheric plates. They do so for two main reasons.

First, the world's highest mountain chains lie at or near plate boundaries, and on steep mountain slopes the safety factor often lies close to 1. The rocks of many mountain ranges consist of well-jointed strata that have been strongly fractured and deformed as they were uplifted. Both the joint planes and the bedding surfaces are potential zones of failure. Furthermore, along these belts lie the world's high stratovolcanoes, the slopes of which also tend to lie close to the angle of repose.

Second, it is along the boundaries between plates, where plate margins slide past or over one another, that most large earthquakes occur. Earthquakes also are associated with upward-moving magma that feeds volcanic eruptions at the Earth's surface. All of the major landslides listed in Tables 8.2 and 8.3 occurred in active tectonic zones near plate margins, and several are known to have been directly related to major earthquakes.

Nanga Parbat (8125 m), the tenth highest mountain in the world, lies at the boundary where the northward-moving Indian Plate meets the vast Eurasian Plate. The impact of the converging plates has led to uplift of the Himalaya, the highest mountain range in the world. Geologists have determined that the present uplift rate at Nanga Parbat is about 5 mm/yr. At this rate, the mountain should increase in altitude by 5000 m every million years. However, high mountains mean high erosion rates, and erosion is tearing down the mountain almost as rapidly as it is rising.

At the base of Nanga Parbat, only 21 km from the summit, lies the deep gorge of the Indus River, one of the two largest streams draining the Himalaya. From the river to the top of the mountain, the vertical *relief* (the difference in altitude between highest and lowest points on a landscape) measures nearly 7000 m (Fig. B8.1). High relief, steep slopes, fractured rocks, persistent undercutting by active streams and glaciers, and frequent large earthquakes all make Nanga Parbat an obvious place for landsliding. In fact, landsliding is probably the most effective of the agents involved in tearing down the mountain. The record of one such landslide, and its resulting human impact, has been vividly recorded.

In early June 1841, a Sikh army was camped upstream from the town of Attock, which lies beyond the mouth of the steep gorge where the Indus flows out of the Himalaya. Suddenly, in midafternoon, a huge wall of muddy debris came rushing out of the gorge and overwhelmed the army. An eyewitness described the

SUMMARY

1. Mass-wasting causes rock debris to move downslope under the pull of gravity without a transporting medium. Mass-wasting occurs both on land and beneath the sea.

2. The composition and texture of debris, the amount of air and water mixed with it, and the steepness of slope influence the type and velocity of slope movements.

3. Mass-wasting processes include sudden slope failures (slumps, falls, and slides) and downslope flow of mixtures of sediment, water, and air.

4. Failures occur when the shear stress reaches or exceeds the shear strength of slope materials.

High water pressure in rock voids or sediment reduces shear strength and increases the likelihood of failure.

5. Slumps involve a rotational movement along a concave-up surface that results in backward-tilted blocks of rock or regolith.

6. Falling and sliding masses of rock and debris are common in mountains where steep slopes abound.

7. Rockfall debris accumulates at the base of a cliff to produce a talus with slopes that stand at the angle of repose.

FIGURE B8.1 Nanga Parbat, seen from the deep gorge of the Indus River, some 7000 m below the mountain's summit. Steep slopes and frequent earthquakes make this region particularly susceptible to large-scale landsliding.

terrible scene: "It was a horrible mess of foul water, carcasses of soldiers, peasants, war-steeds, camels, prostitutes, tents, mules, asses, trees, and household furniture, in short every item of existence jumbled together in one flood of ruin . . . As a woman with a wet towel sweeps away a legion of ants, so the river blotted out the army of the Raja."

The cause of the disaster lay 400 km up the river canyon where, in January of that same year, an earthquake caused a spur of Nanga Parbat to collapse, triggering a massive landslide that rushed downslope toward the river. The slide debris quickly dammed the Indus, forming a lake that steadily grew until it was 150 m deep and more than 30 km long. As the rising water reached the top of the landslide dam, it overflowed and rapidly cut through the unconsolidated debris. The gigantic flood of water that was released rushed swiftly downstream, sweeping everything before it, toward the unsuspecting army on the plains below.

8. Slurry flows involve dense moving masses of water-saturated sediment that form nonsorted deposits when flow ceases. Flow velocities range from very slow (solifluction) to rapid (debris flows).

9. In granular flows, sediment is in grain-to-grain contact or grains constantly collide. The sediment may be largely dry or it may be saturated with water that can escape easily.

10. Although creep is imperceptibly slow, it is widespread and therefore quantitatively important in the downslope transfer of debris.

11. Large, rapidly moving debris avalanches are rela-tively infrequent but potentially hazardous to humans.

12. In regions of perennially frozen ground, frost heaving, creep, and gelifluction are quantitatively important mass-wasting processes.

13. Large areas of seafloor on the continental slopes show evidence of widespread slumps, slides, and flows. Mass-wasting on submarine slopes was especially active during glacial ages when sea level was lower and large quantities of stream sediment were transported to the edge of the continental shelves.

14. Slope failures can be triggered by earthquakes,

undercutting by streams, heavy or prolonged rains, or volcanic eruptions. Subaqueous slope failures are frequently related to rapid deposition of sediments, oversteepening of slopes, and earthquake shocks.

15. Loss of life and property from mass-wasting events can be prevented or mitigated by adequate assessment and planning based on geologic studies of previous occurrences.

IMPORTANT TERMS TO REMEMBER

angle of repose (p. 220)

colluvium (p. 225)
creep (p. 223)

debris avalanche (p. 227)
debris fall (p. 219)
debris flow (p. 221)
debris slide (p. 219)

earthflow (p. 225)

frost heaving (p. 228)

gelifluction (p. 229)
granular flow (p. 221)

landslide (p. 218)
liquefaction (p. 227)

mass-wasting (p. 216)
mudflow (p. 222)

rockfall (p. 219)
rock glacier (p. 229)
rockslide (p. 219)

shear strength (p. 216)
shear stress (p. 216)
sediment flows (p. 221)
slump (p. 218)
slurry flow (p. 221)
solifluction (p. 221)

talus (p. 220)

QUESTIONS FOR REVIEW

1. How does mass-wasting differ from weathering and from stream erosion?

2. What primary factors influence shear stress and shear strength on hillslopes? How can the shear strength of sediment or a rock mass on a hillslope suddenly change?

3. Why can the presence of water in rock or regolith promote downslope movement?

4. What distinctive landscape features might enable you to identify an area where numerous slumps have occurred?

5. What conspicuous type of deposit generally is found at the base of a cliff subject to frequent rockfalls? How would you expect the particles in the deposit to be sorted?

6. Why is a deposit that has been artificially excavated so that its surface exceeds the natural slope angle likely to fail?

7. How might one prove that creep is occurring on a slope and how can its rate be measured?

8. Why are lava flows erupted from stratovolcanoes largely restricted to the volcanic cones, while mudflows originating on the same mountains are often distributed for many tens of kilometers beyond the volcanoes along adjacent valleys?

9. Explain why large-scale mass-wasting on the continental slopes apparently was far more active during the last glacial age than it is today.

10. Why are large debris avalanches and volcanic mudflows generally far more dangerous to people than are lava flows?

11. How might prolonged and heavy rainfall affect the shear strength of a body of regolith and make it susceptible to failure?

12. What geologic conditions make high mountains especially prone to landslide activity?

STREAMS IN THE LANDSCAPE

Almost anywhere we travel over the land surface, we can see evidence of the work of running water. Even in places where no rivers flow today, we are likely to find deposits and landforms that tell us water has been instrumental in shaping the landscape. Most of these features can be related to the activity of streams that are part of complex drainage systems.

A **stream** is a body of water that flows downslope along a clearly defined natural passageway, in the process transporting detrital particles and dissolved substances. The passageway is called the stream's **channel,** and the detritus constitutes the bulk of its **load,** which is the sediment and dissolved matter the stream transports. The quantity of water passing by a point on the streambank in a given interval of time is a measure of the stream's **discharge.** As a stream moves sediment from place to place, its channel is continually being altered. A stream and its channel are closely related and form an ever-changing, interrelated system.

Streams play an important role in our lives. Large streams, like the Amazon, the Rhine, and Mark Twain's Mississippi, are important avenues of transportation. Many of the world's great cities are built in stream valleys; New Orleans, Cairo, London, Paris, Rome, and Moscow are examples (Fig. 9.1). People choose to live near streams because valley floors are flat and easy to build on, soils tend to be deep and fertile, and water is available. But stream valleys have drawbacks as well. They can be threatened by floods, and as cities grow, human and industrial wastes begin to pollute the water. How to achieve an acceptable balance between human needs and the capacities of streams to maintain a safe, clean water supply is one of the major issues facing society.

In addition to their immediate practical and esthetic importance, streams are vital geologic agents:

Streams carry most of the water that goes from land to sea and so are an essential part of the hydrologic cycle.

Streams transport billions of tons of sediment to the oceans each year; there, the sediment is deposited and eventually becomes part of the rock record.

Streams carry to the sea small amounts of soluble salts released by weathering that play an essential role in maintaining the saltiness of seawater.

Streams shape the surface of the Earth. Most landscapes consist of stream valleys separated by higher ground and are the result of weathering, mass-wasting, and stream erosion working in combination.

FIGURE 9.1   Paris, like many of the world's great cities, was founded along the banks of a major river. The Seine provides water for human and industrial use, is an avenue of transportation, and has great esthetic and recreational value. However, under the stress of a growing population, the Seine, like other urban rivers throughout the world, is increasingly susceptible to pollution.

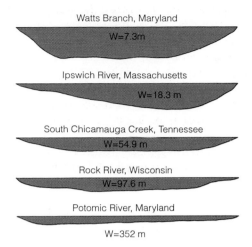

FIGURE 9.2 Cross sections of some natural streams, drawn so that their widths (W) are at the same scale. In general, the wider the channel, the larger the ratio of width to depth.

STREAM CHANNELS

A stream's channel is an efficient conduit for running water. The discharge varies both along the channel and through time, mainly because of changes in precipitation. In response to varying discharge and load, the channel continuously adjusts its shape and orientation. Therefore, a stream and its channel are dynamic elements of the landscape.

Cross-Sectional Shape

The size and shape of any particular channel cross section reflect the typical stream conditions at that place. Very small streams may be as deep as they are wide, whereas very large streams usually have widths many times greater than their depths (Fig. 9.2). Because the volume of water moving through a channel generally increases downstream, it follows that the ratio of channel width to channel depth is likely to change downstream as the volume of water increases.

Long Profile

If we measure the vertical distance that a stream channel falls between two points along its course, we will have obtained a measure of the stream's **gradient.** The average gradient of a steep mountain stream, such as the Sacramento River of California, may reach 60 m/km or even more, whereas near the mouth of a large stream like the Missouri River, the gradient may be as low as 0.1 m/km or even less (Fig. 9.3). The average gradient of a river decreases downstream, and so the stream's **long profile** (a line drawn along the surface of a stream from its source to its mouth) is a curve that decreases in gradient downstream (Fig. 9.3). However, a long profile is not a perfectly smooth curve because irregularities in the gradient occur along the channel. For example, a local change in gradient may occur where a channel passes from a bed of resistant rock into one that is more erodible, or where a landslide or lava flow forms a temporary dam across the channel. A hydroelectric

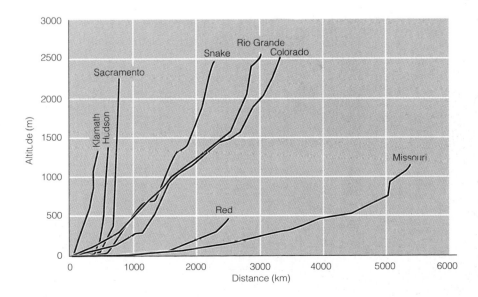

FIGURE 9.3 Long profiles of some streams in the United States. The Klamath, Hudson, and Sacramento are relatively short, steep-gradient streams, whereas the Missouri is a long stream with a low average gradient.

dam also introduces an irregularity in the long profile of a stream channel and may create an extensive reservoir upstream. Abrupt changes in the long profile cause water to flow rapidly and turbulently through a stretch of rapids or to plunge over a steep drop as a waterfall.

DYNAMICS OF STREAMFLOW

The average annual rainfall on the area of the United States is equivalent to a layer of water 76 cm thick covering this same land surface. Of this layer, 45 cm returns to the atmosphere by evaporation and transpiration (Fig. 1.23) and 1 cm infiltrates the ground; the remaining 30 cm forms **runoff,** the portion of precipitation that flows over the land surface. By standing outside during a heavy rain, you can see that water initially tends to move down slopes in broad, thin sheets, a process called **overland flow.** You will also notice, however, that after traveling a short distance overland flow begins to concentrate into well-defined channels, thereby becoming **streamflow.** Runoff is a combination of overland flow and streamflow.

Factors in Streamflow

Several basic factors control the way a stream behaves: (1) *gradient,* expressed in meters per kilometer; (2) *channel cross-sectional area* (width × average depth), expressed in square meters; (3) *average velocity* of water flow, expressed in meters per second; (4) *discharge,* expressed in cubic meters per second; and

(5) *load,* expressed in kilograms per cubic meter. Unlike the sediment forming the bulk of a stream's load, dissolved matter generally does not affect stream behavior.

Because gravity pulls on water, just as it pulls on rock and regolith, the steepness of a stream's gradient is a factor in stream behavior. A stream plunging over a waterfall or cascading down a series of steep rapids obviously is behaving differently than the same stream where it reaches level terrain.

The relationship among discharge, velocity, and channel shape can be expressed by the equation

$$Q \quad = \quad A \quad \times \quad V$$

Discharge	Cross-sectional area	Average
(m^3/s)	of channel (width × average depth) (m^2)	velocity (m/s)

This equation tells us that when discharge changes, one or more of the factors on the right side of the equation also must change if equilibrium is to be maintained. For example, if discharge increases, both velocity and channel cross-sectional area are likely to increase so the channel can accommodate the added flow. Conversely, a decrease in discharge can lead to a corresponding decrease in channel dimensions and velocity.

A dramatic example of changes in stream factors can be seen when floods occur. During 1956, the channel of the Colorado River at Lees Ferry, Arizona, experienced a major change in dimensions as discharge increased and then fell (Fig. 9.4). Prior to the flood, the channel averaged about 2 m deep and 100 m wide. As discharge increased in late spring, the wa-

FIGURE 9.4 Changes in the cross-sectional area of the Colorado River at Lees Ferry, Arizona, during 1956. As discharge increased from February to June, the channel floor was scoured and deepened and the water level rose higher against the banks. During the falling-water phase, the river level fell and sediment was deposited in the channel, decreasing its depth.

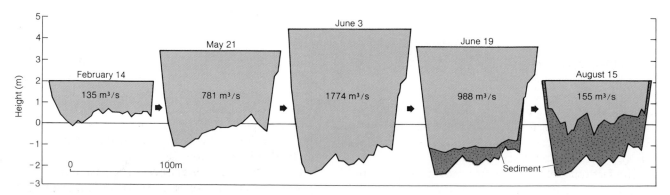

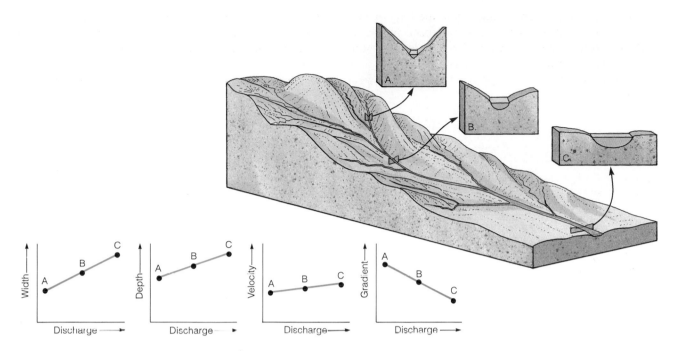

FIGURE 9.5   Changes in the downstream direction along a stream system. Discharge increases as new tributaries join the main stream. Width and depth of the channel are shown by cross sections A, B, and C. Graphs show the relationship of discharge to channel width and depth, to velocity, and to channel gradient at the same three cross sections.

ter rose in the channel and erosion scoured the bed until at peak flow the channel was about 7 m deep and 125 m wide. Together with an increase in velocity, the enlarged channel was now able to accommodate the increased flood discharge and carry a greater load. As discharge fell, the stream was unable to transport as much sediment, and the excess load was dropped in the channel, causing its floor to rise. At the same time, decreasing discharge caused the water level to fall, thereby returning the cross-sectional area to its preflood dimensions.

As we can see from this example, a stream and its channel are intimately related. The channel is so responsive to changes in discharge that the system, at any point along the stream, is always close to a balanced condition.

Changes Downstream

Traveling down a stream from its head to its mouth, we can see that orderly adjustments occur along the channel: (1) discharge increases; (2) channel cross-sectional area increases; (3) velocity increases slightly; and (4) gradient decreases (Fig. 9.5).

The fact that velocity increases downstream seems to contradict the common observation that water rushes turbulently down steep mountain slopes and flows smoothly over nearly flat lowlands. However, the physical appearance of a stream is not a true indication of its velocity. Discharge is low in the headward reaches of a stream, and therefore average velocity is also low. Discharge increases downstream as each **tributary** (a stream joining a larger stream) introduces more water. To accommodate the greater volume of water, velocity increases accordingly, together with the cross-sectional area of the channel.

Floods

The uneven distribution of rainfall through the year causes many streams to rise seasonally in flood. A *flood* occurs when a stream's discharge becomes so great that it exceeds the capacity of the channel, therefore causing the stream to overflow its banks (Fig. 9.6). People affected by floods are frequently surprised and even outraged at what a rampaging stream has done to them. Geologists, however, tend to view floods as normal and expectable events, for the geologic record shows that floods have been occurring as long as rain has been falling on the Earth.

Unusually large discharges associated with floods appear as major peaks on a *hydrograph,* a graph that

FIGURE 9.6 During a spring flood in 1982, the White River (in upper right corner of view) inundated the small town of Jacksonport, Arkansas, which is built on the river's floodplain.

plots stream discharge against time. In the example shown in Figure 9.7, a passing storm generates a brief interval of intense rainfall. As the runoff moves into the stream channel, the discharge quickly rises. The crest of the resulting flood, when peak flow is reached, passes the point where the discharge is being measured about 2 hours after the storm passes. It takes an additional 8 hours before the flood runoff passes through the channel and discharge decreases to the normal nonflood level.

As discharge increases during a flood, so does velocity. This velocity increase has the double effect of enabling a stream to carry not only a greater load, but also larger particles. The collapse of the large St. Francis Dam in southern California in 1928 provides an extreme example of the exceptional force of floodwaters. When the dam gave way, the water behind it rushed down the valley as a spectacular flood, moving blocks of concrete weighing as much as 9000 metric tons through distances of more than 750 m. Because natural floods are also capable of moving very large objects as well as great volumes of sediment, they are able to accomplish considerable geologic work.

Flood Prediction

Major floods can be disastrous events, causing both loss of life and extensive property damage (Table 9.1), and so it is highly desirable to be able to predict their occurrence. By plotting the frequency of occurrence of past floods of different sizes on a probability graph, a *flood-frequency curve* can be produced (Fig. 9.8). The measure of how often a flood of a given magnitude is likely to recur is called the *recurrence interval*.

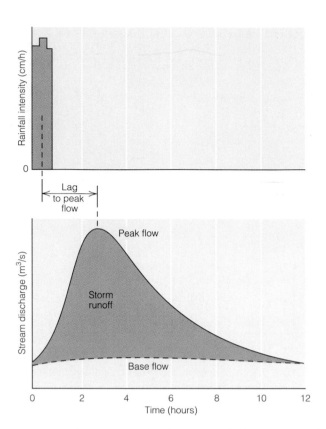

FIGURE 9.7 Hydrograph of a stream following a brief storm. An interval of intense rainfall causes a rise in stream discharge as the runoff passes through the stream. Peak discharge lags about 2.5 hours behind the peak rainfall. It takes an additional 8 hours for the storm runoff to pass and the discharge to fall to its prestorm value (the stream's base flow).

We can use a flood-frequency curve to estimate the probability that a flood of a certain magnitude will occur in any one year. In the case of the Skykomish River at Gold Bar, Washington, a flood having a discharge of 1750 m³/s has a 1 in 10 (10 percent) probability of occurring in any given year, whereas a larger flood of 2500 m³/s has a 1 in 50 (2 percent) probability. We refer to a flood having a recurrence interval of 10 years as a 10-year flood, whereas if the interval is 50 years, it is a 50-year flood, and so forth.

One potential problem with using flood-frequency curves to estimate the probability of future floods is that the curves are based on floods of the recent past. If the Earth's climate changes during the next several decades (chapter 19), present flood-frequency curves will be of little value in predicting future floods.

Catastrophic Floods

Exceptional floods—those well outside a stream's normal range—occur very infrequently, perhaps only once in several centuries. Even-greater floods, evidence for which we can find in the geologic record,

TABLE 9.1   Fatalities from some disastrous floods

River	Date	Fatalities	Remarks
Huang He, China	1887	ca. 900,000	Flood inundated 130,000 km² and swept many villages away
Johnstown, Pennsylvania	1889	2200	Dam failed. Wave 10–12 m high rushed downvalley
Yangtze, China	1911	ca. 100,000	Formed lake 130 km long and 50 km wide
Yangtze, China	1931	ca. 200,000	Flood extended from Hankow to Shanghai (>800 km), leaving tens of millions homeless
Vaiont, Italy	1963	2000	Landslide into lake caused wave that overtopped dam and inundated villages below

Source: *Encyclopedia Americana* (1983).

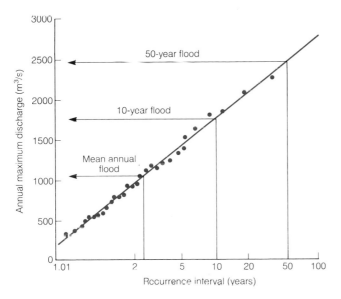

FIGURE 9.8   Curve of flood frequency for Skykomish River at Gold Bar, Washington, plotted on a probability graph. A flood with a discharge of 1750 m³/s has a 1 in 10 probability of occurring in any year (a 10-year flood), whereas a flood of 2500 m³/s has a 1 in 50 probability (a 50-yr flood).

can be viewed as catastrophic events that occur very rarely even on geologic time scales.

In the 1920s, geologist J Harlen Bretz began a study of a curious landscape in eastern Washington State called the Channeled Scabland. An array of dark, channel-like features mark places where bare lava flows lie exposed at the surface, stripped of the fertile topsoil that makes this a prime wheat-producing region (Fig. 9.9A). Bretz carefully documented the character and distribution of a variety of landforms that provide evidence of the Scabland's origin: dry coulees (canyons) with abrupt cliffs marking sites of former huge waterfalls, deep rock basins carved in the basalt, massive piles of gravel containing enormous boulders, linear deposits of gravel in the form of huge current

ripples (Fig. 9.9B), and upper limits of water-eroded land that lie hundreds of meters above valley floors.

Bretz considered different hypotheses to explain this array of features, but he was led inescapably to conclude that they could be accounted for only by a catastrophic event—a truly gigantic flood, far larger than any historic flood. The source of the enormous volume of flood water was resolved with the discovery that the continental ice sheet covering western Canada during the last glaciation had advanced across the Clark Fork River and dammed a huge lake in the vicinity of Missoula, Montana. The glacier-blocked lake contained between 2000 and 2500 km³ of water when it was filled and remained in existence only so long as the ice dam was stable. When the glacier retreated or began to float in the rising lake water, the dam failed, and water was released rapidly from the basin, as through a plug had been pulled from a giant bathtub. The only possible exit route lay across the Channeled Scabland region and down the Columbia River to the sea. Recent geologic studies have shown that the lake formed and drained repeatedly, creating numerous floods. The array of features scattered throughout the Scabland region thus provides us with dramatic evidence that the geologic work accomplished by catastrophic floods can be prodigious.

Base Level

As a stream flows downslope, its potential energy decreases and finally falls to zero as it reaches the sea. It now no longer has the ability to deepen its channel. The limiting level below which a stream cannot erode the land is called the **base level** of the stream. The base level for most streams is global sea level (Fig. 9.10). Exceptions are streams that drain into closed interior basins having no outlet to the sea. Where the floor of a tectonically formed basin lies below sea level (e.g., Death Valley, California), base level coincides with the basin floor. A stream flowing into such

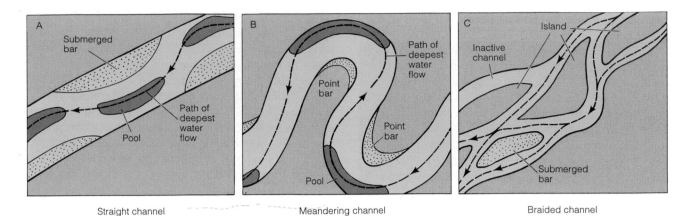

FIGURE 9.11 Features associated with (A) straight, (B) meandering, and (C) braided channels. Pools are places along a channel where water is deepest. Arrows indicate the direction of stream flow and trace the path of the deepest water.

the Menderes River (in Latin, *Meander*) in southwestern Turkey, which is noted for its winding course. Meanders are not accidental. They occur most commonly in channels that lie in fine-grained stream sediments and have gentle gradients. The meandering pattern reflects the way in which a river minimizes resistance to flow and dissipates energy as uniformly as possible along its course.

Try to wade or swim across a meandering stream and it quickly becomes apparent that the velocity of the flowing water is not uniform. Velocity is lowest along the bed and walls of the channel because here the water encounters maximum frictional resistance to flow. The highest velocity along a straight channel segment usually is found near the surface in midchannel. However, wherever the water rounds a bend, the zone of highest velocity swings toward the outside of the channel (Fig. 9.13).

Over time, meanders migrate slowly down a valley. As water sweeps around a meander bend, the zone of highest velocity swings toward the outer stream bank. Strong turbulence causes undercutting and slumping of sediment where the fast-moving water meets the steep bank. Meanwhile, along the inner side of each meander loop, where water is shallow and velocity is low, coarse sediment accumulates to form a *point bar* (Figs. 9.11B and 9.12). As a result, meanders slowly change shape and shift position along a valley as sediment is subtracted from and added to their banks.

The behavior of artificial streams and their channels has been studied in laboratory experiments using large sedimentation tanks. The experiments show that

FIGURE 9.12 A meandering stream near Pnom Penh, Cambodia, flows past agricultural fields that cover the river's floodplain. Light-colored point bars, composed of gravelly alluvium, lie opposite cutbanks on the outside of meander bends. Two oxbow lakes, the product of past meander cutoffs, lie adjacent to the present channel.

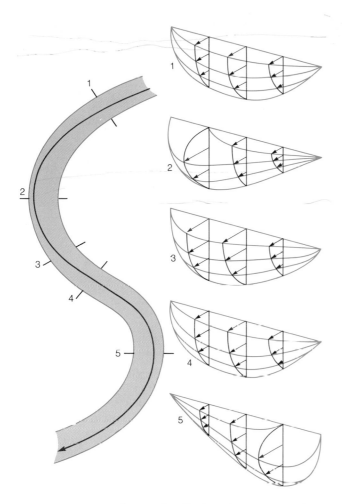

FIGURE 9.13 Velocity distribution in cross sections through a sinuous channel (lengths of arrows indicate relative flow velocities). The zone of highest velocity (red arrow) lies near the surface and toward the middle of the stream where the channel is relatively straight (sections 1 and 4). At bends, the maximum velocity swings toward the outer bank and lies below the surface (sections 2 and 5).

if the stream-bank sediment has a uniform grain size, meanders are symmetrical and all tend to migrate downstream at the same rate. In nature however, bank sediment generally is not uniform, as shown in Figure 9.14, which is an example from the lower Mississippi River. Wherever the downstream part of a meander that is cutting into sandy sediment encounters a less-erodible sediment, such as clay, its migration can be slowed. Meanwhile the next meander upstream, migrating more rapidly, may intersect the slower-moving meander (Fig. 9.14A). The water in the channel now can take a shorter route downstream over a steeper gradient, so the stream bypasses the meander which has been cut off. As sediment is deposited along the margin of the new channel route, the cutoff meander

is blocked and converted into a curved *oxbow lake* (Figs. 9.12 and 9.14A).

Nearly 600 km of the Mississippi River channel has been abandoned through cutoffs since 1776. However, contrary to the analysis of these channel changes offered by Mark Twain at the beginning of this chapter, the river has not been shortened appreciably because the loss of channel due to cutoffs has been balanced by lengthening of the channel as other meanders have enlarged.

Braided Channels

The intricate geometry of a **braided stream** resembles the pattern of braided hair, for the water repeat-

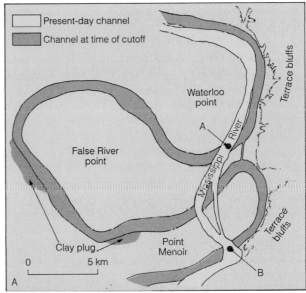

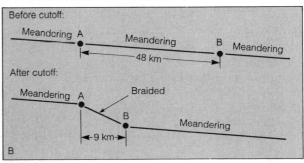

FIGURE 9.14 Cutoff of a meander loop of the Mississippi River in Louisiana. A. The downvalley migration of the river was halted when the channel encountered a body of clay in the floodplain sediments. This allowed the next meander loop to advance and finally cut off the river segment surrounding False River Point. B. The new, shorter channel had a steeper gradient than the abandoned course, and a braided pattern developed.

edly divides and reunites as it flows through two or more adjacent but interconnected channels separated by bars or islands (Fig. 9.11C). The cause of braiding is related to the ability of a stream to transport sediment. If a stream is unable to move all the available load, it tends to deposit the coarsest sediment as a bar which locally divides the flow and concentrates it in the deeper segments of channel to either side. As the bar builds up, it may emerge above the surface as an island and become stabilized by vegetation that anchors the sediment and inhibits erosion.

A braided pattern tends to form in streams having highly variable discharge and easily erodible banks that can supply abundant sediment load to the channel system. Streams of meltwater issuing from glaciers generally have a braided pattern because the discharge varies both daily and seasonally, and the glacier supplies the stream with large quantities of sediment. The braided pattern, therefore, seems to represent an adjustment by which a stream increases its efficiency in transporting sediment.

Large braided rivers typically have numerous constantly shifting shallow channels (Fig. 9.15). Although at any moment the active channels may cover no more than 10 percent of the width of the entire channel system, within a single season all or most of the surface sediment may be reworked by the laterally shifting channels.

EROSION BY RUNNING WATER

Erosion by water begins even before a distinct stream has formed. It occurs in two ways: by impact as raindrops hit the ground and by overland flow during heavy rains, a process known as **sheet erosion.** As raindrops strike bare ground they dislodge small particles of loose soil, spattering them in all directions. On a slope the result is net displacement downhill. One raindrop has little effect, but the number of raindrops is so great that together they can accomplish a large amount of erosion.

The effectiveness of raindrops and overland flow in eroding the land is greatly diminished by a protective cover of vegetation. The leaves and branches of trees break the force of falling raindrops and cushion their impact upon the ground. More important, the intricate network of roots forms a tight mesh that holds soil in place, greatly reducing erosion. The root network also holds water, letting it percolate slowly down through the soil. As a consequence, in vegetated areas there is less runoff than in areas of bare ground.

The ability of streams to erode is related to the way water moves through a stream channel. If the velocity is very slow, the water particles travel in parallel layers, a motion called **laminar flow** (Fig. 9.16). With increasing velocity, the movement becomes more erratic and complex, giving rise to the swirls and eddies that characterize **turbulent flow.** The velocity in stream channels is sufficiently high for turbulent flow to dominate. Only in a very thin zone along the bed and channel walls, where frictional drag is high, is velocity low enough for laminar flow to occur.

The ability of a stream to pick up particles of sediment from its channel and move them along depends largely on the turbulence and velocity of the water. Figure 9.17, based on experimental data, shows the velocities required to erode particles of different size from a stream bed, the range of velocity in which particles can be transported, and the velocities at which particles can no longer be moved and will settle to the bottom. In general, as velocity increases, so does the ability of the turbulent water to lift ever

FIGURE 9.15 Intricate braided pattern of Brahmaputra River where it flows out of the Himalaya en route to the Ganges delta. Noted for its huge sediment load, the river may be 8 km wide during the rainy monsoon season.

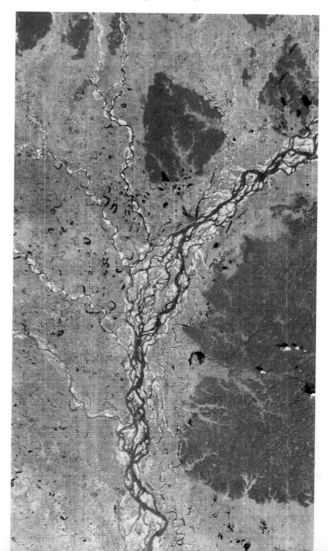

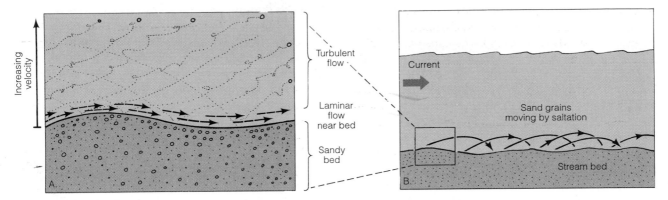

FIGURE 9.16 Laminar and turbulent flow in a stream channel. Laminar flow is confined to a thin zone along the bed and the channel walls where frictional drag is high and velocity is very low. Turbulent flow occurs as velocity increases above the bed and is responsible for moving bedload sand by saltation. In the turbulent zone, sand grains are carried up into the current where turbulent eddies locally reach the bottom or where suspended grains impact other grains on the bed.

larger particles of sediment. Silt and clay are an exception, for when very cohesive, they are difficult to erode except under conditions of high velocity.

THE STREAM'S LOAD

The solid portion of a stream's load consists of two parts. The first part is the coarse particles that move along the streambed (the **bed load**), while the second is the fine particles that are suspended in the water (the **suspended load**). Wherever they are dropped, these solid particles constitute **alluvium,** which is any detrital sediment deposited by a stream.

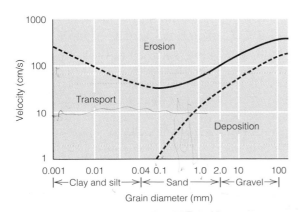

FIGURE 9.17 Graph showing how stream velocity controls erosion, transport, and deposition of sediment grains of different sizes.

Streams also carry dissolved substances (the **dissolved load**) that are chiefly a product of chemical weathering.

Bed Load

The bed load generally amounts to between 5 and 50 percent of the total load of most streams. Bed-load particles move at a lower velocity than the stream water, for the particles are not in constant motion. Instead, they move discontinuously by rolling or sliding. Where forces are sufficient to lift a particle, it may move short distances by saltation, a motion that is intermediate between suspension and rolling or sliding. **Saltation** involves the progressive forward movement of a particle in a series of short intermittent jumps along arcuate paths (Fig. 9.16). Saltation continues as long as currents are turbulent enough to lift particles and carry them downstream.

The distribution of bedload sediment in a stream channel is related to the velocity distribution (Fig. 9.18). Coarse-grained sediment is concentrated where the velocity is high, whereas finer-grained sediment is relegated to zones of progressively lower velocity.

Placer Deposits

The famous California gold rush of 1849 followed the discovery that the sand and gravel in the bed of a small stream contained bits of gold. Similar gold-bearing gravels are found in many other parts of the world. The gravels themselves are sometimes rich enough to

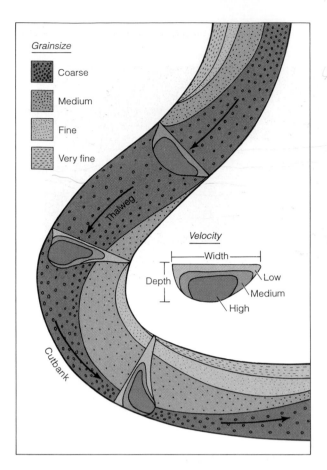

Grainsize

■ Coarse
▨ Medium
▦ Fine
▨ Very fine

Width
Depth
Low
Medium
High

Velocity

Thalweg

Cutbank

FIGURE 9.18 Relationship of bed load grain size to velocity in a section of meandering channel. The coarsest sediment is associated with the zone of highest velocity; both lie on the outside of a bend adjacent to the cutbank, but in the center of the channel between bends. The finest sediment is associated with the zone of lowest velocity and lies on the inside of a meander bend, opposite the cutbank.

be mined, but even when they are too lean, the gold is a clue that a source must lie upstream. In fact, many mining districts have been discovered by following trails of gold and other minerals upstream to their sources in veins in bedrock.

Because pure gold is heavy (specific gravity = 19), it is deposited from the bed load of a stream very quickly, while quartz, with a specific gravity of only 2.65, is washed away. As most silicate minerals are light by comparison with gold, grains of gold become mechanically concentrated in places where the velocity of stream flow is high enough to remove the light particles but not high enough to remove the heavy ones. Such concentration occurs, for example, behind rock bars or in bedrock holes along the channel, below waterfalls, on the inside of a meander bend, and downstream from the point where a tributary enters a main stream. A deposit of heavy minerals concentrated mechanically is a **placer.**

Many heavy, durable metallic minerals other than gold also form placers. These include minerals that occur as pure metals, such as platinum and copper, as well as tinstone (cassiterite, SnO_2) and nonmetallic minerals such as diamond, ruby, and sapphire. Even if a vein contains a low percentage of a mineral, the placer it yields may be quite rich. In order to become concentrated in placers, the minerals must be not only dense but also resistant to chemical weathering and not readily susceptible to cleaving as the mineral grains are tumbled in the stream.

Every phase of the conversion of gold in a fissure vein to placer gold has been traced. Chemical weathering of the exposed vein releases the gold, which then moves slowly downslope by· mass-wasting. In some places mass-wasting alone concentrates the gold sufficiently to justify mining. More commonly, however, the mineral particles get into a stream, which concentrates them more effectively than mass-wasting.

Most placer gold occurs as grains the size of silt particles, the "gold dust" of miners. Some of it is coarser. A lump of pebble or cobble size is a *nugget* (Fig. 9.19). The largest nugget ever recorded weighed 80.9 kg.

In following placers along stream channels, prospectors have learned that rounding and flattening (by pounding) of nuggets increase downstream. Therefore, when they find rough, angular nuggets, prospectors know the primary source is close.

Suspended Load

The muddy character of many streams is due to the presence of fine particles of silt and clay moving in suspension (Fig. 9.20). Most of the suspended load is derived from fine-grained regolith washed from areas unprotected by vegetation and from sediment eroded and reworked by the stream from its own banks. The Yellow River (or Huang He) of China is yellow because of the great load of yellowish silt it erodes and transports seaward from widespread deposits of eolian sediment that underlie much of its basin.

Because upward-moving currents within a turbulent stream exceed the velocity at which particles of silt and clay can settle toward the bed under the pull of gravity, such particles tend to remain in suspension longer than they would in nonturbulent waters. They settle and are deposited only where velocity decreases and turbulence ceases, as in a lake or in the sea.

Dissolved Load

Even the clearest stream water contains dissolved chemical substances that constitute part of its load. Only seven ions make up the bulk of the dissolved

254

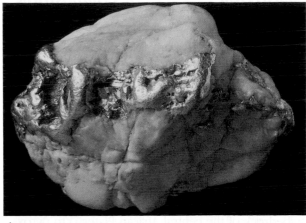

A.

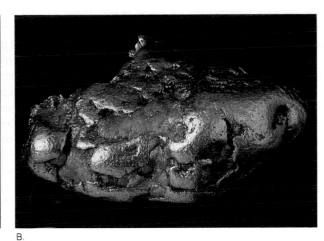B.

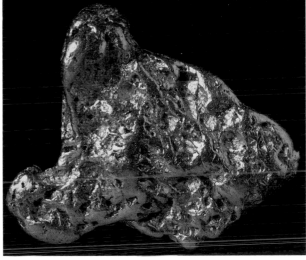C.

FIGURE 9.19 Formation of a nugget. A. A vein of metallic gold cutting through a pebble of vein quartz. Stream abrasion causes the brittle quartz to chip and be reduced in size, while the malleable gold deforms but is not reduced. B. The ratio of gold to quartz increases as the quartz is abraded away. Eventually a nugget of almost solid gold forms. C. A nugget of metallic gold from California. No quartz remains. Each of the specimens has a diameter of about 4 cm.

content of most rivers: bicarbonate (HCO_3^{1-}), calcium (Ca^{2+}), sulfate (SO_4^{2-}), chloride (Cl^{1-}), sodium (Na^{1+}), magnesium (Mg^{2+}), and potassium (K^{1+}).

Although in some streams the dissolved load may represent only a few percent of the total load, in others it amounts to more than half. Streams that receive large contributions of underground water generally have higher dissolved loads than those whose water comes mainly from surface runoff.

Downstream Changes in Grain Size

The size of the particles a stream can transport is mainly related to the flow velocity. Therefore, we might expect the average size of sediment to increase in the downstream direction as velocity increases. In fact, the opposite is true; sediment normally decreases in coarseness downstream. The explanation for this unexpected result involves both sorting and abrasion.

FIGURE 9.20 A large suspended load, eroded from extensive deposits of wind-blown silt, gives the Huang He a very muddy appearance and its English name (Yellow River).

The profile of most fans, from top to base in any direction, has the same curved form characteristic of the long profiles of streams. The exact form of the profile depends chiefly on discharge and on the size of particles in the bed load. Hence, no two fans are exactly alike. A small stream carrying a load of coarse particles builds a shorter, steeper fan than a larger stream carrying a load of finer particles. The area of a fan generally is closely related to the size of the area upstream from which its sediments are derived.

Deltas

When a stream enters the standing water of the sea or a lake, its speed drops rapidly, its ability to transport sediment decreases markedly, and it deposits its load. As we learned in chapter 4, a sedimentary deposit that forms where a stream flows into standing water is a **delta,** so-named because it may develop a crudely triangular shape that resembles the Greek letter delta (Δ) (Fig. 9.27).

Deltas built by streams transporting coarse sediments are of two types. A gravel-rich delta that is formed where an alluvial fan is building outward into a standing body of water (a *fan delta,* Fig. 9.28A) typically is built adjacent to a mountain front. Stream-channel, sheet-flood, and debris-flow sediments characteristic of alluvial fans form the upper parts of such deltas. The sediments in a fan delta show evidence of highly variable currents and abrupt changes of facies.

A *braid delta* is a coarse-grained delta constructed by a braided stream that builds outward into a standing body of water (Fig. 9.28A). Its upper part displays features characteristic of braided streams. Braid deltas are especially conspicuous where braided glacial meltwater streams flow into lakes or the sea.

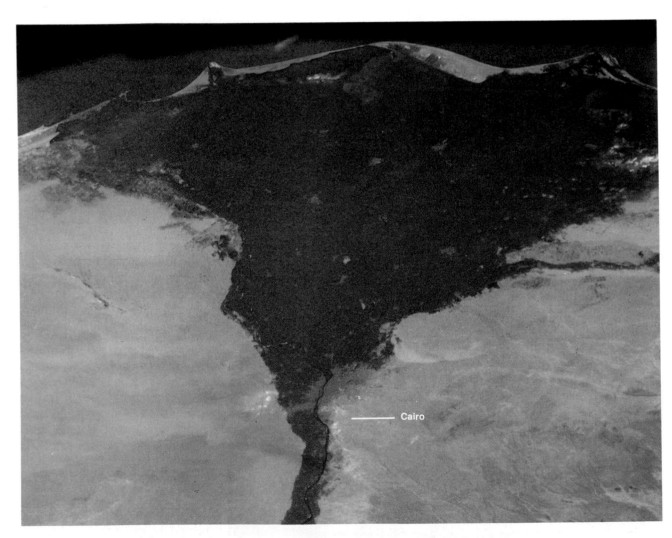

FIGURE 9.27 View of the Nile delta as seen from an orbiting spacecraft. Darker-colored, well-watered agricultural land of the delta and Nile Valley contrasts with barren desert landscape beyond.

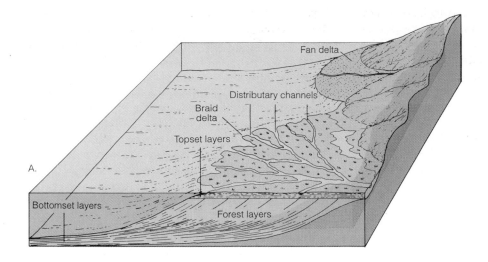

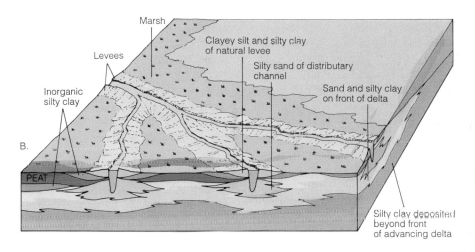

FIGURE 9.28 Main features of deltas. A. A braid delta built into a lake displays topset, foreset, and bottomset layers. A nearby fan delta is an alluvial fan that is building out into the body of water. B. Part of a large fine-grained delta built into the sea shows the intertonguing relationship of coarse channel deposits and finer sediments deposited on the delta front and beyond.

As a stream enters standing water, particles of the bed load are deposited first, in order of decreasing weight. Then the suspended sediments settle out. A layer representing one depositional event (such as a single flood) therefore grades from coarse sediment at the stream mouth to finer sediment offshore. The accumulation of many successive layers creates an embankment that grows progressively outward (Fig. 9.28A). The coarse, thick, steeply sloping part of a depositional layer in a delta is a *foreset layer*. Traced away from the shore, the same layer changes facies, becoming rapidly thinner and finer, and covering the bottom over a wide area. This part of a depositional layer in a delta is called a *bottomset layer*.

As deposition proceeds and the delta builds outward, the coarse foreset layers progressively overlap the bottomset layers. Thus, the stream gradually extends its channel outward over the growing delta. Coarse channel deposits and finer sediment deposited between channels, together called *topset layers*, overlie the foreset layers in a delta (Fig. 9.28A).

Many of the world's largest streams, among them the Ganges-Brahmaputra, the Huang He, the Amazon, and the Mississippi, have built massive deltas at their mouths. Each delta has its own peculiarities, determined by such factors as the stream's discharge, the character and volume of its load, the shape of the bedrock coastline near the delta, the offshore topography, and the intensity and direction of currents and waves.

Most major streams transport large quantities of fine suspended sediment, the bulk of which is carried seaward as the fresh stream water overrides denser salt water at the coast. The fine sediment then settles out to form a gently sloping delta front.

Where strong currents and wave action redistribute sediment as quickly as it reaches the coast, delta formation may be inhibited. However, if the rate of sediment supply exceeds the rate of coastal erosion, then a delta will be built seaward. The Mississippi River delivers a huge sediment load to the margin of the Gulf of Mexico each year. Much of the load is deposited along and around numerous *distributaries,* which are long fingerlike channels that branch from the main channel. The coarsest sediment is found along the channels. Finer sediment reaches the front of the delta and also accumulates between distributary channels during floods. The result is a complex intertonguing of facies and an intricate delta margin (Fig. 9.28B).

DRAINAGE SYSTEMS

To city dwellers used to a structured, artificial environment, nature sometimes seems to lack any obvious organization or pattern. But organization does exist, even though we may have to look for it. Streams are not distributed randomly across the landscape, but are organized into intricate drainage systems, the geometry of which can provide clues about the underlying geology and the evolution of continents.

Drainage Basins and Divides

Every stream is surrounded by its **drainage basin,** the total area that contributes water to the stream. The line that separates adjacent drainage basins is a **divide.** Drainage basins range in size from less than a square kilometer to vast areas of subcontinental dimension (Fig. 9.23). The vast drainage basin of the Mississippi River encompasses an area that exceeds 40 percent of the area of the contiguous United States (Fig. 9.29). Not surprisingly, the area of any drainage basin is proportional to both the length and the mean annual discharge of the stream that drains the basin.

Stream Order

The arrangement and dimensions of streams in a drainage basin tend to be orderly. This can be verified by examining a stream system on a map and numbering the observed stream segments according to their position, or *order,* in the system. The smallest segments lack tributaries and are classified as first-order streams. Where two first-order streams join they form a second-order stream, which has only first-order tributaries. Third-order streams are formed by the joining of two second-order streams and can have first-order and second-order tributaries, and so forth for successively higher stream orders (Fig. 9.30).

If, for any drainage basin, the number of streams assigned to each order is counted, we quickly see that the sums increase with decreasing stream order. In its pattern of a main stream joined by increasingly greater numbers of successively smaller tributaries, a stream system resembles a tree with its trunk and increasing numbers of sucessively smaller branches. This orderliness is like that inherent in a stream's long profile, in which gradient decreases systematically from head to mouth, while discharge, velocity, and channel dimensions increase. All these relationships imply that in response to a given quantity of runoff, stream systems develop with just the size and spacing required to move the water off each part of the land with greatest efficiency.

Evolution of Drainage

A stream system can develop quickly, as indicated by the following example. In August 1959, an earthquake

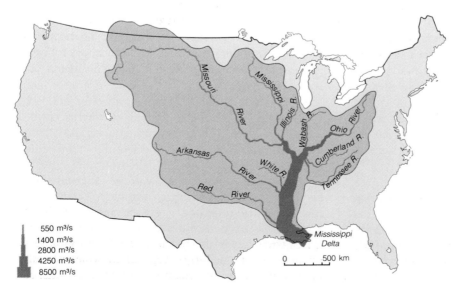

FIGURE 9.29 The drainage basin of the Mississippi River encompasses a major portion of the central United States. In this diagram, the width of the river and its major tributaries reflect discharge values.

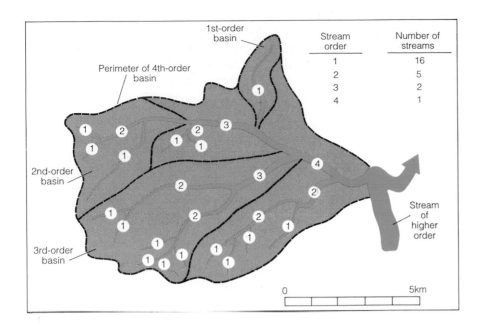

Stream order	Number of streams
1	16
2	5
3	2
4	1

FIGURE 9.30 Drainage basin of a 4th-order stream in the Appalachian region showing tributary channels numbered according to stream order. Both the number of tributaries and their length are related to stream order. Basins are classified according to the order of the largest stream they contain.

raised and tilted the bed of Hebgen Lake, near West Yellowstone, Montana, exposing a large area of silt and sand. With the first rain, small stream systems began to develop on the newly exposed lakebed. Sample areas were surveyed and mapped one and two years after the earthquake. The results showed the same basic geometry that characterizes much larger and older stream systems. The small, newly formed valleys, together with the areas between them, were disposing of the available runoff in a highly systematic way, and all within a period of two years after the surface had emerged from beneath the lake.

As a system of drainage develops, details of its pattern change. Streams acquire new tributaries, and some old tributaries are lost as a result of *stream capture,* which is the interception and diversion of one stream by another stream that is expanding its basin by erosion in the headward direction. When stream capture occurs, some stream segments are lengthened and others are shortened. Just as the hydraulic factors within a stream are constantly adjusting to changes, so too is the drainage system constantly changing and adjusting as it grows. Like a stream channel, a drainage system is a dynamic system tending toward a condition of equilibrium.

Drainage Patterns, Rock Structure, and Stream History

One of the best ways to view stream systems on the landscape is from the window of an orbiting spacecraft; next best is from an airplane. From an altitude of 8 or 9 kilometers, stream patterns can tell us a great deal about geologic structure and landscape history.

The ease with which a formation is eroded by streams depends chiefly on its composition and structure. The course a stream takes across the land therefore bears a close relationship to these factors. Thus, drainage patterns we can see from an airplane or those that can be traced on a topographic map give us information about underlying rock type and structure. Figure 9.31 shows some of the most common drainage patterns and the geologic factors that control them. An experienced geologist can use these drainage patterns to infer such things as rock type, the direction and degree of slope of a dipping rock unit, the manner in which the rocks are folded or offset, and the orientation and spacing of joints.

A close relationship between streams and the rock units across which they flow can provide important insights regarding the structure and geologic history of an area. Geologists classify such relationships into several categories that reflect distinctive stream histories (Fig. 9.32).

Continental Divides

Every continent except ice-covered Antarctica can be divided into large regions in each of which all through-flowing streams enter one of the world's major oceans. The line separating any two such regions is a continental divide. North America, for example, is divided into three major regions in which streams drain into the Pacific, Atlantic, and Arctic oceans, re-

Dendritic — Irregular branching of channels ("treelike") in many directions. Common in massive rock and in flat-lying strata. In such situations, differences in rock resistance are so slight that their control of the directions in which valleys grow headward is negligible.

Parallel — Parallel or subparallel channels that have formed on sloping surfaces underlain by homogenous rocks. Parallel rills, gullies, or channels are often seen on freshly exposed highway cuts or excavations having gentle slopes.

Radial — Channels radiate out, like the spokes of a wheel, from a topographically high area, such as a dome or a volcanic cone.

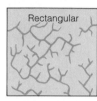

Rectangular — Channel system marked by right-angle bends. Generally results from the presence of joints and fractures in massive rocks or foliation in metamorphic rocks. Such structures, with their cross-cutting patterns, have guided the directions of valleys.

Trellised — Rectangular arrangement of channels in which principal tributary streams are parallel and very long, like vines trained on a trellis. This pattern is common in areas where the outcropping edges of folded sedimentary rocks, both weak and resistant, form long, nearly parallel belts.

Annular — Streams follow nearly circular or concentric paths along belts of weak rock that ring a dissected dome or basin where erosion has exposed successive belts of rock of varying degrees of erodibility.

Centripetal — Streams converge toward a central depression, such as a volcanic crater or caldera, a structural basin, a breached dome, or a basin created by dissolution of carbonate rock.

Deranged — Streams show complete lack of adjustment to underlying structural or lithologic control. Characteristic of recently deglaciated terrain whose preglacial features have been remodeled by glacial processes.

FIGURE 9.31 Some common stream patterns and their relationship to rock type and structure.

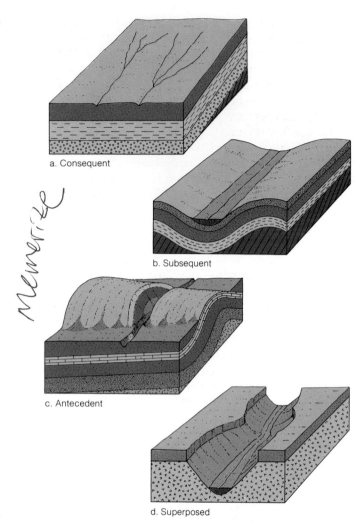

a. Consequent

b. Subsequent

c. Antecedent

d. Superposed

FIGURE 9.32 The relationship of streams to geology provides information about the structural history of an area. A. The course of a *consequent stream* is determined by the slope of land surface. B. A *subsequent stream* occupies belts of weak rock or takes a course determined by geologic structure. C. An *antecedent stream* maintains its course across rocks that have been raised across its path; typically, it occupies a gorge that crosses a structural ridge, rather than taking a course around the obstruction. D. A *superposed stream* has cut down through strata until its channel lies in underlying rock of different lithology or structure; the initial course of the stream was not determined by the rocks across which it is now flowing.

spectively (Fig. 9.33). In South America, a single continental divide extends along the crest of the Andes from Venezuela to the Strait of Magellan and divides the continent into two regions of unequal size. Streams draining the western (Pacific) slope of the Andes are steep and short, whereas to the east the streams take much longer routes along more gentle gradients to reach the Atlantic shore.

Major stream systems originate in the highest mountains of the continents. Because continental divides tend to coincide with the crests of mountain ranges, and mountain ranges result from uplift related to the interaction of tectonic plates, a close relationship must exist between plate tectonics and the location of primary stream divides and drainage basins.

The continental divide in western North America follows the crest of the Rocky Mountains northward from southern Mexico to northwestern Alaska (Fig. 9.33). The mountain building that produced this complex of ranges began near the end of the Mesozoic Era. We can reasonably infer that the present continental divide probably originated as the Rocky Mountains were uplifted. Although younger mountain ranges have risen near the west coast of the continent, major through-flowing rivers like the Yukon, Fraser, Columbia, and Colorado all head farther east at the continental divide. Thus, the imprint of this important mountain-building event many tens of millions of years ago is still recorded in the continental drainage pattern.

The position of the Arctic/Atlantic divide stretching across the middle of North America (Fig. 9.33) has been strongly influenced by glaciation. The landscape of central Canada was eroded by a succession of huge ice sheets during the last 2 million years. These glaciers repeatedly disrupted drainage systems and caused the Arctic/Atlantic drainage divide to shift back and forth as the ice sheets waxed and waned. At present, the divide lies near the southern limit of the former glaciers and close to a major geologic boundary that separates relatively erodible sedimentary rocks to the south from more resistant metamorphic rocks in central Canada.

An even older collision zone strongly influences the drainage systems of the eastern United States. The Appalachian Mountains date to the end of the Paleozoic Era when converging plates of lithosphere produced a major mountain chain that now stretches from Canada to the Gulf of Mexico. The ancient mountain system, which likely was a primary continental divide hundreds of millions of years ago, still forms a major drainage divide separating streams flowing directly into the Atlantic Ocean from those reaching the Atlantic via the continental interior and the Gulf of Mexico.

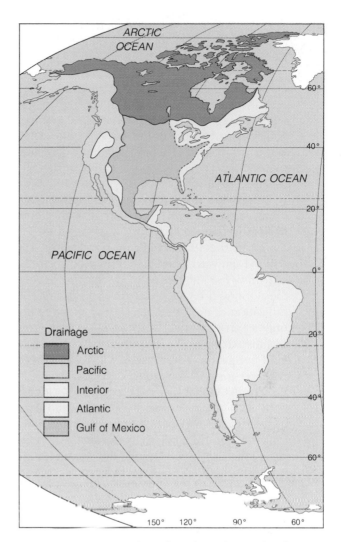

FIGURE 9.33 Map of North and South America showing location of continental drainage divides. The dashed line in eastern North America shows the location of the divide that separates streams draining directly into the Atlantic Ocean from those draining into the Gulf of Mexico.

TAMPERING WITH THE NILE

For more than seven millennia people have lived along the banks of the lower Nile River, a ribbon of green that crosses the vast desert of Egypt. As far back as the age of the pyramids, people learned to live with the Nile's annual cycle. During part of each year the river flowed peacefully within its banks, but by late summer it began to rise during the annual period of flooding. Although floods could devastate those living along the stream's margins, at the same time the floods were beneficial, for they carried an abundant load of silt that added annual nourishment to agricultural fields.

In the early nineteenth century, when a French mission sent by Napoleon conducted a census, the estimated population of Egypt was less than 2,500,000. By the 1920s the number had risen to 14 million, and since then the population has tripled. This burgeoning population placed a severe strain on the ability of the Nile to supply basic water needs.

To provide an adequate water supply and stabilize flow through the lower, densely populated Nile basin, a high dam was constructed at Aswan in the 1960s (Figs. B9.1 and 9.22). The Aswan Dam was designed to reduce downstream the immense difference in discharge between the flood and low-flow seasons, and permit full utilization of the Nile water. It was intended not only to provide water for human consumption, but for irrigation, hydroelectric power, and inland navigation. These primary goals have been realized, for the dam led to a marked change in the river's annual discharge pattern. No longer is there a large seasonal flood below the dam; instead, the annual hydrograph reflects the more uniform, controlled discharge of water from the reservoir (Fig. B9.2). Despite the success of the project, some important geologic consequences

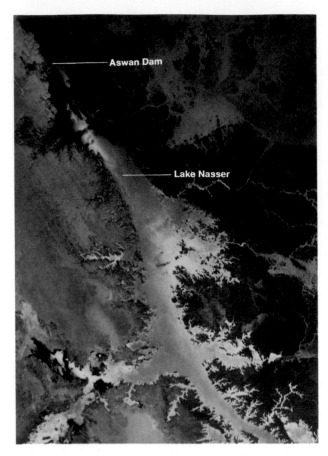

FIGURE B9.1   Vertical satellite view of the Aswan Dam and its reservoir, Lake Nasser. Sediment formerly carried northward to the Mediterranean Sea and now settling out in the lake will eventually fill the reservoir and make it unusable.

SUMMARY

1. Streams are part of the hydrologic cycle and the chief means by which water returns from the land to the sea. They help shape the Earth's surface and transport sediment to the oceans.

2. A stream's long profile decreases in gradient downstream.

3. Discharge, velocity, and cross-sectional area of a channel are interrelated such that when discharge changes, the product of the other two factors also changes.

4. As discharge increases downstream, channel width and depth increase, and velocity increases slightly.

5. Streams experiencing large floods are capable of transporting large loads and moving large boulders. Exceptional floods can do a great deal of

geologic work, but they have a low recurrence interval.

6. World sea level constitutes the base level for most streams. A local base level, such as a lake, may temporarily halt downward erosion upstream.

7. Straight channels are rare. Meandering channels form where gradients are low and the load is fine-grained. Braided patterns develop in streams with highly variable discharge and a large load to transport.

8. Stream load is the sum of bed load, suspended load, and dissolved load. Bed load may amount to as much as 50 percent of the total load. Most suspended load is derived from erosion of fine-grained regolith or from stream banks.

resulted from tampering with the Nile's hydrology.

The Nile, like all large streams, is a complex natural system, and its behavior reflects a delicate balance between discharge and sediment load. Ninety-eight percent of the Nile's load is suspended sediment. Prior to construction of the Aswan Dam, an average of 125 million metric tons of sediment passed downstream each year, but the dam reduced this value to only 2.5 million metric tons. Nearly 98 percent of the suspended sediment is now deposited in the reservoir behind the dam; at the present sedimentation rate, the reservoir will be filled with silt and no longer usable in 500 years. Under natural conditions, this sediment was carried downstream by floodwaters where much of it was deposited over the floodplain and delta, thus adding to the rich agricultural soils at a rate of 6 to 15 cm/century. With this natural source of nourishment eliminated, farmers must now resort to artificial fertilizers and soil additives to keep the land productive.

Before the Aswan Dam was constructed, the annual Nile flood transported at least 90 million metric tons of sediment to the Mediterranean Sea, adding it to the front of the delta. The shoreline at that time reflected a balance between sediment supply and the attack of waves and currents that redistributed the sediment along the coast. Because the annual discharge of sediment has now been cut off, the coast has become increasingly vulnerable to erosion. Over the long term, we can expect to see continuing changes to the delta as the shoreline adjusts to the reduction in Nile sediment and a new balance is reached.

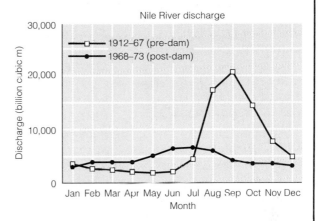

FIGURE B9.2 Prior to construction of the Aswan Dam, discharge varied seasonally, with peak discharge coming during the late summer and early fall interval of flooding. Controlled release of water after the dam was built greatly reduced the seasonal variability in discharge.

Streams that receive large contributions of underground water commonly have higher dissolved loads than those deriving their discharge principally from surface runoff.

9. 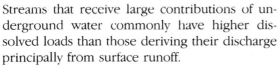 Sediment size decreases downstream due to sorting and abrasion of particles. The composition of a stream's load changes downstream as sediments of different composition are introduced.

10. Sediment yield is influenced by rock type and structure, climate, and topography. The greatest sediment yields are recorded in mountainous terrain with steep slopes and abundant runoff, and in small basins that are transitional from desert to grassland conditions. In moist climates, vegetation anchors the surface, thereby inhibiting erosion.

11. During floods, streams overflow their banks and construct natural levees, which grade laterally into silt and clay deposited on the floodplain. Terraces are due to the abandonment of a floodplain as a stream erodes downward.

12. Alluvial fans are constructed where a stream experiences a sudden decrease in gradient. The area of a fan is closely related to the size of the area upstream that supplies sediment to the fan.

13. A delta forms where a stream enters a body of standing water and loses its ability to transport sediment. The shape of a delta reflects the balance between sedimentation and erosion along the shore.

14. A drainage basin encompasses the area supplying

water to the stream system that drains the basin. Its area is proportional to the stream's length and annual discharge.

15. Stream systems possess an inherent orderliness, with the number of stream segments increasing with decreasing stream order.

16. Drainage patterns are related to underlying rock type and structure, and often can reveal information about a stream's history.

17. Most continental divides are associated with major tectonic alignments, such as the boundaries between present and former lithospheric plates.

IMPORTANT TERMS TO REMEMBER

alluvial fan (p. 259)
alluvium (p. 253)

base level (p. 247)
bed load (p. 253)
braided stream (p. 251)

channel (p. 242)

delta (p. 260)
discharge (p. 242)
dissolved load (p. 253)
divide (p. 262)
drainage basin (p. 262)

floodplain (p. 258)

gradient (p. 243)

laminar flow (p. 252)
load (p. 242)
long profile (p. 243)

meander (p. 249)

natural levee (p. 258)

overland flow (p. 244)

placer (p. 254)

runoff (p. 244)

saltation (p. 253)
sheet erosion (p. 252)
stream (p. 242)
streamflow (p. 244)
suspended load (p. 253)

terrace (p. 259)
tributary (p. 245)
turbulent flow (p. 252)

QUESTIONS FOR REVIEW

1. What evidence leads us to think that streams must be a major force in shaping the Earth's landscapes?

2. Describe how overland flow differs from streamflow.

3. Why does vegetation decrease the effectiveness of sheet erosion?

4. How do a stream's channel dimensions (depth, width) and velocity adjust in response to changes in discharge?

5. What controlling factors would have to change in order for a braided channel system to change to a meandering system?

6. Why does stream velocity generally increase downstream, despite a decrease in stream gradient?

7. When a large dam is built across a stream to create a reservoir upstream, the channel immediately below the dam commonly experiences erosion. Why should this erosion occur?

8. What is meant by a "200-year flood"? What is

the likely effect of such a flood on landscape evolution compared to the cumulative effect of annual floods?

9. What factors in a stream would have to change to cause fine gravel being moved as bedload to be transported as suspended load? In what way would the forward movement of such particles also be likely to change?

10. Why, if velocity increases downstream, does sediment in transport typically decrease in size in that direction?

11. How does increasing urbanization affect the amount of sediment eroded from a drainage basin, and why?

12. How might internal stratification and sedimentary characteristics permit you to distinguish between a delta and an alluvial fan that are preserved in the stratigraphic record?

13. What is a reasonable explanation for the orderly arrangement of tributaries that is typical of most drainage systems?

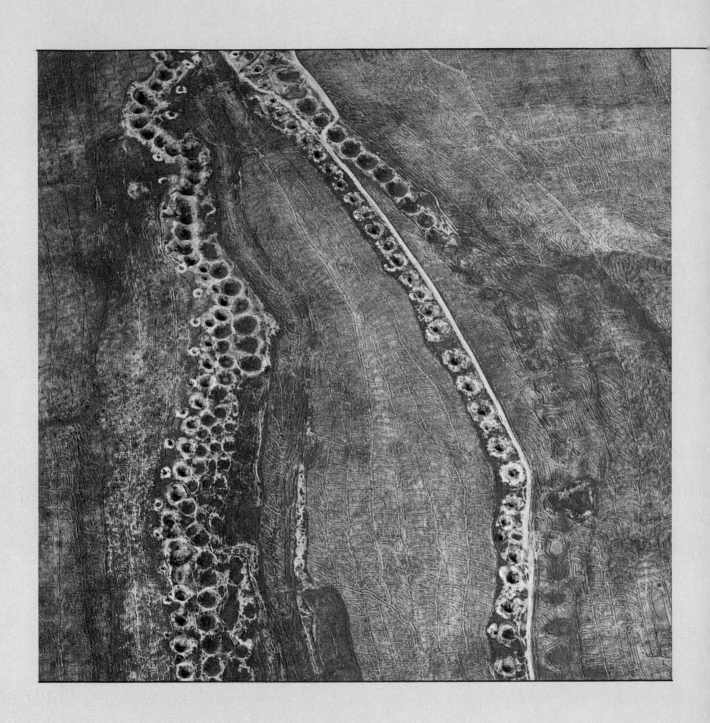

Aerial view of qanat system in southern Iraq, built to carry underground water beneath the surface of an alluvial fan toward agricultural fields in a desert basin.

CHAPTER 10

Groundwater

M any alluvial fans in the deserts of the Near East and central Asia are pockmarked by craterlike holes several meters in diameter that call to mind the landscape of a combat zone. However, unlike a heavily shelled battlefield where craters are scattered at random, these holes are aligned down the slope of the land. Each hole was dug by hand, and the discarded dirt was distributed around the resulting depressions. If we were to climb down one of these holes, we would find a nearly horizontal shaft that connects underground with other holes, both upslope and downslope. What we have here is an example of ancient aqueduct systems (called *karez* in western China and *qanats* in Iran) that were laboriously dug centuries ago to carry underground water to towns and agricultural fields on the floors of desert basins. High evaporation, a result of intense desert heat, causes surface streams to run dry. However, porous alluvium beneath the surface contains water that can be artificially channeled underground, thereby reducing loss by evaporation.

The qanats of the Near East are believed to have first been dug more than 2000 years ago in Persia (present-day Iran) and are among the earliest examples of extensive exploitation of underground water for human use. Vast reservoirs of underground water throughout the world are now routinely tapped to support agriculture, livestock, cities, and towns. Through improved techniques, the ease and rate of water withdrawal have increased to such a degree that today many supplies of underground water are being rapidly depleted. An added concern is the increasing pollution of such water supplies by human and industrial waste. The solution to these environmental problems lies in an improved understanding of how water moves beneath the land surface, how this underground water is replenished, and how we can avoid polluting it with human and industrial wastes.

WATER IN THE GROUND

Access to water, whether from streams, lakes, springs, and direct rainfall or from underground, is a vital human need. Most early cities and towns were founded close to streams that would provide a reliable source of water. As the population of these towns and cities grew, the streams often became insufficient. People then either resorted to bringing water from a more distant source through canals, or they dug wells to obtain water from underground.

As society has become increasingly more populous and industrialized, communities have generated ever-larger amounts of human and industrial wastes, a good deal of which has inevitably found its way into the very water that people must rely on for their existence. In many places water is dwindling in both quantity and quality, creating important questions for the communities involved: Will there be enough clean water to sustain future needs? Is the quality adequate for the uses to which we put this water? Is the water being used with a minimum of waste?

Origin of Groundwater

Less than 1 percent of the water on the Earth is **groundwater,** defined as all the water contained in spaces within bedrock and regolith. Although the volume of groundwater sounds small, it is 40 times larger than the volume of all the water in freshwater lakes or flowing in streams and nearly a third as large as the water contained in all the world's glaciers and polar ice.

Most groundwater originates as rainfall. Rainwater that soaks into the ground becomes part of the groundwater system and moves slowly toward the ocean, either directly through the ground or by flowing out onto the surface and joining streams (Fig. 1.23).

That groundwater comes from rain was finally established on a quantitative basis in the seventeenth century, when Pierre Perrault, a French physicist, measured the mean annual rainfall for a part of the drainage basin of the Seine River in eastern France and the mean annual stream runoff in the same basin area. After allowing for loss by evaporation, Perrault concluded that the difference between the amounts of rainfall and runoff was ample enough, over a period of years, to account for the amount of water in the ground.

Depth of Groundwater

Water is present everywhere beneath the land surface, but more than half of all groundwater, including most

of what is usable, occurs within about 750 m of the Earth's surface. The volume of water in this zone is estimated to be equivalent to a layer of water approximately 55 m deep spread over the world's land areas. Below a depth of about 750 m, the amount of groundwater gradually, though irregularly, diminishes. Holes drilled for oil have found water as deep as 9.4 km, and one deep experimental hole drilled on the Kola Peninsula by Soviet scientists encountered water at more than 11 km. However, even though water may be present in crustal rocks at such depths, the pressure exerted by overlying rocks is so high and openings in rocks are so small that it is unlikely that much water is present or that it can move freely through the enclosing rocks. Thus, for our purposes, we shall think of groundwater as the water to be found between the land surface and a depth of about 750 m.

The Water Table

Much of what we know about the occurrence of groundwater has been learned from the accumulated experience of generations of people who have dug or drilled millions of wells. This experience tells us that a hole penetrating the ground ordinarily passes first through a layer of moist soil and then into a zone in which open spaces in regolith or bedrock are filled mainly with air (Fig. 10.1). This is the **zone of aeration** (also called the *unsaturated zone,* for although water may be present, it does not saturate the ground). The hole then enters the **saturated zone,** a zone in which all openings are filled with water. We call the upper surface of the saturated zone the **water table.** Normally the water table slopes toward the nearest stream or lake, but in deserts, for example, it may lie far underground. Just as water is present at some depth everywhere beneath the land surface, so too is the water table always present.

In fine-grained sediment, a narrow fringe as much as 60 cm thick immediately above the water table is kept wet by capillary attraction that pulls water from the top of the water table a small distance into the zone of aeration (the capillary fringe shown in Fig. 10.1). *Capillary attraction* is the adhesive force between a liquid and a solid that causes water to be drawn into small tubelike openings. This is the same force that draws ink through blotting paper and kerosene through the wick of a lamp.

In humid regions, the water table is a subdued imitation of the land surface above it (Fig. 10.1). It is high beneath hills and low beneath valleys because water tends to move toward low points in the topography, where the pressure on it is least. If all rainfall were to cease, the water table would slowly flatten and gradually approach the levels of the valleys. Water seepage

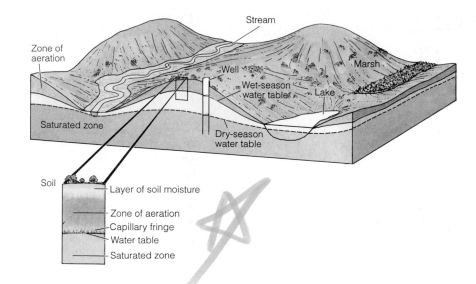

FIGURE 10.1 In a typical ground-water system, the water table separates the zone of aeration from the saturated zone and fluctuates in level with seasonal changes in precipitation. Corresponding fluctuations are seen in the water level in wells that penetrate the water table. Lakes, marshes, and streams occur where the water table intersects the land surface. A layer of moisture coincides with the surface soil, and a thin capillary fringe lies immediately above the water table. In shape, the water table is a subdued imitation of the overlying land surface.

into the ground would diminish and then cease, and streams would dry up as the water table fell beneath valleys. In times of drought, when rain may not fall for several weeks or even months, we can sense the flattening of the water table in the drying up of wells. When a well becomes dry, we know that the water table has dropped to a level below the bottom of the well. It is repeated rainfall, dousing the ground with fresh supplies of water, that maintains the water table at a normal level.

Whatever its depth, the water table is a significant surface, for it represents the upper limit of all readily usable groundwater. For this reason, a major aim of groundwater geologists and well drillers alike is to determine the depth and the shape of the water table.

HOW GROUNDWATER MOVES

Groundwater operates continuously as a small but integral part of the hydrologic cycle (Fig. 1.23). Water, evaporated mainly from the oceans and falling on the land as rain, seeps into the ground and enters the groundwater reservoir. Some of this slowly moving underground water reaches stream channels and contributes to the water they carry to the ocean, where the cycle begins again.

Most of the groundwater within a few hundred meters of the surface is in motion. Unlike the swift flow of rivers, however, which is measurable in kilometers per hour, groundwater moves so slowly that velocities are expressed in centimeters per day or meters per year. The reason for this contrast is simple. Whereas the water of a stream flows unimpeded through an open channel, groundwater must move through small, constricted passages, often along a tortuous route.

Therefore, the flow of groundwater to a large degree depends on the nature of the rock or sediment through which the water moves.

Porosity and Permeability

Porosity is the percentage of the total volume of a body of regolith or bedrock that consists of open spaces, called *pores*. It is porosity that determines the amount of water that a given volume of sediment or rock can contain.

The porosity of sediments is affected by the sizes and shapes of the rock particles, as well as by the compactness of their arrangement (Fig. 10.2A and B). In some well-sorted sands and gravels, the porosity may reach 20 percent, while some very porous clays have porosities as high as 50 percent.

The porosity of a sedimentary rock is affected not only by the sorting and arrangement of the particles, but also by the extent to which the pores become filled with cement (Fig. 10.2C). The porosity of igneous and metamorphic rocks, on the other hand, generally is low. However, if such rocks have many joints and fractures, their porosity will be higher.

Permeability is a measure of how easily a solid allows fluids to pass through it. A rock of very low porosity is also likely to have low permeability. However, a high porosity does not necessarily mean a correspondingly high permeability, because the size and continuity of the pores influence permeability in an important way.

The relationship between pore size and the molecular attraction of rock surfaces plays a large role in determining the permeability of a rock. Molecular attraction is a force that exists between a solid surface and a film of water. This force causes a thin film of

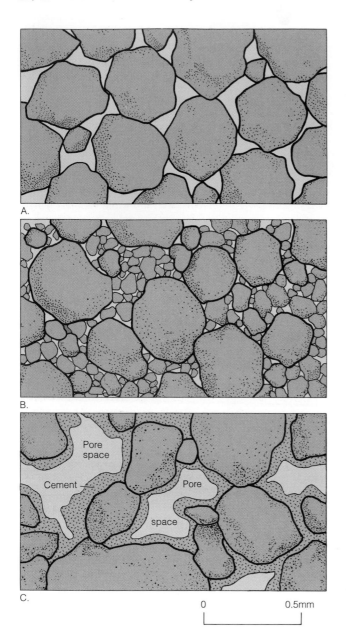

FIGURE 10.2 Porosity in different sediments. A. A porosity of 30 percent in a reasonably well-sorted sediment. B. A porosity of 15 percent in a poorly sorted sediment in which fine grains fill spaces between larger grains. C. Reduction in porosity in an otherwise very porous sediment due to cement that binds grains together.

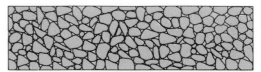

FIGURE 10.3 Effect of molecular attraction on permeability in the intergranular spaces of a fine, silty sediment (top) and a coarse, sandy sediment (bottom) of equal porosity.

water to adhere to a rock surface despite the force of gravity. If the open space between adjacent particles in a rock is small enough, the films of water that adhere to the particles will come into contact with each other. The force of molecular attraction therefore extends right across the pore (Fig. 10.3). At ordinary pressures, this water is held firmly in place and so permeability is low. An example of such a sediment is clay, the particles of which have diameters of less than 0.005 mm (Table 4.1). Although clay may have a high porosity, because the pores are very small, the permeability is low.

By contrast, in a sediment with grains at least as large as sand (grain diameters of 0.06 to 2 mm), the pores commonly are wider than the combined thickness of the films of water adhering to adjacent grains. Therefore, the water in the centers of the pores is free to move (Fig. 10.3). Such sediment is permeable. As the diameters of the pores increase, permeability increases. Gravel, with very large pores, is more permeable than sand and can yield large volumes of water to wells.

Movement in the Zone of Aeration

Water from a rain shower soaks into the soil, which usually contains clay resulting from the chemical weathering of bedrock. Because of these fine clay particles, the soil is generally less permeable than underlying coarser regolith or rock. The low permeability and the fine clay particles cause part of the water to be retained in the soil by forces of molecular attraction. This is the layer of soil moisture shown in Figure 10.1. Some of this moisture evaporates directly into the air, but much of it is absorbed by the roots of plants that later return it to the atmosphere through transpiration (Fig. 1.23).

Because of the pull of gravity, water that cannot be held in the soil by molecular attraction seeps downward until it reaches the water table. With every rainfall, more water enters the ground, but apart from soil moisture and the capillary fringe, the zone of aeration is likely to be nearly dry between rains.

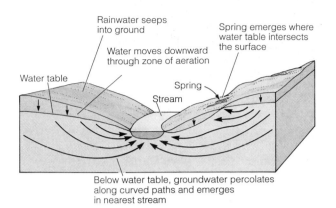

FIGURE 10.4 Paths of groundwater flow in a humid region in uniformly permeable rock or sediment. Long, curved arrows represent only a few of many possible paths. Springs are located where the water table intersects the land surface.

Movement in the Saturated Zone

The movement of groundwater in the saturated zone, termed **percolation,** is similar to the flow of water when a saturated sponge is squeezed gently. Water moves slowly by percolation through very small pores along parallel, threadlike paths. Movement is easiest through the central parts of the spaces but diminishes to zero immediately adjacent to the sides of each space because there the water is held in place by molecular attraction.

Responding to gravity, water percolates from areas where the water table is high toward areas where it is lowest. In other words, it generally percolates toward

surface streams or lakes (Fig. 10.4). Only part of the water moves directly down the slope of the water table by the shortest route. Much of it flows along innumerable long, curving paths that go deeper through the ground. Some of the deeper paths turn upward against the force of gravity and enter the stream or lake from beneath. This upward flow is possible because, at any given altitude, the water in the saturated zone is under greater pressure beneath a hill than beneath a stream. The water therefore tends to flow toward points where pressure is least. However, most of the groundwater entering a stream travels along shallow paths not far beneath the water table.

Recharge and Discharge Areas

Replenishment, or **recharge,** of groundwater occurs as rainfall and snowmelt enter the ground in **recharge areas,** which are areas of the landscape where precipitation seeps downward beneath the surface and reaches the saturated zone (Fig. 10.5). The water continues to move slowly toward **discharge areas,** which are areas where subsurface water is discharged to streams or to lakes, ponds, or swamps. The surface extent of recharge areas is invariably larger than that of discharge areas.

The time water takes to move through the ground from a recharge area to the nearest discharge area depends on rates of flow and on the travel distance. It may take only a few days, or possibly thousands of years in cases where water moves through the deeper parts of a groundwater body (Fig. 10.5).

In humid regions, recharge areas encompass nearly all the landscape beyond streams and their adjacent

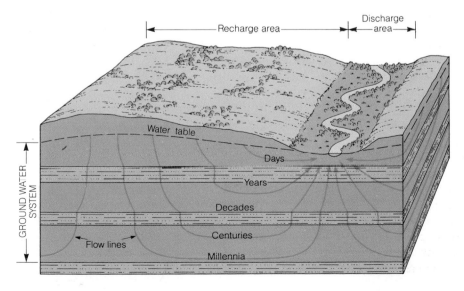

FIGURE 10.5 Distribution of recharge and discharge areas in a humid landscape. The time required for groundwater to reach the discharge area from the recharge area depends on the path and distance of travel. Downward and upward percolation is faster and more direct in the most porous strata.

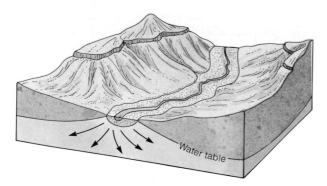

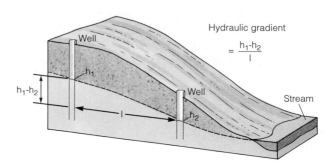

FIGURE 10.6 In arid regions, direct recharge is minimal and the water table lies below the bed of a river. During times of low flow, large through-flowing streams and intermittent streams lose water, which seeps downward to resupply groundwater in the saturated zone.

FIGURE 10.7 The hydraulic gradient is the slope of the water table. In the example illustrated, the hydraulic gradient is found by subtracting the altitude of h_2, the level of water in the downslope well, from h_1, the level of the water in the upslope well, and dividing the difference by the distance (l) between the two wells.

floodplains (Fig. 10.5). In more arid regions, recharge occurs mainly in mountains and in the alluvial fans that border them. In such regions recharge also occurs along the channels of major streams that are underlain by permeable alluvium; the water leaks downward through the alluvium and recharges the groundwater (Fig. 10.6).

Discharge and Velocity

Groundwater does not everywhere flow at a constant rate. We can demonstrate this by injecting colored dye at some point in a recharge area and measuring the time it takes for the dye to appear in nearby springs or wells. Experiments conducted with materials of uniform permeability have shown that the velocity of groundwater flow increases as the slope of the water table increases; in other words, the steeper the slope, the faster the water moves. As shown in Figure 10.7, the slope can be determined by measuring the difference in altitude of two points (h_1 and h_2) on the water table and dividing this figure by the horizontal distance (l) between the points. The resulting slope value is generally referred to as the **hydraulic gradient.** Thus, the velocity of groundwater (V) is proportional to the hydraulic gradient:

$$V \propto \frac{h_1 - h_2}{l}$$

In 1856, Henri Darcy, a French engineer, concluded that the velocity of groundwater must be related not only to the slope of the water table (the hydraulic gradient), but also to the permeability of the

rock or sediment through which the water is flowing. He proposed an equation in which permeability, together with the acceleration due to gravity and the viscosity of water, is expressed as a coefficient (K). This coefficient, referred to as the *coefficient of permeability,* is simply a measure of the ease with which water moves through a rock or sediment. The equation Darcy proposed can thus be expressed as

$$V = \frac{K(h_1 - h_2)}{l} \qquad (1)$$

In chapter 9 we learned that discharge (Q) in streams varies as a function of both stream velocity (V) and cross-sectional area (A). The discharge of groundwater through a rock or sediment also depends on the velocity of flow and cross-sectional area of flow; in this case, however, the cross-sectional area is not that of an open channel but of an interconnected system of pores. We can express discharge by using an equation we learned in Chapter 9: $Q = AV$. If we substitute the value for V given in equation 1 in this equation from chapter 9, we arrive at a new equation,

$$Q = \frac{AK(h_1 - h_2)}{l} \qquad (2)$$

that is known as **Darcy's Law.** If we take the cross-sectional area (A) as constant for any given situation, by measuring any two of the remaining three variables [discharge (Q), coefficient of permeability (K), and hydraulic gradient ($h_1 - h_2)/l$], we can calculate the value of the third.

Because percolating groundwater encounters a

large amount of frictional resistance, flow rates tend to be very slow. Normally, velocities range between half a meter a day and several meters a year. The highest rate yet measured in the United States, in exceptionally permeable material, is only about 250 m/yr.

SPRINGS AND WELLS

People generally obtain supplies of groundwater either from natural springs or by excavating wells that reach a body of water underground.

Springs

A **spring** is a flow of groundwater emerging naturally at the ground surface. The simplest kind of spring is one that issues from a place where the land surface intersects the water table (Fig. 10.4). Small springs are found in all kinds of rocks, but almost all large springs issue from lava, limestone, or gravel.

A vertical or horizontal change in permeability is a common reason for the localization of springs (Fig. 10.8). Often this change involves the presence of an **aquiclude,** a body of impermeable or distinctly less permeable rock adjacent to a permeable one (Fig. 10.8A). If a porous sand overlies a relatively impermeable clay aquiclude, water percolating downward through the sand will flow laterally when it reaches the underlying clay and will emerge as a spring where the stratigraphic boundary between the two units intersects the land surface, as along the side of a valley or a coastal cliff (Fig. 10.8B). Springs may also issue from lava flows, especially where a jointed lava bed overlies an aquiclude (Fig. 10.8C), or along the trace of a fault (Fig. 10.8D).

Wells

A well will supply water if it intersects the water table (Fig. 10.1). Figure 10.9 shows that a shallow well can become dry at times when the water table is low, whereas a nearby deeper well may yield water throughout the year.

When water is pumped from a new well, the rate of withdrawal initially exceeds the rate of local groundwater flow. This imbalance in flow rates creates a conical depression in the water table immediately surrounding the well, called a **cone of depression** (Fig. 10.9). The locally steepened slope of the water

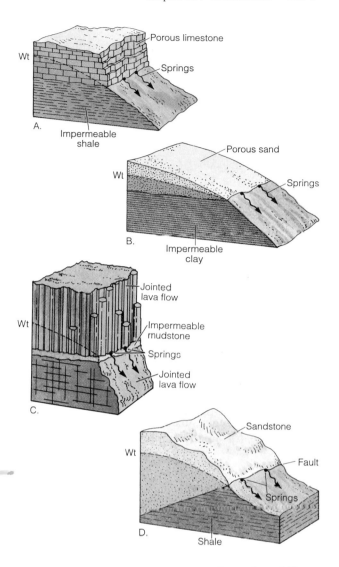

FIGURE 10.8 Examples of springs formed in different geologic conditions. A. A spring discharges water at the contact between a porous limestone and an underlying impermeable shale. B. Springs lie at the contact between a porous sandy unit and an underlying impermeable clay. C. Springs issue along the contact between a highly jointed lava flow and an underlying impermeable mudstone. D. Springs issue along the trace of a fault where it intersects the land surface.

table increases the flow of water to the well, consistent with Darcy's Law. Once the rate of inflow balances the rate of withdrawal, the hydraulic gradient stabilizes, but it will change if either the rate of pumping or the rate of recharge changes. In most small domestic wells the cone of depression is hardly discernible. Wells pumped for irrigation and industrial uses, however, withdraw so much water that the cone can become very wide and steep and can lower the water table in all wells of a district.

If the source of a groundwater supply is inhomogeneous rock or sediment, water yields from wells may

The placement of hazardous wastes underground, even far underground, immediately raises concerns about groundwater. Water is a nearly universal solvent, and the weakly acidic character of most groundwater means that any container of toxic or radioactive substances eventually is likely to corrode, so that the contents will then dissolve and be transported away from the storage site. Water is present in crustal rocks to depths of many kilometers, and in many of these rocks it is circulating at rates of 1 to 50 m/yr. Over tens or hundreds of thousands of years, even such slow rates can move dissolved substances over great distances and introduce them to more rapidly flowing parts of the hydrologic system.

There is general agreement among geologists that the ideal underground storage site for radioactive wastes should possess the following characteristics:

1. The enclosing rock should have few fractures and low permeability.
2. The enclosing rock should have no present or future economic mineral potential.
3. Local groundwater flow should be away from the biosphere.
4. Only very long paths of groundwater flow should be directed toward places accessible to humans.
5. The area should have low rainfall.
6. The zone of aeration should be thick.
7. The rate of erosion should be very low.
8. The probability of earthquakes or volcanic activity should be very low.
9. Future change of climate in the region should be unlikely to affect groundwater conditions substantially.

The safe long-term storage of toxic and nuclear wastes at underground sites provides a major challenge for geologists. Historically, geologists have studied past events, but now they are being asked to predict possible future events. To do so with any confidence requires considerable knowledge of local and regional groundwater conditions. It also demands a much better understanding of how complex groundwater systems might respond to crustal movements, local and global climatic change, and other natural factors that can affect the stability of a storage site.

GEOLOGIC ACTIVITY OF GROUNDWATER

In regions underlain by rocks that are highly susceptible to chemical weathering, groundwater creates distinctive landscapes that are among the most interesting and picturesque on our planet.

Dissolution

As soon as rainwater infiltrates the ground, it begins to react with minerals in regolith and bedrock and weathers them chemically. An important part of chemical weathering involves minerals and rock materials passing directly into solution; this process is known as **dissolution.** Among the rocks of the Earth's crust, the carbonate rocks are the most readily attacked by dissolution (Fig. 10.21).

Limestone, dolostone, and marble are the most common carbonate rocks and underlie millions of square kilometers of the Earth's surface. Although carbonate minerals are nearly insoluble in pure water, they are readily dissolved by carbonic acid (Table 7.1, reaction 1) in the form of downward-percolating rainwater. As a result, groundwater becomes charged with calcium cations and bicarbonate anions (Table 7.1, reaction 5).

The weathering attack occurs mainly along joints and other partings in the carbonate bedrock. The result is impressive. When granite is weathered chemically, quartz and other resistant minerals are little affected and remain as part of the regolith. However, when limestone weathers, nearly all its volume can be dissolved away in slowly moving groundwater.

FIGURE 10.21 A marble balustrade of the Forbidden City in Beijing, China, shows the effects of more than 300 years of dissolution. The original sharply carved design has become smooth and indistinct as chemical weathering, enhanced by acid rainfall, has dissolved the stone.

By measuring over a period of time the amount of dissolution observed on small, precisely weighed limestone tablets placed at open sites in various areas, geologists have obtained estimates of the average rate at which limestone landscapes are being lowered by dissolution. In temperate regions with high rainfall, a high water table, and a nearly continuous cover of vegetation, carbonate landscapes are being lowered at average rates of up to 10 mm/1000 yr. In dry regions with scanty rainfall, low water tables, and discontinuous vegetation, rates are far lower. Measured rates of dissolution by groundwater in carbonate terrains of the United States show that the dissolution rate can exceed the average erosional reduction of the surface by mass-wasting, sheet erosion, and streams.

Chemical Cementation and Replacement

The conversion of sediment into sedimentary rock is primarily the work of groundwater. A body of sediment lying beneath the sea is generally saturated with water, as is sediment lying in the saturated zone beneath the land. Substances in solution in the water are precipitated as cement in the spaces between rock and mineral particles of the sediment. As we learned in chapter 4, this diagenetic process transforms the loose sediment into firm rock. Calcite, quartz, and iron compounds (mainly hydroxides such as limonite) are, in that order, the chief cementing substances.

Less common than the deposition of cement between the grains of a sediment is **replacement,** the process by which a fluid dissolves matter already present and at the same time deposits from solution an equal volume of a different substance. Evidently, replacement takes place on an approximately volume-for-volume basis because the new material preserves the most minute textures of the material replaced. Both mineral and organic substances can be replaced. *Petrified wood* is a common example of the replacement of organic matter (Fig. 10.22).

Carbonate Caves and Caverns

People have long been interested in caves. The earliest evidence we have of human dwellings comes from limestone caves in Europe and Asia that provided shelter for paleolithic peoples during the Pleistocene glacial ages. The walls of these caves were the rocky canvases of prehistoric artists, whose polychrome paintings provide us with superb renditions of the prey of ice-age big-game hunters.

Caves come in many sizes and shapes. Although most caves are small, some are of exceptional size. A very large cave or system of interconnected cave

FIGURE 10.22 Logs of petrified wood weather out of mudstone layers in Petrified Forest National Park, Arizona.

chambers is often called a *cavern.* The Carlsbad Caverns in southeastern New Mexico include one chamber 1200 m long, 190 m wide, and 100 m high. Mammoth Cave, in Kentucky, consists of interconnected caverns with an aggregate length of at least 48 km. The recently discovered Good Luck Cave on the tropical island of Borneo has one chamber so large that into it could be fitted not only the world's largest previously known chamber (in Carlsbad Caverns), but also the largest chamber in Europe (in Gouffre St. Pierre Martin, France), and the largest chamber in Britain (Gaping Ghyll).

Cave formation is mainly a chemical process involving the dissolution of carbonate rock by circulating groundwater. The usual sequence of development is thought to involve (1) initial dissolution along a system of interconnected open joints and bedding planes by percolating groundwater, (2) enlargement of a cave passage along the most favorable flow route by water that fully occupies the opening, (3) deposition of carbonate formations on the cave walls while a stream occupies the cave floor, and (4) continued deposition of carbonate on the walls and floor of the cave after the stream has stopped flowing. Although geologists have argued for years as to whether caves form in the zone of aeration or in the saturated zone, available evidence favors the idea that most caves are excavated in the shallowest part of the saturated zone, along a seasonally fluctuating water table.

The rate of cave formation is related to the rate of dissolution. Where the water is acidic, the rate of dissolution increases with increasing flow velocity.

Therefore, as a passage increases in size, and the flow changes from very slow laminar flow to more rapid turbulent flow, the rate of dissolution will rise. The development of a continuous passage by slowly percolating waters has been estimated to take up to 10,000 years, while the further enlargement of the passage by more rapidly flowing water into a fully developed cave system may take an additional 10,000 to 1 million years.

Although limestone caves are generally believed to involve dissolution by carbonic acid, chemical evidence suggests that at least some caves, including Carlsbad Caverns, may involve dissolution by sulfuric acid. The proposed agent is hydrogen-sulfide-bearing solutions derived from petroleum-rich sediments. The solutions rise along joints where they meet and interact with oxygenated water to form sulfuric acid, which then dissolves the limestone.

Cave Deposits

Some caves have been partly filled with insoluble clay and silt, originally present as impurities in limestone and gradually concentrated as the limestone was dissolved. Others contain partial fillings of **dripstone** and **flowstone,** deposits chemically precipitated from dripping and flowing water, respectively, in an air-filled cavity. Both kinds of deposits are commonly composed of calcium carbonate. The carbonate precipitates take on many curious forms, which are among the chief attractions for cave visitors. Among the most common shapes are *stalactites* (iciclelike forms of dripstone hanging from ceilings), *stalagmites* (blunt "icicles" of dripstone projecting upward from cave floors), *columns* (stalactites joined with stalagmites, forming connections between the floor and roof of a cave), and crenulated or curtainlike formations of flowstone (Fig. 10.23).

As its name implies, dripstone is deposited by successive drops of water. As each drop forms on the ceiling of a cave, it loses a tiny amount of carbon dioxide gas and precipitates a particle of calcium carbonate (Fig. 10.24). This chemical reaction is the reverse of the one by which calcium carbonate is dissolved by carbonic acid.

Dripstone and flowstone can be deposited only in caves that are at least partially filled with air and therefore lie at or above the water table. Yet many, perhaps most, caves are believed to have formed below the water table, as is suggested by their shapes and by the fact that some caves are lined with crystals, which can form only in an aqueous environment. How can we reconcile these apparently conflicting observations? The answer lies partly in the observation that both

FIGURE 10.23   A caver surveys the spectacular dripstone and flowstone formations of Lechuguilla Cave, a limestone cave in the Carlsbad Caverns region of New Mexico.

dissolution and deposition are known to take place in the zone of aeration. In some cases, the answer probably also lies in a change in the level of the water table. This change in level can result when uplift of the land causes a stream to cut downward into the landscape or when a change of climate causes a lowering of the regional water table (Fig. 10.25). Caves formed in the upper part of the saturated zone when the water table is high would shift into the zone of aeration as the water table falls. Dissolution could then give way to deposition of dripstone and flowstone.

Sinkholes

In contrast to a cave, a **sinkhole** is a large dissolution cavity that is open to the sky. Some sinkholes are caves whose roofs have collapsed; others are formed at the surface, where rainwater is freshly charged with carbon dioxide and is most effective as a solvent. Many sinkholes located at the intersections of joints, where

FIGURE 10.24 A drop of water collects at the end of a growing stalactite in Carlsbad Caverns, New Mexico. As the water loses carbon dioxide, a tiny amount of calcium carbonate precipitates from solution and is added to the end or sides of the stalactite.

downward movement of water is most rapid, are funnel shaped.

Sinkholes of the Yucatan Peninsula in Mexico, which are locally called *cenotes* (a word of Mayan origin), have high, vertical sides and contain water because their floors lie below the water table (Fig. 10.26). The cenotes were the primary source of water for the ancient Maya and formerly supported a considerable population in Yucatan. A large cenote at the ruined city of Chichen Itza was sacred and dedicated to the rain gods. Remains of more than 40 human sacrifices, mostly young children, have been recovered from the cenote, together with huge quantities of jade, gold, and copper offerings.

In the carbonate landscape of Florida, new sinkholes are forming constantly. In one small area of about 25 km², more than 1000 collapses have occurred in recent years. In this case, lowering of the water table brought on by drought and excessive pumping of local wells has led to extensive collapse of cave roofs.

Some sinkholes form catastrophically. An account of one such event in rural Alabama describes how a resident was startled by a rumble that shook his house. He then distinctly heard the sound of trees snapping and breaking. A short time later, hunters walking through nearby woods discovered a huge 50-m-deep sinkhole that measured 140 m long and 115 m wide.

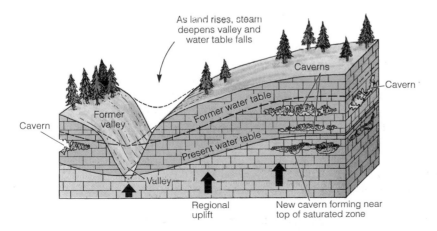

FIGURE 10.25 One possible history of a cavern containing carbonate deposits. The cavern formed in the saturated zone when the water table lay at level 1. Uplift of the region caused streams to deepen their valleys. The water table then lowered in response to valley deepening, leaving the cavern above the lowered water table (level 2). Stalactites and stalagmites could then form in the air-filled cavern.

FIGURE 10.26 The sacred well at Chichen Itza, a ruined Mayan city on the Yucatan Peninsula of Mexico. This cenote, formed in flat-lying limestone strata, contained a rich store of archaeological treasures that were cast into the water with human sacrifices centuries ago.

Karst Topography

In some regions of exceptionally soluble rocks, caves and sinkholes are so numerous that they form a peculiar topography characterized by many small, closed basins and a disrupted drainage pattern. Streams disappear into the ground and eventually reappear elsewhere as large springs. Such terrain is called **karst topography** after the Karst region of Yugoslavia, a remarkable landscape of closely spaced sinkholes. While most typical of carbonate landscapes, karst can also develop in areas underlain by dolomite, gypsum, and salt.

Several factors control the development of karst landscapes. First, the topography must produce a steep enough hydraulic gradient to permit the flow of groundwater through soluble rock under the pull of gravity. Second, precipitation must be adequate to supply the groundwater system, soil and plant cover must supply an adequate amount of carbon dioxide, and temperature must be high enough to promote dissolution. Although karst terrain is found throughout a wide range of latitudes and at varied altitudes, it often is best developed in moist temperate to tropical regions underlain by thick and widespread soluble rocks.

Several distinctive kinds of karst landscape are recognized. The most common is *sinkhole karst,* a landscape dotted with sinkholes of various sizes and shapes. Such landscapes are seen in southern Indiana, south-central Kentucky, central Tennessee, and Jamaica, among other places.

Cone karst and *tower karst* occur in thick, well-jointed limestone that separates into isolated blocks as it weathers. Cone karst consists of many closely spaced conical- or pinnacle-shaped hills separated by deep sinkholes (Fig. 10.27). Tower karst, by contrast, consists of isolated towerlike limestone hills separated by expanses of alluvium. If sediment being deposited in the bottom of sinkholes is not removed by the local drainage system, the depressions slowly fill, and areas between towers merge to form a flat alluvial surface. Cone and tower karst landscapes are found in parts of Mexico, Central America, the Caribbean islands of Cuba and Puerto Rico, and in the South Pacific.

One of the most famous and distinctive of the world's tower karst regions lies near Guilin in southeastern China. There, vertical-sided peaks of limestone rise up to 200 m high. The dramatic landscape of this region has inspired both classical Chinese painters and present-day photographers (Fig. 10.28).

Pavement karst consists of broad areas of bare limestone in which joints and bedding planes have been etched and widened by dissolution, creating a distinctive land surface. Pavement karsts are especially common in high-latitude regions where continental glaciation has stripped away regolith and left carbonate bedrock exposed to weathering. These rocky, intricately etched landscapes, nearly devoid of vegetation, are found in such places as Spitzbergen, Greenland, and the Burren region of western Ireland (Fig. 10.29).

FIGURE 10.27 Cone karst near Arecibo in northwest Puerto Rico, The Arecibo radiotelescope (lower right) occupies a circular depression in the midst of a vast limestone landscape of closely spaced conical hills separated by deep sinkholes.

FIGURE 10.28 Steep limestone pinnacles surrounded by flat expanses of alluvium form a distinctive tower karst landscape around the Li River near Guilin, China.

FIGURE 10.29 The burren landscape in western Ireland is an extensive platform karst developed on limestone. Crevices have developed by solution along prominent joints, giving the terrain a regular, geometric texture.

TOXIC GROUNDWATER IN THE SAN JOAQUIN VALLEY

The only time most people ever hear of selenium is in their high school chemistry course when they learn it is thirty-fourth in the periodic list of the elements. Selenium is a naturally occurring element and a necessary trace nutrient in our diet as well as in the diet of livestock and many wild animals. However, if excessive amounts are ingested, selenium can prove toxic.

In 1983, selenium attracted national attention when the U. S. Fish and Wildlife Service reported fish kills and high incidences of mortality, birth defects, and decreased hatching rates in nesting waterfowl at the Kesterson National Wildlife Refuge in California's San Joaquin Valley (Fig. B10.1). Laboratory studies showed high concentrations of selenium in fish from Kesterson Reservoir in the wildlife refuge, while birds using the reservoir were found to have high concentrations of selenium and obvious symptoms of selenium poisoning. Federal agencies quickly began investigations to learn how these toxic levels of selenium were entering the ecological system.

Geologists have now shown that the poisoning of wildlife at Kesterson resulted from a combination of geological, hydrological, and agricultural factors. The western San Joaquin Valley is a prime agricultural area, but because it lies in a zone of arid climate, the land is irrigated. The irrigation artificially raised the water table to such a degree that a system of subsurface drains was established to remove excess water. The drainage system carries subsurface water to a surface canal that funnels the water northward along the valley to Kesterson Reservoir in the wildlife refuge.

A natural source of selenium lies in the Coast Ranges immediately to the west of the San Joaquin Valley. Rainwater falling in these mountains dissolves selenium-bearing salts from marine sedimentary strata. Surface runoff then carries the dissolved salts to broad alluvial fans in the valley where the water seeps into the ground and recharges a shallow regional aquifer. Evaporation in this arid region, where an annual rainfall of less than 250 mm is greatly exceeded by an annual evaporation rate of about 2300 mm, concentrates the salts in the soil. An irrigation system on the fan surfaces is designed to supply water to crops and also to flush salts out of the soil and into the drainage canal that leads toward Kesterson Reservoir. Because the reservoir has no outlet, the selenium is concentrated there and now has reached toxic levels.

Groundwater conditions in the western San Joaquin Valley contribute to the Kesterson Reservoir problem. Most of the shallow groundwater in this area is alkaline and slightly to highly saline. Under these conditions, selenium is very soluble and is carried with the flowing groundwater. This mobility greatly increases the ease with which selenium moves from the soils and into the artificial drainage system.

A.

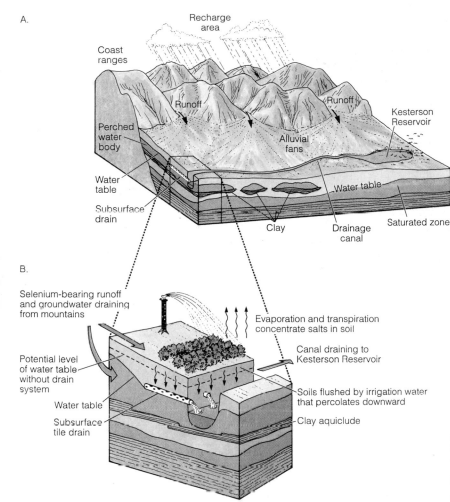

Recharge area

Coast ranges

Runoff

Runoff

Kesterson Reservoir

Perched water body

Alluvial fans

Water table

Subsurface drain

Water table

Clay

Drainage canal

Saturated zone

B.

Selenium-bearing runoff and groundwater draining from mountains

Evaporation and transpiration concentrate salts in soil

Canal draining to Kesterson Reservoir

Potential level of water table without drain system

Water table

Subsurface tile drain

Soils flushed by irrigation water that percolates downward

Clay aquiclude

FIGURE B10.1 Geologic setting of Kesterson Reservoir in western San Joaquin Valley, California. Runoff carries dissolved selenium to alluvial fans where irrigation water flushes it into a drainage system that concentrates it in the Kesterson Reservoir. Subsurface drains lower the water table, which otherwise would lie at shallow depths because the groundwater is perched above a clay aquiclude.

A further condition is the presence of a clay layer 3 to 23 m beneath the land surface. Being impermeable, the clay layer restricts the downward percolation of selenium-bearing groundwater and produces a perched water body. It is this perched water body above the clay aquiclude that necessitates the drain system, which in turn creates the environmental hazard at Kesterson Reservoir.

Although we are constantly alerted to the environmental impact of pesticides and other manufactured poisons that are introduced into nat-ural ecosystems, the Kesterson saga illustrates how natural substances that pose no special hazard under normal conditions reach toxic levels through human intervention. The solution to such problems, which are increasing in number as an expanding human population places greater demands on limited natural resources, rests on an understanding of the complex interrelationship of the factors involved. In seeking these solutions, geologists have an increasingly important role to play.

SUMMARY

1. Groundwater is derived mostly from rainfall and occurs everywhere beneath the land surface.

2. The water table is the top of the saturated zone.

In humid regions, the form of the water table is a subdued imitation of the overlying land surface.

3. Groundwater moves chiefly by percolation, at

rates far slower than those of surface streams. In rock or sediment of constant permeability, the velocity of groundwater increases as the slope of the water table increases.

4. In moist regions, groundwater in recharge areas percolates downward under the pull of gravity. It moves away from hills toward valleys, where it may emerge to supply streams (groundwater discharge areas). In dry regions, the groundwater is recharged by water percolating downward beneath surface streams.

5. According to Darcy's Law, the discharge of water in a groundwater system is equal to the product of the cross-sectional area of flow, the coefficient of permeability, and the hydraulic gradient.

6. Springs often occur at places where either the water table or an aquiclude intersects the land surface.

7. Groundwater flows into most wells directly by gravity. Pumping of water from wells creates cones of depression in the water table.

8. Major supplies of groundwater are found in aquifers, among the most productive of which are porous sand, gravel, and sandstone.

9. If the top of a well that penetrates an artesian aquifer lies below the altitude of the water table in the recharge area, hydrostatic pressure will allow water to rise in the well and flow out at the surface without pumping.

10. The water table defines the upper surface of an unconfined aquifer, whereas a confined aquifer is bounded by aquicludes.

11. Excess withdrawal of groundwater can lead to lowering of the water table and to land subsidence.

12. Water quality is influenced by the content of natural dissolved substances, seawater intrusion, and pollution by human and industrial wastes that percolate into groundwater reservoirs.

13. Hazardous (toxic and radioactive) wastes should be stored underground only if geologic conditions imply little or no change in groundwater systems over geologically long intervals of time.

14. Groundwater dissolves mineral matter from rock. It also deposits substances as cement between grains of sediment, thereby reducing porosity and converting the sediments to sedimentary rock.

15. In carbonate rocks, groundwater not only creates caves and sinkholes by dissolution but also deposits calcium carbonate as dripstone and flowstone.

16. Karst topography forms in areas of porous carbonate or other soluble rocks where the relief is great enough to permit gravitational flow of groundwater.

IMPORTANT TERMS TO REMEMBER

aquiclude (p. 277)
aquifer (p. 278)
artesian aquifer (p. 279)

cone of depression (p. 277)
confined aquifer (p. 279)

Darcy's Law (p. 276)
discharge area (p. 275)
dissolution (p. 286)
dripstone (p. 288)

flowstone (p. 288)

groundwater (p. 272)

hydraulic gradient (p. 276)

karst topography (p. 290)

percolation (p. 275)
permeability (p. 273)
porosity (p. 273)

recharge (p. 275)
recharge area (p. 275)
replacement (p. 287)

saturated zone (p. 272)
sinkhole (p. 288)
spring (p. 277)

unconfined aquifer (p. 279)

water table (p. 272)

zone of aeration (p. 272)

QUESTIONS FOR REVIEW

1. What is the ultimate source of groundwater?

2. Why does a thin zone immediately above the water table remain continuously moist?

3. Why do the flow paths of groundwater moving beneath a hill tend to turn upward toward a stream in an adjacent valley?

4. What variables determine how long it takes water to move from a recharge area to a discharge area?

5. Explain why it is possible to determine the discharge, velocity, or coefficient of permeability of groundwater passing through an aquifer provided two of these three factors are known.

6. What is the hydraulic gradient and what importance does it have in determining the rate of flow of groundwater?

7. Why are sandstones generally better aquifers than siltstones or shales?

8. What features in igneous and metamorphic rocks promote the flow of groundwater through them?

9. How are springs related to the water table?

10. What causes a cone of depression to form around a producing well?

11. What causes water to rise to or above the ground surface in an artesian well?

12. Why is sand especially effective in purifying water flowing through it?

13. What is the origin of "hard" water in regions of carbonate bedrock?

14. Why do dripstone and flowstone formations not form in a cave that is in the saturated zone and therefore completely filled with water?

15. Why are karst landscapes not well developed in limestone terrain at high latitudes?

16. A large area on a hillside has been suggested as a landfill site for garbage generated by a small nearby city. You are asked for a geologic appraisal of the site to determine if local subsurface water supplies might be affected. What geologic factors would you investigate and why?

17. What geologic and biologic factors are likely to disqualify a site from being selected for underground storage of high-level radioactive waste?

Radiocarbon dates of peat exposed by the recent retreat of Ruitor Glacier in the Italian Alps show that between 8500 and 5500 years ago the glacier was smaller than it has been during the past 5000 years.

CHAPTER 11

Glaciers and Glaciation

A popular science fiction film of the 1980s recounts the discovery of the body of a prehistoric ice-age hunter that had been frozen and preserved in an arctic glacier for the last 20 millennia. Resuscitated in a laboratory, the ancient hunter faces the prospect of surviving in the modern world, a world as different to him as his own world would be to us.

While this tale is fictitious, fossils preserved in and under glaciers do provide us with important clues about environments of the distant past. In the European Alps, receding glaciers have disgorged the remains of trees and bog vegetation that once flourished at sites subsequently covered by ice for many millennia. Dated by radiocarbon, the fossil trees and other plant remains tell us of an interval of mild climate during the early Holocene, a time when glaciers retreated high up in the mountains and plants invaded the upper reaches of alpine valleys.

Another kind of fossil preserved in glaciers has revolutionized our understanding of the glacial ages. As a glacier forms, tiny amounts of air are preserved as bubbles in the ice. These samples of "fossil" atmosphere are trapped in layers of ice as a glacier grows. When a sample of old ice is melted in the laboratory and the chemistry of its trapped air is analyzed, we are able to measure the composition of the ancient atmosphere and calculate the air temperature when the ice formed. Armed with such information, geologists can test theories about the causes of glacial ages and learn many details about the history of the Earth's changing climate.

GLACIERS

At any place where more snow falls on the land during each winter than is melted during the following summer, the snow will gradually grow thicker. As the snow accumulates, the increasing weight of overlying snow causes the basal layers to recrystallize, forming a solid mass of ice. This process is analogous to the way sedimentary rocks, buried under a deep pile of strata within the Earth's crust, recrystallize to form metamorphic rocks. When the accumulating snow and ice become so thick that the pull of gravity causes the frozen mass to move, a glacier is born. Accordingly, we define a **glacier** as a permanent body of ice, consisting largely of recrystallized snow, that shows evidence of downslope or outward movement due to the pull of gravity.

Glaciers are found in areas where average temperature is so low that water can exist year round in the frozen state. As we might expect, therefore, most glaciers are found in high latitudes, the coldest parts of our planet. However, because low temperatures also occur at high altitudes, many small glaciers are found in middle and low latitudes on high mountains.

Because glaciers vary considerably in their physical characteristics, we can distinguish several kinds based on their shape and size (Table 11.1), as well as on their internal temperature.

TABLE 11.1 Principal Types of Glaciers, Classified According to Form

Glacier Type	Characteristics
Cirque glacier	Occupies a cirque on a mountain side
Valley glacier	Flows from cirque(s) onto and along the floor of a valley
Fjord glacier	Valley glacier that occupies a fjord. Base lies below sea level. May have a steep front that recedes rapidly as icebergs break off and float away
Piedmont glacier	Broad lobe of ice that terminates on open slopes beyond a mountain front. Fed by one or more large valley glaciers
Ice cap	Dome-shaped body of ice and snow that covers mountain highlands (or lower lying lands at high latitudes) and displays generally radial outward flow
Ice sheet	Continent-sized mass of ice that overwhelms nearly all land within its margins
Ice shelf	Thick slablike glacier that floats on the sea and is fed by one or more glaciers on land. Commonly located in large bays

Mountain Glaciers and Ice Caps

The shape and direction of movement of most mountain glaciers is determined by the surrounding bedrock topography (Fig. 11.1). A *cirque glacier* occupies a **cirque,** a protected bowl-shaped depression on a mountainside, open downslope and bounded upslope by a steep cliff called a *headwall.* A growing cirque glacier that spreads outward and downward along a valley becomes a *valley glacier* (Fig. 11.2). Many of the Earth's high mountain ranges (e.g., the Alaska Range and the Himalaya) contain glacier systems that include valley glaciers tens of kilometers long (Fig. 11.3).

Valley glaciers in some coastal mountain ranges at middle to high latitudes occupy deep glacier-carved valleys that are filled by an arm of the sea. Such a valley is called a **fjord,** and a glacier that occupies it is a *fjord glacier* (Fig. 11.1).

A very large mountain glacier may spread out onto a gentle piedmont slope beyond the mountain front. It then becomes a *piedmont glacier,* forming a broad lobe of ice that resembles an inverted spoon (Fig. 11.1).

An *ice cap* covers a mountain highland or lower-lying land at high latitude, and displays generally radial outward flow (Figs. 11.1 and 11.4).

Ice Sheets and Ice Shelves

An *ice sheet* is the largest type of glacier on the Earth. These continent-sized masses of ice overwhelm nearly all the land surface within their margins. Modern ice sheets, confined to Greenland and Antarctica, include about 95 percent of the ice in existing glaciers. If all the ice in these vast ice sheets were to melt, their combined volume—close to 24 million km^3—would be sufficient to raise world sea level by nearly 66 m.

The Greenland Ice Sheet has about the same area as the United States west of the Rocky Mountains (Fig. 11.5). It is so thick (about 3000 m) that its weight is enough to cause the land surface beneath much of it to be depressed below sea level.

Antarctica is covered by two large ice sheets that meet along the lofty Transantarctic Mountains (Fig. 11.6). The East Antarctic Ice Sheet is the larger of the two and covers the continent of Antarctica. Because of its ice sheet, Antarctica has the highest average altitude and the lowest average temperature of all the continents. The smaller West Antarctic Ice Sheet overlies numerous islands of the Antarctic archipelago. Parts of it rest on land that rises above sea level, but extensive portions cover land lying below sea level.

An *ice shelf* is a thick, nearly flat sheet of floating

FIGURE 11.1 Common types of mountain glaciers, classified according to shape and size. A small cirque glacier occupies a bowl-shaped hollow at the head of a valley, while in the adjacent valley two cirque glaciers have expanded and merged to form a valley glacier. In the next valley a fjord glacier sheds icebergs into a long fjord. Nearby, several tributary glaciers merge into a single ice stream that terminates in a broad piedmont glacier. In the distance, a high mountain plateau is mantled by a broad ice cap.

FIGURE 11.2 Dark bands of rock debris mark the boundaries between adjacent tributary ice streams that have merged to form Kaskawulsh Glacier, a large valley glacier in Yukon Territory, Canada.

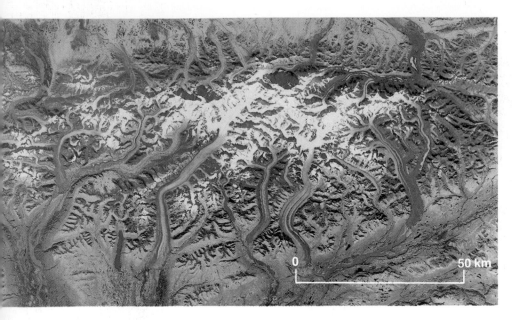

FIGURE 11.3 A vertical satellite view of the valley–glacier complex that covers much of Denali National Park in south-central Alaska. Mount McKinley, the highest peak in North America, lies near the center of the glacier-covered region.

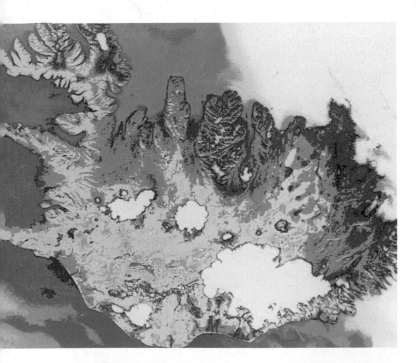

FIGURE 11.4 Several ice caps mantle the areas of highest land on Iceland in the north Atlantic Ocean. Vatnajökull, lying in the southeastern part of the island, is the largest and covers 8300 km².

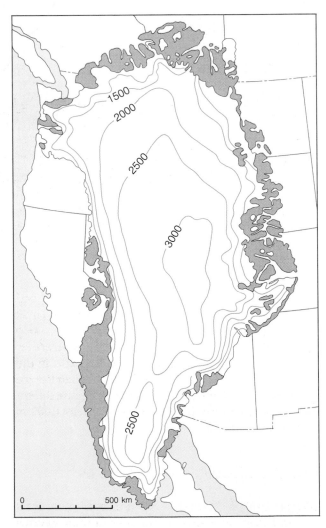

FIGURE 11.5 Map of Greenland superimposed on a map of western United States showing that the Greenland Ice Sheet covers an area approximating that of the United States west of the Great Plains. At the crest of the ice sheet, the ice thickness is more than 3000 m.

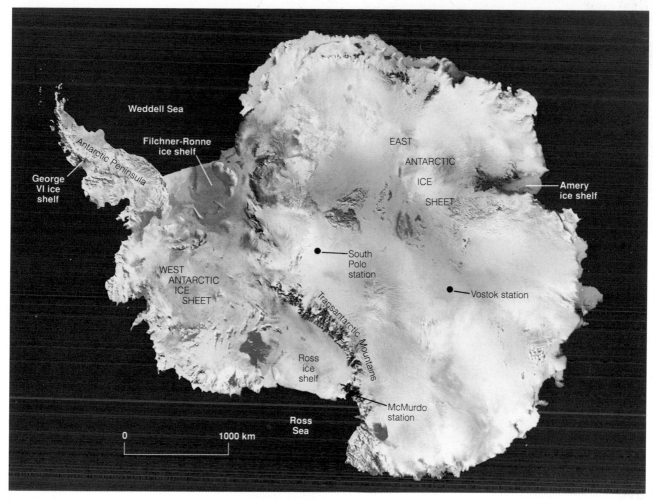

FIGURE 11.6 Satellite view of Antarctica. The East Antarctic Ice Sheet overlies the continent of Antarctica, whereas the much smaller West Antarctic Ice Sheet overlies a volcanic island arc and surrounding seafloor. Three major ice shelves occupy large bays. The ice-covered regions of Antarctica nearly equal the combined area of Canada and the conterminous United States.

ice that is fed by one or more glaciers on land and terminates seaward in a steep ice cliff as much as 50 m high. Ice shelves are found at several places along the margins of the Antarctic ice sheets, where they are confined in large bays (Fig. 11.6), and also in the Canadian Arctic islands. The largest Antarctic ice shelves extend hundreds of kilometers seaward from the coastline and reach a thickness of at least 1000 m near their landward margins.

Temperate and Polar Glaciers

Glaciers can be classified not only according to their size and shape, but also on the basis of their internal temperature. Temperature is an important criterion for classification because it helps determine how glaciers move and how they shape the landscape.

Ice in a **temperate glacier** is at the **pressure melting point**, which is the temperature at which ice melts at a particular pressure (Fig. 11.7A). In such glaciers, which are restricted mainly to low and middle latitudes, meltwater and ice exist together in equilibrium. At high latitudes and high altitudes, where the mean annual air temperature lies below freezing, the temperature in a glacier remains below the pressure melting point and little or no seasonal melting occurs. Such a glacier, in which the ice remains below the pressure melting point, is called a **polar glacier** (Fig. 11.7B).

Since the temperature of snow crystals falling to the surface of a temperate glacier is below freezing, how can ice throughout the glacier reach the pressure melting point? The answer lies in the seasonal fluctuation of air temperature and in what happens when water freezes to form ice. In summer, when air temperature rises above freezing, solar radiation melts the glacier's surface snow and ice. The meltwater percolates downward, where it encounters freezing temperatures and therefore freezes. When changing state

A.

TEMPERATURE GLACIER

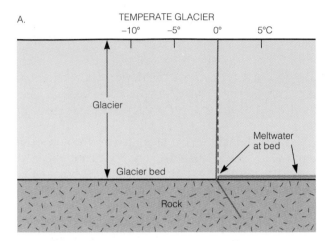

B.

POLAR GLACIER

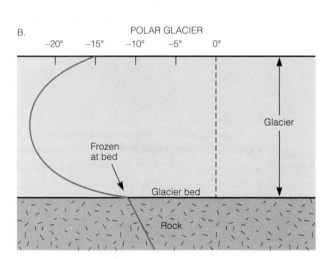

FIGURE 11.7 A. Ice in a temperate glacier is at the pressure melting point from surface to bed. The terminus is rounded, as shown in this view of Pré de Bar Glacier in the Italian Alps, because melting occurs at the surface. B. Ice in a polar glacier remains below freezing, and the ice is frozen to its bed. Subfreezing temperatures inhibit melting at the terminus, which forms a steep cliff of ice, as shown in this view of Commonwealth Glacier in Antarctica.

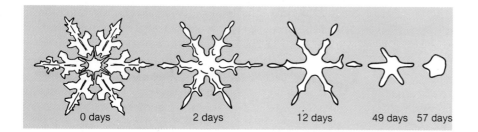

FIGURE 11.8 Conversion of a snowflake into a granule of old snow. The delicate points of a snowflake disappear as melting and evaporation occur. The resulting meltwater refreezes and vapor condenses near the center of the crystal, making it denser.

from liquid to solid, each gram of water releases 335 J of heat. This released heat warms the surrounding ice and, together with heat flowing upward from the solid earth beneath the glacier, keeps the temperature of the ice at the pressure melting point.

Glaciers and the Snowline

Glaciers can form only at or above the **snowline,** which is the lower limit of perennial snow. The snowline is sensitive to local climate, especially temperature and precipitation, and rises in altitude from near sea level in polar latitudes to as much as 6000 m in the tropics. It also rises inland from moist coastal regions toward the drier interiors of large islands and continents.

A glacier potentially can form at any place where the snowline intersects the land surface and the topography will permit a glacier to exist. This explains why glaciers are found not only at sea level in the polar regions, where the snowline is low, but also near the equator, where some high peaks in New Guinea, East Africa, and the Andes rise above the snowline.

Conversion of Snow to Glacier Ice

Glacier ice is a metamorphic rock that consists of interlocking crystals of the mineral *ice* and owes its characteristics to deformation under the weight of overlying snow and ice.

Newly fallen snow is very porous and has a density less than a tenth that of water. Air easily penetrates the pore spaces, and the delicate points of snowflakes gradually disappear by evaporation. The resulting water vapor condenses, mainly in constricted places near the centers of snowflakes. In this way, the fragile ice crystals slowly become smaller, rounder, and denser, and the pore spaces between them disappear (Fig. 11.8).

Snow that survives a year or more gradually increases in density until it is no longer permeable to air, at which point it becomes glacier ice. Although

now a rock, such ice has a far lower melting temperature than any other naturally occurring rock, and its density of about 0.9 g/cm^3 means that it will float in water.

Further changes in ice take place with increasing depth in a glacier. Figure 11.9 shows a core obtained by Russian glaciologists who drilled to the base of the East Antarctic Ice Sheet at Vostok Station (Fig. 11.6). As snowfall adds to the glacier's thickness, the increasing pressure causes initially small grains of glacier ice to grow in size until, near the base of the ice sheet, they reach a diameter of 1 cm or more. This increase in grain size is analogous to what happens in a fine-grained rock that is carried deep within the Earth's crust: subjected to high pressure over a long time, large mineral grains slowly develop (chapter 5).

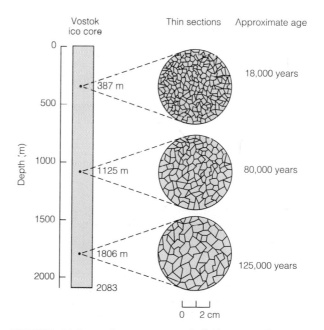

FIGURE 11.9 A deep ice core drilled at Vostok Station penetrates through the East Antarctic Ice Sheet to bedrock at a depth of 2083 m. Thin-section samples of ice taken from different depths in the core show a progressive increase in the size of ice crystals, a result of slow recrystallization as the weight of overlying ice increases with time.

Why Glaciers Change in Size

The pioneer Scottish glaciologist James D. Forbes visited the Alps in the mid-nineteenth century and carefully mapped the extent of long valley glaciers flowing from Mont Blanc, the highest peak in Europe. Since Forbes's day, these glaciers have shrunk greatly, exposing extensive areas of valley floor that only a century ago were buried beneath thick ice. Over this same period, changes in glacier size have been recorded worldwide. Like those in the Alps, most other glaciers have receded dramatically since the end of the last century, although some have remained relatively unchanged and others have expanded. To understand why glaciers advance and retreat, and why glaciers in any region may behave in different ways, we must examine how a glacier responds to a gain or loss of mass.

Mass Balance

The mass of a glacier is constantly changing as the weather varies from season to season and, on longer time scales, as local and global climates change. These ongoing environmental changes cause fluctuations in the amount of snow added to the glacier surface and in the amount of snow and ice lost by melting.

We can think of a glacier as being like a checking account at the bank. The balance in a bank account at the end of the year is the difference between the amount of money added to the account and the amount removed during the year. The balance of a glacier's account is measured in terms of the amount of snow added, mainly in the winter, and the amount of snow (and ice) lost, mainly during the summer. Additions to the glacier's account are collectively called **accumulation,** and losses are termed **abla-**

tion. The total in the account at the end of a year—in other words, the difference between accumulation and ablation—is a measure of the glacier's **mass balance** (Fig. 11.10). The account may have a surplus (a positive balance) or a deficit (a negative balance), or it may be in exact balance (accumulation = ablation).

If a glacier is viewed at the end of the summer ablation season, two zones are generally visible on its surface (Fig. 11.11). An upper zone, the *accumulation area,* is the part of the glacier covered by remnants of the previous winter's snowfall and is an area of net gain in mass. Below it lies the *ablation area,* a region of net loss where bare ice and old snow are exposed because the previous winter's snow cover has melted away.

The **equilibrium line** marks the boundary between the accumulation area and the ablation area (Fig. 11.11). It lies at the level on the glacier where net mass loss equals net mass gain. The equilibrium line on temperate glaciers coincides with the local snowline, which marks the lower limit of fresh snow at the end of the ablation season. Being very sensitive to climate, the equilibrium line fluctuates in altitude from year to year and is higher in warm, dry years than in cold, wet years (Fig. 11.12).

Fluctuations of the Glacier Terminus

If, over a period of years, a glacier's mass balance is positive more often than negative, the mass of the glacier increases. The front, or *terminus,* of the glacier is then likely to advance as the glacier grows. Conversely, a succession of years in which negative mass balance predominates will lead to retreat of the terminus. If no net change in mass occurs, the glacier is in a balanced state. If this condition persists, the terminus is likely to remain relatively stationary.

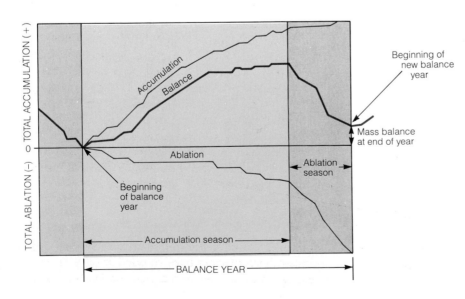

FIGURE 11.10   Graph showing how accumulation and ablation determine glacier mass balance (heavy line) over the course of a balance year. The balance curve, obtained by summing values of accumulation (positive values) and ablation (negative values), rises during the accumulation season as mass is added to the glacier, then falls during the ablation season as mass is lost. The mass balance at the end of the balance year reflects the difference between mass gain and mass loss.

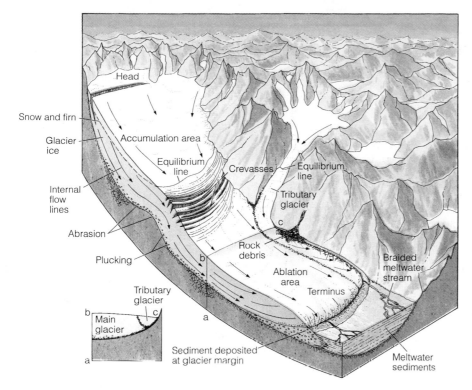

FIGURE 11.11 Main features of a valley glacier. The glacier has been cut away along its center line; only half is shown. Crevasses form where the glacier bed has a steeper slope. Arrows show directions of ice flow. A band of rock debris marks the boundary between the main glacier and a tributary glacier that joins it from a lateral valley.

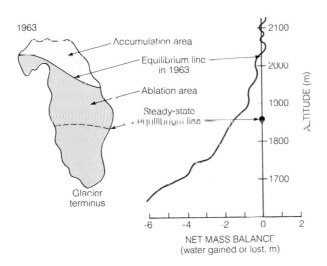

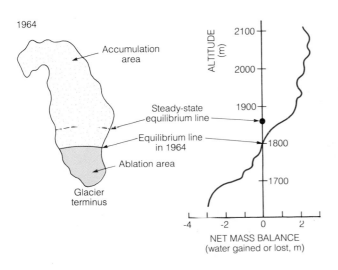

FIGURE 11.12 Maps of South Cascade Glacier in Washington State at the end of the 1963 and 1964 balance years showing the position of the equilibrium line relative to the position it would have under a balanced (steady-state) condition. The curves show values of mass balance as a function of altitude. During 1963, a negative balance year, the glacier lost mass and the equilibrium line was high (2025 m). In 1964, a positive balance year, the glacier gained mass and the equilibrium line was low (1800 m).

Response Lags. Advance or retreat of a glacier terminus does not necessarily give us an accurate picture of changing climate because a lag occurs between climatic change and the response of the glacier terminus to that change. The lag reflects the time it takes for the effects of an increase or a decrease in accumulation above the equilibrium line to be transferred through the slowly moving ice to the glacier terminus. The length of the lag depends both on the size of a glacier and the way the ice flows; the lag will be longer for large glaciers than for small ones, and longer for polar glaciers than for temperate ones. Temperate glaciers of modest size (like those in the European Alps) have response lags that range from several years to a decade or more. This lag time can explain why, in any area having glaciers of different sizes, some glaciers may be advancing while others are stationary or retreating.

Calving. During the last century and a half, many Alaskan fjord glaciers have receded at rates far in excess of typical glacier retreat rates on land. Their dra-

305

FIGURE 11.13 Icebergs break away from the calving front of a fjord glacier in southwest Greenland. This fjord was filled by an arm of the sea as the glacier retreated back rapidly by frontal calving.

matic recession is due to frontal **calving,** defined as the progressive breaking off of icebergs from the front of a glacier that terminates in deep water (Fig. 11.13). Although the base of a fjord glacier may lie far below sea level along much of its length, its terminus can remain stable as long as it is resting (or "grounded") against a shoal (Fig. 11.14). However, if the terminus retreats off the shoal, water will replace the space that had been occupied by ice. With the glacier now terminating in water, conditions are right for calving. Because a fjord glacier increases in thickness in the upfjord direction, the water becomes progressively deeper as the calving terminus retreats. The deepening water leads to faster retreat, because the greater the water depth, the faster the rate of calving. Once started, calving will continue rapidly and irreversibly until the glacier front recedes into water too shallow for much calving to occur, generally near the head of the fjord.

Icebergs produced by calving glaciers constitute an ever-present hazard to ships in subpolar seas. When the S.S. *Titanic* sank after striking a berg in the North Atlantic in 1912, the detection of approaching icebergs relied on sailors' vision. Today, with sophisticated electronic equipment, large bergs can generally be identified well before an encounter. Nevertheless, ice has a density of 0.9, so that 90 percent of an iceberg lies under water, making it difficult to detect. In coastal Alaska, where calving glaciers are commonplace, icebergs pose a potential threat to huge oil tankers. For

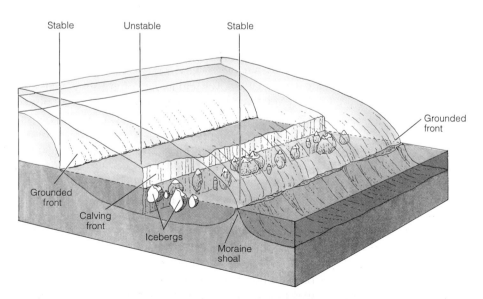

FIGURE 11.14 The terminus of a fjord glacier remains stable if it is grounded against a shoal, but if it retreats into deeper water, calving will begin. The unstable terminus then retreats, at a rate that depends on the water depth. Once it becomes grounded farther up the fjord, the terminus is stable once again.

this reason, Columbia Glacier, which lies adjacent to the main shipping lanes from Valdez at the southern end of the Alaska Pipeline, is being closely monitored as its terminus pulls steadily back and multitudes of bergs are released.

How Glaciers Move

We can easily prove to ourselves that glaciers move. One way is to visit a glacier near the end of the summer and carefully measure the position of a boulder lying on the ice surface with respect to some fixed point beyond the glacier margin. If we return a year later, we will find that the boulder has moved at least several meters in the downglacier direction. Actually, it is the ice that has moved, carrying the boulder along for the ride.

What causes a glacier to move may not be immediately obvious, but we can find clues by examining the ice and the terrain over which it lies. These clues tell us that ice moves in two ways: by internal flow and by sliding of the basal ice over underlying rock or sediment.

Internal Flow

When an accumulating mass of snow and ice on a mountainside reaches a critical thickness, the mass will begin to deform and flow downslope under the pull of gravity. The flow takes place mainly through movement within individual ice crystals, which are subjected to higher and higher stress as the weight of the overlying snow and ice increases. Under this stress, ice crystals are deformed by slow displacement (termed creep) along internal crystal planes in much the same way that cards in a deck of playing cards slide past one another if the deck is pushed from one end (Fig. 11.15). As the compacted, frozen mass begins to move, stresses between adjacent ice crystals cause some to grow at the expense of others, and the resulting larger crystals end up with their internal planes oriented in the same direction. This alignment of crystals leads to increased efficiency of flow, for the internal creep planes of all crystals now are parallel.

In contrast to deeper parts of a glacier where the ice flows as a result of internal creep, the surface portion of a glacier has relatively little weight on it and is brittle. Where a glacier passes over an abrupt change in slope, such as a bedrock cliff, the surface ice cracks as tension pulls it apart. The cracks open up and form crevasses. A **crevasse** is a deep, gaping fissure in the upper surface of a glacier, generally less than 50 m deep (Fig. 11.11). At depths greater than about 50 m, continuous flow of ice prevents crevasses from forming. Because it cracks at the surface yet flows

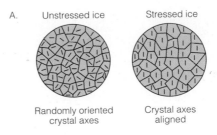

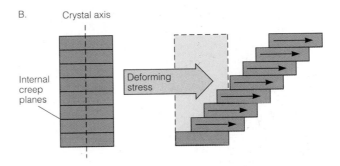

FIGURE 11.15 Internal creep in ice crystals of a glacier. A. Randomly oriented crystals of ice in the upper layers of a glacier are transformed under stress so that their axes are aligned. B. When a stress is applied to an ice crystal, creep along internal planes results in slow deformation of the crystal.

at depth, a glacier is similar to the upper layers of the Earth, which include a surface zone that cracks and fractures (the lithosphere) and a deeper zone (the asthenosphere) that can flow slowly.

Basal Sliding

Ice temperature is very important in controlling the way a glacier moves and its rate of movement. Meltwater at the base of a temperate glacier acts as a lubricant and permits the ice to slide across its *bed* (the rocks or sediments on which the glacier rests). In some temperate glaciers such sliding accounts for up to 90 percent of the total observed movement (Fig. 11.16). By contrast, polar glaciers are so cold they are frozen to their bed. Their motion largely involves internal deformation rather than basal sliding, and so their rate of movement is greatly reduced.

Velocities and Directions of Flow

Measurements of the surface velocity across a valley glacier show that the uppermost ice in the central part of the glacier moves faster than ice at the sides, similar to the velocity distribution in a river (Figs. 11.16 and 9.11). The reduced rates of flow toward the margins are due to frictional drag of the ice against the valley

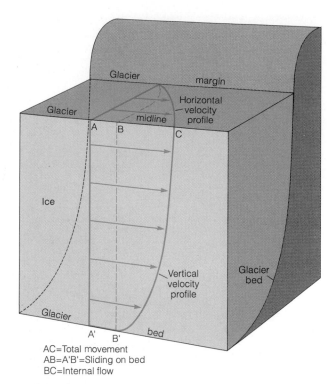

AC=Total movement
AB=A'B'=Sliding on bed
BC=Internal flow

FIGURE 11.16 Three-dimensional view through half of a glacier showing horizontal and vertical velocity profiles. A portion of the total observed movement is due to internal flow within the ice, whereas part is due to sliding of the glacier along its bed, lubricated by a film of meltwater.

walls. A similar reduction in flow rate toward the bed is observed in a vertical profile of velocity (Fig. 11.16).

Snow continues to pile up in the accumulation area each year, while melting removes snow and ice from the ablation area. The surface profile of a glacier does not change much, however, because ice is transferred from the accumulation area to the ablation area. In the accumulation area, the mass of accumulating snow and ice is pulled downward by gravity, and so the dominant flow direction is toward the glacier bed. However, the ice does not build up to ever greater thickness because a downglacier component of flow is also present. Ice flowing downglacier replaces ice being lost from the glacier's surface in the ablation area, and so in this area the flow is upward toward the surface (Fig. 11.11). Ice crystals falling as snowflakes on the glacier near its head therefore have a long path to follow before they emerge near the terminus. Those falling close to the equilibrium line, however, travel only a short distance through the glacier before reaching the surface again.

Even if the mass balance of a glacier is negative and the terminus is retreating, the downglacier flow of ice is maintained. Retreat does not mean that the ice-flow direction reverses; instead, it means that the rate of flow downglacier is insufficient to offset the loss of ice at the terminus.

In most glaciers, flow velocities range from only a few centimeters to a few meters a day, or about as fast as the rate at which groundwater percolates through crustal rocks. Hundreds of years have elapsed since ice now exposed at the terminus of a very long glacier fell as snow near the top of its accumulation area.

Glacier Surges

Although most glaciers slowly grow or shrink in size as the climate fluctuates, some glaciers experience episodes of very unusual behavior marked by rapid movement and dramatic changes in size and form. Such an event, called a **surge,** is unrelated, or only secondarily related, to a change in climate. When a surge occurs, a glacier seems to go berserk. Ice in the accumulation area begins to move rapidly downglacier, producing a chaos of crevasses and broken pinnacles of ice in the ablation area. Medial moraines, which are bands of rocky debris marking the boundaries between adjacent tributary glaciers (Fig. 11.11), are deformed into intricate patterns (Fig. 11.17). The termini of some glaciers have advanced up to several kilometers during surges. Rates of movement as great as 100 times those of nonsurging glaciers and averaging as much as 6 km a year have been measured.

The cause of surges is still imperfectly understood, but available evidence points to a reasonable hypothesis. We know that the weight of ice can produce high hydrostatic pressure in water at the base of a glacier.

FIGURE 11.17 Surging tributary glaciers flowing from the mountains on the right of this view have deformed the medial moraines of Alaska's Yanert Glacier into a series of complex folds.

Over a period of years, steadily increasing hydrostatic pressure in water trapped beneath the ice may lead to widespread separation of the glacier from its bed. The resulting effect is similar to the hydroplaning of an automobile on a wet road surface or the displacement of a beverage can on a sheet of glass in the experiment described in chapter 8. According to this hypothesis, as the ice is floated off its bed, its forward mobility is greatly increased and it moves rapidly forward before the escape of water brings the surge to a halt.

GLACIATION

Glaciation, the modification of the land surface by the action of glacier ice, has occurred so recently over large areas of Europe, Asia, and North and South America that weathering, mass-wasting, and erosion by running water have not yet had time to alter the landscape appreciably. Except for a cover of vegetation, the appearance of these glaciated landscapes has remained nearly unchanged since they emerged from beneath the ice.

Like the geologic work of other surface processes, glaciation involves the erosion, transport, and deposition of sediment.

Glacial Erosion and Sculpture

In changing the surface of the land over which it moves, a glacier acts collectively like a plow, a file, and a sled. As a plow it scrapes up weathered rock and soil and plucks out blocks of bedrock; as a file it rasps away firm rock; and as a sled it carries away the load of sediment acquired by plowing and filing, along with additional rock debris that falls onto the glacier from adjacent slopes.

Small-Scale Features of Glacial Erosion

The base of a temperate glacier is studded with rock fragments of various sizes which are all carried along with the moving ice. Small fragments of rock embedded in the basal ice scrape away at the underlying bedrock and produce long, nearly parallel scratches called *glacial striations* (Fig. 11.18). Larger rock fragments that the ice drags across the bedrock abrade *glacial grooves* aligned in the direction of glacier flow. Grains of fine sand and silt in the basal ice act like sandpaper and polish the rock until it has a smooth, reflective surface.

Because striations and grooves are aligned with the direction of ice flow, geologists use these and other aligned erosional features to reconstruct the flow

FIGURE 11.18 Recently deglaciated bedrock surface beyond Findelen Glacier in the Swiss Alps. Debris carried at the base of the glacier produced grooves, striations, and polish on bedrock as the ice flowed forward in the direction of the Matterhorn.

paths of former glaciers. Debris-laden ice striates and polishes the upglacier sides of small bedrock knobs or hills over which it moves, and plucks blocks of rock from the downglacier sides. The resulting asymmetrical landforms, smooth and gently sloping upglacier and steeper and angular downglacier, provide clear evidence of the direction in which the glacier was moving (Fig. 11.19).

Landforms of Glaciated Mountains

Skiers racing down the steep slopes at Alta, Mammoth, or Whistler, and rock climbers inching their way up the cliffs of Yosemite Valley, the granite spires of Mont Blanc, or the icy monoliths of the southern Andes owe a debt to the ancient glaciers that carved these mountain playgrounds. The scenic splendor of most of the world's high mountains is the direct result of glacial sculpturing that has produced a distinctive assemblage of alpine landforms.

FIGURE 11.19 Asymmetrical glacially sculptured rock surface beyond the terminus of Franz Josef Glacier in New Zealand's Southern Alps. The glacier flowed from right to left. Slopes facing toward the glacier are smooth and polished. Scarps facing downvalley result from the plucking of bedrock blocks by flowing ice.

Cirques.   Cirques are among the most common and distinctive landforms of glaciated mountains. The characteristic bowl-like shape of a cirque is the combined result of frost-wedging, glacial plucking, and abrasion (Fig. 11.20). Many cirques are bounded on their downvalley side by a bedrock threshold that impounds a small lake (a *tarn*).

A cirque probably begins to form beneath a large snowbank or snowfield at or just above the snowline. As meltwater infiltrates rock openings under the snow, it refreezes and expands, disrupting the rock and dislodging fragments. Small rock particles are carried away by snowmelt runoff during periods of thaw. A shallow depression in the land is thereby created, and the depression gradually becomes larger and larger.

As snow continues to accumulate in the deepening hollow, the snowbank thickens and becomes a glacier. Plucking then helps to enlarge the cirque still more, and abrasion at the bed further deepens it.

As cirques on opposite sides of a mountain grow larger and larger, their headwalls intersect to produce a sharp-crested ridge called an *arête* (Fig. 11.20). Where three or more cirques have sculptured a mountain mass, the result can be a high, sharp-pointed peak (a *horn*), examples of which include the Matterhorn in the Swiss/Italian Alps (Fig. 11.18) and many of the steep, high peaks surrounding Mount Everest in the Himalaya (Fig. 11.20).

Glacial Valleys.   Valleys that were shaped by former glaciers differ from ordinary stream valleys in several ways. The chief characteristics of glacial valleys include a U-shaped cross profile and a floor that lies below the floors of smaller tributary valleys (Fig. 11.21). Streams commonly descend as waterfalls, or cascades, as they flow from the tributary valleys into the main valley. The glaciated Cascade Range of western United States derives its name from such streams. The long profile of a glaciated valley floor may possess steplike irregularities and shallow basins. These usually are related to the spacing of joints in the rock, which influences the ease of glacial plucking, or to changes in rock type along the valley. Finally, the valley typically heads in a cirque or group of cirques.

Fjords.   Fjords are common features along the mountainous, west-facing coasts of Norway, Alaska, British Columbia, Chile, and New Zealand, as well as in northern Canada. The floors of many fjords contain elongate basins that reach depths of 300 m or more

FIGURE 11.20 Sharp-crested arêtes flank the Western Cwm, a deep cirque on the west side of Mount Everest in the central Himalaya. Sharp-crest peaks in the distance are horns that have resulted from headward growth of flanking cirque and valley glaciers.

and are evidence of deep glacial erosion (Fig. 11.22). As a result, fjords typically are shallow at their seaward end and deepen inland. Sognefjord in Norway reaches a depth of 1300 m, yet near its seaward end the water depth is only about 150 m.

Landforms Produced by Ice Caps and Ice Sheets

Abrasional Features. Landscapes that were shaped by overriding ice sheets display the same small-scale erosional features typical of valley glaciation. Striations have been especially helpful to geologists in reconstructing the flow lines of long-vanished ice sheets. Striations also demonstrate that basal ice in the thick central zone of former ice sheets was at the pressure melting point, a necessary condition for the sliding action required to produce these features. In peripheral zones, evidence of glacial erosion is sometimes less obvious, leading to the conclusion that the thinner ice there was cold and largely frozen to its bed.

Wherever ice sheets overwhelmed mountainous terrain, as in the high ranges of northwestern North America during the most recent glacial age, the highest evidence of glaciation can frequently be seen as the level where smooth, ice-abraded slopes pass abruptly upward into rugged, frost-shattered peaks and mountain crests that stood above the glacier surface. Some divides between adjacent drainages are broad and smooth and show evidence of glacial abrasion and plucking where they were overridden by ice.

Streamlined Forms. In many areas inside the limits of former ice sheets, the land surface has been molded into smooth, nearly parallel ridges, some of which are several kilometers long. Among the most distinctive of these landforms is the **drumlin,** a streamlined hill consisting largely of glacially deposited sediment and elongated parallel to the direction of ice flow (Fig. 11.23). Glacially molded drumlin-shaped hills of bedrock (called *rock drumlins*) also owe their shape to erosion by flowing ice. Drumlins and rock drumlins, like the streamlined bodies of supersonic airplanes which are designed to reduce air resistance, offer minimum resistance to glacier ice flowing over and around them.

Transport of Sediment by Glaciers

One way a glacier differs from a stream is the way in which it carries its load of rock particles. Unlike a stream, part of a glacier's coarse load can be carried

FIGURE 11.21 Repeated glaciations by a thick valley glacier carved the deep, U-shaped Yosemite Valley in California's Sierra Nevada. The valley glacier was nourished by an extensive mountain ice cap that covered the undulating upland surface of the range, seen in the distance.

FIGURE 11.22 A deep fjord indents the northeast coast of Baffin Island in northeastern Canada. If the fjord were drained of water, its cross-section would resemble that of Yosemite Valley (Fig. 11.21).

FIGURE 11.23   A field of elongate drumlins formed during the last glaciation occupies the floor of Beagle Channel along the south coast of Tierra del Fuego in southernmost Argentina. The glacier, flowing from left to right, produced these smooth, elongate ridges of till with their long axes oriented in the direction of ice flow.

at its sides and even on its surface. A glacier can carry far larger pieces of rock, and it can transport large and small pieces side by side without segregating them according to size and density into a bed load and a suspended load. Because of these differences, sediments deposited directly by a glacier are neither sorted nor stratified.

The load of a glacier typically is concentrated at its base and sides because these are the areas where glacier and bedrock are in contact and where abrasion and plucking are effective. Much of the rock debris on the surface of valley glaciers arrived there by rockfalls from adjacent cliffs. If a rockfall reaches the accumulation area, the flow paths of the ice (Fig. 11.11) will carry the debris downward through the glacier and then upward to the surface in the ablation area. If rocks fall onto the ablation area, the debris will remain at the surface and be carried along by the moving ice. Where two glaciers join, rocky debris at their margins merges to form a distinctive, dark-colored medial moraine (Figs. 11.2 and 11.11).

Much of the load in the basal ice of a glacier consists of very fine sand and silt grains informally called *rock flour*. These particles have sharp, angular surfaces that are produced by crushing and grinding (Fig. 4.24A).

Glacial Deposits

Glaciers are efficient agents of erosion and transport. Many of the sediments and landforms they produce are distinctive, making it relatively easy for geologists to interpret the record of past glacier variations.

Sediments deposited by a glacier or by streams produced by melting glacier ice are collectively called **glacial drift**, or simply **drift.** The term *drift* dates from the early nineteenth century when it was vaguely conjectured that all such deposits had been "drifted" to their resting places during the biblical flood of Noah or in some other ancient body of water. Glacial drift includes sediments associated both with moving ice and with stagnant ice. Several types of sediment are recognized. They form a gradational series ranging from nonsorted to sorted deposits.

Ice-laid Deposits

Till and Erratics.   At one end of the range is **till,** which is nonsorted drift deposited directly from ice. The term was used by Scottish farmers long before the origin of the sediment was understood. The rock particles in a body of till lie just as they were released from the ice (Fig. 11.24). Most tills are a random mixture of rock fragments in which a matrix of fine-grained sediment surrounds larger stones of various sizes. The till matrix consists largely of sand and silt derived by abrasion of the glacier bed and from reworking of preexisting fine-grained sediments. Pebbles and larger rock fragments in till often have smoothed and abraded surfaces, and some are striated. Both the stones and the coarser matrix grains in till tend to lie with their longest axis aligned with the direction of ice flow.

As we learned in chapter 4, a *tillite* is an ancient till that has been converted to rock. Tillites constitute a primary line of evidence for pre-Pleistocene glacial ages.

In many cases, not all the boulders and smaller rock fragments in a till are the same kind of rock as the underlying bedrock, indicating that these components of the till were carried to their present site from somewhere else. A glacially deposited rock or rock

FIGURE 11.24 Till deposited by the Laurentide Ice Sheet in the James Bay Lowland of southern Canada contains numerous igneous and metamorphic clasts resting in a fine-grained matrix. The well-developed jointing probably is related to the release of stress as the 2-km-thick glacier thinned and disappeared at the end of the last glaciation.

fragment with a lithology different from that of the underlying bedrock is called an **erratic** (Latin for wanderer) (Fig. 11.25). Some huge erratics weigh many tons and are found tens or even hundreds of kilometers from their sources. In areas of ice-sheet glaciation, erratics derived from distinctive bedrock sources may have a fanlike distribution, spreading out from the area of outcrop and reflecting the diverging pattern of ice flow.

FIGURE 11.25 Erratic boulders of Precambrian granite rest on eroded Ordovician shale on the south shore of the Gulf of St. Lawrence in eastern Canada. The boulders were carried southward across the gulf by the most recent Pleistocene ice sheet and left behind when the glacier retreated at the end of the glacial age.

Glacialmarine Drift. Closely resembling till, *glacialmarine drift* is sediment deposited on the seafloor from ice shelves or bergs. As an iceberg or the base of an ice shelf slowly melts, the contained sediment is released and settles to the seafloor where it forms a nonsorted deposit. Individual stones released from icebergs that have drifted into the open sea fall to the seafloor and plunge into unconsolidated marine sediments. The impact causes any laminated structure in the uppermost sediment layers to be deformed. Such *dropstones* are also common in the sediments of lakes that are ponded along glacier margins.

Moraines. A moving glacier carries with it rock debris eroded from the land over which it is passing or dropped on the glacier surface from adjacent cliffs. As the debris is transported past the equilibrium line and ablation reduces ice thickness, the debris begins to be deposited. Some of the basal debris is plastered directly onto the ground as till. Some also reaches the glacier margin where it is released by the melting ice and either accumulates there or is reworked by meltwater that transports it beyond the terminus.

An accumulation of drift having a surface form that is unrelated to the underlying bedrock is called a **moraine.** If a body of drift is widespread, has a relatively smooth surface topography, and consists of gently undulating knolls and shallow, closed depressions, we call it *ground moraine*; commonly, it is a blanket of till 10 m or more thick that was deposited beneath a glacier. By contrast, an *end moraine* is a ridgelike accumulation of drift deposited along the margin of a glacier (Fig. 11.26). An end moraine deposited at the glacier terminus is a *terminal moraine,* whereas a similar deposit along the side of a valley glacier is a *lateral moraine.* Normally terminal and lateral moraines are segments of a single, continuous landform deposited below the equilibrium line.

End moraines can form in several ways: as sediment is bulldozed by a glacier advancing across the land, as loose surface debris on a glacier slides off and piles up along a glacier margin, or as debris melts out of ice and accumulates at the margin of a glacier. End moraines range in height from a few meters to hundreds of meters. The great thickness of some lateral moraines results from the repeated accretion of sediment from debris-covered glaciers during successive ice advances.

Stratified Drift

In contrast to nonsorted till and glacialmarine drift, some glacial drift is both sorted and stratified. This kind of drift is not deposited directly by glacier ice but rather by meltwater flowing from the ice. *Stratified drift* ranges from coarse, very poorly sorted sandy

FIGURE 11.26 A bouldery end moraine surrounds the terminus of Suess Glacier in Taylor Valley, Antarctica. The terminal moraine system, which rises above a frozen lake on the valley floor, passes upslope into a lateral moraine along the glacier margin.

gravels deposited by turbulent streams of meltwater, to fine-grained, well-sorted silts and clays deposited in quiet water.

Outwash. Stratified sediment deposited by meltwater streams as they flow away from a glacier margin is called **outwash** (because the sediment is washed "out" beyond the ice). Such streams typically have a braided pattern because of the large sediment load they are moving. If the streams are free to swing back and forth widely beyond the glacier terminus, they deposit outwash to form a broad *outwash plain.* Meltwater streams confined by valley walls build an outwash body called a *valley train* (Figs. 11.11 and 11.28A).

During glacier retreat, a stream's sediment load is reduced and the underloaded stream cuts down into its outwash deposits to produce *outwash terraces* (Fig.

11.28B). A succession of terraces commonly is found in valleys that have experienced repeated glaciations (Fig. 11.27). Generally, each prominent terrace can be traced upstream to an end moraine or to the limit of a former glacier.

Deposits Associated with Stagnant Ice. When rapid ablation greatly reduces a glacier's thickness in its terminal zone, ice flow may virtually cease. Sediment carried by meltwater flowing over or beside such stagnant ice is deposited as stratified drift that slumps and collapses as the supporting ice slowly melts away. Such sediment, called **ice-contact stratified drift,** is recognized by abrupt changes in grain size; distorted, offset, and irregular stratification; and extremely uneven surface form. Bodies of ice-contact stratified drift have many distinctive forms and are classified according to their shape (Fig. 11.27). Among the landforms most likely to be seen are the *kame,* a small hill of ice-contact stratified drift, and the *kettle,* a basin in drift created by the melting away of a mass of underlying glacier ice. Extremely uneven terrain underlain by ice-contact stratified drift and marked by numerous kettles and kames is clear evidence of former stagnant-ice conditions.

THE GLACIAL AGES

As early as 1821, European scientists began to recognize features characteristic of glaciation in places far from any existing glaciers. They drew the then remarkable conclusion that glaciers must once have covered extensive regions. The concept of a glacial age with widespread effects was first proposed in 1837 by Louis Agassiz, a Swiss scientist who achieved considerable fame through his hypothesis. Although at first many people regarded Agassiz's idea as outrageous, gradually, through the work of many geologists, the concept gained widespread acceptance. Today, the study of the glacial ages provides us with dramatic evidence of rapid global climatic changes on the Earth and with clues about how natural physical and biological systems responded to these changes. It also gives us important information about how glaciers behave and helps us understand some of the basic physical processes of the crust and upper mantle.

Ice-Age Glaciers

Over tens of millions of years the climate has slowly grown cooler as the Earth moved into a late Cenozoic glacial era. During the last few million years, the planet has experienced numerous glacial-interglacial

FIGURE 11.27 Flat-topped outwash terraces related to several phases of glaciation border meandering Cave Stream on South Island, New Zealand.

cycles superimposed on the long-term cooling trend. At present, we find ourselves near a time of maximum warmth in such a cycle and poised to begin the slow, but inexorable decline into the next glacial age which will culminate many thousands of years in the future.

About 30,000 years ago, late in the Pleistocene Epoch, an extensive ice sheet that had formed over eastern Canada began to spread south toward the United States and west toward the Rocky Mountains. Simultaneously, another great ice sheet that originated in the highlands of Scandinavia spread southward across northwestern Europe and overwhelmed the landscape (Fig. 11.29). Other large ice sheets grew over arctic regions of North America and Eurasia, including some areas now submerged by shallow polar seas, and over the mountain ranges of western Canada. The ice sheets in Greenland and Antarctica grew larger and advanced across areas of the surrounding continental shelves that were exposed by falling sea level. Glaciers also developed in the world's major mountain ranges, including the Alps, Andes, Himalaya, and Rockies, as well as in numerous smaller ranges and on isolated peaks scattered widely through all latitudes.

On a global scale, the areas of former glaciation add up to an impressive total of more than 44 million km², which is about 29 percent of the Earth's present

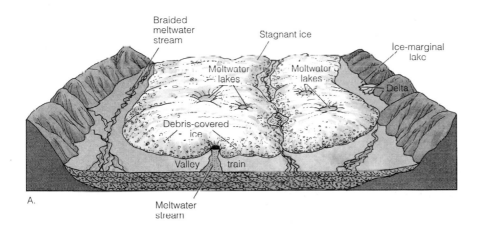

A.

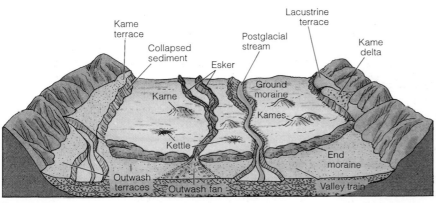

B.

FIGURE 11.28 Origin of ice-contact stratified drift and related landforms associated with stagnant ice. A. Ablating stagnant ice furnishes temporary retaining walls for bodies of stratified sediment deposited by meltwater streams and in meltwater lakes. B. As ice melts, bodies of sediment slump, creating kettle-and-kame topography.

land area. Today, by comparison, only about 10 percent of the world's land area is covered with glacier ice; of this area, 84 percent lies in the Antarctic region.

Drainage Diversions and Glacial Lakes

The growth of ice sheets over the continents caused disruption of major stream systems. In North America, repeated encroachment of Pleistocene ice sheets displaced the Missouri and Ohio rivers into new courses beyond the ice margin. When glaciers blocked preglacial drainage paths, water was ponded to form ice-dammed lakes. Vast ice-marginal lakes that developed beyond the edges of the expanding ice sheet in eastern North America shifted in location and changed size as the glacier receded. Major disruption of drainage also produced large ice-marginal lakes in north-

ern Asia as ice expanding southward in western Siberia blocked the courses of major north-flowing rivers.

Lowering of Sea Level

Whenever large glaciers formed on the land, the moisture needed to produce and sustain them was derived primarily from the oceans. As a result, sea level was lowered in proportion to the volume of ice on land. During the most recent glacial age, world sea level fell at least 100 m, thereby causing large expanses of the shallow continental shelves to emerge as dry land. At that time, the Atlantic coast of the United States south of New York lay as much as 150 km east of its present position. At the same time, lowering of sea level joined Britain to France where the English Channel now lies, and North America and Asia formed a continuous landmass across what is now the Bering Strait (Fig. 11.29). These and other land connections

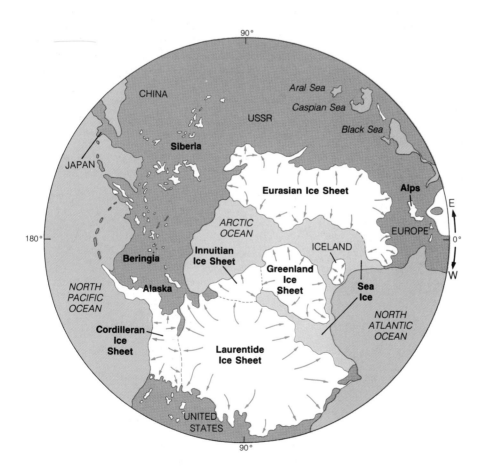

FIGURE 11.29 Areas of the Northern Hemisphere that were covered by glaciers during the last glacial age. Arrows show the general direction of ice flow. Coastlines are shown as they were at that time, when world sea level was at least 100 m lower than present. Sea ice, shown covering the Arctic Ocean, extended south into the North Atlantic. Some scientists postulate that thick ice shelves, rather than sea ice, covered these portions of the ocean. The extent of former glacier ice over shallow continental shelves of northern Eurasia, as well as in parts of northern North America, is controversial.

allowed plants and animals, as well as humans, to pass freely between land areas that now are separated by ocean waters.

Deformation of the Crust

The weight of the massive ice sheets caused the crust of the Earth to subside beneath them, a process described further in chapter 15. The difference in density between crustal rocks (about 2.7 g/cm³) and glacier ice (about 0.9 g/cm³) means that an ice sheet 3 km thick could cause the crust to subside by as much as 1 km. The Hudson Bay region of Canada, which 20,000 years ago lay near the center of the vast Laurentide Ice Sheet (Fig. 11.29), is still rising as the crust adjusts to the removal of this ice load. Using glacial-geologic evidence, we can measure accurately the rate at which the crustal rocks have risen over many thousands of years. Such measurements provide us with important information about how the lithosphere and asthenosphere behave when subjected to changing loads.

Earlier Glaciations

Until recently, it was thought that the Earth had experienced four glacial ages during the Pleistocene Epoch. This assumption was based on studies of ice-sheet and mountain-glacier deposits and had its roots in early studies of the Alps, where geologists mapped moraines and outwash terraces and interpreted them as evidence of four Pleistocene glaciations. This traditional view had to be modified, however, when studies of deep-sea sediments disclosed a long succession of glaciations, the most recent of which was shown by radiocarbon dating to equate with the youngest extensive glacial drift on land. Paleomagnetic dating of these marine sediments showed that during the last 800,000 years the length of the glacial-interglacial cycles averaged about 100,000 years. For the Pleistocene Epoch as a whole, more than 20 glacial ages are recorded, rather than the traditional four. Geologists now realize that the glacial record on land is incomplete and marked by numerous unconformities, whereas many areas of the deep sea contain a record of continuous sedimentation.

Seafloor Evidence

Deep-sea sediments provide some of the best evidence we have of glacial/interglacial cycles. The biologic component of seafloor sediments shows repeated shifts in the composition of surface-water animal and plant populations—from warm interglacial forms to cold glacial forms and back to warm interglacial forms—as we sample down a core. The ratio of the isotopes ^{18}O to ^{16}O in layers of calcareous ooze in these cores also fluctuates with a similar pattern. These $^{18}O/^{16}O$ variations in Pleistocene marine sediments are thought primarily to represent changes in global ice volume. When water is evaporated from the oceans and precipitated on land to form glaciers, water containing the light isotope ^{16}O is more easily evaporated than water containing the heavier ^{18}O. As a result, Pleistocene glaciers contained more of the light isotope, while the oceans became enriched in the heavy isotope. Isotope curves derived from the seafloor sediments therefore provide a continuous reading of changing ice volume on the planet (Fig. 11.30). Because glaciers wax and wane in response to climatic changes, the isotope curves also give a generalized view of global climatic change.

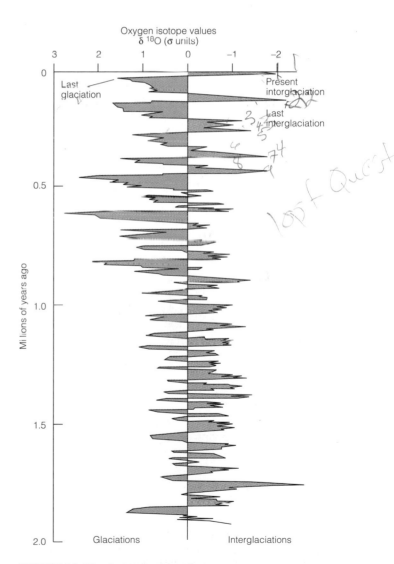

FIGURE 11.30   Average curve of oxygen-isotope variations in deep-sea cores representing global changes in ice volume during the last 2 million years. During the last million years, each glacial-interglacial cycle was about 100,000 years long; earlier cycles were about 40,000 years long.

Pre-Pleistocene Glaciations

Ancient glaciations, identified mainly by tillites and striated rock surfaces, are known from older parts of the geologic column (Fig. 4.8). The earliest recorded glaciation dates to about 2.3 billion years ago, in the early Proterozoic. Evidence of other glacial episodes has been found in rocks of late Proterozoic, early Paleozoic, and late Paleozoic ages. During the late Paleozoic, 50 or more glaciations are believed to have occurred.

Little Ice Ages

Old documents, lithographs, and paintings of Alpine valleys dating to the sixteenth and seventeenth centuries describe and depict glaciers that advanced over small villages and became larger than at any time in human memory (Fig. 11.31). Similar ice advances took place in other parts of the world and in many cases led to the greatest expansion of glaciers since the end of the last glacial age 10,000 years ago. During this recent interval of generally cool climate, which started in the mid-thirteenth century and lasted until the mid-nineteenth century, mountain glaciers expanded worldwide. This *Little Ice Age*, as it is commonly known, was similar to other brief episodes of glacier expansion that were superimposed on the much longer glacial-interglacial cycles.

WHAT CAUSES GLACIAL AGES?

Ever since geologists first became convinced that the Earth had experienced a succession of ice ages, the problem of their cause has been a subject of continuing debate and investigation. We are now much closer to resolving this problem than ever before, but many answers still elude us, for climatic change is a complex phenomenon that involves not only the atmosphere, but also the solid earth, the oceans, and the biosphere, as well as extraterrestrial factors. The ultimate solution to the problem of the ice ages, therefore, will involve the cooperative efforts of specialists from many different fields of science.

Glacial Eras and Shifting Continents

Successions of glacial ages, each lasting tens of millions of years, can be identified in the geologic record. What seems to be the only reasonable explanation for their pattern is suggested by slow but important geographic changes that affect the crust of the planet. These changes include (1) the movement of conti-

A

B

FIGURE 11.31   A. A lithograph made in the 1850s shows Rhone Glacier in the Swiss Alps close to its maximum extent during the Little Ice Age. The glacier terminates in a broad lobe that crosses the floor of the upper Rhone valley. B. A photograph of the same area taken 110 years later shows the terminus of Rhone Glacier perched high up in the headward part of the valley. Extensive thinning and recession of this and other glaciers worldwide marked the end of the Little Ice Age.

nents as they are carried along with shifting plates of lithosphere; (2) the large-scale uplift of continental crust where continents collide; (3) the creation of mountain chains where one plate overrides another; and (4) the opening or closing of ocean basins and seaways between moving landmasses.

The effect of such earth movements on climate is illustrated by the fact that low temperatures are found, and glaciers tend to form and persist, in two kinds of situations: (1) on landmasses at high latitudes and (2) at high altitudes. Furthermore, glaciers are particularly common in places where winds can supply abundant moisture evaporated from a nearby ocean. Today, about 85 percent of the Earth's glacier ice lies in Antarctica where temperatures are constantly below freezing and the land is surrounded by ocean. The only glaciers found at or close to the equator lie at extremely high altitudes.

Abundant evidence leads us to conclude that the positions, shapes, and altitudes of landmasses have changed with time (chapter 16), in the process altering the paths of ocean currents and atmospheric circulation. Where evidence of ancient ice-sheet glaciation is now found in low latitudes, we infer that such lands were formerly located in higher latitudes where large glaciers could be sustained. The most dramatic case concerns glacial deposits of late Paleozoic age that are exposed in South America, southern Africa, India, Australia, and Antarctica. Tillites and related rocks have been interpreted as deposits of continental glaciers that repeatedly covered large portions of a vast southern continent (called Gondwanaland) which was located near the South Pole. As a result of the breakup and subsequent northward movement of large fragments of Gondwanaland, many of the ancient glacial rocks are now found in low latitudes (Fig. 16.1).

The absence of widespread glacial deposits in rocks of Mesozoic age implies that during this era most of the world's landmasses had moved away from polar latitudes and that climates were mild. By the early Cenozoic, slowly shifting landmasses had once again moved into polar latitudes and tectonic movements were beginning to uplift large areas of the western United States and central Asia to high altitudes. By middle Cenozoic time, the Earth was again poised to enter another lengthy glacial era.

Ice Ages and the Astronomical Theory

As initially discovered through studies of glacial deposits and later verified by studies of deep-sea cores, glacial and interglacial ages have alternated for at least 2 million years (Fig. 11.30). Determining their cause has long been a fundamental challenge to the development of a comprehensive theory of climate. A preliminary answer was provided by Scottish geologist John Croll, in the mid-nineteenth century, and later elaborated by Milutin Milankovitch, a Serbian astronomer of the early twentieth century.

Croll and Milankovitch recognized that minor variations in the Earth's orbit around the Sun and in the tilt of the Earth's axis cause slight but important variations in the amount of radiant energy reaching any given latitude. Three movements are involved (Fig. 11.32).

First, the *eccentricity* of the orbit, which is a measure of its circularity, changes over periods of 100,000 and 400,000 years (Fig. 11.32A). About 50,000 years ago, the orbit was more circular (lower eccentricity) than it has been for the last 10,000 years.

Second, the *tilt* of the axis, which now averages 23.5°, shifts about 1.5° to either side during a span of about 41,000 years (Fig. 11.32B).

Third, the axis of rotation, which now points in the direction of the North Star, wobbles like a spinning top (Fig. 11.32C). The wobbling movement causes the North Pole to trace a cone in space, completing one full revolution every 26,000 years. Viewed from a point above the North Pole, this movement would appear to be counterclockwise. At the same time, the axis of the Earth's elliptical orbit is also rotating, but much more slowly, in the opposite direction. These two motions together cause a progressive shift in the position of the four cardinal points of the Earth's orbit (spring and autumn equinoxes and winter and summer solstices). As the equinoxes move slowly around the orbital path, a motion called *precession of the equinoxes,* they complete one full cycle in about 23,000 years.

The slow but predictable changes in precession, tilt, and eccentricity cause long-term variations of as much as 10 percent in the amount of radiant energy that reaches any particular place on the Earth's surface in a given season (Fig. 11.33). By reconstructing and dating the history of climatic variations during the Quaternary Period, geologists have shown that fluctuations of climate on glacial-interglacial time scales match those of tilt and precession. This persuasive evidence supports the theory that astronomical changes which determine the amount of radiation reaching the Earth's surface control the *timing* of the glacial-interglacial cycles.

Atmospheric Composition

Although orbital factors can explain the timing of the glacial-interglacial cycles, the variations in solar radiation reaching the Earth's surface are too small to account for average global temperature changes of 4 to 10°C that are implied by geologic and biologic evidence. We therefore must conclude that other factors are also involved. Somehow, the slight temperature decrease caused by orbital changes must be translated into a temperature change sufficiently large to generate and maintain the huge Pleistocene ice sheets. We do not yet know how this is accomplished, but some of the factors involved are likely to be changes in the chemical composition and dustiness of the atmosphere, and changes in the reflectivity of the Earth's surface.

Air bubbles in glacier ice of the present Antarctic and Greenland ice sheets are samples of the Earth's

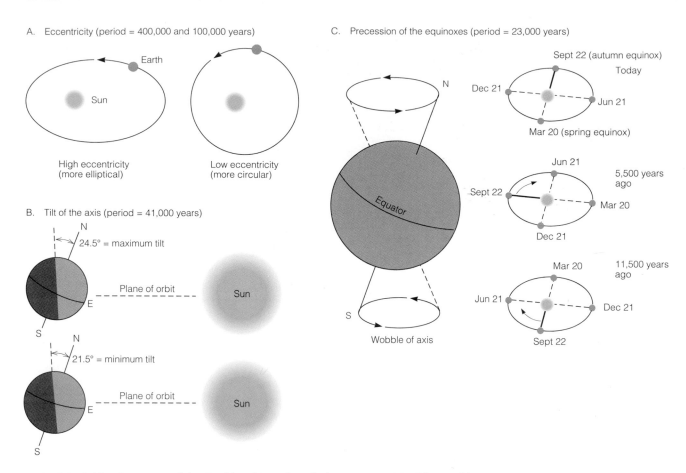

FIGURE 11.32   Geometry of the Earth's orbit and axial tilt. A. *Eccentricity*. The Earth's orbit is an ellipse with the Sun at one focus. The shape of the orbit changes from almost circular (low eccentricity) to more elliptical (high eccentricity) over periods of 100,000 and 400,000 years. The higher the eccentricity, the greater is the seasonal variation in radiation received at any point on the Earth's surface. B. *Tilt*. The tilt of the Earth's axis, which now is about 23.5°, ranges from 21.5 to 24.5°, with each cycle lasting about 41,000 years. Increasing tilt means a greater difference, for each hemisphere, between the amount of solar radiation received in summer and that received in winter. C. *Precession*. The Earth wobbles on its axis like a spinning top, making one revolution every 26,000 years. The axis of the Earth's elliptical orbit also rotates, though more slowly, in the opposite direction. These motions together cause a progressive shift, or precession, of the spring and autumn equinoxes, with each cycle lasting about 23,000 years.

ancient atmosphere. Studies of the chemical composition of trapped air that dates back to the most recent ice age indicate that during glacial times the atmosphere contained less carbon dioxide and methane than it does today (Fig. 11.34). These two gases are important "greenhouse" gases (chapter 19). If their concentration in the atmosphere is high, they trap radiant energy emitted from the Earth's surface that would otherwise escape to outer space. As a result, the lower atmosphere heats up and the Earth's climate becomes warmer. If the concentration of these gases is low, as it was during glacial times, surface air temperatures are reduced. Calculations suggest that the low levels of these two important atmospheric gases during glacial times can account for nearly half of the total ice-age temperature lowering. Therefore, the greenhouse gases likely play a significant role in explaining the *magnitude* of past global temperature changes. Although we know that the percentages of these gases fell during glacial times, we do not yet know for certain what caused them to drop.

Ice core studies have also shown that the amount of dust in the atmosphere was unusually high during glacial times. The fine dust was picked up by strong winds blowing across outwash deposits and dry desert basins. So much dust was delivered to the atmosphere

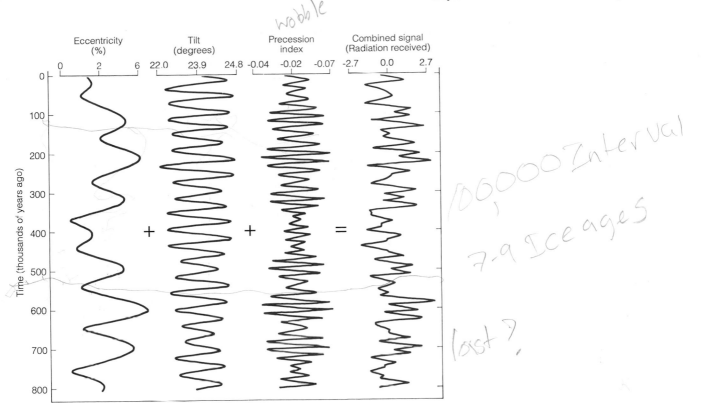

FIGURE 11.33 Curves showing variations in orbital eccentricity, tilt, and precession during the last 800,000 years. Summing these factors produces a combined signal that shows the amount of radiation received on the Earth at a particular latitude through time.

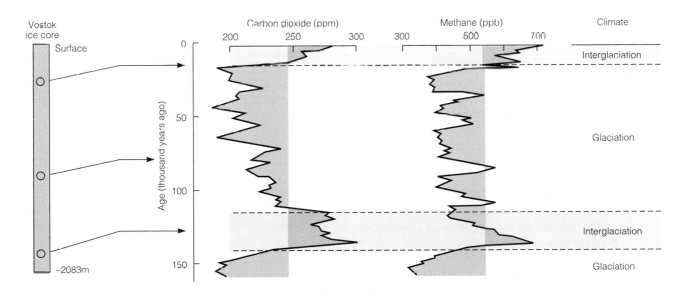

FIGURE 11.34 Curves showing changes in the amount of carbon dioxide and methane in samples of the atmosphere obtained from bubbles in a deep ice core from Vostok Station, Antarctica. Concentrations of these greenhouse gases were high during the last interglaciation, just as during the present interglaciation, but they were lower during glacial times. The curves are consistent with the concept that these gases contributed to warm interglacial climates and cold glacial climates.

that the sky must have appeared hazy much of the time. The fine atmospheric dust scattered incoming radiation back into space, which would have further cooled the earth's surface.

Whenever the world enters a glacial age, large areas of land are progressively covered by snow and glacier ice. The highly reflective surfaces of snow and ice scatter incoming radiation back into space, further cooling the lower atmosphere. Together with lower greenhouse gas concentrations and increased atmospheric dust, this additional cooling would favor the expansion of glaciers.

Solar Variations, Volcanic Activity, and Little Ice Ages

Climatic fluctuations measured in centuries or decades were responsible for the Little Ice Age and similar episodes of glacier expansion. However, such fluctuations are too brief to be caused either by movements of continents or variations in the Earth's orbit, and require us to seek other explanations for their cause. Two have received special attention.

One hypothesis regarding the cause of glacial events like the Little Ice Age is based on the concept that the energy output of the Sun fluctuates over time.

The idea is appealing because it might explain climatic variations on several different time scales. However, although correlations have been proposed between weather patterns and rhythmic fluctuations in the number of sunspots appearing on the surface of the Sun, as yet there has been no clear demonstration that solar variations are responsible for climatic changes on the scale of the Little Ice Age.

Large explosive volcanic eruptions can eject huge quantities of fine ash into the atmosphere to create a veil of fine dust that circles the globe. Like other types of dust, the fine ash particles tend to scatter incoming solar radiation, resulting in a slight cooling at the Earth's surface. Although the dust settles out rather quickly, generally within a few months to a year, tiny droplets of sulfuric acid, produced by the interaction of volcanically emitted SO_2 gas and water vapor, also scatter the Sun's rays, and such droplets remain in the upper atmosphere for several years. During a large eruption, volcanic dust and gases in the atmosphere can lower average surface air temperature by 0.5 to 1.0C°, sufficient to influence the mass balance of glaciers. A close association between intervals of glacier advance and periods of sustained volcanic activity during the last several centuries lends support to the hypothesis that volcanic emissions can produce detectable changes of climate on the scale of decades.

⎾── GLACIAL WATER FOR ARID LANDS ──

As world population increases and our insatiable demand for fresh water continues to rise, a critical problem we face is how to meet this demand. Some imaginative people think the answer may lie in a vast unutilized resource in the frigid polar seas: Antarctic icebergs.

Captain James Cook was among the first to recognize the potential for using fresh water locked up in polar icebergs. In 1773, while his ship lay off the coast of Antarctica, his sailors hoisted 15 tons of berg ice on board, and Cook noted in his log that "this is the most expeditious way of watering I ever met with."

Today, scientists visualize towing large icebergs far to the north where they could supply water to the arid coasts of South America, Africa, Australia, and the southwestern United States. Although icebergs are shed from arctic glaciers as well, these icebergs are far fewer and on average much smaller than those of the Antarctic region. A French engineering firm, commissioned by Saudi Arabia, has already studied the feasibility

of towing Antarctic icebergs weighing as much as 85 million tons northward more than 8000 km to the Red Sea where they could supply drinking and agricultural water.

As impressive as such proposals sound, the technological problems are formidable. How can blocks of ice hundreds of meters thick and several square kilometers in area be towed or pushed thousands of kilometers through warm and often stormy seas and still have enough mass remaining to make the venture economically feasible? Large icebergs are plentiful around Antarctica, where an estimated 1000 km³ of glacier ice breaks off as icebergs each year (Fig. B11.1). (One exceptionally large berg about the size of Rhode Island floated 2000 km in three years.) Those measuring 1 to 2 km long may be optimal for towing. A huge berg towed 3000 km northward through relatively cool waters to Australia at an average speed of 0.5 m/s (1 knot) would require 70 days to make the transit, but about 60 m of ice would melt around

FIGURE B11.1 A small tabular iceberg near Antarctica has calved off the margin of a nearby ice shelf. Scientists visualize towing bergs many times this size to supply fresh water to arid lands in Australia, the American Southwest, and the Middle East.

its sides. An unprotected iceberg of this size is unlikely to survive the long trip (up to a year) through much warmer waters to southern California or the Middle East. To avoid excessive loss, the ice would have to be insulated against melting and evaporation and protected against collapse and disintegration. In addition, the towing vessels or propulsion systems large enough to move huge icebergs over great distances have yet to be developed. Although these technological challenges appear immense, the day may come when ice that has been stored in Antarctic glaciers for tens of thousands of years will bring life-sustaining waters to the Earth's low-latitude arid lands.

SUMMARY

1. Glaciers are permanent bodies of moving ice that consist largely of recrystallized snow.

2. Based on geometry, glaciers are classified as cirque glaciers, valley glaciers, fjord glaciers, piedmont glaciers, ice caps, ice sheets, and ice shelves.

3. Ice in a temperate glacier is at the pressure melting point, and liquid water exists at the base of the glacier; in a polar glacier ice is below the pressure melting point and is frozen to the rock on which it rests.

4. Glaciers can form only at or above the snowline, which is close to sea level in polar regions and rises to high altitudes in the tropics.

5. The mass balance of a glacier is measured in terms of accumulation and ablation. The equilibrium line separates the accumulation area from the ablation area and marks the level on the glacier where net gain is balanced by net loss.

6. Temperate glaciers move as a result of internal flow and basal sliding. In polar glaciers, which are frozen to their bed, motion is much slower and involves only internal flow. Surges involve extremely rapid flow, probably related to excessive amounts of water at the base of a glacier.

7. Glaciers erode rock by plucking and abrasion. Rock debris, transported chiefly at the base and sides of a glacier, includes fragments of all sizes, from fine rock flour to large boulders.

8. Mountain glaciers erode stream valleys into U-shaped glacial valleys with cirques at their heads. Fjords are excavated far below sea level by glaciers in high-latitude coastal regions.

9. Glacial drift is sediment deposited by glaciers and glacial meltwater. Till is deposited directly by glaciers, while glacialmarine drift is deposited on the seafloor from floating glacier ice. Stratified drift includes outwash deposited by meltwater streams and ice-contact stratified drift deposited on or against stagnant ice.

10. Ground moraine is built up beneath a glacier, whereas end moraines (both terminal and lateral) form at a glacier margin.

11. During glacial ages, huge ice sheets repeatedly covered northern North America and Eurasia, causing the crust beneath the ice to subside and world sea level to fall.

12. Glacial ages have alternated with interglacial ages in which temperatures approximated those of today. Studies of marine cores indicate that more than 20 glacial-interglacial cycles occurred during the Pleistocene Epoch.

13. Glacial eras in Earth history probably are related to the favorable positioning of continents and ocean basins, brought about by movements of lithospheric plates. The timing of glacial-interglacial cycles appears to be closely controlled by changes in the Earth's precession, tilt, and orbital eccentricity, three factors that affect the distribution of solar radiation received at the Earth's surface. Changes in the atmospheric concentration of carbon dioxide, methane, and dust may help explain the magnitude of global temperature lowering during glacial ages.

14. Climatic variations on the scale of centuries and decades have been ascribed to fluctuations in energy output from the Sun or to injections of volcanic dust and gases into the atmosphere.

IMPORTANT TERMS TO REMEMBER

ablation (p. 304)
accumulation (p. 304)

calving (p. 306)
cirque (p. 298)
crevasse (p. 307)

drift (p. 312)
drumlin (p. 311)

equilibrium line (p. 304)
erratic (p. 313)

fjord (p. 298)

glacial drift (p. 312)
glaciation (p. 309)
glacier (p. 298)

ice-contact stratified drift (p. 314)

mass balance (p. 304)
moraine (p. 313)

outwash (p. 314)

polar glacier (p. 301)
pressure melting point (p. 301)

snowline (p. 303)
surge (p. 308)

temperate glacier (p. 301)
till (p. 312)

QUESTIONS FOR REVIEW

1. What distinguishes temperate glaciers from polar glaciers?

2. What is the snowline and how are glaciers related to it?

3. Describe the steps in the conversion of snow to glacier ice.

4. Why does the position of the equilibrium line provide a rough estimate of a glacier's mass balance?

5. Why is there a lag in time between a change of climate and the response of a glacier's terminus to the change?

6. In what ways does ice temperature influence the way a glacier moves?

7. Describe the unique motions of surging and calving glaciers.

8. Illustrate how small-scale and large-scale erosional features can be used to infer directions of flow of former glaciers?

9. How might you distinguish till from stratified drift in a roadside outcrop?

10. In what different ways are moraines built at a glacier margin?

11. What can you say about the condition of a glacier in association with which kettles and kames are developing?

12. Why, and by approximately how much, does world sea level fall and rise during glacial-interglacial cycles?

13. What evidence obtained from deep-sea cores indicates that glacial-interglacial cycles have occurred repeatedly during the Pleistocene Epoch?

14. What natural factors explain the recurrence of glacial events on time scales of tens of thousands of years?

15. How might large volcanic eruptions influence climate and cause glaciers to grow or shrink?

WIND AS A GEOLOGIC AGENT

If we lived on Mars instead of the Earth, a substantial percentage of this book would likely be devoted to the theme of wind action and deserts, for Mars is an arid, windy, and dusty planet. When the Mariner 9 spacecraft approached Mars in 1971, for instance, a dust storm of major proportions enveloped much of the planet and continued unabated for several months. Photographs taken both during this Mariner mission and during subsequent Viking missions revealed a planetary surface that has been extensively modified by wind action.

Wind is also an important agent of erosion and sediment transport on the Earth, but its effects are mainly visible in desert regions where few people live. Most of the world's population is concentrated in the relatively moist parts of the temperate and tropical latitudes (0–66° N and S), where a protective cover of vegetation makes wind an ineffective geologic agent. Nevertheless, even in these populated regions the occurrence of ancient sand dunes and widespread deposits of dust show us that wind has been important in shaping the landscape at times when the continents were drier and windier places than they are today.

Planetary Wind System

To help us understand why winds are effective geologic agents in some regions but not in others, we need to see how surface winds are related to the global circulation of the atmosphere.

Circulation of the Atmosphere

The atmosphere consists of a mixture of gases that together we call air. The atmosphere is always moving, a fact we are well aware of whenever we feel a gentle breeze or a strong wind blowing. The basic reason the atmosphere is always in motion is that more of the Sun's heat is received per unit of land surface near the equator than near the poles. This unequal heating gives rise to convection currents. The heated air near the equator expands, becomes lighter, and rises. High up, it spreads outward in the direction of both poles. As the upper air travels both northward and southward, it gradually cools, becomes heavier, and sinks. On reaching the Earth' surface, this cool, descending air then flows back toward the equator, warms up, and rises, thereby completing a convective cycle.

The Coriolis Effect

The Earth's rotation modifies what would otherwise be a simple convective circulation pattern in the atmosphere. The **Coriolis effect,** named after the nineteenth-century French mathematician who first analyzed it, causes any body that moves freely with respect to the rotating Earth to veer to the right in the Northern Hemisphere and to the left in the Southern Hemisphere. This is true regardless of the direction in which the body may be moving. Both flowing water (such as an ocean current) and flowing air (wind) respond to the Coriolis effect. So too do small bodies such as projectiles, if they travel over a sufficiently long path.

The Coriolis effect breaks up the simple flow of air between the equator and the poles into belts (Fig. 12.1). The result, in both the Northern and Southern hemispheres, is a large cell of circulating air lying between the equator and about 30° latitude. In these low-latitude cells, the prevailing winds are northeasterly in the Northern Hemisphere (that is, they flow *from* the northeast toward the southwest), while in the Southern Hemisphere they are southeasterly. These wind systems are called the *trade winds,* for their direction and consistent flow carried trade ships across the tropical oceans at a time when winds were the chief source of power.

In each hemisphere, a second cell of circulating air lies poleward of the low-latitude cell. In these second, middle-latitude cells, westerly winds prevail (blowing *from* the west). In these cells, cold equatorward-flowing upper air descends near 20–30° latitude in both hemispheres, and northward-moving surface air rises in higher latitudes where it meets dense, cold air flowing from the polar regions.

A third cell of circulating air lies over the polar regions. In each cell, cold, dry, upper air descends near the pole and moves equatorward in a wind system called the polar easterlies. As this air slowly warms and encounters the belt of westerlies, it rises along the *polar front* and returns toward the pole.

The cold upper air converging where the low- and mid-latitude cells meet cannot hold as much moisture as warm air, so where this cold air descends, dry conditions are created at the land surface. As a result, much of the arid land in both the Northern and Southern hemispheres is centered between latitudes 20 and 30°. By contrast, abundant moisture in the warm air surrounding the equator condenses as the air rises and becomes cooler and denser, creating clouds that release their moisture as tropical rains.

Climate

The global pattern of air flow, which is determined by the nonuniform heating of the Earth's surface, the Coriolis effect, the distribution of land and sea, and the topography of the land, ultimately controls the variety and pattern of the Earth's climates. **Climate** is the average weather of a place, together with the de-

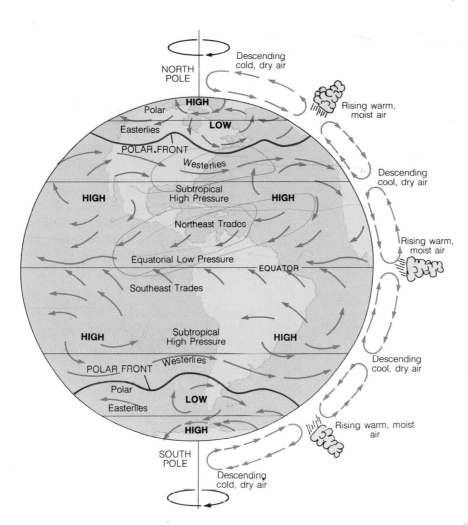

FIGURE 12.1 The Earth's planetary wind system, shown schematically. Moist air, heated in the warm equatorial zone, rises convectively and forms clouds that produce abundant rain. Cool, dry air descending at latitudes 20–30° produces a belt of subtropical high pressure in which lie many of the world's great deserts.

gree of variability of that weather, over a period of years. It is measured by such factors as temperature, precipitation, cloudiness, and windiness. If the Earth had no mountains and no oceans to affect the moving atmosphere, the major climatic zones would all lie parallel to the equator. However, the pattern of climatic zones is distorted by the distribution of oceans, continents, high mountains, and plateaus. As a result, average temperature, precipitation, cloudiness, and windiness vary greatly from one place to another and give rise to an array of distinct climatic regions.

Movement of Sediment by Wind

We have all seen pictures of the tremendous destruction done by hurricanes and typhoons when winds achieve speeds of at least 120 km/h and sometimes exceed 500 km/h. The force of such a wind is so strong that trees are uprooted, houses are ripped

apart, and large objects are thrown substantial distances. Fortunately, hurricane winds are the exception, but they provide a glimpse of the potential power of wind as a geologic agent.

Because the density of air at sea level (1.22 kg/m³) is far less than that of water (1000 kg/m³), air cannot move as large a particle as water can flowing at the same velocity. In extraordinary wind storms, when wind speeds locally reach or exceed 300 km/h, coarse rock particles up to several centimeters in diameter can be lifted to heights of a meter or more. Pebbles swept aloft by exceptional winds have been found lodged in buildings, trees, and cracks in telephone poles. In most regions, however, wind speed rarely exceeds 50 km/h, a velocity described as a strong wind. In a strong wind, the largest particles of sediment that can be suspended in the air stream are grains of sand. Larger particles settle out too quickly to remain aloft. At lower wind speeds, sand moves along close to the ground surface, and only finer grains of dust move in suspension.

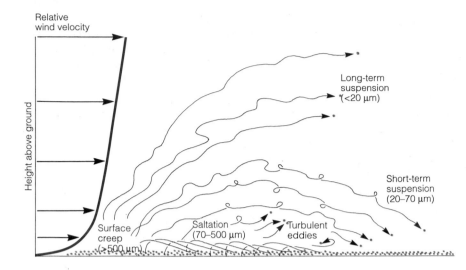

FIGURE 12.2 Under conditions of moderate wind, sand grains larger than 500 μm (0.5 mm) move by surface creep, while smaller grains (70–500 μm) saltate across the ground surface. Still finer particles (20–70 μm) are carried aloft in turbulent eddies and encounter faster moving air that transports them downwind as they slowly settle to the ground. The finest dust (less than 20 μm) reaches greater heights and is swept along in suspension as long as the wind is blowing.

Wind-Blown Sand

If a wind blows across a bed of sand, the grains begin to move when the wind speed reaches about 4.5 m/s (16 km/h). The resulting forward rolling motion of the sand is called *surface creep* (Fig. 12.2). With increasing wind speed, turbulence near the ground lifts moving sand grains into the air, where they travel along arcuate paths, landing a short distance downwind. This is the same process (*saltation*) we see in a stream, where grains of sand move close to the bottom, also following arcuate paths (see chapter 9).

Saltation. Saltation accounts for at least three-quarters of the sand transport in areas covered by sand dunes. Measurements of the rate of sand movement in deserts of the Middle East indicate a rapid increase in sand movement with increasing wind speed. For example, a strong wind blowing at 58 km/h will move as much sediment in one day as it would take a wind blowing at 29 km/h to move in three weeks.

If a wind is strong enough, it can start a grain rolling along the ground where it may impact another grain and knock it into the air. As this second grain falls to the ground it will impact other grains, some of which are thrown upward into the airstream. Within a very short time the air close to the ground may contain a very large number of saltating sand grains, all moving along with the wind and moving in arclike paths somewhat similar to those of a Ping-Pong ball bouncing across a table (Fig. 12.3). However, even in strong winds saltating sand grains seldom rise far off the ground, as shown by abrasion marks on utility poles and fence posts that are sandblasted up to a height of about a meter.

Sand Ripples. Sheets of well-sorted sand that have accumulated on the land surface are inherently unstable, even under gentle winds. As the wind passes across such an accumulation, saltation moves the smaller, most easily transported grains. Sand grains too large to be moved are left behind. As the saltating

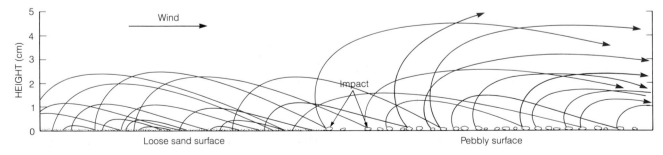

FIGURE 12.3 Strong wind causes movement of sand grains by saltation. Impacted grains bounce into the air and are carried along by the wind as gravity pulls them back to the land surface where they impact other particles, repeating the process.

FIGURE 12.4 Sand ripples cross the surface of a desert sand sheet on the floor of Monument Valley, Arizona.

finer grains impact the surface at some average distance downwind, additional fine particles are set in motion, and another accumulation of coarse grains develops as the fine sand moves onward. By this process, the coarse grains form a series of small, linear ridges of sand called **sand ripples**. Sand ripples tend to be aligned in a regular pattern with their crests oriented perpendicular to the wind direction (Fig. 12.4). Under very strong winds, the ripples disappear because then all grains can be moved and sorting is less likely to occur.

Wind-Blown Dust

Sand grains blown across the land surface travel slowly and are deposited quickly when wind velocity subsides. Fine particles of dust (silt- and clay-size sediment), on the other hand, travel faster, longer, and much farther before settling to the ground.

As you might guess, the dustiest places on Earth tend to coincide with some of the world's major desert regions. Estimates of annual world dust production, which are imprecise at best, range as high as 5 billion tons. Among the many types of terrain that can give rise to large quantities of dust, the following are especially important: dry lake and stream beds, alluvial fans, outwash plains of glacial streams, and regions underlain by deposits of wind-blown dust that have lost their vegetation cover as a result either of climatic change or human disturbance.

Mobilization and Transport of Dust. As a result of frictional drag, the velocity of moving air decreases sharply near the ground surface. Right at the surface lies a layer of quiet air less than 1 mm thick (Fig. 12.5). Sand grains that protrude above this layer of quiet air can be swept aloft by rising turbulent eddies. Dust particles, however, are so small and often so closely packed that they form a very smooth surface that does not protrude above the quiet air. Even a strong wind blowing over such a surface may not disturb the dust. Mobilization of the fine sediment may require the impact of saltating sand grains or other physical disruption of the smooth surface. We can see how such dust is set in motion by looking at a dusty desert road covered by dry, compact silt on a windy day. The wind blowing across the road generates little or no dust, but a vehicle driving over the road creates a choking cloud, which is blown away downwind before settling once more to the ground. The passing wheels have broken up the surface of the powdery dust that was too smooth to be disturbed by the wind.

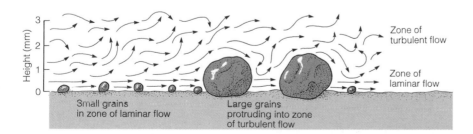

FIGURE 12.5 Particles of fine sand and silt at the ground surface lie within a zone of laminar air flow less than 0.5 mm thick where wind velocity is extremely low. As a result, it is difficult for the wind to dislodge and erode these small grains. Larger grains protrude into a zone of faster moving, turbulent air. The turbulence, which exerts a greater push on the top of the grains than does the laminar flow at their base, makes it easy to start the grains moving.

Once in the air, dust constitutes the wind's suspended load. The grains of dust are continually tossed about by eddies, like particles in a stream of turbulent water, while gravity tends to pull them toward the ground (Fig. 12.2). Meanwhile the wind carries the dust forward. Although in most cases suspended sediment is deposited fairly near its place of origin, strong winds associated with large dust storms are known to carry very fine dust into the upper atmosphere, where it can be transported thousands of kilometers.

Dust Storms.   Dust storms are the chief events leading to large-scale transport of dust. In a dust storm, the visibility at eye level is reduced to 1000 m or less by dust raised from the ground surface. Such storms are most frequent in the vast arid and semiarid regions of central Australia, western China, Soviet Central Asia, the Middle East, and North Africa shown in Figure 12.6. In the United States, blowing dust is especially common in the southern Great Plains and in the desert regions of California and Arizona.

Dust storm frequency commonly is related to cycles of drought, with a marked rise in atmospheric dust concentrations coinciding with severe droughts. The frequency also has risen with increasing agricultural activity, especially in semiarid lands. Another example of how human activities can contribute to an increase in dustiness is seen in records from the western desert of Egypt in the 1930s and 1940s: the number of dust storms rose from three or four per year before the Second World War to more than 40 between 1939 and 1941, when tank action and artillery bombardment were at a peak, and then declined to four per year in 1944 after military activity ceased.

Deposition of Dust.   Dust can be deposited if (1) wind velocity and air turbulence decrease so that particles can no longer remain in suspension, (2) the particles collide with rough or moist surfaces that trap them, or with surfaces having a weak electrical charge that attracts them, (3) the particles accumulate to form aggregates, which then settle out because of their greater mass, or (4) the particles are washed out of the air by rain.

Vegetation acts as a trap for descending dust particles because wind velocity is reduced over vegetated landscapes. Forest is more efficient at trapping dust than is low-lying vegetation because of the greater effect trees have on reducing wind velocity in the critical zone above the ground.

Deposition also occurs where a topographic obstacle causes a divergence of air flow, thereby leading to reduced wind velocity behind the obstruction (Fig. 12.7). This explains why deposits of dust are generally

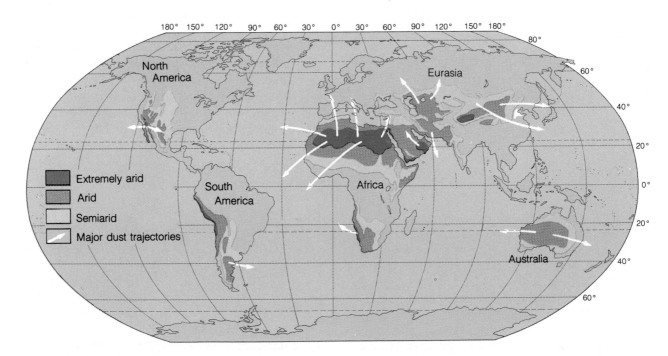

FIGURE 12.6   Major dust storms are most frequent in arid and semiarid regions that are concentrated in the belts of subtropical high pressure north and south of the equatorial zone. Arrows show the most common trajectories of dust transported during major storms.

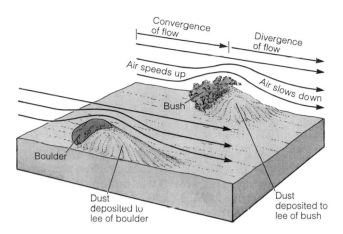

FIGURE 12.7 Obstacles in the path of wind produce a convergence of air flow, causing the air to move faster. As the air passes across the obstacle, the flow paths diverge, leading to lower velocity. The lower velocity reduces the wind's carrying capacity, causing the air to drop some of its dust load, which accumulates on the lee side of the obstruction.

thick on the *lee,* or downwind, side of obstacles (the side away from the wind) and either thin or absent on the *windward,* or upwind, side (the side from which the wind is blowing).

Coarse and medium-grained dust particles, which are carried at relatively low altitudes, are likely to settle out first. Finer particles are carried higher in the atmosphere and therefore may remain suspended for long periods.

Detrimental Effects of Wind-Blown Sediment

Wind action can create many problems for people. Each year the material losses can range into millions of dollars, and the direct and indirect effects of wind can lead to a significant loss in human lives.

Blowing sand and dust can severely damage crops and other vegetation. Large wind storms are reported to have caused major damage to wheat fields, citrus orchards, and vineyards, as well as to root crops. In one severe California storm in 1977, cattle were asphyxiated by dense dust, and hair and skin were sandblasted from their hind quarters. A strong, dense dust storm can quickly turn a car's clear windshield into a sheet of frosted glass, and blowing sand can pit and strip away the shiny paint of a new car.

The engines of vehicles operating in dusty areas are particularly susceptible to damage. Dust can contaminate fuel, clog air filters, abrade cylinders, and cause electrical short circuits. In the Second World War, higher-than-average engine failure among vehicles operating in desert areas was attributed to such

abnormal cylinder wear. Dust similarly created a problem for military vehicles operating in the deserts of Saudi Arabia and Kuwait during the Middle East War of 1991.

Blowing dust can severely reduce visibility on roads and highways. Near Tucson, Arizona, clouds of dust caused so many accidents that a Dust Storm Alert system was introduced in 1976 to warn motorists of hazardous conditions.

Airborne dust also can pose a hazard to aircraft. In 1973 a Royal Jordanian Airlines plane crashed at an airport in northern Nigeria while flying through dense dust, killing 176 persons. An attempt to rescue U.S. hostages in Iran in 1979 was aborted after helicopter engine failure, attributed to dense airborne dust, caused a fatal crash in the desert staging area. Helicopter pilots operating in the Saudi Arabian desert in 1990–91 reported that dense dust raised while flying close to the ground produced a bright electrostatic glow around their rotor blades, thereby greatly reducing the effectiveness of light-sensitive night-vision goggles.

Inhalation of dust can lead to various medical problems. When too many fine mineral particles enter the lungs, tissue damage can cause emphysema. Quartz inhalation can lead to silicosis, a debilitating disease common among unprotected miners working in dusty conditions. In addition, disease-causing organisms may be carried in dust, where they may survive for long periods. Among the deadly germs that can be transported in wind-blown dust are anthrax and tetanus. In central China, a close correlation has been found between deaths due to cancer of the esophagus and the distribution of dust deposits, with the death rate increasing as the average grain size of the dust decreases. Outside China, the disease is mainly present in the dusty, arid regions of Iran, Soviet central Asia, Mongolia, and Siberia, suggesting that the cancer may somehow be related to persistent inhalation of fine dust.

Wind Erosion

Wind is an important agent of erosion wherever winds are strong and persistent and either the land is too dry to support vegetation or the influx of airborne sediment is so rapid that vegetation cannot gain a foothold and thereby stabilize the ground surface. Flowing air erodes in two ways. **Deflation** (from the Latin word meaning to blow away) is the picking up and removal of loose rock fragments, sand, and dust by the wind. This process provides most of the wind's load. The second process, *abrasion,* results when rock is impacted by wind-driven grains of sediment.

FIGURE 12.26 A flash flood has just passed through this steep-walled arroyo on the Navajo reservation in northeastern Arizona. As the floodwater subsides, sediment is deposited across the flat alluvial floor of the canyon.

Flash Floods

The sparse vegetation cover in deserts presents no great impediment to surface runoff, which can readily erode loose, dry regolith. A major rainstorm is likely to be accompanied by a **flash flood,** a sudden, swift flood that can transport large quantities of sediment. The debris from such floods forms fans at the base of mountain slopes and on the floors of wide valleys and basins.

Often streams in flood pass rapidly through desert canyons where they erode preexisting alluvium and undercut valley sideslopes, causing the slopes to cave in. As a flood subsides, its load is deposited rapidly, creating a flat alluvial surface (Fig. 12.26). The stratigraphy of such alluvial fills often discloses a complex history of cutting and filling.

Fans and Bajadas

Alluvial fans develop under a wide range of climatic conditions, but they are especially common in arid and semiarid lands where they typically are composed of both alluvium and debris-flow deposits. They are a characteristic landform of deserts and can be a major source of groundwater for irrigation. In some semiarid regions entire cities have been built on alluvial fans or fan complexes (for example, San Bernardino, California, and Teheran, Iran). As we learned in chapter 10 (frontispiece), alluvial fans in Iran, Afghanistan, and Pakistan are dotted with mounds of debris that mark the sites of deep artificial shafts connecting horizontal tunnel systems. The shafts were designed to collect water within the upper reaches of fans for use in surface irrigation. Some such systems date back a thousand years or more.

In desert basins of the southwestern United States, the Middle East, and central Asia, alluvial fans form a prominent part of the landscape. In these regions, the fans border highlands, with the top of each fan lying at the mouth of a mountain canyon. Where a mountain front is straight and its canyons widely spaced, each fan will encompass an arc of about 180° (Fig. 9.26). If canyons are closely spaced along the base of a mountain range, coalescing adjacent fans form a broad alluvial apron, or **bajada** (Spanish for *slope*), that has an undulating surface due to the convexities of the component fans (Fig. 12.27).

Desert Lakes and Playas

Runoff in arid regions is rarely abundant enough to sustain permanent lakes. Instead, the floor of a desert basin may contain a dry lake bed, called a **playa** (Spanish for *beach*) (Fig. 12.27). Following a major rainstorm, runoff may be sufficient to form a temporary playa lake that will last for up to several weeks. White or grayish salts at the dry surface of a playa, left by the repeated formation and evaporation of temporary lakes, can accumulate to thicknesses of tens of meters and constitute an important source of industrial chemicals (see chapter 4).

Pediments

One of the most characteristic landforms of dry regions is the **pediment,** a broad, relatively flat surface, eroded across bedrock and thinly or discontinuously veneered with alluvium, that slopes away from the

FIGURE 12.27 A vast salt-encrusted playa occupies the floor of Death Valley in California. The playa is bordered by a bajada, composed of coalescing alluvial fans constructed beyond the mouths of adjacent mountain valleys.

base of a highland (Fig. 12.28). Although from a distance a pediment may resemble a bajada, it is a bedrock surface rather than a thick alluvial fill. Rock debris scattered over a pediment is carried by running water from adjacent mountains and is also derived by weathering of the pediment surface. Downslope, the scattered rock debris gradually forms a continuous cover of alluvium that merges with the thick alluvial fill of an adjacent valley.

The long profile of a pediment, like that of an alluvial fan, is concave upward, becoming progressively steeper toward a mountain front. Such a profile is typically associated with the work of running water (see chapter 9). Faint, shallow channelways on pediment surfaces show that water is involved in their formation, and it is generally agreed that pediments are slopes across which sediment is transported by mass-wasting and running water. Eyewitness accounts of sheetflood activity and laterally eroding floodwaters associated with intense desert storms have led geologists to suggest that both these processes may be involved in pediment formation. However, the exact manner in which pediments form is still not firmly established.

The surface of a pediment rises toward a mountain that it meets at an abrupt angle. This suggests that desert mountain slopes do not become gentler with time, as they tend to do in moist regions, where chemical weathering and creep are dominant. Instead, the slopes apparently achieve an angle determined by the resistance of the bedrock and maintain that angle as they gradually retreat under the attack of weathering and mass-wasting (Fig. 12.29). In this way, as a mountain slope retreats, a pediment will increase in size by expanding at its upslope edge. The growth of the pediment, at the expense of the mountain, may continue until the entire mountain has been consumed.

Inselbergs

Among the most impressive of the Earth's landforms are steep-sided mountains, ridges, or isolated hills that rise abruptly from adjoining plains like rocky islands standing above the surface of a broad, flat sea. Ayers Rock in central Australia is a famous example (Fig. 12.30). Called **inselbergs** (German for *island mountain*), these landforms appear in many environmental settings, ranging from coastal to interior and arid to

FIGURE 12.28 A pediment in the Mojave Desert of southeastern California has left only a few residual hills near the crest of a former mountain ridge. The flat bedrock surface cut across crystalline rocks passes downslope beneath a thin cover of alluvium.

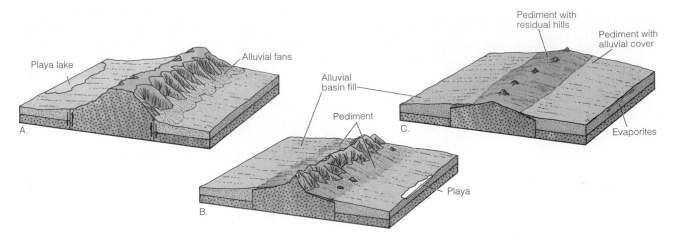

FIGURE 12.29 Stages in the formation of a pediment. A. A mountain block, uplifted along bordering faults, is eroded by streams that contribute sediment to a growing alluvial basin fill. B. A pediment is cut across the margins of the uplifted block and grows headward into the mountains as sheetfloods and running water transport sediment across the planar rock surface toward the basin fill. C. Growing headward from both sides of the upland, the pediment slowly consumes the mountains, leaving only a few residual hills rising above the eroded bedrock surface.

humid. However, they are especially common and well developed in semiarid grasslands in the middle of tectonically stable continents. Numerous examples can be found in southern and central Africa, northwestern Brazil, and central Australia.

Field evidence suggests that inselbergs form in areas of relatively homogeneous, resistant rock (most commonly granite or gneiss, but also sedimentary rocks such as conglomerate and sandstone) that are surrounded by rocks more susceptible to weathering. Differential weathering over long time intervals lowers adjacent terrain, leaving these resistant rock masses standing high. Once formed, the bare rock hills tend to shed water, whereas surrounding debris-

FIGURE 12.30 Ayers Rock, a massive inselberg, rises about 360 m above the surrounding flat plain in central Australia.

mantled plains absorb water, causing the underlying rocks to weather more rapidly. For this reason, inselbergs may remain as stable parts of a landscape and persist for tens of millions of years or more. Some may even date back to the Mesozoic Era. If true, they have remained prominent landscape features since the time of the dinosaurs.

DESERTIFICATION

In the region south of the Sahara lies a belt of dry grassland known as the Sahel (Arabic for *border*). There the annual rainfall is normally only 100 to 300 mm, most of it falling during a single brief rainy season. In the early 1970s, the drought-prone Sahel experienced the worst drought of this century (Fig. 12.31). For several years in succession the annual rains failed to appear, causing adjacent desert to spread southward—according to one estimate as much as 150 km. The drought extended from the Atlantic to the Indian Ocean, a distance of 6000 km, and affected a population of at least 20 million people, many of them seminomadic herders of cattle, camels, sheep, and goats. The results of the drought were intensified by the fact that between about 1935 and 1970 the human population had doubled, and the number of livestock had also increased dramatically. This increase in the number of people and animals led to severe overgraz-

ing, so that with the coming of the drought the grass cover almost completely failed. About 40 percent of the cattle—a great many millions—died. Millions of people suffered from thirst and starvation, and many died as vast numbers migrated southward in search of food and water. In the mid-1970s the rains returned briefly. Then, in the 1980s, drought conditions resumed. Ethiopia and the Sudan were especially hard hit, and experienced widespread famine. Mass starvation was alleviated only by worldwide relief efforts.

Such invasion of desert into nondesert areas, referred to as **desertification**, can result from natural environmental changes as well as from human activities. The major symptoms are declining groundwater tables, increasing saltiness of water and topsoil, reduction in supplies of surface water, unnaturally high rates of soil erosion, and destruction of native vegetation. Although we can find evidence of natural desertification events in the geologic record, there is increasing concern that human activities, regardless of natural climatic trends, can in themselves help promote widespread desertification.

Desertification in North America

The impact of desertification on human life in North America is less severe than in more densely populated regions of the world, but it nevertheless has important and far-reaching implications for the continent's food,

FIGURE 12.31 Overgrazing during years of drought killed most of the vegetation around wells in the Azaouak Valley, Mali. Without vegetation, soil blows away and the desert advances.

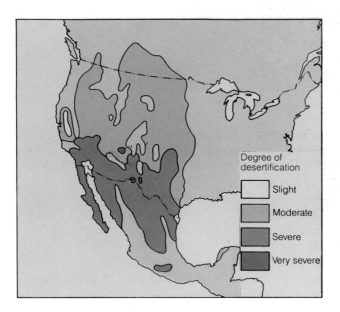

FIGURE 12.32 Vast areas of the American West show evidence of recent moderate to severe desertification.

water, and energy supplies, as well as its natural environment. Nearly 37 percent of the dry regions of the continent have experienced "severe" desertification (Fig. 12.32). In the southwestern United States, about 10 percent of the land area—approximately the same size as the original 13 states—has been severely affected by desertification over the past century. Large shifting sand dunes have formed, erosion has virtually denuded the landscape of vegetation, numerous gullies have developed, and salt crusts have accumulated on nearly impermeable irrigated soils. Mostly, this desertification has been brought about by overgrazing, by excessive withdrawal of groundwater, and by unsound water-use practices, in part allied with a population increase and expanded agricultural production.

Countermeasures

How can the detrimental effects of desertification be halted and even reversed? The answer lies largely in understanding the geologic principles involved and in intelligent application of measures designed to reestablish a natural balance in the affected areas. Elimination of incentives to exploit arid lands beyond their natural capacity, coupled with long-range planning aimed at minimizing the negative effects of human activity, should help in reaching the desired goal. Because arid lands of the western United States supply about 20 percent of the total agricultural output of the nation, the long-term benefits could be substantial.

TECTONIC DESERTIFICATION

Western China contains two of the world's major deserts. One, the Takla Makan, is a hot desert with vast regions of shifting sand dunes (Fig. 12.17), while the second encompasses the western Tibetan Plateau, a high, cold desert with sparse steppe vegetation underlain by perennially frozen ground (Fig. B12.1). Both are relatively young, geologically speaking, and their origin is tied to the tectonics of converging lithospheric plates.

Evidence of the long-term natural desertification of western China has been assembled by Chinese paleontologists who have studied the fossil remains of plants and animals in Tertiary strata. The fossil sites lie scattered about the high Tibetan Plateau and adjacent mountain ranges, mostly at altitudes of 4000 to 6000 m where the landscape now consists of alpine steppe and cold desert. Fossil plants and pollen of Eocene and Oligocene age disclose a flora that consisted of evergreen broadleaf forests and included eucalyptus, magnolia, and fig. Today,

these warmth-loving trees are found in moist, tropical environments at low altitudes. Remains of a giant rhinoceros, a larger relative of the rhinoceros that lives in tropical southeast Asia today, have also been found in the Oligocene strata. From this evidence, we can infer that western China during the early Tertiary must have had a relatively low altitude (perhaps 500 to 1000 m) and a warm climate, and that there were no mountain barriers to block the passage of moist oceanic air from the south. The widespread distribution of *Hipparian* supports this view, for it implies that this primitive horse could migrate freely over the continent unimpeded by major mountain chains.

During the Miocene and Pliocene, changing plant and animal assemblages imply progressive uplift of the region, accompanied by cooling and drying of the climate. By the beginning of the Pleistocene, nearly 2 million years ago, the Tibetan Plateau had reached an altitude of 1000 to 2000 m, and alpine coniferous forests had re-

FIGURE B12.1 The snow- and ice-capped Himalaya rising high above the dense green vegetated plains of India (lower left) keep moisture-bearing winds from reaching the high Tibetan Plateau. As a result, the plateau is a brown and barren alpine desert, receiving less than 250 mm of annual precipitation.

placed subtropical trees. During the last half of the Pleistocene, these forests gave way to dry alpine steppe as the rate of uplift accelerated, bringing the plateau and the adjacent Himalaya to high altitudes.

The progressive desertification of western China during the Tertiary coincides with the ongoing collision of India with Asia and resulting uplift of the Tibetan Plateau and associated mountain ranges (B12.2). Prior to this collision, the area of Tibet and the Takla Makan desert stood at low altitude and lay much closer to an ancient seaway that separated India and Asia. As the two continents converged, the seaway narrowed and then disappeared, placing these future desert regions closer to the center of Asia and therefore in a more continental climatic environment. Uplift associated with continental collision further intensified the desertification process by raising Tibet into successively drier and colder climatic zones and simultaneously raising the lofty Himalaya across the path of northward-flowing monsoonal air, thereby blocking off the primary source of precipitation. The cold desert of the lofty plateau and the adjacent hot Takla Makan desert therefore owe their existence both to their mid-continental position and the rainshadow formed by the high mountain barrier that separates the hot, humid plains of India from the frigid wastes of Tibet.

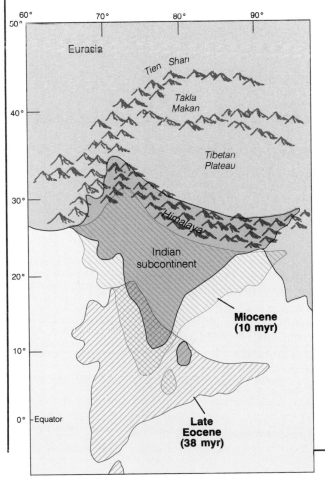

FIGURE B12.2 As India drifted northward during the Tertiary period, the seaway separating it from Asia narrowed and disappeared. Successive positions of the continents during the late Eocene and the late Miocene are shown relative to the present continent of Asia. The collision of the two land masses resulted in uplift of the Himalaya and associated high mountain ranges, and isolation of the Tibetan Plateau and Takla Makan desert from moisture sources to the south.

SUMMARY

1. Unequal heating of the Earth by solar radiation sets up convective circulation in the atmosphere. The Coriolis effect breaks the equator-to-poleward circulation in the northern and southern hemispheres into cells dominated, respectively, by the tradewinds, westerlies, and polar easterlies.

2. Climate is the average weather of a place over a period of years, together with the degree of variability of that weather. The Earth's climates are influenced by the distribution of land and oceans, and by surface topography.

3. Wind moves saltating sand grains close to the ground and suspended dust particles at higher levels. Sorting of sediment results.

4. Through deflation and abrasion, winds create deflation basins, desert pavement, ventifacts, and yardangs.

5. Dunes originate where obstacles distort the flow of air. Dunes have steep slip faces and gentler windward slopes. They migrate in the direction of wind flow, forming cross strata that dip downwind.

6. Loess is deposited chiefly downwind from deserts and from the floodplains of glacial meltwater streams. Once deposited, it is stable and is little affected by further wind action.

7. Airborne tephra deposited during explosive volcanic eruptions decreases in thickness and grainsize away from the source vent.

8. Hot deserts constitute about a quarter of the world's nonpolar land area and are regions of slight rainfall, high temperature, excessive evaporation, relatively strong winds, and sparse vegetation. Polar deserts occur at high latitudes where descending cold, dry air creates arid conditions.

9. Mechanical weathering, flash floods, and winds are especially effective geologic agents in deserts.

10. Fans, bajadas, and pediments are conspicuous features of many deserts. Pediments are probably shaped by running water and are eroded surfaces across which sediment is transported.

11. Inselbergs form in relatively homogeneous, resistant rocks and may remain as persistent landforms for millions of years.

12. Recurring natural droughts can lower the water table, cause high rates of soil erosion, and destroy vegetation, thereby leading to the invasion of deserts into nondesert areas. Overgrazing, excessive withdrawal of groundwater, and other human activities can promote desertification. Desertification can be halted or reversed by measures that restore the natural balance.

IMPORTANT TERMS TO REMEMBER

bajada (p. 346)

climate (p. 328)
Coriolis effect (p. 328)

deflation (p. 333)
desert (p. 341)
desert pavement (p. 334)
desert varnish (p. 345)

desertification (p. 349)
dune (p. 336)

flash flood (p. 346)

inselberg (p. 347)

loess (p. 339)

pediment (p. 346)
playa (p. 346)

sand ripples (p. 331)
sand sea (p. 338)

ventifact (p. 335)

yardang (p. 335)

QUESTIONS FOR REVIEW

1. What atmospheric factors cause most of the world's large hot deserts to be concentrated in belts lying between 20° and 30° from the equator?

2. Explain what controls the depth to which deflation is effective in arid regions.

3. Why are the erosional effects of blowing sand generally confined to a zone within about a meter of the ground surface?

4. Explain the origin of the internal stratification of a sand dune. How do sand dunes migrate downwind?

5. How might you tell the previous direction of the prevailing wind from the form and internal stratification of an ancient, inactive sand dune? from a ventifact? from a tephra deposit?

6. What measures could you recommend a farmer in the southwestern United States to take that would halt the migration of sand dunes now threatening his agricultural fields?

7. How might you tell a deposit of loess from an alluvial silt having a similar range of particle sizes?

8. Why do hillslopes in arid landscapes tend to be steeper and sharper than those in humid landscapes?

9. What evidence can you cite that points to streams being effective agencies of erosion and sediment transport in desert regions?

10. How would you tell a bajada from a pediment in the field? What process(es) are involved in the formation of each?

11. Why do playas often have a distinctive deposit of salts at their surface?

12. What factors influence the formation of inselbergs? Why are inselbergs likely to remain persistent features of a semiarid landscape?

13. What are some of the obvious symptoms of desertification? How might they be retarded or reversed by human intervention?

Coastline recession, the result of storm waves pounding against a bluff of erodible sandy sediments, has undermined the foundations of this house.

CHAPTER 13

The Ocean Margins

S horeline homes command premium prices in the real estate market. If built on solid rock, they can prove to be a lasting investment; however, coastlines are among the most dynamic places on the Earth's surface, and a house built on a coast composed of erodible sediment can prove to be a poor bargain.

A recent government assessment of coastal erosion problems in the United States showed that moderate to severe beach erosion is taking place along more than 80 percent of the shoreline. The most rapid erosion is occurring along sandy stretches of the Atlantic and Gulf coasts and the north coast of Alaska. Because these coasts consist of erodible sand, they are especially vulnerable to the attack of waves and currents.

A primary reason why erosion and other shoreline problems are receiving increasing attention is that coastal zones of the United States now have population densities that are five times the national average. At the time of the 1990 census, 50 percent of all Americans lived within 75 km of a coast. Projections indicate that this number will increase to 75 percent by 2010. The nation's fragile shorelines, already severely stressed by the existing population, will come under increased pressure as continuing growth leads to demands for further development, additional fresh water, and adequate waste disposal.

Unlimited development, however, could well prove disastrous, for coastlines are places that change constantly under the attack of wind and the sea. The natural risk of living in a coastal zone may increase still further if a warming climate causes world sea level to rise. The combination of natural risk and increasing population pressure has led many well-informed scientists to conclude that we are rapidly approaching a coastal crisis. Like other environmental crises of our own making, the solution to coastal problems must rest on balancing human needs and expectations with a clear understanding of how natural processes control coastal environments.

OCEAN CURRENTS AND TIDES

The shoreline, where land and ocean meet, invariably is a dynamic place. Pay an annual visit to almost any shore, and you will see changes that have taken place since your previous visit. Sometimes the changes are small, but often they are substantial. The energy driving these changes comes from ocean currents and waves, which in turn derive their energy from winds and tides.

Surface Ocean Currents

Surface ocean currents are broad, slow drifts of surface water. They are set in motion by the prevailing surface winds. Air that flows across a water surface drags the water slowly forward, creating a current of water as broad as the current of air, but rarely more than 50 to 100 m deep.

In low latitudes, surface seawater moves westward with the trade winds. The general westerly direction of the North and South Equatorial currents (Fig. 13.1), driven by these winds, is reinforced by the Earth's rotation. Both north and south of the equator, the westerly moving currents are deflected wherever they encounter a coast, and by the Coriolis effect (chapter 12). On reaching middle latitudes, the currents travel eastward, moved by the prevailing westerly winds. The result is a circular motion of water in each major ocean basin, both north and south of the equator.

Along most continental margins, the predominant flow of water roughly parallels the coast. Warm surface water originating in the equatorial region moves north or south along the eastern margins of continents to about latitude 45°, while cold, east-flowing water at higher latitudes encounters the western margins of continents and is deflected south (Fig. 13.1). At still higher latitudes, cold currents generally prevail.

Tides

Tides are the rhythmic rise and fall of ocean waters resulting from the gravitational attraction of the Moon and (to a lesser degree) the Sun acting on the Earth. Gravitational attraction of the Moon causes the ocean to bulge upward on opposite sides of the Earth. The bulges are caused chiefly by the difference in the Moon's attraction for the ocean and for the solid Earth, which in turn is related to differences in their respective distance from the Moon. On the side toward the Moon, ocean water is closer to the Moon than is the solid Earth beneath, and so the water is attracted more strongly; on the far side, the water is farther from the moon than the underlying solid Earth and so it is pulled less strongly. The result is two *tidal bulges* on

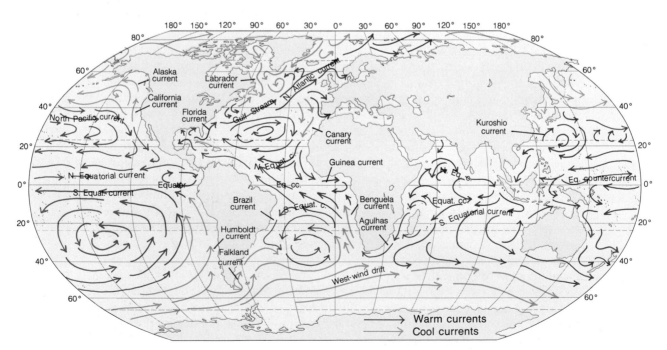

FIGURE 13.1 Surface ocean currents form a distinctive pattern, curving to the right in the Northern Hemisphere and to the left in the Southern Hemisphere. The westward flow of tropical Atlantic and Pacific waters is interrupted by continents that deflect the water poleward. The flow then returns as eastward-moving currents in higher latitudes.

A.

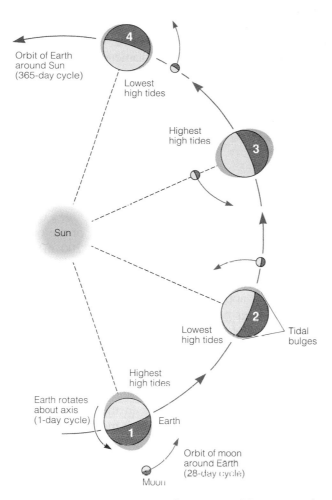

NW

Sand and clay

Depth (m)

B.

and nearsho
file, and the

Organic
Many of the
stone reefs
principally
Such organ
the temper
are built or
istic of low

Three p
ing reef is
ing the adj
a lagoon.
surface as
plunges st₆

A **barri₆
a lagoon th
(Fig. 13.22
lie far off t
Great Barr

An **atol**
shallow la₆
cal volcani

FIGURE 13.2 The gravitational attraction of the Moon and
Sun on the Earth raises tidal bulges in the ocean on opposite
sides of the planet. When the Earth, Moon, and Sun are
aligned (positions 1 and 3), tides of highest amplitude are
observed. When the Moon and Sun are pulling at right
angles to each other (positions 2 and 4), tides of lowest
amplitude are experienced.

opposite sides of the planet. Because of the Earth's
daily rotation about its axis, and the moon's much
slower 28-day revolution around the Earth, these
bulges move continually across the Earth's surface
(Fig. 13.2).

To visualize how tides work, lets consider the tidal
bulges oriented with their maximum amplitude lying
along a line extending through the center of the Earth
and the center of the moon. As the Earth rotates about
its axis, any given point of land moves westward
through both tidal bulges each day (in other words,
during each complete rotation). Every time a land
mass encounters a tidal bulge, the water level along
the coast rises. As the Earth continues to rotate, the
coast passes through the highest point of the tidal

bulge (high tide), and then the water level begins to
fall.

At most places on the ocean margins, two high tides
and two low tides are observed each day. There are
two tidal cycles because during every complete rotation of the Earth, any point on a coast passes across
both tidal bulges.

Because the distance from the Earth to the Sun is
so much greater than the distance from the Earth to
the Moon, the Sun is only about half as effective as
the Moon in producing tides. Therefore, the opposing
tidal effects never entirely cancel each other.

In the open sea, the effect of the tides is small.
However, in bays, straits, estuaries, and other narrow
places along coasts, tidal fluctuations of as much as 16
m are known. At the Medieval abbey of Mont St. Michel on the coast of Normandy, the coastline changes
dramatically between high and low tide (Fig. 13.3). At
high tide, the small, rocky hill is nearly surrounded
by the sea, being connected to the mainland only by
a narrow artificial causeway. As the tidal flow reverses
and the water level falls, boats become grounded on
the broad expanse of surrounding tidal flat.

Tides also can generate rapid currents. Some tidal
currents approach 25 km/h, and the incoming tide
locally can create a wall of water a meter or more
high (called a *tidal bore*) that moves up estuaries and
the lower reaches of streams. Such fast-moving currents, although restricted in extent, can easily move
sediment around. Large linear sand ridges can be built
paralleling such currents, as well as sand waves, which

FIGURE 13.3 Mont St. Michel, a Medieval abbey on a
rocky islet amid expansive tidal flats at the Normandy coast,
was accessible only at low tide until a causeway was built a
century ago.

FIGU
sach
worl
side
sout
of th

and
buil
dro
cur
cre:
cur
A
isla
ton
a s]
,
and
dir
Pac
sec
is I
acr
ori
(sc
rei

Protection of Beaches

Because of their great recreational value, beaches in densely populated regions justify greater expense for maintenance than most headlands do. A beach, however, presents a special sort of problem. As a result of beach drift, what happens on one part of a beach affects all other parts that lie in the downdrift direction. This is true because any beach is in a state of delicate balance between forces of erosion and deposition. For example, a seawall, dock, or other structure built at the updrift end of a beach reduces the amount of sand available for beach drift. The surf becomes underloaded and makes good the loss by eroding sand from along the beach. Small beaches have been completely destroyed by this process in only a few years.

Breakwaters and Groins

A **breakwater** is an offshore barrier designed to protect a beach or boat anchorages from incoming waves. However, a breakwater upsets the natural balance of the adjacent beach, leading to shoreline changes. Breakwaters constructed along the shore at Tel Aviv in Israel (Fig. 13.29) protect the beach from the onslaught of waves, but they have turned a straight coastline into a scalloped one. Arriving waves now are refracted around each breakwater. The currents created by the refracted waves transport sand toward the zone of protected water, causing the beach to expand seaward behind each barrier.

Beach erosion can be checked to some extent by building groins at short intervals along the beach. A **groin** is a low wall built out into the water at a right angle to the shoreline (Fig. 13.30). A groin acts as a check on the rate of beach drift because it traps sand carried to it by longshore drift. However, erosion tends to occur on the downdrift side of a groin where the beach has been deprived of its constant supply of sand. The net effect, therefore, is to protect one part of a beach at the expense of another part.

Artificial Nourishment

Another way of protecting an eroding beach is to haul in sand and pile it on the beach at the updrift end. Surf then erodes the pile and drifts the new sand down the length of the beach. Sand for artificially nourishing a beach must be continuously replenished, however. As can be imagined, the feeding of a beach, like the construction and maintenance of groins, can be expensive.

Effects of Human Interference

Many beaches around the world are deteriorating because of human interference. In southern California, for example, most of the sand on beaches is supplied not by erosion of wave-cut cliffs but by alluvium carried to the sea at times of flood. However, because buildings and other structures in stream valleys are vulnerable to floods, dams have been built across the stream courses to control flooding. Of course, the dams also trap the sand and gravel carried by the streams, thus preventing the sediment from reaching the sea. As a consequence of halting the through-flow

FIGURE 13.29 Breakwaters constructed offshore from Tel Aviv, Israel, protect the beach zone from incoming waves. Wave refraction around the barriers has led to progradation of the beach behind each breakwater, producing a scalloped coastline.

FIGURE 13.30 Groins have been built perpendicular to the shoreline of Miami Beach, Florida, to prevent excessive loss of sand by longshore drift at this popular resort area.

of sand, the natural balance among the factors involved in the longshore transport of sediment has been upset. The result has been significant erosion of some beaches.

A further dramatic example of human interference can be seen along the Black Sea coast of the Soviet Union. Of the sand and pebbles that form the natural beaches there, 90 percent used to be supplied by rivers as they entered the sea. During the 1940s and 1950s, three things occurred: large resort developments were built at the beaches, large breakwaters were constructed so that two major harbors could be extended into the sea, and dams were built across some rivers inland from the coast. All this construction interfered with the steady state that had existed among the supply of sediment to the coast, longshore currents and beach drift, and deposition of sediment on beaches. By 1960, it was estimated that the combined area of all beaches along the coast had decreased by 50 percent. Then beachfront buildings began to sag or collapse as the surf ate away at their foundations. An ironic twist to the chain of events lies in the fact that large volumes of sand and gravel were removed from beaches and used as concrete aggregate, not only to construct the resort buildings but also to build the dams that cut off the supply of sediment to the coast.

HOW TO MODIFY AN ESTUARY

San Francisco Bay is one of the world's best-known estuaries (Fig. B13.1). When first discovered by Europeans, its shores and productive waters supported as many as 20,000 Native Americans. The population boom that followed the discovery of gold in California in 1848, and which has continued unabated to the present day, has led to major changes in the river systems that enter the bay and thereby has greatly affected the estuary environment.

The need for water is a continuing concern for Californians, who require it not only for human consumption and industry but also to support huge agricultural enterprises. Diversion of water from natural streams has reduced freshwater inflow to San Francisco Bay to less than half what it was in 1850; by the year 2000, the projected flow will decline still further—to only 30 percent of the post-1850 average. At the same time, increasing urbanization has led to a loss of 95 percent of the bordering tidal marshes as a result of filling and diking. Hydraulic mining for gold in the foothills of the Sierra Nevada following the discovery of gold in 1848 produced vast quantities of fine sediment that choked streams, destroyed fish spawning grounds, obstructed navigation, and ultimately reached the bay, where it reduced the area and volume of the estuary and modified tidal circulation.

The unforeseen consequences of this human tampering with a natural stream and estuarine system have been many. Most commercial fisheries have disappeared. Reduced freshwater flushing of the bay has concentrated agricultural, domestic, and industrial waste products in estuarine sediments, contaminating both them and the organisms that feed on them and raising increasing concerns about human health. Natural habitats of migrating birds have been extensively

FIGURE B13.1 Vertical satellite image of the shrinking San Francisco Bay estuary. Filling and diking of tidal marshes to create farmland, evaporation ponds, and residential and industrial developments has reduced 2200 km^2 of marsh land that existed before 1850 to less than 130 km^2 today.

destroyed, with presumed major effects on bird populations.

Such changes are not unique to San Francisco Bay. Many other large estuaries, which are favored sites of urban development and industrialization, have similar problems (the Rhine River in the Netherlands, the Thames River in England, and the Susquehanna and Potomac rivers in the United States are examples). All are sensitive to human-induced changes and susceptible to steady deterioration. The best hope for reversing their decline is through an improved understanding of how human activities affect the physical, chemical, and biological processes in river–estuarine systems, and of the measures that must be taken to curtail potentially destructive actions.

SUMMARY

1. Huge wind-driven ocean currents that circulate clockwise in the Northern Hemisphere and counterclockwise in the Southern Hemisphere carry warm equatorial water toward the polar regions.

2. Twice-daily ocean tides are generated as the surface of the rotating Earth passes through tidal bulges on opposite sides of the planet produced by the gravitational attraction of the Moon and the Sun.

3. Fast-moving tidal currents cause movement of sediment in bays, straits, estuaries, and other restricted places along coasts.

4. In deep water, waves have little or no effect on the seafloor. At a depth equal to half its wavelength, a wave can begin to stir up the bottom sediments.

5. Most of the geologic work of waves is performed by surf at depths of 10 m or less.

6. Wave refraction tends to concentrate wave erosion on headlands and to diminish it along the inner shores of bays.

7. Longshore currents and beach drift can transport great quantities of sand along coasts.

8. On gentle, sandy coasts the beach typically consists of a foreshore, berm, and backshore. On rocky coasts the shore profile includes a wave-cut cliff, a wave-cut bench, and a beach.

9. Depositional shore features include beaches, marine deltas, spits, bay barriers, and barrier islands. Barrier islands form offshore in areas where rising sea level causes a shoreline to advance across a gently sloping coastal plain.

10. Organic reefs form in tropical seas where water temperature averages at least 18°C. A fringing reef on a volcanic island can be transformed into a barrier reef as the island subsides, and eventually into an atoll.

11. The shape of coasts partly reflects the amount of energy available to erode and deposit sediment. Rock structure and degree of erodibility help dictate the form of rocky coasts.

12. Nearly all coasts have experienced recent submergence due to postglacial rise of sea level. Some coasts have experienced more complicated histories of emergence and submergence due to tectonic and isostatic movements on which are superimposed the worldwide sea-level rise.

13. Infrequent but powerful storms, tsunamis, and large landslides can pose significant threats to people and structures in the coastal zone.

14. Erosional damage to a shore cliff can be minimized by a seawall or an armor of boulders. Beach erosion is a serious problem along many inhabited coasts, but beaches can be temporarily protected by a series of groins or by importation of sand.

IMPORTANT TERMS TO REMEMBER

atoll (p. 369)

barrier island (p. 368)

barrier reef (p. 369)
bay barrier (p. 368)
beach (p. 363)
beach drift (p. 361)

emergence (of a coast) (p. 373)

fringing reef (p. 369)

groin (p. 376)

longshore current (p. 361)

spit (p. 367)
submergence (of a coast)
   (p. 372)

surf (p. 359)

tsunami (p. 373)

wave base (p. 358)
wave-cut bench (p. 364)
wave-cut cliff (p. 364)

wavelength (p. 358)
wave refraction (p. 359)

QUESTIONS FOR REVIEW

1. Explain why there are two high tides and two low tides each day and suggest why their time of occurrence is not always the same.

2. Describe the motion of a parcel of water below a passing wave in the open ocean, and explain how and why the motion changes as the wave moves into shallow water.

3. How is wave base related to wavelength?

4. What causes a wave to "break" near the shore?

5. What is the effective depth of erosion by surf, and what determines this depth?

6. How does wave refraction explain why rocky headlands are more vigorously eroded by surf than bays are?

7. What causes longshore currents to develop and, in some cases, to shift direction seasonally?

8. Why are the sediments in a beach generally well sorted and well stratified?

9. Describe how you think an atoll would look during the peak of a glacial age, and why.

10. How might you demonstrate whether an emerged marine terrace was related to (a) a higher world sea level or (b) a local tectonic uplift of the land.

11. Describe measures that can be taken to reduce the impact of erosion (a) along a cliffed coast and (b) along a sandy beach. What negative effects might the measures have, even though they stop erosion?

12. How might the construction of a dam along the lower part of a large river affect the coast where the river enters the ocean?

PART III

The Evolving Earth

Windmill farm, California

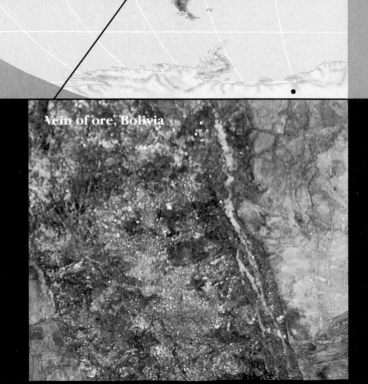

Vein of ore, Bolivia

Postglacial uplift, Svalbard, Norway

Earthquake damage, Armenia

Air bubbles in glacier, Antarctica

Lateritic landscape, Western Australia

Thin sedimentary strata, once horizontal, have been complexly tilted and bent as a result of deformation during high-grade metamorphism in the Limpopo Mobile Belt, Sand River, South Africa.

CHAPTER 14

Deformation of Rock

G reece is well known for its earthquakes. One destroyed Sparta, the famous city of antiquity, in the fifth century B.C. and there have been many more since. Recently, a team of Greek and British scientists discovered that the tectonic forces that cause the earthquakes are also stretching Greece and slowly making it grow larger.

A century ago the distances between a series of Greek survey monuments were measured very accurately. In 1988 the scientific team remeasured the distances and found that Greece is now a meter longer. They also discovered that Greece is being twisted so that the southern end, the Peloponnesus, is moving to the southwest relative to the rest of Greece. The reason for the stretching and twisting is plate tectonics. Africa is moving north and slowly forcing a slice of Mediterranean seafloor under Greece. The resulting forces are stretching and twisting the overlying continental crust.

Because Greece is being stretched, we can conclude that rock in the Greek crust is being deformed. There is nothing unique or unusual about such a conclusion. Evidence that rocks can be deformed is easy to find. If you look at a photograph of the Alps, the Rockies, the Appalachians, or any other mountain range, you will see strata, once horizontal, that are now tilted and bent. Enormous forces are needed to deform such huge masses of rock.

Tectonic forces continuously squeeze, stretch, bend, and break rock in the lithosphere. The source of energy for tectonic forces is the Earth's heat energy. As we discussed in chapter 1, convection converts the Earth's heat energy to mechanical energy. The huge, slow, convective flows of hot rock in the mesosphere and asthenosphere continuously buckle and warp the lithosphere. It is those convective forces that are ultimately the cause of the rock deformation we observe in mountain ranges and that are stretching Greece.

Rarely can we see rock being twisted and bent. Most deformation happens deep in the crust or in the mantle, and we see the bent rock only when it is exposed by erosion. We are then forced to infer how the deformation occurred and the origin of the deforming forces.

Some of the evidence on which we base our inferences as to how deformation occurs comes from laboratory studies, and some from direct studies of deformed rock. In this chapter we first discuss how rocks can be deformed, and then we turn to the kind of evidence visible in the crust and the terms geologists use to describe the geometry of deformation.

HOW ROCK IS DEFORMED

In order to discuss deformation in rocks, such as bending, twisting, and fracture, it is helpful to review some of the elementary properties of solids. Knowledge of the factors controlling rock deformation comes largely from laboratory experiments in which cylinders or cubes of rock are squeezed and twisted under controlled conditions.

Stress and Strain

We learned in chapter 5 that the term **stress** is used rather than pressure when rock deformation is being discussed. We also learned that **uniform stress** describes that situation where the stress is equal in all directions, such as the stress on a small body immersed in a liquid or gas. Uniform stress in rocks is also called **confining stress** because any body of rock in the lithosphere is confined by the rocks around it and is uniformly stressed by the weight of the overlying rocks. **Differential stress,** by contrast, is stress that is not equal in all directions. The stress that deforms rocks is usually differential stress. The three kinds of differential stress are shown in Figure 14.1. **Tensional stress** stretches rocks, **compressional stress** squeezes them, while **shear stress** causes slippage and translation. Differential stresses are caused by tectonic forces. An example of the difference between uniform and differential stress is illustrated in Figure 5.2.

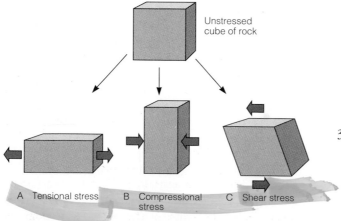

FIGURE 14.1 Shape of a cube of rock deformed by differential stress. Arrows indicate directions of maximum stress. A. Tension stress. B. Compressional stress. C. Shear stress.

The term used to describe the deformation of a rock is **strain,** which is defined as the change in size or shape, or both, in a solid as a result of stress. Uniform stress causes a solid to change size but not shape. Differential stress causes a solid to change shape, but it may or may not also cause a change in size.

Stages of Deformation

When a rock is subjected to increasing stress, it passes through three stages of deformation in succession:

1. **Elastic deformation** is a reversible, or nonpermanent, change in the volume or shape of a stressed rock. When the stress is removed, the rock returns to its original size and shape. There is a limiting stress, called the **elastic limit,** beyond which a solid suffers permanent deformation and does not return to its original size or shape once the stress is removed.

   A famous British scientist, Sir Robert Hooke (1635–1703), was the first to demonstrate that, for all materials, provided the elastic limit is not exceeded, a plot of stress versus strain is always a straight line. Hooke proved his point by using a spring, as in Figure 14.2A. However, Hooke's law is equally true for rocks, as illustrated in Figure 14.2B.

2. **Ductile deformation** is an irreversible change in shape and/or volume of a rock that has been stressed beyond the elastic limit. If a cylinder of rock is stressed by a compressional stress applied parallel to the long axis of the cylinder, an interesting result is obtained. As shown in Figure 14.3, the stress–strain curve for the cylinder rises first through the elastic region, and then, at the elastic limit (point Z), the curve flattens and additional stress causes ductile deformation. If the stress is removed at point X', the cylinder partially returns to its original shape—the strain decreases along the curve $X'Y$. A permanent strain, equal to $XY,$ has been induced in the rock. The permanent strain XY is due to ductile deformation.

3. **Fracture** occurs in a solid when the limits of both elastic and ductile deformation are exceeded. Consider again the stress–strain curve in Figure 14.3. If, instead of releasing the stress at point X', we continued to increase the stress, the stress–strain curve would continue to point $F,$ where the cylinder breaks by fracturing. Obviously, fracture, like ductile deformation, is an irreversible kind of deformation.

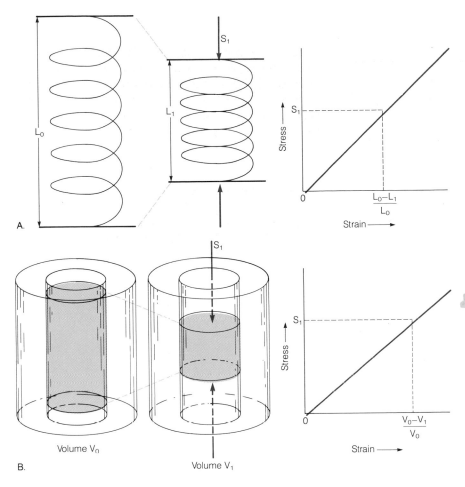

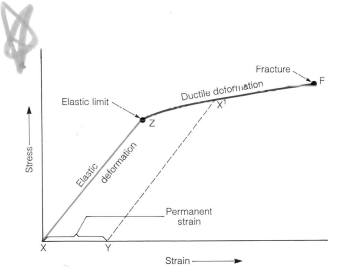

FIGURE 14.2 Hooke's law states that for elastic solids, strain is proportional to stress. A. A spring, shortened by a compressional stress S_1, has its length reduced from L_0 to L_1. A plot of the strain $(L_0 - L_1)/L_0$ as a function of stress produces a straight line. B. A cylinder of rock subjected to a confining stress by a tight metal jacket, and shortened by a compressional stress S_1, has its volume reduced from V_0 to V_1. A plot of strain $(V_0 - V_1)/V_0$ as a function of stress produces a straight line.

FIGURE 14.3 Typical stress-strain curve for a cylinder of rock tested in the laboratory. Following elastic deformation (X to Z), the elastic limit Z marks the onset of ductile deformation. If, at point X^1, the stress is removed, the solid will return to an unstressed state at point Y by following the path X^1Y. The distance XY is a measure of the permanent, irreversible strain produced by ductile deformation. If the stress is not released at X^1, but is increased and maintained, the strength of the solid is exceeded when rupture occurs at the point of fracture, F.

Ductile Deformation Versus Fracture

A brittle substance tends to deform by fracture, while a ductile substance deforms by a change of shape. Drop a piece of chalk on the floor and it will fracture, but if you drop a piece of cheese it will bend but not break. Chalk is brittle; cheese is ductile.

A typical stress–strain curve for a brittle substance is shown in Figure 14.4A. Note that Z, the elastic limit, is very close to F, the point of fracture. Thus, very little ductile deformation occurs in a brittle substance. In contrast, in a ductile substance the elastic limit and fracture point are far apart, as shown in Figure 14.4B.

Examples of deformation of brittle and ductile rocks are shown in Figure 14.5A and B. The strata in Figure 14.5A have fractured cleanly with little or no evidence of bending due to ductile deformation. The strata in Figure 14.5B, by contrast, have been intensely twisted and bent by ductile deformation.

To evaluate deformation in rocks, it is necessary to estimate the relative importance of brittle properties versus ductile properties. The essential conditions controlling the relative importance of the two kinds of properties are (1) temperature; (2) confining stress; (3) time and strain rate; and (4) composition.

385

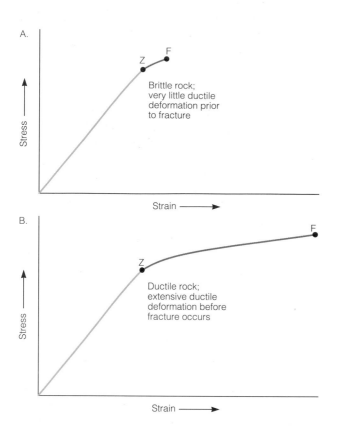

A.

FIGURE 14.4 Comparison of stress-strain curves. A. Curve for a brittle substance. B. Curve for a ductile substance.

Temperature

The higher the temperature, the more ductile and less brittle a solid becomes. A rod of glass is difficult to bend at room temperature; if we try too hard, it will break because it is brittle. However, a glass rod becomes ductile and can be readily bent if it is heated to redness over a flame. Rocks are like glass rods. They are brittle at the Earth's surface, but at depth, where temperatures are high because of the geothermal gradient, rocks become ductile.

B.

FIGURE 14.5 Examples of rock deformation. A. Fracture of strata by brittle deformation. B. Bending of strata by ductile deformation.

Confining Stress

The effect of confining stress on deformation is not familiar in common experience. Confining stress is a uniform squeezing of rock owing to the weight of all the overlying strata. High confining stress hinders the formation of fractures and so reduces brittle properties. At high confining stress it is easier for a solid to bend and flow than to fracture. Reduction of brittleness by high confining stress is a second reason why solid rock can be bent and folded by ductile deformation.

Time and Strain Rate

The effect of time on rock deformation is vitally important, but as with confining stress, it is not obvious from common experience. Stress applied to a solid is transmitted by all the constituent atoms of the solid. If the stress exceeds the strength of the bonds between atoms, either the atoms must move to another place in the crystal lattice in order to relieve the stress or the bonds must break, which means fracture occurs. Atoms in solids cannot move rapidly. Nevertheless, if the stress builds up slowly and gradually and is maintained for a long period, the atoms have time to move, and the solid can slowly readjust and change shape by ductile deformation. The important point to appreciate is that the *rate* at which a solid is strained is just as significant as how long a stress is active.

The term used for time-dependent deformation of a rock is **strain rate,** which is the rate at which a rock is forced to change its shape or volume. Strain rates are measured in terms of change of volume per unit volume per second. For example, a strain rate that is sometimes used in laboratory experiments is 10^{-6}/s, by which is meant a change in volume of one millionth of a unit volume per unit volume per second. Strain rates in the Earth are much lower than this—about 10^{-14} to 10^{-15}/s. The lower the strain rate, the greater the tendency for ductile deformation to occur.

A comparison of the influences of temperature, confining pressure, and strain rate on rock properties can be seen in Figure 14.6. Low temperature, low confining stress, and high strain rate enhance brittle properties. These conditions are characteristic of the crust (especially the upper crust), and as a result failure by fracture is common in upper-crustal rocks. High temperature, high confining stress, and low strain rates, which are characteristic of the deeper crust and mantle, reduce brittle properties and enhance the ductile properties of rock. Down to a depth of about 15 km rocks are brittle, and they will readily fracture. Below 15 km fractures become less common because rocks become increasingly ductile. The depth in the crust where ductile properties start to predominate over brittle properties is known as the *brittle-ductile transition.*

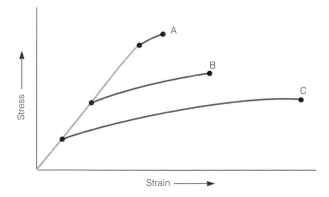

FIGURE 14.6 The effects of temperature and strain rate on rock deformation. Curve A: low temperature, high strain rate. Curve B: high temperature, high strain rate. Curve C: high temperature, low strain rate.

Composition

The composition of a rock has a pronounced effect on its properties. Composition has two aspects. First, the kinds of minerals in a rock exert a strong influence on properties because some minerals (such as quartz, garnet, and olivine) are very brittle, while others (such as mica, clay, calcite, and gypsum) are ductile. Second, the presence of water in a rock reduces brittleness and enhances ductile properties. Water affects properties by weakening the chemical bonds in minerals and by forming films around mineral grains, thereby reducing the friction between grains. Thus, wet rocks have a greater tendency to be deformed in a ductile fashion than do dry rocks.

Rocks that readily deform by ductile deformation are limestone, marble, shale, slate, phyllite, and schist. Rocks that tend to be brittle rather than ductile are sandstone and quartzite, granite, granodiorite, and gneiss.

DEFORMATION IN PROGRESS

Most deformation in the crust is too slow or too deeply buried to be observed. Large movements happen so slowly, which means at such low strain rates, that they can be measured only over a hundred or more years. The deformation occurring in Greece, discussed at the beginning of this chapter, is an example. Nevertheless, deformation does sometimes happen fast enough to be detected and measured. For convenience we divide large-scale, observable deformation of the crust into two groups: abrupt movement that involves fracture, in which blocks of the crust suddenly move a few centimeters or a few meters in a matter of minutes or hours; and gradual movement involving ductile deformation, in which slow, steady motions occur without any abrupt jarring.

Abrupt Movement

Abrupt movement involves the fracture of brittle rocks and movement along the fractures. A fracture in a rock along which movement occurs is a **fault.** Once fracturing has started, friction inhibits continuous slippage. Instead, stress again builds up slowly until friction between the two sides of the fault is overcome. Then abrupt slippage occurs again. If the stresses persist, the whole cycle of slow buildup followed by an abrupt movement repeats itself many times. Although movement on a large fault may eventually total many kilometers, this distance is the sum of numerous small, sudden slips. Each sudden movement may

FIGURE 14.7 An orange grove in southern California planted across the San Andreas Fault. Movement on the fault displaced the originally straight rows of trees. The direction of motion is such that trees in the background moved from left to right relative to the trees in the foreground.

cause an earthquake and, if the movement occurs near the Earth's surface, may disrupt and displace surface features (Fig. 14.7).

Figure 14.7 is an example of horizontal abrupt movement. Abrupt vertical movements are also well documented. The largest abrupt vertical displacement ever observed occurred in 1899 at Yakutat Bay, Alaska, during an earthquake. A stretch of the Alaskan shore (including the beach, barnacle-covered rocks, and other telltale features) was suddenly lifted as much as 15 m above sea level. This visible vertical displacement may be less than the total amount, because the fault is hidden offshore and the block of crust on the other side of it, entirely beneath the sea, may have moved downward, thus adding to the total displacement of the stretch of beach.

Abrupt movements in the lithosphere are commonly accompanied by earthquakes and can therefore be hazardous to people living nearby. Earthquakes and earthquake hazards are discussed in chapter 15.

Gradual Movement

Movement along faults is usually, but not always, abrupt. Measurements along the San Andreas Fault in California reveal places where gradual slipping occurs, sometimes reaching a rate as high as 5 cm a year. Because the San Andreas Fault cuts through the entire lithosphere—that is, well below the brittle-ductile transition—it is probable that deformation of rocks on either side of the fault is ductile at depth but brittle near the surface.

Possibly no spot on the Earth is completely stationary. Measurements by U.S. government surveyors over

100 years, for example, reveal great areas of the United States where the land is slowly sinking and other places where it is slowly rising (Fig. 14.8). The causes of these vast, slow vertical movements are not well understood, but the movements do prove that the solid Earth is not as rigid as it seems at first sight and that great internal forces are continuously deforming its crust.

EVIDENCE OF FORMER DEFORMATION

With such convincing evidence of present-day deformation of the Earth's crust, we might reasonably expect to find a great deal of evidence of former deformation. Studies of land and sea-bottom topography provide abundant evidence of vertical movements. In some areas the distribution of various kinds of rock provides clear evidence that horizontal movements have occurred through distances as great as several hundred kilometers.

Not all evidence of movement and deformation observed in bedrock is as obvious as the examples cited. But once we learn to recognize it, evidence of deformation is seen to be very widespread—so much so that a special branch of geology, *structural geology,* has the study of rock deformation as its primary focus.

Strike and Dip

The law of original horizontality (chapter 6) tells us that sedimentary strata and lava flows were initially

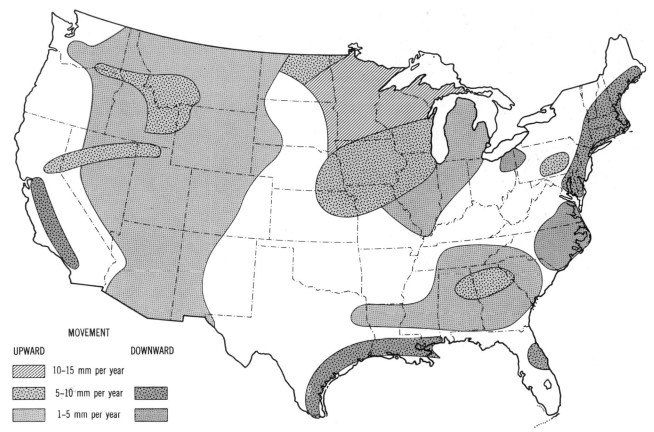

FIGURE 14.8 Accurate measurements over a 100-year period show that in large areas of the United States, the surface is slowly moving either upward or downward. Subsidence along the coasts of California and the Gulf of Mexico is believed to have been caused in part by withdrawal of gas, oil, and water, which allows subsurface reservoirs to collapse. Uplift near the Great Lakes is a rebound effect following the melting of the last ice sheet. The causes of movements in other areas are not known with certainty. Those areas in which no movement is shown are not necessarily stationary. They are simply areas in which measurements are very few or movement is so slow it has not yet been detected.

horizontal. Where we observe such rocks to be tilted, we can conclude that deformation has occurred. In order to decipher and explain this deformation, a geologist starts by measuring the angle and direction of tilting.

In order to measure the orientation of a tilted plane, we need to remember the two principles of geometry shown in Figure 14.9: (1) the intersection of two planes defines a line: and (2) in an inclined plane all horizontal lines are parallel. The line formed by the intersection of an inclined plane with a horizontal plane is always horizontal. Such a line can be visualized as the waterline on an inclined stratum along the shore of a lake, as shown in Figure 14.10. The lake surface is a convenient horizontal plane. The waterline marks the **strike,** which is the compass direction of the horizontal line formed by the intersection of a horizontal plane and an inclined plane.

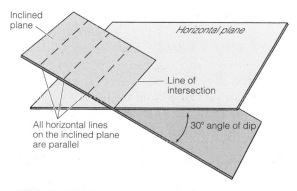

FIGURE 14.9 The geometric principles used to measure the direction and angle of tilt of an inclined plane.

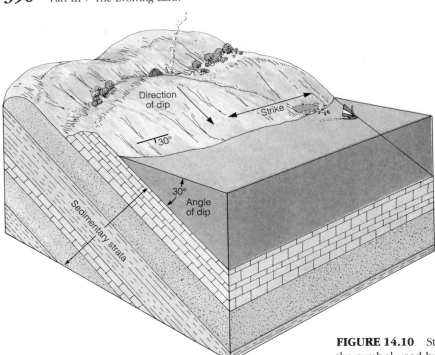

FIGURE 14.10   Strike and direction and angle of dip. Note the symbol used by geologists to indicate strike and dip.

Once we know the strike, we need only one more measurement to fix the orientation of an inclined plane. That is the **dip,** the angle in degrees between a horizontal plane and the inclined plane, measured down from horizontal. The direction and angle of dip are indicated in Figures 14.9 and 14.10.

Geologic Maps

Rarely is it possible for a geologist to see all the structural details of deformed rocks in a given area. Soil and vegetation usually cover much of the evidence. A geologist must therefore decipher the structures by making observations and measurements at a number of individual *outcrops* (places where bedrock is exposed at the surface). When the observations made at each outcrop are plotted on a map, and inferences made as to what has happened beneath the cover of soil and vegetation, the result is a geologic map. An example of a simple geologic map is shown in Figure 14.11. Note the way the geologist recorded the strike and the angle and direction of dip of the contacts between strata at each outcrop.

Deformation by Fracture

Rock in the crust, especially rock close to the surface, tends to be brittle. As a result, rock at or near the Earth's surface tends to be cut by innumerable frac-

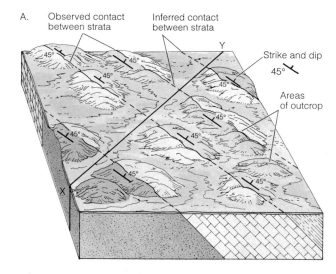

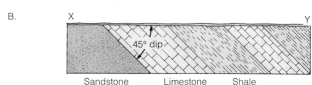

FIGURE 14.11   Geologic map. A. Dotted lines denote areas of outcrop where the geologist identified rock types and strata. Dashed lines indicate inferred contacts. B. Section along the line *XY* in part A.

tures called either *joints* or *faults*. As discussed in chapter 7, a joint is a special kind of fracture because there has not been any movement along it. A fault, as we learned earlier, is a fracture along which visible displacement has occurred.

Relative Displacement

Generally, it is not possible to tell how much movement has occurred along a fault, nor which side of the fault has moved. In an ideal case—for example, if a single mineral grain or a single pebble in a conglomerate has been cut through by the fault and the halves have been carried apart a measurable distance—the amount of movement can be determined. Yet even then it is not possible to say whether one block stood still while the other moved past it, or whether both sides moved. In classifying fault movements, therefore, geologists can determine only relative displacements; that is, one side of a fault has moved in a given direction relative to the other side. For example, in Figure 14.5A it is apparent that there has been displacement on a fault, but all we can say about the movement is that the left-hand side moved down relative to the right-hand side. Similarly, in Figure 14.9 all we can say is that the orange trees in the rear moved to the right relative to those in the front.

Hanging Wall and Footwall

Most faults are inclined like the faults in Figure 14.5A. (That is, to use a geological term, they dip.) To describe the inclination, geologists have adopted two old mining terms. From a miner's viewpoint, the rocks above an inclined vein overhang him, while the rocks below the vein are beneath his feet (Fig. 14.12). Because veins occupy openings created by faults, we use the old miner's terms in the following way. The **hanging wall block** is the block of rock above an inclined

fault; the block of rock below an inclined fault is the **footwall block.** These terms, of course, do not apply to vertical faults.

Classification of Faults

Faults are classified according to (1) the dip of the fault and (2) the direction of relative movement. The common classes of faults, together with the changes in local topography they sometimes create, are listed in Figure 14.13. The standard planes of reference in classifying faults are the vertical and the horizontal. Along many faults, movement is entirely vertical or entirely horizontal, but along some faults combined vertical and horizontal movements occur.

Normal Faults

Normal faults are caused by tensional stresses that tend to pull the crust apart, as well as by stresses created by a push from below that tend to stretch the crust. Movement on a normal fault is such that the hanging wall block moves down relative to the footwall block. The faults shown in Figure 14.5A are normal faults.

Commonly, two or more normal faults with parallel strikes but opposite dips enclose an upthrust or down-dropped segment of the crust. As shown in Figure 14.14, a down-dropped block is a **graben, or rift,** if it is bounded by two normal faults and a **half-graben** if subsidence occurs along a single fault. An upthrust block is a **horst.** The central, steep-walled valley that runs down the center of the mid-Atlantic Ridge and cuts through Iceland (Fig. 1.14) is a graben. Perhaps the world's most famous system of grabens and half-grabens is the African Rift Valley seen in Figure I.6, which runs north–south through the countries of East Africa for more than 6000 km. Within parts of the Rift Valley magma has followed channels that lead upward along the fault surfaces, creating volcanoes. Five volcanoes are visible in Figure I.6.

Normal faults are innumerable. Therefore, horsts and grabens are also very common, although none is more spectacular than the African Rift Valley. The north–south valley of the Rio Grande in New Mexico is a graben. The valley in which the Rhine River flows through Western Europe follows a series of grabens. A spectacular example of normal faulting is found in the Basin and Range Province in Utah, Nevada, and Idaho. There, movement on a series of parallel and nearly parallel north–south striking, normal faults has formed horsts and half-grabens. The horsts are now mountain ranges, and the half-grabens are sedimentary basins. As seen in Figure 14.15, the province, which is bounded in the east by the western edge

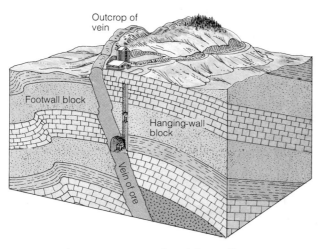

FIGURE 14.12 Hanging wall and footwall.

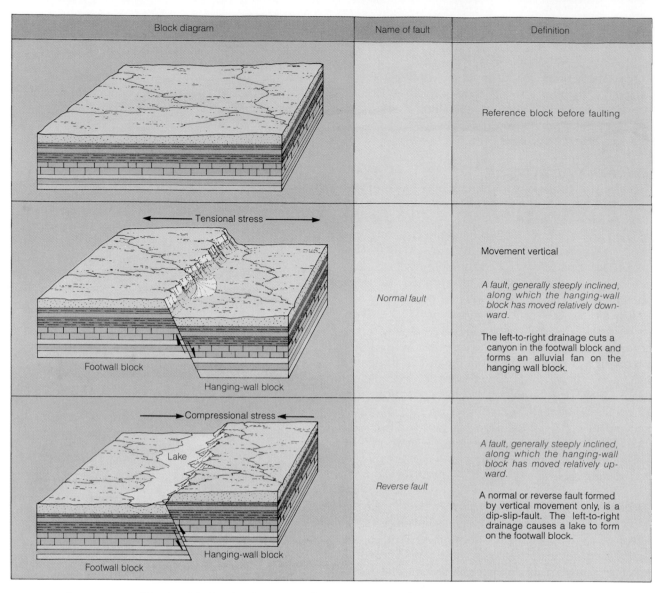

Block diagram	Name of fault	Definition
		Reference block before faulting
Normal fault	Movement vertical *A fault, generally steeply inclined, along which the hanging-wall block has moved relatively downward.* The left-to-right drainage cuts a canyon in the footwall block and forms an alluvial fan on the hanging wall block.
Reverse fault	*A fault, generally steeply inclined, along which the hanging-wall block has moved relatively upward.* A normal or reverse fault formed by vertical movement only, is a dip-slip-fault. The left-to-right drainage causes a lake to form on the footwall block.

FIGURE 14.13 Principal kinds of faults, the directions of maximum stress that cause them, and some of the topographic changes they produce.

of the Wasatch Range and continues westward to the eastern edge of the Sierra Nevada, contains some spectacularly beautiful scenery.

Reverse Faults and Thrust Faults
Reverse faults arise from compressional stresses. Movement on a reverse fault is such that a hanging wall block moves up relative to a footwall block. Reverse fault movement shortens and thickens the crust.

A special class of reverse faults, called **thrust faults,** are low-angle reverse faults with dips less than 15° (Fig. 14.16). Such faults, common in great mountain chains, are noteworthy because along some of them the hanging wall block has moved many kilometers over the footwall block. In most cases the hanging wall block, thousands of meters thick, consists of rocks much older than those adjacent to the thrust on the footwall block (Fig. 14.17).

Strike-slip Faults
Strike-slip faults are those in which the principal movement is horizontal and therefore parallel to the strike of the fault (Fig. 14.13). Strike-slip faults arise from shear stresses. One strike-slip fault is so famous almost everyone has heard of it—the San Andreas Fault. In Figure 14.7 it is strike-slip movement on the San Andreas that is offsetting the rows of orange trees.

Horizontal fault movement is designated as follows; to an observer standing on either fault block, the movement of the other block is *left lateral* if it is to the left and *right lateral* if it is to the right. The sense of relative motion is the same regardless of which block the observer is standing on. The San Andreas is a right-lateral strike-slip fault. Apparently, movement has been occurring along it for at least 65 million years. The total movement is not known, but some evidence suggests that it now amounts to more than 600 km.

392

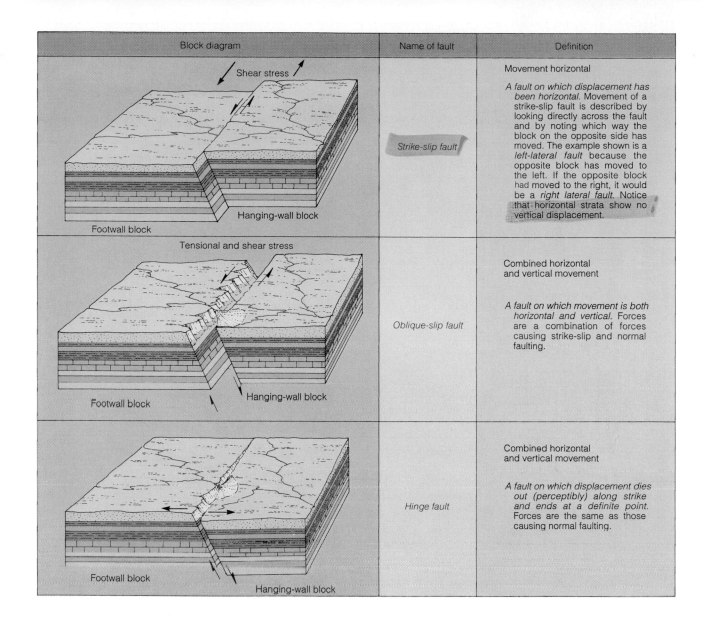

Block diagram	Name of fault	Definition
	Strike-slip fault	**Movement horizontal** *A fault on which displacement has been horizontal.* Movement of a strike-slip fault is described by looking directly across the fault and by noting which way the block on the opposite side has moved. The example shown is a *left-lateral fault* because the opposite block has moved to the left. If the opposite block had moved to the right, it would be a *right lateral fault*. Notice that horizontal strata show no vertical displacement.
	Oblique-slip fault	**Combined horizontal and vertical movement** *A fault on which movement is both horizontal and vertical.* Forces are a combination of forces causing strike-slip and normal faulting.
	Hinge fault	**Combined horizontal and vertical movement** *A fault on which displacement dies out (perceptibly) along strike and ends at a definite point.* Forces are the same as those causing normal faulting.

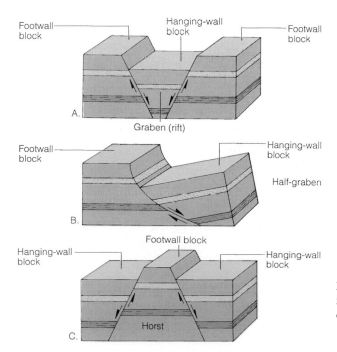

FIGURE 14.14 Horsts and grabens formed when tensional stresses produce normal faults.

393

FIGURE 14.15   Scenery in the northern part of the Basin and Range Province. Looking east toward the Lost River Range, Idaho, from Pioneer Mountain. Valley in the foreground is a graben, the Lost River Range is a horst.

Many of the largest and most active faults are strike-slip faults. This is so because strike-slip faults, spreading edges, and subduction edges are the three kinds of margins that bound tectonic plates (chapter 1). The three plate-margin components link together to form continuous networks encircling the Earth. Where one plate margin terminates, another com- mences; their junction point is called a *transform*. J. T. Wilson, the Canadian scientist who first recog- nized the network relation, proposed that the special class of strike-slip faults that form plate boundaries be called **transform faults.** The three possible configu- rations for transform faults are shown diagrammati- cally in Figure 14.18.

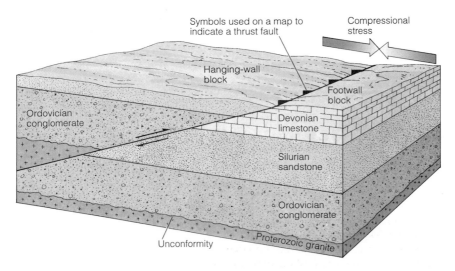

FIGURE 14.16   Example of a thrust fault. Note that thrusting causes older strata in the hanging wall block to lie locally above younger units in the footwall block.

FIGURE 14.17 Keystone Thrust, west of Las Vegas, Nevada. A. Air view northward defines this thrust fault by a color contrast in the strata. B. Section drawn across area shown in the photograph. Light-colored Jurassic sandstone, forming a cliff nearly 600 m high, lies below the fault and forms the footwall block; dark-colored Paleozoic limestone and dolostones lie above the fault and form the hanging wall block.

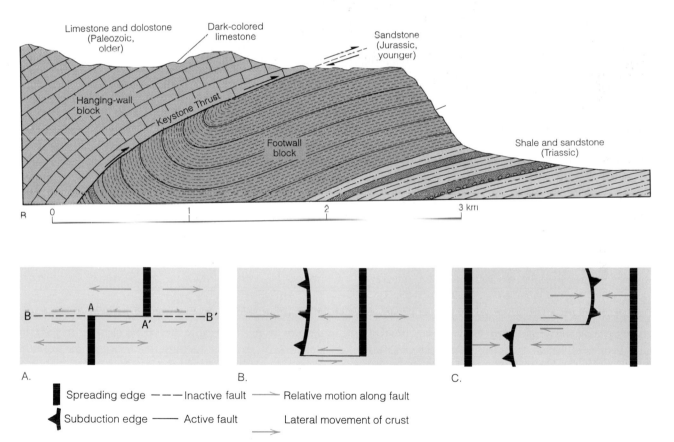

FIGURE 14.18 Transform faults as they appear on a map. A. A spreading edge offset by a transform fault. Crust on both sides of the two spreading edge segments moves laterally away from the edge. Between segments of the edge, along A-A', movement on the two sides of the fault is in opposite directions. Beyond the spreading edge, however, along segments A-B and A'-B', movement on both sides of the projected fault is in the same direction. Therefore, a transform fault does not cause the spreading edge segments to move continuously apart. B. A transform fault joins a spreading edge with a subduction edge. New crust formed at a spreading edge moves laterally away from the edge and plunges back into the mantle at the subduction edge. Note that a subduction edge is a thrust fault. Teeth marks on the thrust fault indicate the hanging wall block. C. Transform fault joining two subduction edges.

A.

B.

FIGURE 14.19 The effects of faulting. A. Slickensides on a fault surface, Dixie Valley, Nevada. The hanging-wall rocks have been removed by erosion, thus exposing the normal fault on which the footwall (left) was raised relative to the hanging wall. B. Fault breccia, Titus Canyon, Death Valley, angular gneiss fragments (dark) broken by faulting set in a matrix of rock flour and calcite.

Evidence of Movement Along Faults

Often we find fractures in rock but cannot tell at first glance whether or not movement has occurred along them—in other words, we don't know whether we are looking at a fault or a joint. For example, in uniform, even-grained rock such as granite, or in a pile of thin-bedded strata no one of which is unique or distinctive, it may not be possible to see the displacement of any obvious features. However, examination of either the fault surface or rock immediately adjacent to it sometimes reveals signs of local deformation, indicating that movement has occurred. Under special circumstances, even the relative direction of movement can be deciphered.

Movement of one mass of rock past another can cause the fault surfaces to be smoothed, striated, and grooved. Striated or highly polished surfaces on hard rocks, abraded by movement along a fault, are called *slickensides*. Parallel grooves and striations on such surfaces record the direction of the most recent movement (Fig. 14.19A).

Not all fault surfaces have slickensides. In many instances fault movement crushes rock adjacent to the fault into a mass of irregular pieces, forming *fault breccia* (Fig. 4.19B). Intense grinding breaks the fragments into such tiny pieces that they may not be individually visible even under a microscope. Some contain such tiny fragments they resemble chert.

Deformation by Bending

Bending may consist of broad, gentle warping that extends over hundreds of kilometers, or it might be close, tight flexing of microscopic size, or anything in between. Regardless of the volume of rock involved or the degree of warping, the bending of rocks is referred to as *folding*. Folding is most easily recognized in layered rocks. An individual bend or warp in layered rock is call a **fold.**

Regardless of their size, folds are formed by ductile deformation as a result of compressional and shear stresses. At very low strain rates, even near-surface rocks that are well above the brittle-ductile transition can be folded. That is apparently what is happening in Greece today. However, the very intense folding that we can observe so widely in mountain ranges probably occurred below the brittle-ductile transition when the rocks were deeply buried and thus were subjected to high temperature and high confining stress.

Types of Folds

The simplest fold is a **monocline,** a local steepening in an otherwise uniformly dipping pile of strata (Fig. 14.20A and B). An easy way to visualize a monocline is to lay a book on a table and then drape a handkerchief over one side of the book and out onto the table. So draped, the handkerchief forms a monocline. Most

A.

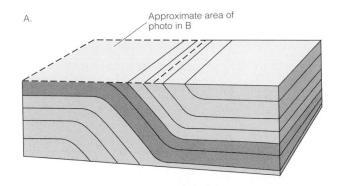

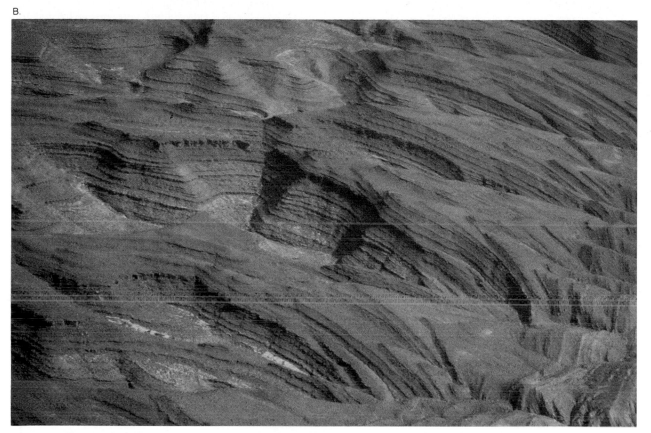

B.

FIGURE 14.20 Monoclines. A. Sketch of a monocline. Note that the two limbs are parallel. B. A monocline in southern Utah that interrupts the generally flat-lying sedimentary strata of the wide Colorado Plateau. In the area of maximum bending, the strata are nearly vertical (right-hand side of photo). The view is looking south.

folds are more complicated than monoclines (Fig. 14.21). An upfold in the form of an arch is an **anticline.** A downfold with a troughlike form is a **syncline.** Anticlines and synclines are usually paired.

Geometry of Folds

As shown in Figure 14.21A, the sides of a fold are the **limbs,** and the median line between the limbs, along the crest of an anticline or the trough of a syncline, is the **axis** of the fold. A fold with an inclined axis is said to be a *plunging fold,* and the angle between a fold axis and the horizontal is the **plunge** of a fold (Fig. 14.21B). An imaginary plane that divides a fold as symmetrically as possible, and that passes through the axis, is the *axial plane.*

Many folds, such as those in Figure 14.21, are nearly symmetrical. Others, however, are not symmetrical; intense stress can create complex shapes. The common forms of folds are shown in Figure 14.22.

An *open fold,* such as that depicted in Figure 14.22A, is one in which the two limbs dip gently and equally away from the axis. The more intense the compressional stress, the less open a fold will be. When stress is very intense, the fold closes up and the limbs become parallel to each other; such a fold is said to be isoclinal (Fig. 14.22B). Intense stress is also the reason why a fold becomes either asymmetric (Fig. 14.22C) in which limbs have unequal dips, or overturned in which limbs dip in the same direction (Figure 14.22D). Eventually, an overturned fold may

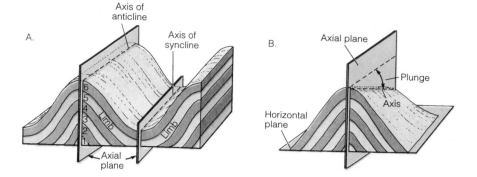

FIGURE 14.21 Features of simple folds. Note that the strata dip away from the axis of an anticline but toward the axis of a syncline. A. Fold axis horizontal. B. Fold axis plunging.

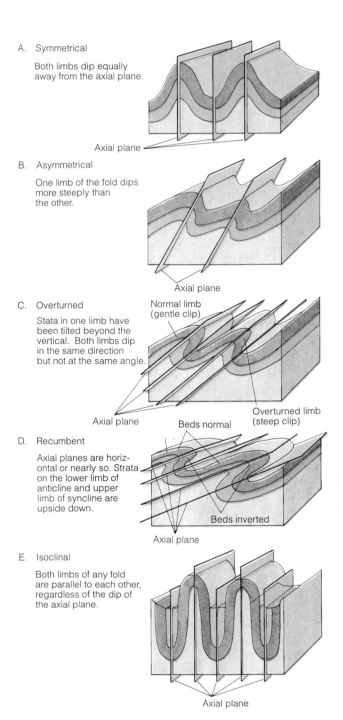

A. Symmetrical

Both limbs dip equally away from the axial plane.

Axial plane

B. Asymmetrical

One limb of the fold dips more steeply than the other.

Axial plane

C. Overturned

Stata in one limb have been tilted beyond the vertical. Both limbs dip in the same direction but not at the same angle.

Axial plane

Normal limb (gentle clip)

Overturned limb (steep clip)

D. Recumbent

Axial planes are horizontal or nearly so. Strata on the lower limb of anticline and upper limb of syncline are upside down.

Beds normal

Beds inverted

Axial plane

E. Isoclinal

Both limbs of any fold are parallel to each other, regardless of the dip of the axial plane.

Axial plane

FIGURE 14.22 Five kinds of folds.

become recumbent, meaning the two limbs are horizontal or nearly so (Fig. 14.22E). Recumbent folds are common in mountainous regions, such as the Alps and the Himalaya, which were produced by continental collisions.

If only fragmentary exposures of bedrock are available, it is apparent that difficulties might arise in deciding whether or not a given fold is overturned. As is apparent in Figures 14.22D and E, it is necessary to know whether a stratum is right-side up or upside down in order to decide which limb of a fold it is in. Figuring this out is not always possible, but in some cases sedimentary structures, such as mud cracks and graded layers, do record whether strata are in their original orientation or inverted (Fig. 6.3). In other examples, only careful, thorough mapping of all bedrock exposures can provide the answer.

Relationship Between Folds and Faults

Folds and faults do not continue forever. Faults tend to die out as folds, as shown in Figure 14.23. Folds die out by becoming smaller and smaller wrinkles, in much the same way as wrinkles in a bedsheet die out.

When two kinds of rock are subjected to the same stresses, one kind may be brittle and deform by fracture, and the other by ductile deformation. Certain monoclines form as a result of such differences. Monoclines commonly result from movement on a fault that causes flat-lying, ductile strata to bend as shown in Figure 14.24.

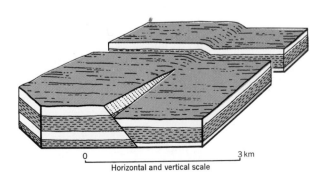

0   3 km
Horizontal and vertical scale

FIGURE 14.23 Normal fault dies out as it passes laterally into a monocline.

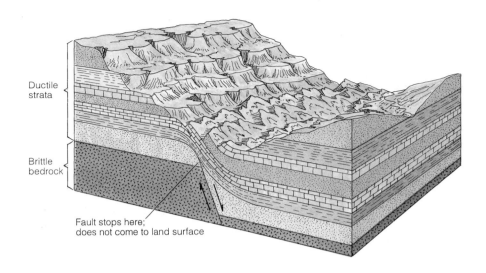

Ductile strata

Brittle bedrock

Fault stops here; does not come to land surface

FIGURE 14.24 Monocline developed in soft, ductile strata due to movement on a fault in the underlying bedrock.

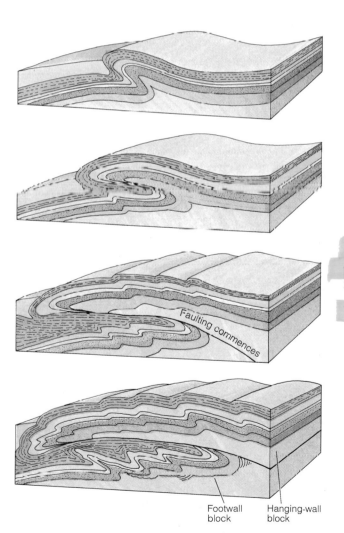

Faulting commences

Footwall block   Hanging-wall block

FIGURE 14.25 The evolution of a recumbent fold into a thrust fault. This kind of structure is important in the Alps and many other mountain ranges.

Some of the great thrust faults in the Alps probably started as recumbent folds. As depicted in Figure 14.25, when stress continues to build up and continues to deform a recumbent fold, the overturned limb may become so stretched and strained that it eventually breaks and becomes a thrust fault. Movement on some of the great recumbent-fold and thrust-fault structures in the Alps is in excess of 50 km.

Folds and Topography

Differences in erodibility of adjacent strata can lead to distinctive topographic forms, and commonly these topographic forms reveal the presence of folds. It is important to remember, however, that anticlines do not necessarily make ridges nor synclines valleys, as Figure 14.26 shows. A particularly striking example of the relationship between plunging folds, erosion, and topography can be seen in the Valley and Ridge province of Pennsylvania (Fig. 14.27). There, a series of plunging anticlines and synclines were created during the Paleozoic Era by a continental collision between North America, Africa, and Europe. The Appalachian mountain chain was the result. Now deeply eroded, the folded rocks determine the pattern of the topography because soft, easily eroded strata (limestones, dolostones, shales) underlie the valleys, while resistant strata (sandstones) form the ridges.

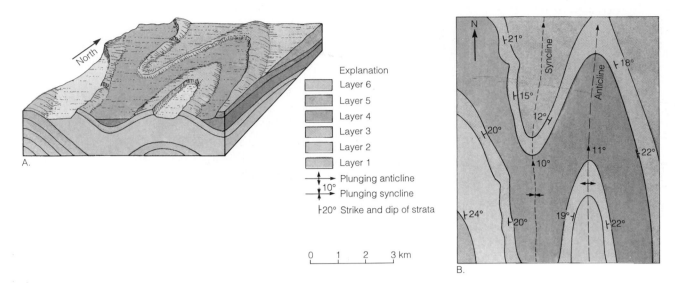

FIGURE 14.26 Distinctive topographic forms and patterns resulting from differing resistance to erosion of different kinds of rock reveal the presence of plunging folds. A. Block diagram showing topographic effects. Note that resistant strata (layers 3 and 5) make topographic highs in both anticlines and synclines, while easily eroded strata (layers 2 and 4) make topographic lows in both anticlines and synclines. B. Geologic map of area shown in Part A.

FIGURE 14.27 Valley and ridge province of the central Appalachians in Pennsylvania. Paleozoic-age strata have been folded into a series of plunging anticlines and synclines. Valleys are developed by erosion of the weakest strata—limestone, dolostone, and shale. The ridges are sandstones. The river flowing across the middle of the photo is the Susquehanna.

ROCK DEFORMATION AND OIL POOLS

Most of the world's largest oil pools are where they are because of the way rocks deform. "Oil pool" is a somewhat confusing term because it is not a pool in the usual sense of a "lake"; the term actually refers to a body of rock in which oil occupies all the pore spaces.

Oil and natural gas occur together, and for a pool to form, five essential requirements must be met:

1. A *source rock* rich in organic matter, such as shale, must provide the oil. Most shales

contain some organic matter from dead plants and animals and can serve as source rocks. Conversion of organic matter to oil and gas happens spontaneously when shale is buried, and, in response to the geothermal gradient, the temperature of the source rock rises.

2. A permeable and porous *reservoir rock* must be present so that the oil can percolate in from the source rock.

3. Oil floats on water, and so oil in a reservoir

Structural traps are far more common than stratigraphic traps.

5. The final, and in some ways most important, requirement for pool formation is that the deformation that forms a trap must occur before all the oil and gas has escaped from the reservoir rock. This is important because conversion of organic matter to oil happens soon after a shale is buried. If the trap were formed long after oil formation, no oil pool would form.

Structural traps

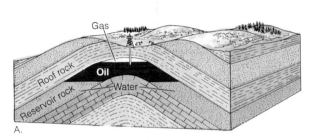

A.

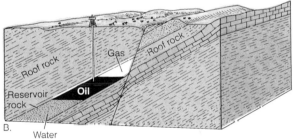

B.

Stratigraphic traps

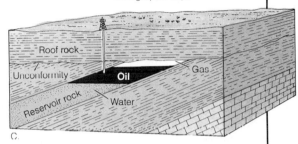

C.

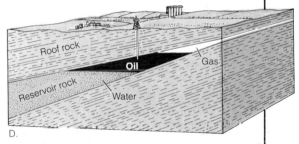

D.

FIGURE B14.1 Four kinds of oil traps. A, B, structural traps; C, D, stratigraphic traps. Folds (part A) are traps formed by ductile deformation. They are the most important of all oil traps. Faults (part B) are traps formed by fracturing of brittle rocks. In C an unconformity marks the top of the reservoir; in D a porous stratum (reservoir) thins out and is overlain by an impermeable roof rock. Gas overlies oil, which floats on groundwater, saturates the reservoir rock, and is held down by a roof of shale. Oil fills only the pore spaces in the rock.

rock will float upward to the water table unless the reservoir rock is covered by an impermeable *roof rock*.

4. Like groundwater in an aquifer, oil will move laterally in a reservoir rock and eventually escape. An important requirement for oil-pool formation is that the reservoir and roof rocks must form a trap that can hold the oil and prevent it from being flushed out by groundwater. As shown in Figure B14.1, traps are either structural—either anticlines or faults—or stratigraphic—either nonconformities or changes in sedimentary facies.

The great oil pools of Saudi Arabia, Kuwait, Iraq, and other parts of the Middle East are in structural traps. The deformation that produced the traps was caused by plate tectonic movements. At the right moment for oil trapping, the sedimentary rocks in the Middle East were deformed by compressional stresses due to Africa banging into Asia.

Iraq invaded Kuwait in 1990 in order to control the gigantic Kuwaiti oil pools; the United Nations went to war in 1991 to expel the Iraqi invaders. How sad it is that we humans find it necessary to kill each other because of tectonic events that happened millions of years ago.

SUMMARY

1. Rocks can be deformed in three ways: by elastic deformation (no permanent change), by ductile deformation (folds), and by fracture (faults, joints).

2. High confining stress and high temperatures enhance ductile properties. Low temperatures and low confining stress enhance elastic properties and failure by fracture when the elastic limit is exceeded.

3. The rate at which a solid is deformed (strained) also controls the style of deformation. High strain rates lead to fractures; low strain rates cause folding.

4. Weak rocks (limestone, marble, slate, phyllite, and schist) enhance ductile properties. Strong rocks (sandstone, quartzite, and granite) enhance brittle properties. Dry rocks are stronger than wet rocks.

5. The orientations of contacts between strata, faults, joints, or any other inclined plane surfaces are determined by the strike (direction of intersection of the inclined plane and a horizontal plane), and the dip (angle between the inclined plane and the horizontal).

6. Fractures in rocks along which slippage occurs are called faults. Normal faults are caused by tensional stresses that tend to pull the crust apart, whereas thrust and other types of reverse faults arise from compressional stresses that squeeze, shorten, and thicken the crust. Strike-slip faults are caused by shear stresses; they are vertical fractures that have horizontal motion.

7. It is usually only possible to determine relative motions of rocks on either side of a fault.

8. Ductile deformation of strata causes bends or warps, which are called folds. Folding is due to compressional stress. An upward arched fold is an anticline; a downward, troughlike fold is a syncline. The sides of a fold are called limbs.

9. Overturned folds (meaning both limbs dip in the same direction) are common in mountain ranges formed by continental collision. In some ranges, such as the Alps, overturned folds have nearly horizontal limbs, in which case the folds are said to be recumbent.

10. Faults die out by becoming folds, and folds die out by becoming smaller and smaller wrinkles.

IMPORTANT TERMS TO REMEMBER

anticline (p. 397)
axis (of a fold) (p. 397)

compressional stress (p. 384)
confining stress (p. 384)

differential stress (p. 384)
dip (p. 390)
ductile deformation (p. 384)

elastic deformation (p. 384)
elastic limit (p. 384)

fault (p. 387)
fold (p. 396)
footwall block (of a fault) (p. 391)
fracture (p. 384)

graben (p. 391)

half-graben (p. 391)
hanging wall block (of a fault) (p. 391)
horst (p. 391)

limb (of a fold) (p. 397)

monocline (p. 396)

normal fault (p. 391)

plunge (of a fold) (p. 397)

reverse fault (p. 392)
rift (p. 391)

shear stress (p. 384)
strain (p. 384)
strain rate (p. 387)
stress (p. 384)
strike (p. 389)
strike-slip fault (p. 392)
syncline (p. 397)

tensional stress (p. 384)
thrust fault (p. 392)
transform fault (p. 394)

uniform stress (p. 384)

QUESTIONS FOR REVIEW

1. What are stress and strain, and what is the relationship between them?

2. Discuss the ways a seemingly rigid solid, such as a rock, can be deformed. In what sequence do deformation properties come into play?

3. What properties determine whether a rock is brittle and fails by fracture or is deformed by ductile deformation?

4. Name three rock types that readily deform by ductile deformation and three that tend to deform by fracture.

5. Identify three prominent topographic features on the Earth made up of a system of grabens and horsts. What kind of stresses cause horsts and grabens to form?

6. Describe the way a transform fault works. Why are they called *transform* faults? Cite an example of a transform fault that is still active.

7. What kind of stresses produce folds? Draw a sketch of an anticline and mark the axis, axial plane, and limbs.

8. How do folds and faults die out?

9. What is a recumbent fold and where would you expect to find large recumbent folds? Describe the way a recumbent fold can become a thrust fault.

10. Draw a geologic map of a plunging anticline. Mark strike and direction of dip at several places around the fold. Draw the fold axis and indicate, if you can, the direction of plunge.

11. Look closely at Figure 14.27 and identify a fold that is a plunging anticline and one that is a plunging syncline. In a sequence of folds, must anticlines and synclines alternate?

More than 3,000 people were killed and over 250,000 made homeless by the magnitude 7.0 earthquake that ravaged southern Italy on November 23, 1980. Among the historic towns devastated by the quake was Teora, an ancient hill town 85 km east of Naples. Severe ground shaking caused old stone buildings to collapse on the unfortunate inhabitants.

CHAPTER 15

Earthquakes and the Earth's Interior

There is an old saying among geologists and engineers that earthquakes don't kill, but buildings do. Shaking ground may make people fall down, but falls don't kill. However, shaking ground can make buildings collapse, and collapsing buildings can definitely kill.

The worst earthquake disaster of the twentieth century happened on July 28, 1976. At 3:42 A.M., when the 1 million inhabitants of T'ang Shan, China, were asleep, a 7.8 magnitude quake leveled the city. Hardly a building was left standing, and the few that did withstand the first quake were destroyed by a second quake, magnitude 7.1, which struck at 6:45 P.M. the same day. When the wreckage of T'ang Shan was cleared, 240,000 were dead. Losses were large because most of the buildings had not been constructed to withstand earthquakes. The buildings had unreinforced brick walls. When the ground started to shake the walls collapsed, the roofs caved in, and the sleeping inhabitants were crushed.

Earthquakes are due to sudden releases of stored elastic energy in the lithosphere. One of the great challenges to scientists is to understand the dynamics of the lithosphere sufficiently well so that earthquakes can be predicted and human safety assured. Chinese scientists have been more successful than most in predicting quakes, but the T'ang Shan quake gave no recognizable warning signs and was completely unexpected. We still have a long way to go.

Not everything about earthquakes is bad. They can be used to study the Earth's interior, for example. The way the Earth vibrates after a large earthquake is controlled in large degree by the properties of the mantle and core. Through measurements of earthquake vibrations, we can find out about the core and mantle. Used in this way, earthquake vibrations are like the X rays a doctor uses to study the inside of a human body—they are the probes we use to sense and measure the world beneath our feet.

In this chapter we discuss how earthquakes occur, how they are measured, the damage they can do, how we use them to study plate tectonics, and how they help us study the unseen parts of the Earth's interior.

EARTHQUAKES

When the Earth quakes, it is as if our world has been struck by a huge hammer. The reason for the quaking is that energy stored in elastically-strained rocks is suddenly released. The more energy released, the stronger the quake. Just how the elastically stored energy is built up and how it is released continue to be subjects for intensive research.

Do an experiment yourself. Have a friend hit one end of a wooden plank or the top of a wooden table with her or his fist or a hammer while you press your hand on the other end. You will feel vibrations set up in the plank or tabletop by the energy of the blow. The harder the blow, the stronger the vibrations. The reason you can feel those vibrations is that some of the energy imparted by the hammer or fist is transferred to your hand by elastic vibrations through the whole length of the solid wood. Fortunately, fists and hammers don't hit the Earth, but a bomb blast or a violent volcanic explosion will serve as an energy source just as well. So, too, will the sudden slipping of rock masses along fault surfaces.

The most widely accepted theory concerning the origin of earthquakes involves slipping faults and the *elastic rebound theory*.

Origin of Earthquakes

Sudden movement along faults seems to be the cause of most earthquakes. But it cannot be that simple. Some earthquakes are millions of times stronger than others. Why? The same energy that in one case is released by thousands of tiny slips and tiny earthquakes is in another case stored and released in a single immense earthquake. The **elastic rebound theory** suggests that if fault surfaces do not slip easily past one another, energy can be stored in elastically deformed bodies of rock, just as in a steel spring that is compressed. Then when the fault finally does slip, the elastically strained bodies of rock rebound to their original shapes. The first evidence supporting the elastic rebound theory came from studies of the San Andreas Fault, which, as discussed in chapter 14, is a right-lateral strike-slip fault. During long-term field observations in central California, beginning in 1874, scientists from the U.S. Coast and Geodetic Survey determined the precise position of many points both adjacent to and distant from the fault (Fig. 15.1). As time passed, movement of the points revealed that the crust was slowly being bent. For some reason, in the area of measurement near San Francisco, the fault was locked and did not slip. On April 18, 1906, the two sides of

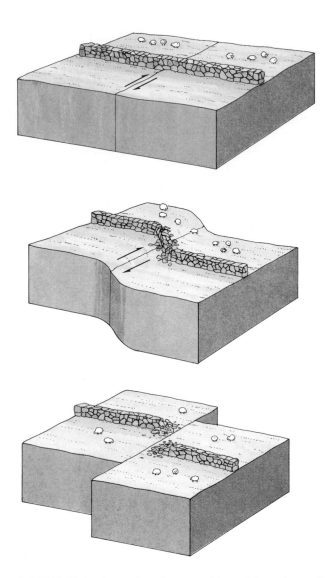

FIGURE 15.1   An earthquake, caused by sudden release of energy. Sketch based on detailed surveys near the San Andreas Fault, California, before and after the abrupt movement that caused the earthquake of 1906. A. Fence crosses the fault and is slowly bent as rock is elastically strained. After the earthquake, two segments of the fence are offset 7 m.

this locked fault shifted abruptly. The elastically stored energy was released as the fault moved and the bent crust snapped back, thereby creating a violent earthquake. Repetition of the survey then revealed that the bending had disappeared.

Most earthquakes occur in the brittle rock of the lithosphere. As discussed in chapter 14, brittleness is the tendency for a solid to fracture when the deforming stress exceeds the limits of elasticity. At great depth, temperatures and pressures are so high that ductile deformation occurs. Rocks can neither fracture nor store elastic energy under such conditions. Rather, they are like putty and undergo permanent changes

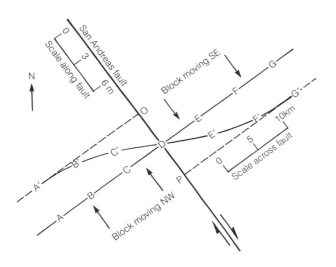

B.

FIGURE 15.1 *(continued)* B. Results of the survey. The seven survey points, *A* to *G*, were originally aligned. Slowly, the line was bent and displaced to the curved line *A'* to *G'*. Suddenly, the frictional lock was broken and the rocks on either side of the fault rebounded. The surveyed points lay along the lines *A'O* and *PG'*.

of shape that remain even after the deforming forces have been removed. Earthquakes, then, are phenomena of the brittle, outer, cooler portion of the Earth.

How Earthquakes Are Studied

The name given to the study of earthquakes is **seismology,** a word that comes directly from the ancient Greek term for earthquakes, *seismos.*

Seismographs

The device used to record the shocks and vibrations caused by earthquakes is a **seismograph.** An ideal way to record the vibrations and motions of the earth would be to put the seismograph on a stable platform that is not affected by the vibrations. But a seismograph must stand on the Earth's vibrating surface, and it will therefore vibrate with the surface. This makes the act of measurement difficult because there is no fixed frame of reference against which to make measurements. The problem is the same one that a sailor in a small boat faces when attempting to measure waves at sea.

To overcome the frame of reference problem, most seismographs make use of **inertia,** which is the resistance of a large stationary mass to sudden movement. If you suspend a heavy mass, such as a block of iron, from a light spring and then suddenly lift the spring, you will notice that because of inertia the block of iron remains almost stationary while the spring stretches (Fig. 15.2). This is the principle used in *inertial seismographs* (Fig. 15.3). Vertical motion is measured by the device shown in Figure 15.3A in which a heavy mass is supported by a spring and the spring is connected to a support which in turn is connected to the ground. When the ground vibrates, the spring expands and contracts but the mass stays almost stationary. Then the distance between the ground and the mass can be used to sense vertical displacement of the ground surface.

Horizontal displacement can be similarly measured by suspending a heavy mass from a string to make a pendulum (Fig 15.3B). Because of its inertia, the mass does not keep up with the horizontal ground motion, and the difference between the pendulum and ground

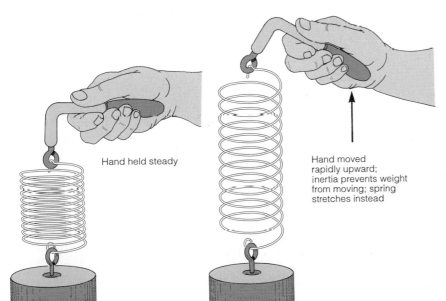

Hand held steady

Hand moved rapidly upward; inertia prevents weight from moving; spring stretches instead

FIGURE 15.2 The principle of the inertial seismograph.

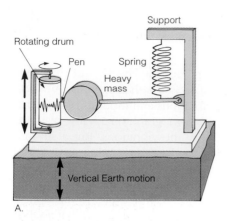

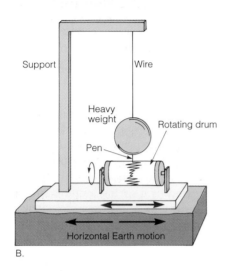

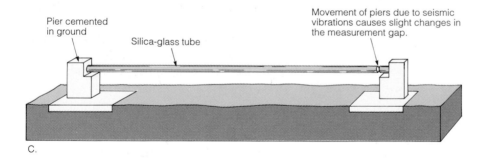

movement records the horizontal ground motion. Inertial seismographs are commonly used in groups so that motions can be measured simultaneously up-down, east–west, and north–south.

Another kind of device, called the *Benioff strain seismograph,* employs two concrete piers in the ground spaced at a distance of about 35 m (Fig. 15.3C). When the Earth vibrates, the two piers move independently of each other. Attached to one pier is a long, rigid, silica-glass tube. The other pier carries a very sensitive detector to measure even the slightest move-

ment in the end of the silica-glass tube. Strain seismographs are frequently installed in mines, tunnels, and other places where temperature is relatively constant and wind disturbance minimal.

Modern seismographs are incredibly sensitive. Vibrational movements as tiny as one hundred millionth (10^{-8}) of a centimeter can be detected. Indeed, many instruments are so sensitive they can detect ground depression caused by a moving automobile several blocks away, and even ocean waves and tides several kilometers from the seashore.

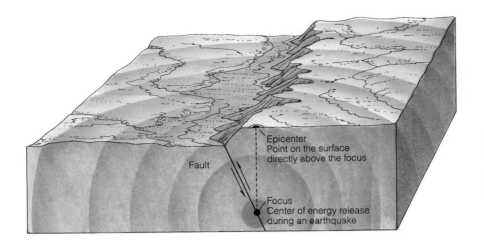

FIGURE 15.4 The focus of an earthquake is the site of first movement on a fault and the center of energy release. The epicenter of an earthquake is the point on the Earth's surface that lies vertically above the focus.

Earthquake Focus and Epicenter

The place where energy is first released to cause an earthquake is called the **earthquake focus.** In reality, because most earthquakes are caused by movement on a fault, the focus is not a point but rather a region that may extend for several kilometers.

An earthquake focus lies at some depth below the Earth's surface. It is more convenient to identify the site of an earthquake from the **epicenter,** which is the point on the Earth's surface that lies vertically above the focus of an earthquake (Fig. 15.4). A good way to describe the location of an earthquake focus is to state the location of its epicenter and its depth.

Seismic Waves

When an earthquake occurs, the elastically stored energy is transmitted from the focus to other parts of the Earth. As with any vibrating body, waves (vibrations) spread outward from the focus. The waves, called **seismic waves,** spread out in all directions, just as sound waves spread in all directions when a gun is fired.

Seismic waves are elastic disturbances, and so unless the elastic limit is exceeded, the rocks through which they pass return to their original shapes after passage of the waves. Seismic waves must therefore be measured and recorded while the rock is still vibrating. For this reason, many continuously recording seismograph stations have been installed around the world.

There are several kinds of seismic waves, and they belong to two families. **Body waves,** of which there are two kinds, corresponding to the two ways a solid can be elastically deformed, travel outward from the focus and have the capacity to travel entirely through the Earth. **Surface waves,** on the other hand, are guided by and restricted to the Earth's surface. Body waves are analogous to either light or sound waves, both of which travel outward in all directions from their point of origin. Surface waves are analogous to ocean waves and are restricted to the vicinity of a free surface, such as the Earth's surface both where it meets the atmosphere and where it meets the ocean.

Body Waves

Rocks can be elastically deformed by either a change of volume or a change of shape. The first kind of body wave, *compressional waves,* deforms rocks by change of volume, and consists of alternating pulses of compression and expansion acting in the direction of wave travel (Fig. 15.5A). Sound waves are also compressional waves. When a sound wave passes through the air, it does so by alternating compressions and expansions of the air. Compression and expansion produce changes in the volume and density of a medium. Compressional waves can pass through solids, liquids, or gases because all three can sustain changes in density. When a compressional wave passes through a medium, the compression pushes atoms closer together. Expansion, on the other hand, is an elastic response to compression, and it causes the distance between atoms to be increased. Movement in a solid subjected to compressional waves is back and forth in the line of the wave motion. Compressional waves have the greatest velocity of all seismic waves—6 km/s is a typical value for the uppermost portion of the crust—and they are the first waves to be recorded by a seismograph after an earthquake. They are therefore called **P (for primary) waves.**

The second kind of body waves are *shear waves.* They deform materials by change of shape. Because gases and liquids do not have the elasticity to rebound to their original shape, shear waves can be transmitted only by solids. Shear waves consist of an alternating

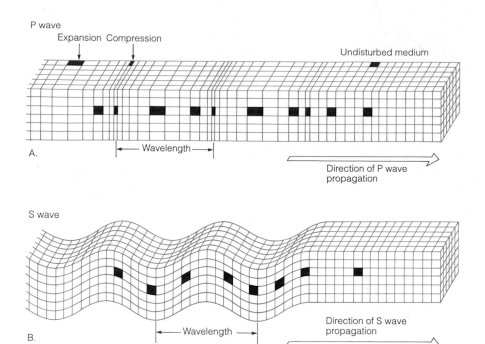

FIGURE 15.5   Seismic body waves of the P (compressional) and S (shear) types. A. P waves cause alternate compressions and expansions in the rock the wave is passing through. An individual point in a rock will move back and forth parallel to the direction of P-wave propagation. As wave after wave passes through, a square will repeatedly expand to a rectangle, return to a square, contract to a rectangle, return to a square, and so on. B. S waves cause a shearing motion. An individual point in a rock will move up and down, perpendicular to the direction of S-wave propagation. A square will repeatedly change to a parallelogram, then back to a square again.

series of sidewise movements, each particle in the deformed solid being displaced perpendicular to the direction of wave travel (Fig. 15.5B). A typical velocity for a shear wave in the upper crust is 3.5 km/s. Because shear waves are slower than P waves and reach a seismograph some time after a P wave arrives, they are called **S (for secondary) waves.**

Seismic body waves behave like light waves and sound waves, which is to say they can be both transmitted through a medium and also *reflected* and *refracted.* Reflection is the familiar phenomenon of light bouncing off the surface of a mirror or a glass of water. Seismic body waves are reflected by numerous surfaces in the Earth. Refraction is a less familiar phenomenon. It occurs whenever a wave velocity changes, and the effect is to cause a bending in the direction the wave is moving. Velocity change and refraction can be either gradual or abrupt. An abrupt change is seen when a ray of light strikes a surface of water as shown in Figure 15.6. Some of the light is reflected, but some also crosses the surface and travels through the water. The velocity of light is different in

water than in air, and the wave path is sharply bent at the surface, making it appear as if the pencil is bent. The ray is said to have been refracted.

Seismic body wave velocities are a function of the density of the medium they are passing through. If the Earth had a homogeneous composition and if density increased steadily with depth as a result of increasing pressure, refraction would cause seismic wave paths to be curved as shown in Figure 15.7A. Measurements show that wave paths are indeed curved, owing to gradual refraction, but measurements reveal, too, that the seismic waves are also refracted and reflected by several zones of sudden density change, such as the core-mantle boundary (Fig. 15.7B).

Surface Waves

To an observer, surface waves appear very similar to P and S waves. However, surface waves travel more slowly than P and S waves, and in addition they pass around the Earth rather than through it. Thus, surface waves are the last to be detected by a seismograph (Fig. 15.8A).

FIGURE 15.6 Refraction of light causes a drinking straw to seem to be bent at the place where it enters the water.

One important difference between body waves and surface waves concerns a property called dispersion. Body waves travel through the Earth at the same velocity regardless of wavelength, but this is not true for surface waves. The longer the wavelength of a surface wave, the deeper the wave motion penetrates the Earth. Because the Earth is compositionally layered, getting denser and denser as depth increases, long wavelength surface waves penetrate to deeper (denser), higher velocity zones, while short wavelength surface waves do not. In this way, surface waves of different wavelengths develop different velocities, and that property is called *dispersion*. Because some wavelengths reach a depth of several hundred kilometers, dispersion of surface waves is a very important way to determine the properties of the outer few hundred kilometers of the Earth. We will return to the importance of dispersion when we discuss the properties of the mantle later in this chapter.

Location of the Epicenter

The location of an earthquake's epicenter can be determined from the arrival times of the P and S waves

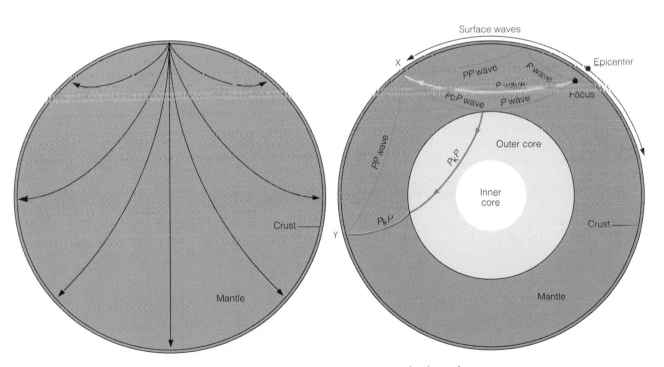

FIGURE 15.7 Refraction and reflection of seismic body waves. A. Wave paths through a planet of homogeneous composition in which density increases with depth due to increasing pressure. Changing density causes wave paths to bend due to refraction. B. Various paths of P waves in the Earth, which is a compositionally layered planet. Seismographs at some places (locations *X* and *Y*, for example) receive both direct P waves (green) as well as reflected (red) and refracted (blue) P waves. A P wave reflected off the Earth's surface is called a PP wave; one reflected off the core-mantle boundary is a PcP wave; one refracted through the liquid outer core is a PkP wave.

411

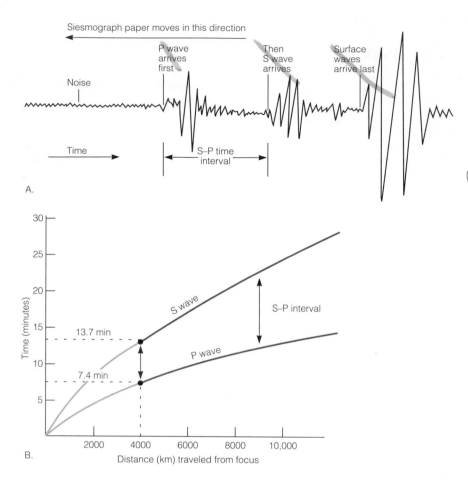

Siesmograph paper moves in this direction

Noise

P wave arrives first

Then S wave arrives

Surface waves arrive last

Time

S–P time interval

A.

S wave

S–P interval

13.7 min

P wave

7.4 min

Time (minutes)

Distance (km) traveled from focus

B.

FIGURE 15.8 Different travel times of P, S, and surface waves. A. Typical record made by a seismograph. The P and S waves leave the earthquake focus at the same instant. The fast moving P waves reach the seismograph first, and some time later the slower moving S waves arrive. The delay in arrival times is proportional to the distance traveled by the waves. The surface waves travel more slowly than either P or S waves. B. Average travel-time curves for P and S waves in the Earth used to locate an epicenter. For example, when seismologists at a station measure the S-P time interval to be 13.7 min − 7.4 min = 6.3 min, they know the epicenter is 4000 km away from their station.

at a seismograph. The farther a seismograph is away from an epicenter, the greater the time difference between the arrival of the P and S waves (Fig. 15.8). After using a graph like that shown in Figure 15.8B to determine how far an epicenter lies from a seismograph, the seismologist draws a circle on a map around the station with a radius equal to the calculated distance to the epicenter. The exact position of the epicenter can be determined when data from the three or more seismographs are available—the center lies where the circles intersect (Fig. 15.9).

The depth of an earthquake focus below an epicenter can also be determined. If a local earthquake is recorded by several nearby seismographs, the focal depth can be determined in the same way that the epicenter is determined, by using the P–S time intervals. For distant earthquakes a different method is employed. Note in Figure 15.7 that a seismograph stationed at position X would record both a direct P wave and a P wave reflected from the Earth's surface (labeled *PP*). The direct P wave, having a shorter path length, would arrive before the *PP* wave. The travel-time difference between them is a measure of the focal depth of the earthquake.

Magnitudes of Earthquakes

Very large earthquakes (of the kind that destroyed San Francisco in 1906, T'ang Shan, China, in 1976, parts of Mexico City in 1985, and the Loma Prieta quake of 1989 that again destroyed parts of San Francisco) are, fortunately, relatively infrequent. In earthquake-prone regions, such as San Francisco and the surrounding area, very large earthquakes occur about once a century. They occur more frequently in some areas and less frequently in others, but a century is an approximate average. This means that the time needed to build up elastic energy to a point where the frictional locking of a fault is overcome is about 100 years. Small earthquakes may occur along a fault during this time as a result of local slippage, but even so, elastic energy is accumulating because most of the fault remains locked. When the lock is broken and an earthquake occurs, the elastic energy is released during a few terrible minutes. By careful measurement of elastically strained rocks along the San Andreas Fault, seismologists have found that about 100 J of elastic energy can be accumulated in 1m^3 of deformed rock. This is not very much—it is equivalent to only about 25 calories

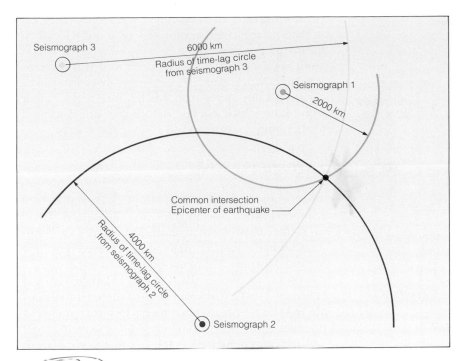

FIGURE 15.9 Locating an epicenter. The effects of an earthquake are felt at three different seismograph stations. The time differences between the first arrival of the P and S waves depend on the distance of a station from the epicenter. The following distances are calculated by using the curves in Figure 15.8B.

	Time Difference	**Calculated Distance**
Seismograph 1	8.8 min − 4.7 min = 4.1 min	2000 km
Seismograph 2	13.7 min − 7.4 min = 6.3 min	4000 km
Seismograph 3	17.5 min − 9.8 min = 7.7 min	6000 km

On a map, a circle of appropriate radius is drawn around each station. The epicenter is where the three circles intersect.

of heat energy—but when billions or trillions of cubic meters of rock are strained, the total amount of stored energy can be enormous. The amount of elastically stored energy released during the Loma Prieta earthquake was about 10^{15}J, and the 1906 San Francisco earthquake released at least 10^{17}J. The energy released by a hydrogen bomb blast is also about 10^{17}J!

Richter Magnitude Scale

Measurements of elastically deformed rocks before an earthquake, and of undeformed rocks after an earthquake, can provide an accurate measure of the amount of energy released. The task is very time consuming, and all too frequently the pre-earthquake measurements are simply not available. Therefore, seismologists have developed a way to estimate the energy released by measuring the amplitudes of seismic waves. The **Richter magnitude scale,** named after the seismologist who developed it, is defined by the maximum amplitudes of the P and S waves (that is, the heights of the waves on a seismogram) 100 km from an epicenter. Because wave signals vary in strength by factors of a hundred million or more, the Richter scale is logarithmic, which means it is divided into steps call magnitudes, starting with magnitude 1 and increasing upward. Each unit increase in magnitude corresponds to a tenfold increase in the amplitude of the wave signal. Thus, a magnitude 2 signal has an amplitude that is ten times larger than a magnitude 1 signal, and a magnitude 3 is a hundred times larger.

We can see from Figure 15.10 how a Richter magnitude is calculated. The energy of a seismic wave is a function of both its amplitude and the duration of a single wave oscillation, T. Divide the maximum amplitude, X, measured in steps of 10^{-4} cm on a suitably adjusted seismograph, by T, measured in seconds. Then add a correction factor, Y, determined from the S-P wave interval. X/T is a measure of the maximum

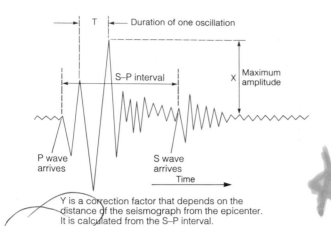

Y is a correction factor that depends on the distance of the seismograph from the epicenter. It is calculated from the S–P interval.

FIGURE 15.10   Measurements used for determining Richter magnitude (M) from a seismograph record.

energy reaching the seismograph. The formula is:

$$\text{Richter Magnitude, } M = \log X/T + Y$$

One Richter magnitude scale unit corresponds to a tenfold increase in X. However, the energy increase is proportional to X^2, which is to say a hundredfold. The duration of a single oscillation differs greatly from one earthquake to another. In particular, the most energetic earthquakes have a higher proportion of long duration waves. As a result, the energy increase corresponding to one Richter scale unit increase, when summed over the whole range of waves in a wave record, is only a thirtyfold increase. Thus, the difference in energy released between an earthquake of magnitude 4 and one of magnitude 7 is $30 \times 30 \times 30 = 27,000$ times!

How big can earthquakes get? The largest recorded to date have Richter magnitudes of about 8.6, which means they release about as much energy as 10,000 atom bombs of the kind that destroyed Hiroshima at the end of World War II. It is possible that earthquakes do not get any larger than this because rocks cannot store more elastic energy. Before they are deformed further, they fracture and so release the energy.

Earthquake Risk

Most people in the United States think immediately of California when earthquakes are mentioned. However, the most intense earthquakes to jolt North America in the past 200 years were centered near New Madrid, Missouri. Three earthquakes of great size occurred on December 16, 1811, and January 23 and February 7, 1812. The exact sizes of these earthquakes are unknown because instruments to record them did not exist at the time. However, judging from the local damage caused, as well as the fact that tremors were felt and minor damage occurred as far away as New York and Charleston, South Carolina, it is estimated that the largest of the quakes was larger than the one that leveled San Francisco in 1906.

Based on known geological structures (mainly faults) and on the location and intensity of past earthquakes, the National Oceanographic and Atmospheric Administration prepared the seismic-risk map shown in Figure 15.11. Although Figure 15.11 is very informative, it is not particularly useful to people who plan and build roadways and public buildings. Builders need to know just how strongly the ground is likely to shake and what the frequency of large earthquakes

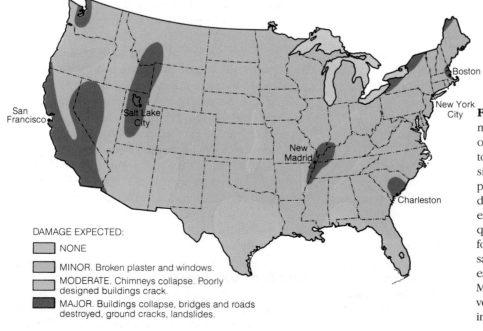

DAMAGE EXPECTED:

☐ NONE

☐ MINOR. Broken plaster and windows.

☐ MODERATE. Chimneys collapse. Poorly designed buildings crack.

■ MAJOR. Buildings collapse, bridges and roads destroyed, ground cracks, landslides.

FIGURE 15.11   Seismic-risk map of the United States based on quake intensity. Zones refer to maximum earthquake intensity and therefore to maximum possible destruction. The map does not indicate frequency of earthquakes. For example, frequency is high in southern California but low in eastern Massachusetts. Nevertheless, when earthquakes occur in eastern Massachusetts, they can be as severe as the more frequent quakes in southern California.

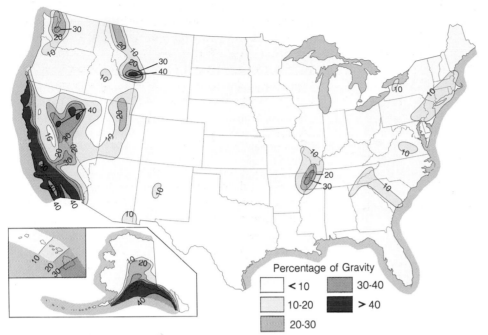

FIGURE 15.12 A seismic-risk map based on horizontal acceleration. Because acceleration during ground movement is the most important factor to be considered in the design of an earthquake-stable building, this type of risk map is preferred by builders and city planners. Numbers on contours refer to the maximum horizontal acceleration during an earthquake, expressed as a percentage of the acceleration due to gravity (980 cm/s^2). The probability that an acceleration of the amount indicated will occur in any given 50-year period is 1 chance in 10.

might be. To answer this need, the U.S. Geological Survey publishes a seismic-risk map of the kind shown in Figure 15.12. The strength of a possible earthquake is compared to the acceleration due to gravity, which is 980 cm/s^2 (that is to say, 1.0G). The maximum acceleration is likely to occur with California quakes and to reach 80 percent of gravity (which would be 0.8G). Damage begins at about 0.1G.

Figure 15.12 includes some additional information. Scientists of the U.S. Geological Survey have calculated that there is only 1 chance in 10 that a given acceleration shown in Figure 15.12 will be exceeded in a 50-year period. That information is of great help to insurance companies.

Earthquake Disasters

Every year the Earth experiences many hundreds of thousands of earthquakes. Fortunately, only one or two are large enough, or close enough to major centers of population, to cause loss of life. Certain areas are known to be earthquake-prone, and special building codes in such places require structures to be as resistant as possible to earthquake damage. However, all too often an unexpected earthquake will devastate an area where buildings are not adequately constructed, as in the T'ang Shan earthquake discussed at the beginning of this chapter. Other examples are the earthquake that destroyed parts of the center of Mexico City and killed 9500 people in 1985 (Fig. 15.13), and the earthquake that struck Soviet Armenia in 1988, killing an estimated 25,000 people (Fig. 15.14).

Seventeen earthquakes are known to have caused 50,000 or more deaths apiece (Table 15.1). The most disastrous one on record occurred in 1556, in Shaanxi Province, China, where an estimated 830,000 people died. Many of those people lived in cave dwellings excavated in loess (chapter 12), which collapsed as a result of the quake. Since 1900, there have been 42 earthquakes worldwide in which 500 or more people have died.

Earthquake Damage

The dangers of earthquakes are profound, and the havoc they can cause is often catastrophic. Their ef-

FIGURE 15.13 A building that was not constructed to withstand expected earthquakes. The Hotel DeCarlo was one of the buildings that collapsed during the earthquake that struck Mexico City in 1985. Proper building design can minimize damage. Nearby buildings of sturdier construction withstood the shaking.

TABLE 15.1 Earthquakes during the past 800 years that have caused 50,000 or more deaths

Place	Year	Estimated Number of Deaths
Silicia, Turkey	1268	60,000
Chihli, China	1290	100,000
Shaanxi, China	1556	830,000
Shemaka, USSR	1667	80,000
Naples, Italy	1693	93,000
Catania, Italy	1693	60,000
Beijing, China	1731	100,000
Calcutta, India	1737	300,000
Lisbon, Portugal	1755	60,000
Calabria, Italy	1783	50,000
Messina, Italy	1908	160,000
Gansu, China	1920	180,000
Tokyo and Yokohama, Japan	1923	143,000
Gansu, China	1932	70,000
Quetta, Pakistan	1935	60,000
Tangshan, China	1976	240,000
Iran	1990	52,000

fects are of six principal kinds. The first two, ground motion and faulting, are primary effects, and they cause damage directly. The other four effects are secondary and cause damage indirectly as a result of processes set in motion by the earthquake.

1. Ground motion results from the movement of seismic waves, especially surface waves, through surface-rock layers and regolith. The motions can

FIGURE 15.14 When a magnitude 6.8 earthquake struck Soviet Armenia on December 7, 1988, poorly constructed buildings with inadequate foundations collapsed like houses of cards. The principal cause of collapse was ground motion.

416

FIGURE 15.15 Fire caused by gas lines broken as a result of the Loma Prieta earthquake in 1989. This shot shows a fire in the Marina district of San Francisco.

damage and sometimes completely destroy buildings. Proper design of buildings (including such features as steel framework and foundations tied to bedrock) can do much to prevent such damage, but in a very strong earthquake even the best buildings may suffer some damage. A lot of the damage from the 1988 Armenian and the 1989 Loma Prieta earthquakes was due to ground motion (Figs. 15.14 and 15.15).

2. Where a fault breaks the ground surface, buildings can be split, roads disrupted, and any feature that crosses or sits on the fault broken apart.

3. A secondary effect, but one that is sometimes a greater hazard than moving ground, is fire. Ground movement displaces stoves, breaks gas lines, and loosens electrical wires, thereby starting fires. Ground motion also breaks water mains so there is no water available to put out fires. In the earthquakes that struck San Francisco in 1906, and Tokyo and Yokohama in 1923, more than 90 percent of the damage to buildings was caused by fire.

4. In regions of steep slopes, earthquake vibrations may cause regolith to slip, cliffs to collapse, and other rapid mass-wasting movements to start. This is particularly true in Alaska, parts of southern California, China, and hilly places such as Iran and Turkey. Houses, roads, and other structures are destroyed by rapidly moving regolith.

5. The sudden shaking and disturbance of water-saturated sediment and regolith can turn seemingly solid ground to a liquidlike mass of quicksand. This process is called liquefaction, and it was one of the major causes of damage during the earthquake that destroyed much of Anchorage, Alaska, on March 27, 1964, and that caused apartment houses to sink and collapse in Niigata, Japan, that same year (Fig. 15.16).

6. Finally, there are **seismic sea waves,** called **tsunami** as discussed in chapter 13, which have been particularly destructive in the Pacific Ocean. About 4.5 h after a severe submarine earthquake near Unimak Island, Alaska, in 1946, a tsunami struck Hawaii. The wave traveled at a velocity of 800 km/h. Although the amplitude of the wave in the open ocean was less than 1 m, the amplitude increased dramatically as the wave approached land. When it hit Hawaii, the wave had a crest 18 m higher than normal high tide. This destructive wave demolished nearly 500 houses, damaged a thousand more, and killed 159 people.

Modified Mercalli Scale

Because damage to the land surface and to human property is so important, the scale of earthquake-damage intensity (called the **Modified Mercalli Scale**) is based on the amount of vibration people feel during low-magnitude quakes and the extent of building damage during high-magnitude quakes. The correspondence between Mercalli intensity, Richter magnitude, and the estimated number of earthquakes is listed in Table 15.2.

FIGURE 15.16 Gaping fissures in a residential area of Anchorage, Alaska, formed as a result of the 1964 earthquake. The fissures are normal faults created as a result of liquefaction and of failure of weak subsurface stratum.

WORLD DISTRIBUTION OF EARTHQUAKES

Although no part of the Earth's surface is exempt from earthquakes, several well-defined **seismic belts** are subject to frequent earthquake shocks (Fig. 15.17). Of these, the most obvious is the *circum-Pacific belt,* for it is here that about 80 percent of all recorded earthquakes originate. The belt follows the mountain chains in the western Americas from Cape Horn to Alaska, crosses to Asia where it extends southward down the coast, through Japan, the Philippines, New Guinea, and Fiji, where it finally loops far southward to New Zealand. Next in prominence, giving rise to 15 percent of all earthquakes, is the Mediterranean-Himalayan belt, extending from Gibraltar to Southeast Asia. Lesser belts follow the midocean ridges.

Seismic belts are places where a lot of the Earth's internal energy is released. Therefore, it might be expected that other manifestations of internal energy would also appear in these belts, and indeed some of them do. Midocean ridges, deep-sea trenches, andesitic volcanoes, and many other features that outline the margins of plates of lithosphere either coincide with or else closely parallel these margins. Compare Figure 15.17 with Figure 1.9 to see that seismic belts outline the plate boundaries.

The depths of earthquake foci around plate edges are also informative. Most foci are no deeper than 100 km, because, as already mentioned, earthquakes occur in brittle rocks and the brittle lithosphere is only 100 km thick. However, a few earthquakes do originate at depths as great as 700 km. It is noteworthy that these deep earthquakes are not associated with either oceanic ridges or transform faults. Rather, they are related to seafloor trenches. Those trenches mark the places where cool, brittle lithosphere sinks down into the mantle.

Benioff Zone

Detailed study of deep-earthquake foci beneath a seafloor trench (Fig. 15.18) shows that the foci follow a well-defined pathway called a **Benioff zone,** named

TABLE 15.2 Earthquake magnitudes, frequencies for the entire Earth, and damaging effects

Richter Magnitude	Number per Year	Modified Mercalli Intensity Scale[a]	Characteristic Effects of Shocks in Populated Areas
<3.4	800,000	I	Recorded only by seismographs
3.5–4.2	30,000	II and III	Felt by some people who are indoors
4.3–4.8	4,800	IV	Felt by many people; windows rattle
4.9–5.4	1,400	V	Felt by everyone; dishes break, doors swing
5.5–6.1	500	VI and VII	Slight building damage; plaster cracks, bricks fall
6.2–6.9	100	VII and IX	Much building damage; chimneys fall; houses move on foundations
7.0–7.3	15	X	Serious damage, bridges twisted, walls fractured; many masonry buildings collapse
7.4–7.9	4	XI	Great damage; most buildings collapse
>8.0	One every 5–10 yr	XII	Total damage, waves seen on ground surface, objects thrown in the air

[a]Mercalli numbers are determined by the amount of damage to structures and the degree to which ground motions are felt. These depend on the magnitude of the earthquake, the distance of the observer from the epicenter, and whether an observer is in or out of doors. *Source:* After B. Gutenberg, 1950.

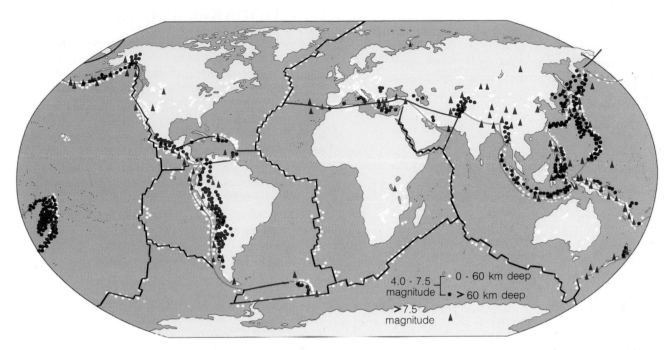

FIGURE 15.17   The Earth's seismicity outlines plate margins. This map shows earthquakes of magnitude 4.0 or greater from 1960 to 1989.

after the scientist who first recognized this phenomenon. This important observation strongly suggests that deep earthquakes originate within the relatively cold, downward-moving plate of lithosphere. Because some earthquake foci can be as deep as 700 km, it must

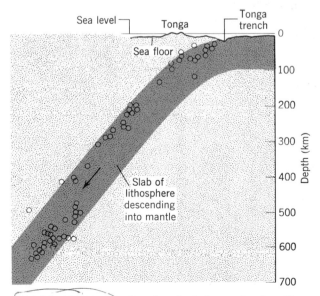

FIGURE 15.18   Earthquake foci beneath the Tonga Trench, Pacific Ocean, during several months in 1965. Each circle represents a single earthquake. The earthquakes define the Benioff zone and are generated by downward movement of a comparatively cold slab of lithosphere.

be concluded that rapidly descending lithosphere can retain at least some brittle properties to that depth. The reason why earthquakes have not been detected at depths below 700 km remains unexplained. It may be that a rapidly sinking slab of lithosphere is sufficiently hot by the time it reaches a depth of 700 km that it has become ductile rather than brittle.

Locations of earthquakes reveal a great deal about the shapes and structures of tectonic plates. But they provide a static picture, a sort of snapshot of the way things are at the moment. In order to discover how the plates move and respond to forces, it is helpful to include other observations that can be obtained from seismic studies—most importantly, first-motion studies.

First-Motion Studies of Earthquakes

By careful study of earthquake waves recorded by seismographs, it is possible to tell the direction of motion of the fault that caused the earthquake. The information that is needed is contained in the arrival records of the seismic body waves.

Consider the P-wave record. If the first arrival is a compressive pulse, the release of elastic energy and the fault motion must be *toward* the seismograph (Fig. 15.19A). If it is an expansion, the fault motion must be *away* from the seismograph. Figure 15.19B shows the effect of an earthquake caused by movement on a strike-slip fault. It is apparent that the first P-wave motion observed depends on the location of the seismo-

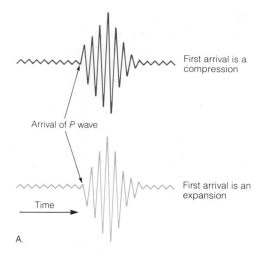

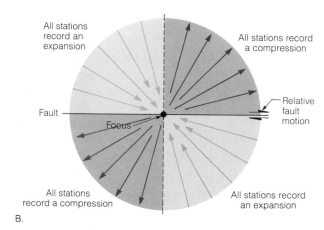

FIGURE 15.19 Initial motion of seismic body waves used to determine the direction of movement on a fault. A. Initial motion of a P wave detected by a seismograph is either a push away from the focus (that is, arrival of a compression) or a pull toward the focus (arrival of an expansion). B. Plotting the first motions detected at a number of seismograph stations allows the direction of movement on a fault to be uniquely determined. The example shown is for right-lateral motion on a strike-slip fault.

graph. The fault movement can be determined by plotting first motions from several seismographs.

The radiation pattern of body waves is in three dimensions. It is not possible to distinguish up-down movement from back-forth movement using P waves alone. A 90° ambiguity is inherent in the P-wave radiation pattern. Because two possibilities exist—up-down versus back-and-forth—two independent measurements are needed. The measurement needed is the strike and dip of the fault, and that can be determined from the locations of several foci on the same fault surface.

S-wave oscillations also carry the signature of the direction of the first motion and can provide independent data that can be used to resolve any ambiguity.

EARTHQUAKES AND PLATE TECTONICS

The combined evidence of quake locations and the first motions of earthquakes provide some of the most detailed proof we have that the plate tectonics theory is correct. As shown in Figure 15.17, plate boundaries are outlined by narrow seismic belts. Plate boundary *motions* can be determined by first-motion studies.

We learned in chapter 1 that there are three kinds of plate boundaries: (1) *divergent boundaries,* or *spreading centers,* which coincide with rift valleys and midocean ridges; (2) *transform fault boundaries;* and (3) *convergent boundaries,* which coincide with oceanic trenches and/or continental collision zones. Each kind of boundary is characterized by earthquakes that are distinctive as to fault motions and to the depth of foci.

Spreading Centers

Along a spreading center, two plates move apart from each other, and the lithosphere is stretched by tensional stresses (Fig. 15.20A). The kinds of faults associated with tensional stresses are normal faults. First-motion studies confirm that earthquakes associated with spreading centers are due to normal faulting.

Earthquakes at spreading centers tend to have low Richter magnitudes and foci that are invariably less than 100 km deep and commonly less than 20 km. This indicates that the lithosphere is thin beneath a spreading center and that the ductile asthenosphere must come close to the surface.

Transform Fault Boundaries

As we learned in chapter 14, transform faults are huge, vertical, strike-slip faults that cut down through the lithosphere. They are boundaries where two plates slide past each other (Fig. 15.20B).

First-motion studies confirm that motion along transform fault boundaries is strike slip. Earthquakes always have shallow foci—that is, they are no deeper than 100 km—and they often have high Richter magnitudes. The location of earthquake foci suggests that when a transform fault cuts continental crust, a system of parallel faults rather than a single fracture can develop. This seems to be the case for the San Andreas Fault.

Convergent Boundaries

As we discussed in chapter 1, convergent plate boundaries are of two kinds: (1) subduction boundaries where lithosphere capped by oceanic crust is subducted into the asthenosphere and mesosphere; and (2) collision boundaries where two continents collide.

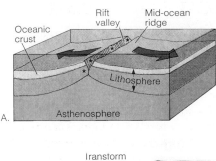

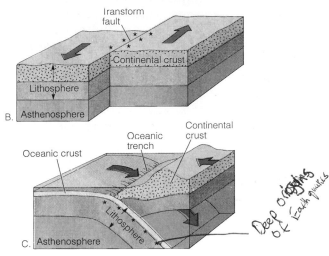

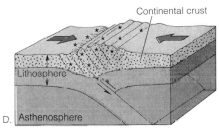

FIGURE 15.20 Relationship between seismicity and plate boundaries. Stars indicate earthquake foci. A. Divergent boundary. Earthquakes have shallow foci, low Richter magnitudes, and normal fault motions. B. Transform faults. Earthquakes have foci down to 100 km, sometimes high Richter magnitudes, and strike-slip motions. C. Subduction boundary. Earthquakes are complex. Earthquakes with low Richter magnitudes, shallow foci, and normal fault motions are observed seaward of the oceanic trench. Deeper earthquakes can have high Richter magnitudes and thrust fault movement. D. Continental collision boundary. Earthquakes have foci down to 300 km, thrust fault motions, and sometimes high Richter magnitudes.

Each kind of convergent boundary has a characteristic pattern of seismic activity, and each is complicated.

When oceanic lithosphere is subducted, it is subjected to complex stresses and different kinds of earthquakes can occur (see Figure 15.20C). Lithosphere bends downward as it is subducted, and the bending causes normal faults in the upper part of the plate. Earthquakes associated with such faults all have very shallow foci, small Richter magnitudes, and normal fault first motions. Subduction involves one plate sliding beneath the other, and so the boundary between the two plates is a thrust fault. Earthquakes to a depth of at least 100 km (the region where the two plates of lithosphere are in contact) often have large Richter magnitudes and invariably have first motions that indicate thrust faulting. Below 100 km, where the subducted lithosphere is sinking through the asthenosphere, earthquakes occur *in* the subducted slab. Some earthquakes indicate tensional stresses (normal faults), whereas others indicate compressional stresses (reverse faults). Exactly why and how such deep earthquakes occur in subducted slabs is something of a puzzle and a matter for research.

Collision boundaries are the places where two continents collide. The Himalayan mountain chain, between India and Asia, is a present-day collision boundary. A zone of collision tends to be a region several hundred kilometers wide, and within it rocks are intensely compressed and thrust-faulted, as shown in Figure 15.20D. Within a collision zone the lithosphere is locally thickened. Earthquakes may have foci as deep as 300 km and may also have high Richter magnitudes and first motions that indicate thrust movements.

EARTHQUAKE PREDICTION

Some of the most dreadful natural disasters have been caused by earthquakes. It is hardly surprising, therefore, that a great deal of research around the world focuses on earthquakes. The hope is that through research we will be able to improve our forecasting ability.

Because China has suffered so many terrible earthquakes, Chinese scientists have tried everything they can think of to predict quakes. They have even observed animal behavior, and on one occasion animals did successfully foretell a quake. On July 18, 1969, zookeepers at the People's Park in Tianjin observed highly unusual animal behavior. Normally quiet pandas screamed, swans refused to go near water, yaks did not eat, and snakes would not go into their holes. The keepers reported their observations to the earthquake prediction office, and at about noon on the same day a magnitude 7.4 earthquake struck.

There have been many informal reports of strange animal behavior before earthquakes, but the Tianjin quake is the only well-documented case. Unfortunately, most quakes do not seem to be preceded by

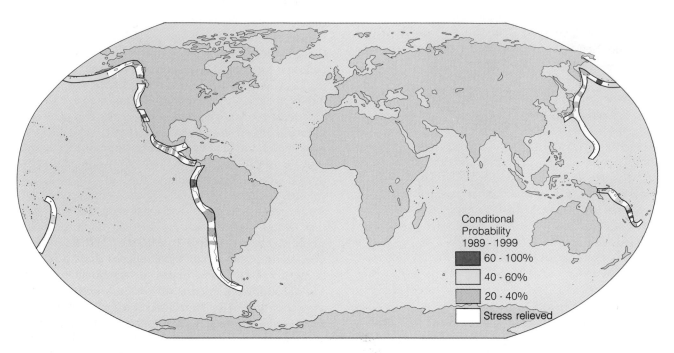

FIGURE 15.21 Seismic gaps in the circum-Pacific belt. In the areas indicated, earthquakes of magnitude 7.0 or greater are known but have not occurred in recent times. Strain is now building up in each seismic gap, raising the probability that a large quake will occur before the year 2000.

anything odd. While scientists haven't given up on animals, they measure lots of other things beside animal behavior.

Most research on earthquake prediction now deals with changes that can be monitored in the properties of elastically strained rocks—properties such as rock magnetism, electrical conductivity, and porosity. Even simple observations, such as the level of well water, might indicate changes in porosity. Tilting of the ground or slow rises and falls in elevation may also indicate that strain is building up. Most significant are the small cracks and fractures that can develop in severely strained rock. These can cause swarms of tiny earthquakes—foreshocks—that may be a clue that a big quake is coming. One of the most successful cases of earthquake prediction, made by Chinese scientists in 1975, was based on slow tilting of the land surface, on fluctuations in the magnetic field, and on the swarms of small foreshocks that preceded the 7.3 Richter magnitude quake that struck the town of Haicheng. Half the city was destroyed, but authorities had evacuated more than a million people before the quake. As a result, only a few hundred were killed.

In places where earthquakes are known to occur repeatedly, such as along plate boundaries, patterns of recurrence can sometimes be discerned. If such a pattern suggests a recurrence interval of, say, a century, it may be possible to predict where and when a large quake may happen. Certainly it is possible to monitor such areas closely when a big quake is thought to be due. Studies of recurrence patterns have identified a number of *seismic gaps* around the Pacific rim (Fig. 15.21). These are places where, for one reason or another, earthquakes have not occurred for a long time and where elastic strain is steadily increasing. Seismic gaps receive a lot of research attention from seismologists because they are considered to be the places most likely to experience large earthquakes.

USING SEISMIC WAVES AS EARTH PROBES

We have seen that P and S waves travel through rock at different velocities. These two types of waves also respond differently to changing rock properties. The arrival times of P and S waves at seismographs around the world provide records of waves that have traveled along many different paths. From such records, it is possible to calculate how the rock properties change and where distinct boundaries occur between layers having sharply different properties. Seismic waves are the most sensitive probes we have to measure the properties of the unseen parts of the crust, the mantle, and the core.

Layers of Different Composition

If the Earth's composition were uniform, and if no polymorphic changes occurred in the minerals present, the velocities of P and S waves would increase smoothly with depth. This is so because higher pressure leads to an increase in the density and the rigidity of a solid, and it is these two properties that control the wave velocities. For an Earth of uniform composition, it is a straightforward matter to predict how long seismic waves would take to pass through. But observed travel times differ greatly from such predictions. These differences can best be accounted for by supposing that velocities do not change smoothly with depth and that neither composition nor physical properties are constant throughout. Distinct boundaries (or discontinuities, as they are more commonly called) can be readily detected by refraction and reflection of body waves, as illustrated in Figure 15.7. From measurements of body waves, two major compositional boundaries have been detected. The first is the boundary between the crust and the mantle, and the second is that between the mantle and the core.

The Crust

Early in the twentieth century, the boundary between the Earth's crust and mantle was demonstrated by a scientist named Mohorovičić, who lived in what today is Yugoslavia. Mohorovičić noticed that, for earthquakes whose focus lay within 40 km of the surface, seismographs about 800 km from the epicenter recorded two distinct sets of P and S waves. He concluded that one pair of waves must have traveled from the focus to the station by a direct path through the crust, whereas the other pair represented waves that had arrived slightly earlier because they had been refracted by a boundary at some depth in the Earth. Evidently, the refracted waves had penetrated a zone of higher velocity below the crust, had traveled within that zone, and then had been again refracted upward to the surface (Fig. 15.22). Mohorovičić hypothesized that a distinct compositional boundary separates the crust from this underlying zone of different composition. Scientists now refer to this boundary as the **Mohorovičić discontinuity** and recognize it as the seismic discontinuity that marks the base of the crust. The feature is commonly called the **M-discontinuity** and in conversation it is shortened still further to **moho.**

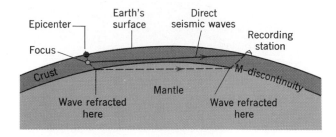

FIGURE 15.22 Travel paths of direct and refracted seismic waves from shallow-focus earthquake to nearby seismograph station.

Thickness and Composition of the Crust

By seismic methods it is possible to determine the thickness of the crust. Seismic-wave velocities can be measured for different rock types in both the laboratory and the field. When the velocities of waves received at a number of seismographs are calculated, laboratory measurements can be used to determine the depth of the moho and to estimate the probable composition of the crust. Beneath ocean basins the crust is less than 10 km thick. Elastic properties of the oceanic crust are those characteristic of basalt and gabbro. But in the continental crust both thickness and composition are very different. The continental crust ranges in thickness from 20 to nearly 60 km and tends to be thickest beneath major mountain masses (Fig. 15.23). Velocities in the continental crust are distinctly different from those in the oceanic crust. They indicate elastic properties like those of rock such as granite and diorite, although at some places just above the moho, velocities close to those of oceanic crust are often observed. These conclusions agree well with what is known about the composition of the crust from other lines of evidence, such as geological mapping and deep drilling. The agreement gives geologists confidence in drawing conclusions about the mantle where these other lines of evidence are scarce.

The Mantle

The mantle is something of an enigma. It is huge and controls much of what happens in the crust, but it cannot be seen. P-wave velocities in the crust range between 6 and 7 km/s. Beneath the moho, velocities are greater than 8 km/s. Laboratory tests show that rocks common in the crust, such as granite, gabbro, and basalt, all have P-wave velocities of 6 to 7 km/s. But rocks that are rich in dense minerals, such as olivine and pyroxene, have velocities greater than 8 km/s. We therefore infer that such rock must be among the principal materials of the mantle. This inference is consistent with what little direct evidence is available

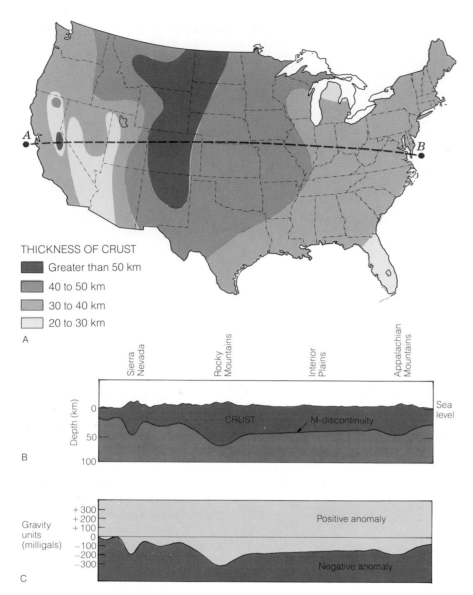

THICKNESS OF CRUST

■ Greater than 50 km
■ 40 to 50 km
■ 30 to 40 km
□ 20 to 30 km

A

Sierra Nevada • Rocky Mountains • Interior Plains • Appalachian Mountains

B

CRUST — M-discontinuity

Depth (km): 0, 50, 100 — Sea level

C

Gravity units (milligals): +300, +200, +100, 0, −100, −200, −300

Positive anomaly

Negative anomaly

FIGURE 15.23 Crust beneath the United States. A. Thickness of crust is determined from measurements of seismic waves. B. Section through the crust along the line *A-B* (above). The crust tends to thicken beneath major mountain masses such as the Sierra Nevada, the Rocky Mountains, and the Appalachians. C. Profile of gravity traverse. The negative gravity anomalies over the Sierra, the Rockies, and the Appalachians are due to the roots of low-density rocks beneath these topographic highs.

FIGURE 15.24 Kimberlite from the Monarch Pipe in South Africa. Fragments of rock from deep in the mantle are carried upward by the forceful intrusion of kimberlite magma. Rounded fragments are the transported mantle rocks; fragmental, grayish background material is the kimberlite.

concerning the composition of the upper part of the mantle. For example, some evidence can be obtained from rare samples of mantle rocks found in *kimberlite pipes,* narrow pipelike masses of igneous rock, sometimes containing diamonds, that intrude the crust but originate deep in the mantle (Fig. 15.24).

The Core

Both P and S waves are strongly influenced by a pronounced boundary at a depth of 2900 km. When P waves reach that boundary, they are reflected and refracted so strongly that the boundary casts a P-wave shadow, which is an area of the Earth's surface, opposite the epicenter, where no P waves are observed (Fig. 15.25). Because this 2900-km boundary is so pronounced, geologists infer that it is the boundary between the mantle and the core. The same boundary casts an even more pronounced S-wave shadow. Here,

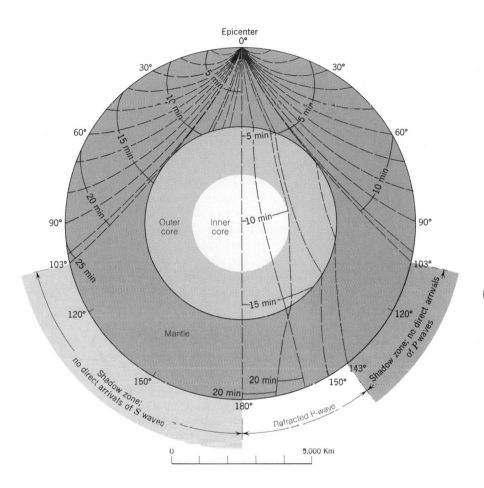

FIGURE 15.25 Paths of P waves from an earthquake focus with epicenter at 0° shown in red in right half of section only. Paths of S waves shown in black in the left half. Reflection and refraction of P waves at the mantle-core boundary create a P-wave shadow zone from 103 to 143°. Because S waves cannot pass through a liquid, an S-wave shadow exists between 103 and 180°.

however, the reason is not reflection or refraction, but the fact that shear waves cannot traverse liquids. Therefore, the huge S-wave shadow lets us conclude that the outer core is liquid.

Seismic waves cannot tell us the composition of the core, but they help us to deduce what it might be. Seismic-wave velocities calculated from travel times indicate that rock density increases slowly from about 3.3 g/cm³ at the top of the mantle to about 5.5 g/cm³ at the base of the mantle. The mean density of the whole Earth is 5.5 g/cm³. Therefore, to balance the less dense crust and the mantle, the core must be composed of material with a density of at least 10 to 11 g/cm³. The only common substance that comes close to fitting this requirement is iron. Suggestive evidence also comes from meteorites. Iron meteorites are samples of material believed to have come from the core of an ancient, tiny planet, now disintegrated. All iron meteorites contain a little nickel, and the

Earth's core presumably does too. Because S waves do not travel beyond the core-mantle boundary, it is inferred that the outer core is molten, that its composition is mostly iron, but that nickel and small amounts of other elements may be present too.

The Inner Core

P-wave reflections indicate the presence of a solid inner core enclosed within the molten outer core. The two cores appear to be essentially identical in composition. The reason for the change from liquid to solid probably relates to the effect of pressure on the melting temperature of iron. As the center of the Earth is approached, pressure rises to a value millions of times greater than atmospheric pressure. Temperature also rises but not steeply enough to offset the effect of pressure. From the base of the mantle (at a depth of 2900 km) to a depth of 5350 km, temperature and pressure are so balanced that iron is molten. But at a depth of 5350 km, another strong reflecting and refracting boundary occurs, one that has properties consistent with a change from a liquid to a solid. Apparently, from 5350 km to the center of the Earth, rising pressure overcomes rising temperature, and iron is solid, creating the solid core.

Layers of Different Physical Properties in the Mantle

As far as we can determine, there are no compositional boundaries within the mantle. Nevertheless, as shown in Figure 15.26, seismic-wave velocities do not increase regularly from the base of the crust to the core-mantle boundary. There are sudden changes in velocity that are apparently due to changes in the physical properties of the mantle.

The Low-Velocity Zone
The P-wave velocity at the top of the mantle is about 8 km/s, and it increases to 14 km/s at the core-mantle boundary. This increase is not smooth and constant, however. From the base of the crust to a depth of about 100 km, the P-wave velocity rises slowly to about 8.3 km/s. However, the velocity then starts to drop slowly to a value just below 8 km/s, and it remains low to a depth of about 350 km. This zone of reduced velocity between 100 and 350 km is not sharply defined and is better developed beneath the oceans than beneath the continents. This low-velocity zone can be seen as a small blip in both the P- and S-wave velocity curves in Figure 15.26. No evidence exists to suggest

that the density decreases or the composition changes in this zone. To account for the velocity changes, therefore, we infer that the zone has a composition similar to that of the mantle immediately above and below, but that it is less rigid, less elastic, and more ductile than the adjacent regions. Evidence to support this idea comes from studies of the dispersion of surface waves. Long-wavelength surface waves that reach down more 100 km are affected by what seems to be a soft, ductile region in the mantle.

A possible explanation for the low-velocity zone is that between 100 and 350 km the geothermal gradient reaches temperatures close to the onset of partial melting of mantle rock. If this explanation is correct, either the rock strength drops sharply at temperatures close to melting, or melting starts and a small amount of liquid develops and forms very thin films around the mineral grains, thus serving as a lubricant. The amount of melting, if it occurs at all, must be very small, because the low-velocity zone does transmit S waves and we know that S waves cannot pass through liquids. Any liquid, like a thick film of oil, merely serves to lubricate the mineral grains in the mantle and at the same time to reduce wave velocities by reducing the elastic properties.

An integral part of the theory of plate tectonics is the idea that plates of lithosphere slide over a somewhat plastic zone in the mantle. The importance of the low-velocity zone for the theory is that such a zone proves the existence of the asthenosphere. The top of the low-velocity zone coincides with the base of the lithosphere. Thus, the low-velocity zone coincides with the asthenosphere.

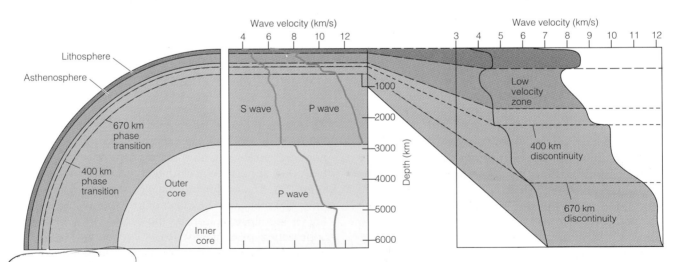

FIGURE 15.26 Variation in seismic-wave velocity within the Earth. Changes occur at the boundaries between crust and mantle and between mantle and core, owing to change in composition. Another change occurs at a depth of 100 km and corresponds to the lithosphere–asthenosphere boundary. Changes also occur at 400 km and 670 km.

The 400 km Seismic Discontinuity

From the P- and S-wave curves in Figure 15.26 it is apparent that the velocities of both P and S waves increase sharply at about 400 km. Sharp though it seems, the increase is not sharp enough to be accounted for by a change in composition; the cause must be something else. A probable explanation is suggested by laboratory experiment. When olivine is squeezed at a pressure equal to that at a depth of 400 km, the atoms rearrange themselves into a denser polymorph. This process of atomic repacking caused by changes in pressure and temperature is called a *polymorphic transition.* In the case of olivine, the repacking involves a change to a structure resembling that found in a family of minerals called the spinels, of which magnetite is a well-known example. The structural repacking involves a 10 percent increase in density. It is likely that the increase in seismic-wave velocities at 400 km is caused by the olivine-spinel polymorphic transition because the density increase, determined from seismic-wave velocities, is almost exactly 10 percent.

The 670 km Seismic Discontinuity

An increase in seismic-wave velocities—particularly P-wave velocity—also occurs at a depth of 670 km. This one is difficult to explain. The observed increase in mantle density is about 10 percent, but the boundary is diffuse. It has not yet been possible to determine from either seismic evidence or laboratory experiment whether the boundary is due to a polymorphic transition, to a compositional change, or to both. Some scientists suggest that the increase at 670 km results solely from a polymorphic transition involving the repacking of atoms in the pyroxene minerals present in mantle rocks. Others suggest that the diffuse boundary indicates a polymorphic change affecting all silicate minerals present. One idea involves the rearrangement of silicon and oxygen to create denser anions in which each silicon atom is surrounded by six oxygen atoms rather than four. An intriguing hypothesis for the origin of the 670 km discontinuity involves evidence from earthquakes. The deepest earthquakes have foci of about 700 km, which is very close to the 670 km seismic discontinuity. Deep earthquakes are associated with sinking slabs of cool, oceanic lithosphere. Perhaps slabs of lithosphere are buoyant at 670 km and form a diffuse compositional boundary.

Opposing the point of view that slabs of lithosphere float buoyantly at 670 km is evidence from a research technique call *seismic tomography.* The method is similar to that used in medicine, where a three-dimensional picture of the interior of the human body is developed from slight differences in the intensities of X rays passing through in different directions. (CAT scan is the common name for X-ray tomography.) In a similar manner, heterogeneities in the mantle can be revealed by measuring slight differences in the velocities of seismic waves (Fig. 15.27). Seismic tomography does not reveal the presence of a compositional boundary at 670 km. For the present, we must conclude that the cause of the 670-km seismic discontinuity is unknown.

GRAVITY ANOMALIES AND ISOSTASY

The Earth is not a perfect sphere, careful measurement reveals that it is an ellipsoid that is slightly flattened at the poles and bulged at the equator.

The radius at the equator is 21 km longer than at the poles. Because the gravitational pull between two objects is inversely proportional to the square of the distance between their centers of mass, the pull exerted by the Earth's gravity on a body at the Earth's surface is slightly greater at the poles than it is at the equator. Thus, a man who weighs 90.5 kg at the North

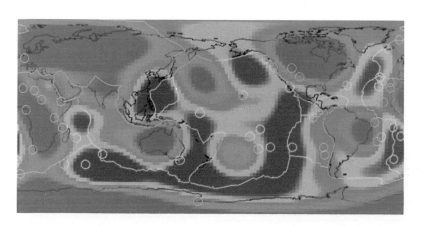

FIGURE 15.27 Lateral heterogeneity in the upper mantle at a depth of 150 km revealed through seismic tomography. Seismic waves travel faster through cooler, more rigid material (shown in blue), and more slowly in hotter, less rigid material (red). White lines show plate boundaries; white circles are centers of long-lived volcanic activity. Note that the red, low-velocity zones lie beneath spreading centers.

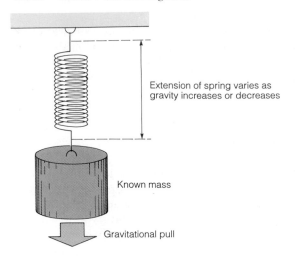

Extension of spring varies as gravity increases or decreases

Known mass

Gravitational pull

FIGURE 15.28 A gravimeter is a heavy mass of metal suspended on a sensitive spring. The mass exerts a greater or lesser pull on the spring as gravity changes from place to place, extending the spring more or less. The mass of metal and the spring are contained in a vacuum together with exceedingly sensitive measuring devices.

Pole would observe his weight decreasing slowly and steadily to 90 kg simply by traveling to the equator. If the weight-conscious traveler made very exact measurements as he traveled, however, he would observe that his weight changed irregularly rather than smoothly. From this he could conclude that the pull of gravity must change irregularly. If the traveler went one step further and carried a sensitive device called a **gravimeter** (or *gravity meter*) for measuring the pull of gravity at any locality, he would indeed find an irregular variation.

Gravity Anomalies

Gravimeters are similar to inertial seismographs. They consist of a heavy mass suspended by a sensitive

spring (Fig. 15.28). When the ground is stable and free from vibrations due to earthquakes, the pull exerted on the spring by the heavy mass provides an accurate measure of the gravitational pull. Modern gravimeters are incredibly sensitive. The most accurate devices in operation can measure variations in the force of gravity as tiny as one part in a hundred million (10^{-8}).

In order to compare the pull of gravity from point to point on the Earth, corrections must be applied to gravimeter measurements for change in latitude and topography. The idea behind the corrections is to know the pull of gravity at a constant distance from the center of the Earth. Then, if the rock mass between the gravimeter and the center of the Earth were everywhere the same, the adjusted figures for the force of gravity might be expected to be the same at every place on the Earth. In fact, the adjusted figures reveal large and significant variations called **gravity anomalies.** The anomalies are due to bodies of rock having differing densities. A simple example of an anomaly is shown in Figure 15.29. A great deal of important information can be derived from the anomalies.

The thickness of the crust beneath the United States, as determined from seismic measurements of the moho, is shown in Figure 15.23. Beneath the three major mountain systems (the Appalachians, the Rockies, and the Sierra Nevada) the crust is thicker than in the nonmountainous regions of the country. In profile (Fig. 15.23B), the crust beneath the mountains resembles icebergs that have high peaks above the waterline but also massive roots below. The accuracy of this analogy is demonstrated by the gravity profile across the United States, shown in Figure 15.23C. Negative gravity anomalies are observed where the crust is thickest. The anomalies are caused by the masses of low-density rock in the mountains, just as the basin of low-density sediment produces the gravity anomaly shown in Figure 15.29.

The reason why a root of low-density rock forms in the first place provides some interesting insights

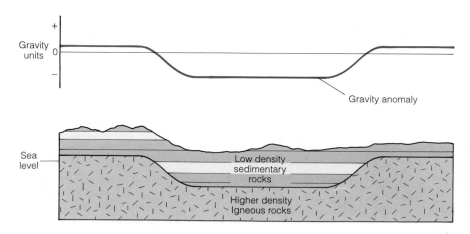

Gravity units

Gravity anomaly

Sea level

Low density sedimentary rocks

Higher density Igneous rocks

FIGURE 15.29 Example of a gravity anomaly; a basin filled with low-density sedimentary rocks sitting on a basement of high-density igneous rocks. Gravity measurements reveal a pronounced gravity low throughout the basin. The magnitude of the anomaly can be used to calculate the thickness of the sedimentary rocks.

into the Earth's physical properties. Mountains stand high and have iceberg-like roots beneath them because they are comprised of low-density rocks and are supported by the buoyancy of weak, easily deformed, but more dense rocks below. Mountains are, in a sense, floating. But it is not the crust that is floating on the mantle. Rather, it is the lithosphere, capped by a mass of thickened crust, that floats on the asthenosphere. Strange as it may seem, the topographic variations observed at the surface of the Earth arise not from the *strength* of the lithosphere but rather from its *weakness* and from the buoyancy of the asthenosphere.

Isostasy

The property of flotational balance among segments of the lithosphere is referred to as **isostasy.** The great ice sheets of the last glaciation provide an impressive demonstration of isostasy. The weight of a large continental ice sheet, which may be 3 to 4 km thick, will depress the lithosphere. When the ice melts, the land surface slowly rises again. The effect is very much like pushing a block of wood into a bucket of thick, viscous oil. When the wood is released, it slowly rises again to an equilibrium position determined by its density. The speed of its rising is controlled by the viscosity of the oil. Just like the block of wood, glacial depression and rebound mean that rock must flow laterally in the asthenosphere when the ice depresses the lithosphere, and then must flow back again when the deforming force is removed (Fig. 15.30). From the fact that the land surface in parts of northeastern Canada and Scandinavia is still rising, even though most of the thick ice sheets that covered these areas during the last glaciation had melted away by 7000 years ago (Fig. 15.31) we infer that the flow must be slow and therefore that the asthenosphere must be extremely viscous.

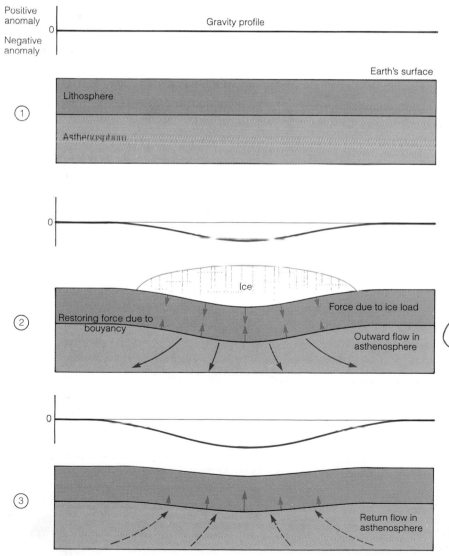

FIGURE 15.30 Depression of the lithosphere by a continental ice sheet. (1) Prior to formation of the ice sheet, there is no gravity anomaly. (2) When the ice sheet forms, it depresses the lithosphere. At some depth in the asthenosphere, material must slowly flow outward to accommodate the sagging lithosphere. (3) When the ice melts, buoyancy slowly restores the lithosphere to its original level. A negative gravity anomaly continues until the depression is removed. The viscosity of the asthenosphere controls the rate of flow and therefore the rate of recovery.

FIGURE 15.31 Beach ridges raised by postglacial uplift of the land in Nordaustlandet, Svalbard, Norway. Such beaches provide clear evidence that the land has emerged from the sea after removal of the ice load.

Continents and mountains are composed of low-density rock, and they stand high because they are thick and light; ocean basins are topographically low because the oceanic crust is composed of denser rock. Isostasy and the fact that the continental crust is less dense than the oceanic crust is the reason why the Earth has two pronounced topographic levels, as shown in Figure 1.10.

The important point to be drawn from this discussion of isostasy is that the lithosphere acts as if it were "floating" on the asthenosphere. (*Floating* is not exactly the correct word because the Earth is solid, but the lithosphere is buoyant and acts as though it were floating.) Sometimes gravity measurements suggest that a mountain has been pushed up so rapidly that it

is top-heavy and has too little root of low-density rock to counterbalance its upper mass. Sometimes, as in the seafloor trenches, it is observed that low-density crust has been dragged down so rapidly that it forms a root without a mountain mass above it. These and many other situations lead to local gravity anomalies. That the anomalies do not seem to become very large suggests that the Earth is always moving toward an isostatic balance. Indeed, isostasy is the principal explanation for vertical motions of the Earth's surface, just as plate tectonics is the principal explanation for lateral motions. In the next chapter we will discuss how the two combine to keep our dynamic Earth ever changing, ever active.

A GREAT EARTHQUAKE IN THE PACIFIC NORTHWEST?

Subduction zones are places of intense seismic activity. Many of the greatest earthquakes ever recorded occurred in subduction zones: the Chilean earthquake of 1960, Richter magnitude 8.5, for instance, and the Alaskan earthquake of 1964, magnitude 8.6. Such high-magnitude earthquakes are totally destructive.

The foci of most earthquakes in subduction zones are within the lithosphere of either the sinking or overriding plate. Such earthquakes tend to be no larger than about Richter magnitude 7.5 and to be caused by the bending and stretching of rock within the plates. The big quakes of magnitude 8 and larger have foci right at the interface between two plates. Presumably

such quakes happen when the downgoing plate sticks to the bottom of the overriding plate. When the lock is broken, a tremendous earthquake occurs.

From northern California to central British Columbia, the Juan de Fuca plate (named for an early Spanish explorer) has been slipping beneath the North American plate at a rate of 3 to 4 cm/yr for the past million years (Fig. B15.1), but for the past 200 years no great earthquakes have occurred on the Cascadia subduction zone. Furthermore, seismographs do not record any current seismic activity along the interface between the two plates. Is the Pacific Northwest a likely spot for a giant quake sometime in the

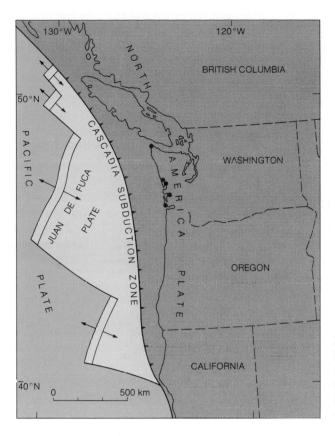

FIGURE B15.1 The Cascadia subduction zone separates the downgoing Juan de Fuca Plate from the overriding North American Plate. Evidence of sudden, catastrophic sinking of near-coastal lowlands has been discovered along the Washington coast at the five sites marked.

future? Is the Juan de Fuca plate locked to the overriding North American Plate, or is it sliding smoothly downward?

Geologists of the U.S. Geological Survey and other institutions are attempting to answer these questions by looking at the stratigraphic record. In general, great earthquakes at subduction zones tend to cause pronounced elevation changes along nearby coastlines. Coastal subsidence up to 2 m or more can change well-vegetated coastal lowlands into intertidal flats. Careful mapping along the coastline of Washington reveals a number of places (labeled in Figure B15.1) where marshes and swamps suddenly became barren tidal flats in the geologically recent past.

Radiocarbon dating of organic matter buried in sediment when lush coastal lowlands were suddenly flooded shows that a catastrophic change in the coastline occurred about 300 years ago in southwestern Washington. Evidence also suggests that catastrophic coastline changes occurred about 1700, 2700, 3100, and 3400 years ago in that area.

Although the evidence so far does not prove that serious earthquakes have ever occurred in the Pacific Northwest, it is likely that they have and—what is more important—that they will happen again. At this point, earth scientists have come to think that the probability of a future large earthquake is low but real. What they cannot do yet is say exactly when it will happen.

SUMMARY

1. Abrupt movements of faults that release elastically stored energy cause earthquakes.

2. Earthquake vibrations are measured with seismographs.

3. Energy released at an earthquake's focus radiates outward as body waves which are of two kinds—P waves (*Primary* waves, which are compressional) and S waves (*Secondary* waves,

which are shear waves). Earthquake energy also causes the surface of the Earth to vibrate because of surface waves.

4. The focus and epicenter of an earthquake can be located by measuring the differences in travel times between P and S waves.

5. The amount of energy released during an earthquake is calculated on the Richter magnitude scale. The calculation is made from seismograph records of the seismic body waves.

6. Ninety-five percent of all earthquakes originate in the circum-Pacific belt (80%) and the Mediterranean-Himalayan belt (15%). The remaining 5 percent are widely distributed along the midocean ridges and elsewhere. Seismic belts outline tectonic plates; seismic-wave first-motion studies are used to determine the directions of movement across a plate boundary.

7. Seismic body waves can be refracted and reflected just as sound and light waves are. From the study of seismic-wave refraction and reflection, scientists infer the internal structure of the Earth by locating boundaries, or discontinuities, in its composition and physical properties. Pronounced compositional boundaries occur between the crust and mantle and between the mantle and outer core.

8. The base of the crust is a pronounced seismic discontinuity called the Mohorovičić discontinuity. Thickness of the crust ranges from 20 to 60 km in continental regions but is only about 10 km beneath oceans.

9. Within the mantle there are two zones, at depths of 400 and 670 km, where sudden density changes produce seismic-wave discontinuities. The change at 400 km is probably produced by a polymorphic transition of olivine. The 670-km change might be due to either a polymorphic transition, a compositional change, or a combination of both.

10. The core has a high density and is inferred to consist of iron plus small amounts of nickel and other elements. The outer core must be molten because it does not transmit S waves. The inner core is solid.

11. From a depth of 100 km to 350 km there is a zone of low seismic-wave velocity that also causes pronounced dispersion of surface waves. This low-velocity zone coincides with the asthenosphere. The lithosphere, which is rigid and on average 100 km thick, overlies the asthenosphere.

12. The outer portions of the Earth are in approximate isostatic balance; in other words, like huge icebergs floating in water, the lithosphere "floats" on the asthenosphere.

IMPORTANT TERMS TO REMEMBER

Benioff zone (p. 418)
body waves (p. 409)

earthquake focus (p. 409)
elastic rebound theory (p. 406)
epicenter (p. 409)

gravimeter (p. 428)
gravity anomaly (p. 428)

inertia (p. 407)
isostasy (p. 429)

M-discontinuity (p. 423)
Modified Mercalli Scale (p. 417)
moho (p. 423)
Mohorovičić discontinuity (p. 423)

P (for primary) wave (p. 409)

Richter magnitude scale (p. 413)

S (for secondary) wave (p. 410)
seismic belt (p. 418)
seismic sea wave (p. 417)
seismic waves (p. 409)
seismograph (p. 407)
seismology (p. 407)
surface waves (p. 409)

tsunami (p. 417)

QUESTIONS FOR REVIEW

1. Explain how most earthquakes are thought to occur and why there seems to be a limit on earthquake magnitudes.

2. What is the relation between an earthquake focus and the corresponding epicenter?

3. How are seismic waves recorded and measured? How would you locate an epicenter from seismic records? Explain how a focus is determined.

4. What are the differences between seismic body waves and surface waves? Identify two kinds of body waves and explain the differences.

5. Explain how seismologists use the Richter magnitude scale to estimate the energy released during an earthquake. In order for us to feel an earthquake, what minimum Richter magnitude can it have?

6. Earthquakes can cause damage in many ways; name four. Where on the Earth was the most disastrous earthquake on record and how did the people die? Where was the biggest known earthquake in the United States?

7. What are reflection and refraction and how do they affect the passage of seismic waves? How can refraction and reflection be used to define the base of the crust? The core-mantle boundary?

8. Briefly describe how seismic waves can be used to infer that the outer core is molten while the inner core is solid. Why is the composition of the core thought to be largely metallic iron?

9. Under what circumstance is it possible to obtain samples of rocks from the mantle? How can such samples be used to check information about the mantle inferred from seismic-wave velocities?

10. What are seismic belts and how are they related to tectonic plates? Explain how seismic-wave records can be used to determine the motions of plate boundaries. How many kinds of plate boundaries are there and what are their characteristic motions?

11. How do gravity anomalies arise and how can they be measured?

12. Describe some evidence that proves that isostasy is operating in the Earth. How is the Earth's surface topography related to isostasy?

13. Draw an east–west profile of the shape of the crust under the United States and indicate how isostasy plays a role in what you have drawn.

Striking rampart of a fold-and-thrust mountain range, the Bourgeau Thrust Sheet in the Southern Canadian Rockies. The ridge of steeply dipping Mississippian and Devonian strata had been thrust over Cretaceous shales and sandstones in the valley to the right. The view is looking north from Riverside Mountain, British Columbia, to Kananaskis Lakes, Alberta.

CHAPTER 16

Global Tectonics

People have wondered for a long time why continents have such irregular shapes and why ocean basins, mountain ranges, earthquake belts, and many other features occur where they do. When the first maps were made of the coastlines on either side of the Atlantic in the sixteenth century, it became apparent that the coasts were approximately parallel. People started to speculate why. They thought about a flood having cut an immense canyon—perhaps the great biblical flood. No realistic answers were forthcoming, but such speculations did get people thinking about why the Earth is the way it is. Scientists eventually began to think that there might be a single, underlying cause for the whole array of the Earth's major features. But what could that cause possibly be?

During the nineteenth century, people favored the idea that the Earth, originally a molten mass, had been cooling and contracting for centuries, with the crust being gradually compressed. These theorists pointed to mountain ranges full of folded strata as the places where past contraction had occurred and to seismic belts as places where contraction might be happening in the present. Contraction did explain some features, but it did not help with questions about the shapes and distribution of continents. Nor did it explain the great rift valleys and other features that are clearly caused by the crust's having been stretched rather than compressed.

When scientists discovered at the beginning of the twentieth century that the Earth's interior is kept hot by radioactive decay, some of them suggested that the Earth might not be cooling but heating up (and therefore expanding). A much smaller Earth, they suggested, could once have been covered largely by continental crust. Heating would cause the Earth to expand, and the continental crust would then crack into fragments. As expansion continued, the cracks would grow into ocean basins, and through the cracks basaltic magma would rise up from the mantle to build new oceanic crust. The theory of an expanding Earth does offer a plausible explanation for the approximately parallel coastlines of adjacent continents, but it does not easily account for mountain ranges formed by compression.

To get around the flaws in both the expansion theory and the contraction theory, geologists began to examine the effects of other forces on the crust. By the middle of the twentieth century, however, all reasonable suggestions concerning the shapes and positions of continents seemed to have been exhausted. The time was ripe for a totally new approach.

The new approach turned out to be plate tectonics. When great slabs of lithosphere—called "plates"—slide sideways across the asthenosphere, some parts of the slabs can be in compression, some in tension (that is, being pulled apart). When a plate splits in two, the broken edges of continental crust match perfectly. The energy needed to move plates turns out to be the Earth's internal heat energy, which causes great convective flows in the mantle.

Plate tectonics is the only hypothesis ever proposed that explains *all* of the Earth's major features.

CONTINENTAL DRIFT

A key proposal on the road to the formulation of the plate tectonics theory was made early in the twentieth century, soon after the contraction theory collapsed. As we learned in chapter 1, the German meteorologist Alfred Wegener proposed in 1912 that continents drift slowly across the surface of the Earth, sometimes breaking into pieces and sometimes colliding with each other. Collisions formed supercontinents, breakages formed smaller continents. According to Wegener, today's continents are the broken fragments of the most recent supercontinent.

Pangaea

Wegener's theory of **continental drift** originated when he attempted, like many before him, to explain the striking match of the shorelines on the two sides of the Atlantic, especially along Africa and South America. Wegener suggested that the most recent supercontinent existed during the Permian period when all the world's landmasses were joined together in a single continent, which he dubbed *Pangaea* (pronounced Pan-jeé-ah, meaning "all lands") (Fig. 16.1A). The northern half of Pangaea is called *Laurasia,* the southern half *Gondwanaland.* Laurasia is a name derived from *Laurentia,* an old name for the Precambrian core of Canada, and from *Eurasia,* a combined term for Europe and Asia. Gondwanaland is a name derived from a distinctive group of rocks found in central India. Similar rocks are found in Africa, Antartica, Australia, and South America—this is one of the bits of evidence that suggest that India and today's southern hemisphere continents were once part of the same landmass. According to Wegener's hypothesis, Pangaea was somehow disrupted during the Mesozoic Era, and its fragments (the continents of today) slowly drifted to their present positions. Proponents of the theory likened the process to the breaking up of a sheet of ice that floats in a pond. The broken pieces, they argued, should all fit back together again, like pieces of a jigsaw puzzle. Figure 16.1A shows that a jigsaw reconstruction indeed works well.

One impressive line of evidence presented by Wegener that supports the former existence of Pangaea is that a continental ice sheet covered parts of South America, southern Africa, India, and southern Australia 300 million years ago (Fig. 16.1B). However, if 300 million years ago continents were in the positions they occupy today, an ice sheet would have had to cover all the southern oceans and in places would even have had to cross the equator. Such a huge ice sheet could mean only that the world climate was exceedingly

cold. Yet if the climate were cold, why has no evidence of glaciation at that time ever been found in the Northern Hemisphere? This dilemma is explained neatly by continental drift: 300 million years ago, the regions covered by ice lay in high, cold latitudes surrounding the South Pole (Fig. 16.1A). No landmass covered the North Pole, however, so there was no northern ice sheet. At that time, therefore, the Earth's climates need not have been greatly different from those of today.

Despite the impressive evidence supporting continental drift, many scientists remained unconvinced by Wegener's ideas, largely because no one could explain how the solid rock of a continent could possibly overcome friction and slide across the oceanic crust. The process is like trying to slide two sheets of coarse sandpaper past each other.

Apparent Polar Wandering

Wegener died in 1930, and although debate continued, its pace slowed down. A turning point came in the 1950s. From the mid-1950s to the mid-1960s, geophysicists made a number of remarkable discoveries. The first arose through studies of paleomagnetism. As we discussed in chapter 6, certain igneous and sedimentary rocks can become weakly magnetized and therefore preserve a fossil record of the Earth's magnetic field at the time and place the rocks formed. Three essential bits of information are contained in that fossil magnetic record. The first is the Earth's polarity—whether the magnetic field was normal or reversed at the time of rock formation. The second is the location of the magnetic poles at the time the rock formed. Just as a free-swinging magnet today will point toward today's magnetic poles, so too does paleomagnetism record the direction of the magnetic poles at the time of rock formation. The third piece of information, and the one that provides the data needed to say how far from the point of rock formation the magnetic poles lay, is the magnetic inclination, which is the angle with the horizontal assumed by a freely swinging bar magnet (Fig. 16.2). Note in Figure 16.2 that the magnetic inclination varies regularly with latitude, from zero at the magnetic equator to 90° at the magnetic pole. The paleomagnetic inclination is therefore a record of the place between the pole and the equator (that is, the **magnetic latitude**) where the rock was formed. Once we know the magnetic latitude of a rock and the direction of the magnetic poles at the time the rock was formed, we can determine the position of the magnetic poles at the time the rock was formed.

Geophysicists studying paleomagnetic pole positions during the 1950s found evidence suggesting that the poles wandered all over the globe. They referred

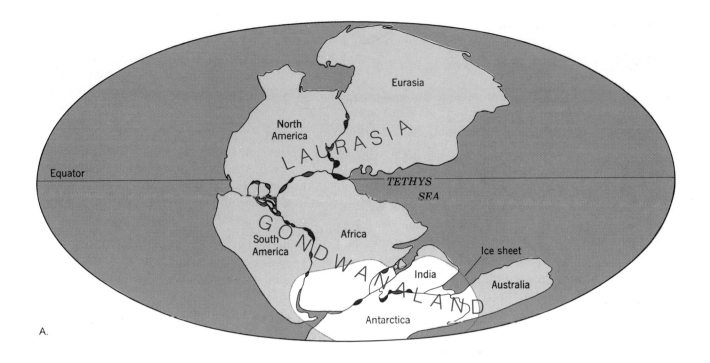

A.

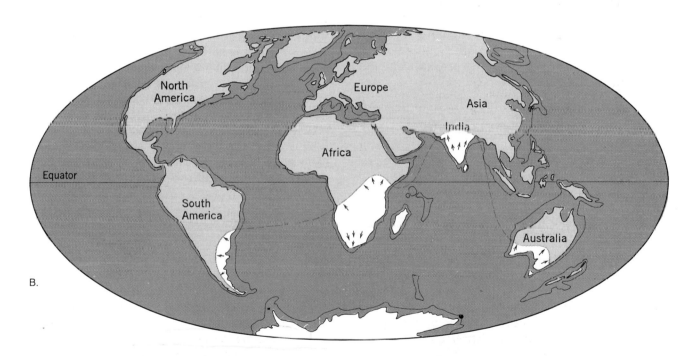

B.

FIGURE 16.1 The continents attained their present shapes when Pangaea broke apart 200 million years ago. A. The shape of Pangaea, determined by fitting together pieces of continental crust along a contour line 2000 m below sea level. (This is the line along which continental crust meets oceanic crust.) In a few places in this drawing, some overlap (black) occurs; elsewhere, small gaps (red) are found. These are places where post break-up events have modified the shapes of the continental margins. The white area is the region affected by continental glaciation 300 million years ago. B. Present continents and the 2000-m contour below sea level. The white areas are where evidence of the old ice sheets exists. Arrows show directions of movement of the former ice. The dashed line joining the glaciated regions indicates how large the ice sheet would have to have been if the continents were in their present positions at the time of glaciation.

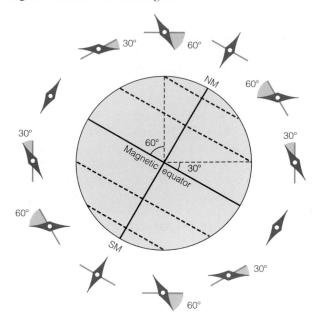

FIGURE 16.2 Change of magnetic inclination with latitude. The solid red diamonds show the magnetic inclinations of a free-swinging magnet. The solid blue line indicates a horizontal surface at each point.

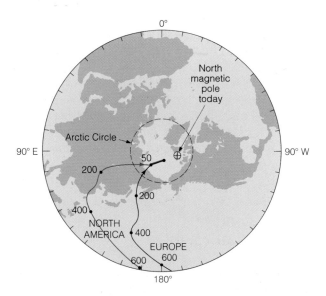

FIGURE 16.3 Curves tracing the apparent path followed by the north magnetic pole through the past 600 million years. Numbers are millions of years before the present. The curve determined from paleomagnetic measurements in North America (red) differs from that determined from measurements made in Europe (black). Wide-ranging movement of the pole is unlikely; therefore, geologists conclude that it was the continents, not the pole, that moved.

to the strange plots of paleopole positions as *apparent polar wandering*. The geophysicists were puzzled by this evidence because the Earth's magnetic poles and the poles of rotation are close together. Determination of the magnetic latitude of any rock should therefore be a good indication to the geographic latitude at which the rock was formed. When it was discovered that the path of apparent polar wandering measured in North America differed from that in Europe (Fig. 16.3), geophysicists were even more puzzled. Somewhat reluctantly, they concluded that, because it is unlikely that the magnetic poles moved, the continents and the magnetized rocks moved. In this way, the hypothesis of continental drift was revived, but a mechanism to explain how the movement occurred was still lacking.

Seafloor Spreading

Help came from an unexpected quarter. All the early debate about continental drift, and even the data on apparent polar wandering, had centered on evidence drawn from the continental crust. But if continental crust moves, why shouldn't oceanic crust move too?

In 1962, Harry Hess of Princeton University hypothesized that the topography of the seafloor could be explained if the seafloor moves sideways, away from the oceanic ridges. His hypothesis came to be called the theory of **seafloor spreading** and was soon proved correct. Once again it was paleomagnetism that provided the proof.

Hess postulated that magma rose from the interior of the Earth and formed new oceanic crust along the mid-ocean ridges. He could not explain what made the crust move away from the ridges, but he nevertheless proposed that it did and that as a consequence the oceanic crust far from any ridge was older than any crust nearer the ridge. A powerful test of the Hess theory was proposed by three geophysicists: Frederick Vine (who was a student at the time), Drummond Matthews (Vine's mentor), and Lawrence Morley (a Canadian scientist who made an independent discovery). The Vine–Matthews–Morley suggestion concerned the magnetism of the oceanic crust.

When lava is extruded at any mid-ocean ridge, the rock it forms becomes magnetized and acquires the magnetic polarity that exists at the time the lava cools. If new lava is continuously making new oceanic crust, and if the crust is continuously moving away from the oceanic ridge, then this crust should contain a continuous record of the Earth's changing magnetic polarity. The oceanic crust is, in effect, a very slowly moving magnetic tape recorder. In fact, two oceanic tape recorders commence at each mid-ocean ridge, one on each side of the ridge, in which successive

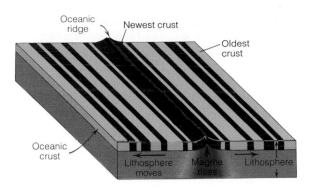

FIGURE 16.4 Schematic diagram of oceanic crust. Lava extruded along an oceanic ridge forms new oceanic crust. As the lava cools, it becomes magnetized with the polarity of the Earth's magnetic field. Successive strips of oceanic crust have alternate normal (black) and reversed (yellowish-green) polarity.

strips of oceanic crust are magnetized with normal and reversed polarity (Fig. 16.4). It was a straightforward matter to match the sort of magnetic pattern observed in Figure 16.4 with a record of magnetic polarity reversals, such as that shown in Figure 6.20. The magnetic striping allowed the age of any place on the seafloor to be determined.

Because the ages of magnetic polarity reversals had been so carefully determined, magnetic striping also provided a means of estimating the speed with which the seafloor had moved. In some places, such movement was found to be remarkably fast, as high as 10 cm/yr.

PLATE TECTONICS: A NEW PARADIGM

Proof that the seafloor moves was the spur needed for the emergence of the theory of plate tectonics. The two essential points in formulating a theory of plate tectonics were, first, that the zone of low seismic wave velocities between 100 and 350 km deep (as discussed in chapter 15) is exceedingly weak and has viscous, fluidlike properties. It was quickly realized that the asthenosphere, which had been postulated many years earlier in order to explain isostasy but had never been proved, and the low-velocity zone must be one and the same. The second point was that the rigid lithosphere is strong enough to form coherent slabs (plates) that can slide sideways over the weak, underlying asthenosphere. These two points answered the main objection to Wegener's ideas—movement must occur with minimal resistance from friction. The lithosphere is much thicker than the crust, however, and so one consequence of plate tectonics is that as the

lithosphere moves, the crust is rafted along as a passenger. Continents move, to be sure, but they do so only as portions of larger plates, not as discrete entities.

Another consequence of the theory of plate tectonics provided a solution for one of the puzzles raised by seafloor spreading. If, as the theory of seafloor spreading required, new oceanic crust is being created along the mid-ocean ridges, either the Earth must be expanding and the ocean basins getting larger, or else an equal amount of old crust must be being destroyed. The answer to the puzzle was provided by the previously unexplained Benioff zones (chapter 15). These slanting zones of deep earthquake foci are the places where old, cold lithosphere is sinking back into the asthenosphere. Destruction of old oceanic crust and the creation of new oceanic crust are in balance.

Structure of a Plate

The surface of the Earth is covered by six large and many small plates of lithosphere, each about 100 km thick (Fig. 1.9). The plates are rigid, or nearly so, and they move as single coherent units; that is, the plates do not crumple and fold like wet paper but act more like semirigid sheets of plywood floating on water. The plates may flex slightly, causing gentle up or down warping of the crust, but the only places where intense deformation occurs are at any edges along which plates impinge on each other. Such plate margins are *active zones*; plate interiors are *stable regions*.

As we learned in chapter 1, plates have three kinds of margins: divergent margins, or spreading centers along which two plates move apart from each other; convergent margins, along which two plates move toward each other; and transform fault margins, along which two plates simply slide past each other. Each type of margin creates distinctive topography in its vicinity and is associated with a distinctive kind of earthquake activity and volcanism. The features are summarized in Table 16.1.

Divergent Margins

As explained in chapter 1, a divergent margin marks the growing edge of plates along which new lithosphere is created. It is here, at the spreading center, that newly formed oceanic crust becomes magnetized.

Magnetic Records and Plate Velocities

The most recent magnetic reversal recorded near the crest of a mid-ocean ridge (in other words, the one closest to the ridge) occurred 730,000 years ago (Fig. 6.20). The oldest reversals so far found in oceanic

TABLE 16.1   Kinds of plate margins and characteristic features

Crust on Each Plate	Feature	Kind of Margin		
		Divergent	Convergent	Transform Fault
Oceanic–Oceanic	Topography	Oceanic ridge with central rift valley	Seafloor trench	Ridges and valleys created by oceanic crust
	Earthquake	All foci less than 100 km deep	Foci from 0 to 700 km	Foci as deep as 100 km
	Volcanism	Basaltic pillow lavas	Andesitic volcanoes in an arc of islands parallel to trench	Volcanism rare; basaltic along "leaky" faults
	Example	Mid-Atlantic Ridge	Tonga-Kermadec Trench; Aleutian Trench	Kane Fracture
Oceanic–Continental	Topography	—	Seafloor trench	—
	Earthquake	—	Foci from 0 to 700 km deep	—
	Volcanism	—	Andesitic volcanoes in mountain range parallel to trench	—
	Example	(No examples)	Western Coast of South America	(No examples)
Continental–Continental	Topography	Rift valley	Young mountain range	Fault zone that displaces surface features
	Earthquake	All foci less than 100 km deep	Foci as deep as 300 km over a broad region	Foci as deep as 100 km throughout a broad region
	Volcanism	Basaltic and rhyolitic volcanoes	No volcanism; intense metamorphism and intrusion of granitic plutons	No volcanism
	Example	African Rift Valley	Himalaya, Alps	San Andreas Fault

crust date back to the middle Jurassic, about 165 million years ago.

From the symmetrical spacing of magnetic time lines on the two sides of a mid-ocean ridge (Fig. 16.5) it appears that both plates move away from a spreading center at equal rates. Appearances can be deceiving, however. The same pattern of magnetic time lines shown in Figure 16.5 would be observed if the African Plate (green) were stationary and both the Mid-Atlantic Ridge and the North American Plate (orange) were

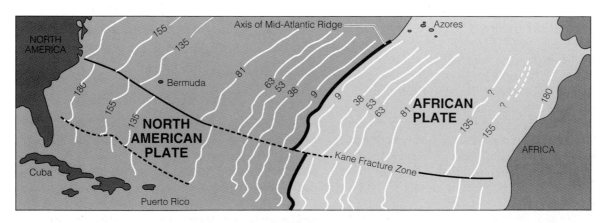

FIGURE 16.5   Age of the ocean floor in the central North Atlantic, deduced from magnetic striping. Numbers give ages in millions of years before the present. The Kane Fracture Zone, which is a transform fault, continues across the Atlantic and causes consistent displacement of the age contours.

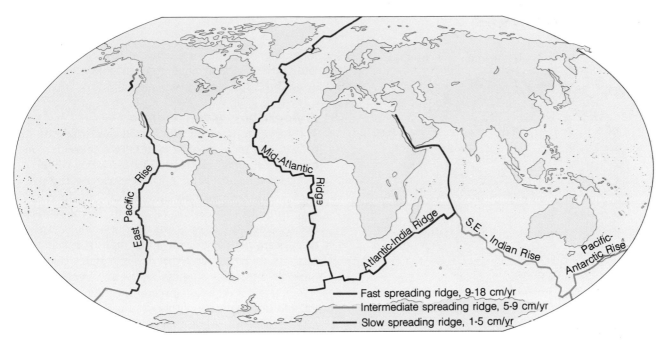

FIGURE 16.6 Spreading rates of principal mid-oceanic ridges. Fast spreading rates mean plates move away from each other between 9 and 18 cm/yr. Intermediate rates are 5 to 9 cm/yr.; slow rates are 1 to 5 cm/yr.

moving westward. (Later in this chapter, evidence will be presented to substantiate the suggestion that mid-ocean ridges do indeed move.) In fact, all that can be deduced from magnetic time lines is the *relative velocity* of two plates. An answer to the question of *absolute velocities* requires more information, and we will return to this question too, later in the chapter.

Variations in Plate Velocities

The relative velocities of some plates are much greater than others (Fig. 16.6). The differences in speed appear to be related to the amount of continental crust sitting on a plate. Plates without any continental crust tend to have high relative velocities. This is the case for the Pacific and Nazca plates. Plates with lots of continental crust, such as the African, North American, and Eurasian plates, have low relative velocities.

A second reason plate velocities vary has to do with the geometry of motion on a sphere. One might think, intuitively, that all points on a plate move with the same velocity, but that is incorrect. Our intuition would be correct only if plates of lithosphere were flat and moved over a flat asthenosphere (like plywood floating on water). If this were true, then all points on the plate would move with the same velocity. However, plates of lithosphere are pieces of a shell on a spherical Earth; they are curved, not flat. In the geometry of a sphere, any movement on the surface is a rotation about an axis of the sphere. A consequence of such rotation is that different parts of a plate move with different velocities, as shown in Figure 16.7.

Plate A in Figure 16.7 moves independently of the

Earth's rotation and rotates instead about an axis of its own, colloquially called a **spreading axis.** In the figure, point P, where the spreading axis reaches the surface, is a **spreading pole.**

The motion of each of the Earth's plates can be described in terms of rotation around a spreading

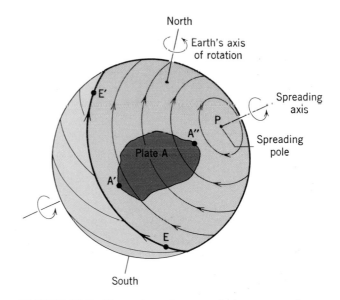

FIGURE 16.7 Movement of a curved plate on a sphere. The movement of each plate of lithosphere on the Earth's surface can be described as a rotation about the plate's own spreading axis. Point P has zero velocity because it is the fixed point around which rotation occurs. Point A', at the edge of the plate closest to the equator EE', has a high velocity. Point A'', closest to the pole, has a low velocity.

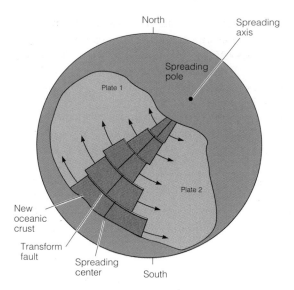

FIGURE 16.8 Relation between spreading axis, oceanic ridge, and transform faults in two adjacent plates. Plates 1 and 2 have a common spreading center (black) displaced by transform faults (red). Each segment of the oceanic ridge lies on a line of longitude that passes through the spreading pole. Each transform fault lies on a line of latitude with respect to the spreading pole. The width of new oceanic crust increases away from the spreading pole.

axis, and the velocity of each point on the plate depends on its distance from the spreading pole. One consequence of different plate velocities is that the width of new oceanic crust bordering a divergent margin increases with distance from the spreading pole (Fig. 16.8). A second consequence is that the projection of a spreading center passes through the spreading pole. Such a projection is analogous to a line of

longitude. A third consequence is that each transform fault lies on a line analogous to a line of latitude around the spreading pole.

Topography of the Seafloor

The topography of the seafloor is controlled by the growth and movement of plates. Two prominent features in particular are related to spreading centers.

The first feature is the mid-ocean ridges. The shape of any ridge is strongly influenced by the rate of spreading. Fast spreading rates, 9 to 18 cm/yr, mean that new oceanic crust is created very rapidly. This in turn means that magma must rise rapidly and continuously from below and that large magma chambers must lie at shallow depths below the center of the ridge. As a result, a fast-spreading center is thermally inflated and stands high above the seafloor. By contrast, a slow-spreading center is cooler and less inflated. The overall ridge at a slow center still stands high above the deep ocean floor, but the central rift is wider and more pronounced than the central rift in a fast spreading center.

The second prominent feature is the ocean floor itself. A large fraction of the heat that escapes from the Earth's interior does so along spreading centers. As a result, not only the mid-ocean ridges but also adjacent portions of the seafloor are high points because the lithosphere beneath them is thermally expanded. As lithosphere moves away from a ridge, it cools and contracts. As contraction occurs, the distance of the seafloor below sea level increases. A constant depth below sea level is reached after about 100 million years, by which time oceanic lithosphere has cooled and reached thermal equilibrium (Fig. 16.9). To a first approximation, therefore, the depth of the

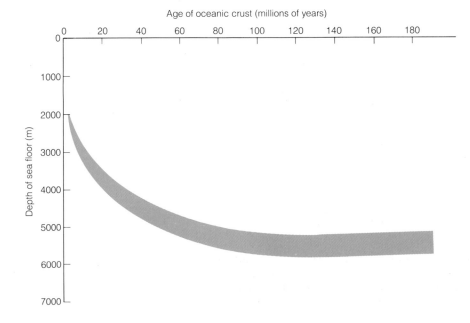

FIGURE 16.9 Depth of the seafloor in the world oceans as a function of the age of the oceanic crust. Near the spreading center, young lithosphere is thermally expanded. As it moves away from the mid-ocean ridge, the lithosphere cools and contracts. The ocean becomes deeper as a result.

ocean floor provides an estimate of the age of the oceanic crust.

Convergent Margins

Subduction zones were defined in chapter 1 as the edges along which plates of lithosphere turn down into the mantle. Because subduction removes a plate from view, what happens as the plate sinks into the asthenosphere can only be inferred. Some of the evidence on which we can base inferences comes from seismic studies, some from the deformed rocks where the two converging plates meet, and a good deal from the kind of volcanism associated with subduction zones.

Magmatic Arcs

As a subducted plate descends, it is heated up, and eventually the water-saturated oceanic crust that caps the plate reaches a temperature at which, as discussed in chapter 3, wet partial melting commences. This process forms andesitic magma. Rising to the surface, the magma forms a chain of stratovolcanoes like those shown in Figures 3.12 and 3.22. The arc-shaped region

of magmatic activity is called a **magmatic arc.** It is parallel to the seafloor trench and separated from it by a distance of 100 to 400 km, the distance depending on the angle of dip of the descending plate. If the stratovolcanoes form on oceanic crust, the magmatic arc is also known as an **island arc,** but if the stratovolcanoes are built on continental crust, the magmatic arc is called a **continental volcanic arc.** The Japanese islands and the Aleutians are modern-day island arcs. The Andes are a continental volcanic arc. Some arcs are curved only slightly because they are parts of a circle with a large radius, and some are highly curved because they are parts of a circle with a smaller radius. The radius of curvature is an indication of the angle at which lithosphere is plunging back into the mantle. If the plunge were perpendicular to the Earth's surface, an island arc would be straight. If the angle of plunge is very shallow, the arc has a pronounced curvature (Fig. 16.10).

Determination of the age of oceanic crust being subducted shows that old, cold, and therefore dense crust forms island arcs that have a large radius of curvature (implying a steep plunge). Young oceanic crust that has not reached thermal equilibrium and is less dense than older, colder crust forms arcs with short

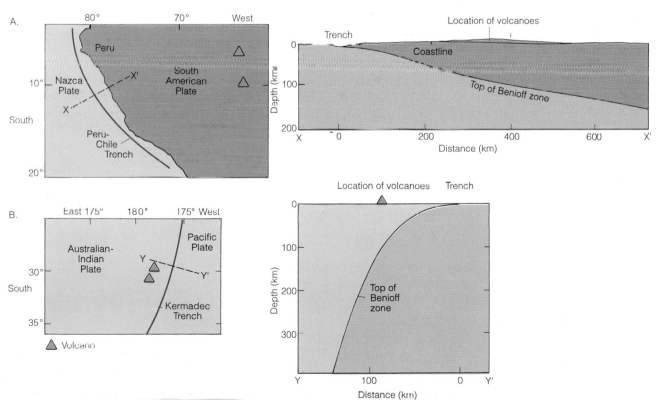

FIGURE 16.10 The steepness of the angle at which lithosphere sinks at a subduction zone controls the curvature of the trench. A. Beneath Peru, the top of the Benioff zone dips at a shallow angle, and so the Peru–Chile Trench has a pronounced curvature. B. The Benioff zone associated with the Kermadec Trench in the southwest Pacific has a steep dip, and so the trench has only a slight curvature.

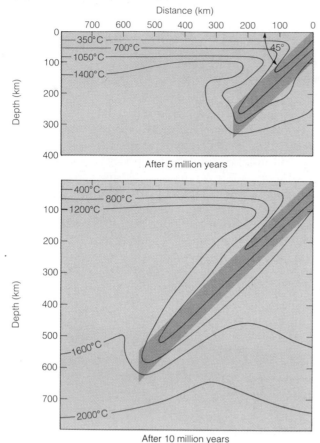

FIGURE 16.11 Computer-aided calculations of the temperature of a descending slab of cool lithosphere. Contours depict the temperature. A plate 100 km thick, descending at an angle of 45° and a rate of 8 cm/yr, will cool the surrounding mantle (as indicated by the bending of the temperature contours), but will slowly become heated as it sinks. Between 600 and 700 km, the temperature of the tip of the descending slab reaches the temperature of the adjacent mantle and earthquakes cease.

radii of curvature (implying a shallow plunge angle). This observation is informative because it indicates that oceanic lithosphere must be sinking under its own weight through the hot, weak asthenosphere. This means that old, cold lithosphere, when capped by oceanic crust, must be more dense than the hot, plastic asthenosphere. The older and colder the lithosphere, the faster the rate of sinking and the steeper the angle of the Benioff zone.

As lithosphere sinks, it must heat up. Earthquakes can occur in the down-going slab as long as it is cool enough to be brittle. Even with a rapid sinking rate of 8 cm/yr, calculations show that lithosphere retains some brittle properties down to a depth of 700 km (Fig. 16.11). This is probably the reason that a few earthquake foci are as deep as 700 km.

Mélange

Many features on the Earth's surface occur as a result of deformation along convergent margins. A distinctive feature of some margins is the development of a **mélange,** a chaotic mixture of broken, jumbled, and thrust-faulted rock. Once a subduction zone forms and a seafloor trench is created, sediment accumulates in the trench. A sinking plate drags the sedimentary rock formed from this accumulated sediment downward beneath the overriding plate. Sedimentary rock has a low density. As a result, it is buoyant and cannot be dragged down very far. Caught between the overriding plate and the sinking plate, the sediment becomes shattered, crushed, sheared, and thrust-faulted to form a mélange (Fig. 16.12). As the mélange thickens, it becomes metamorphosed. The cold sedimentary rocks are dragged down so rapidly that they remain cooler than adjacent rock at the same depth. The kind of metamorphism that is common in many mélange zones, therefore, is that which occurs along curve C

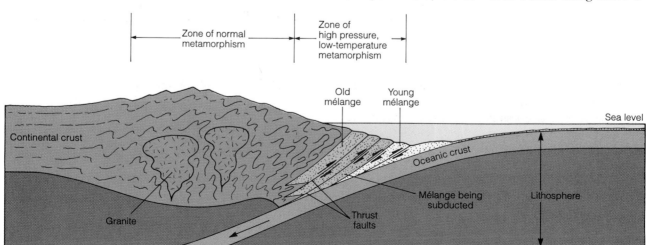

FIGURE 16.12 A mélange is formed when young sediment in a trench is smashed by moving lithosphere and dragged downward in slices bounded by thrust faults. As successive slices are dragged down, older mélange, closer to the overriding plate, is pushed back up. The process is like lifting a deck of cards by adding new cards at the base of the deck.

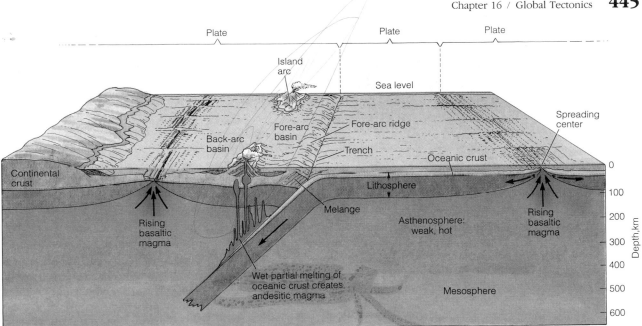

FIGURE 16.13 Structure of a tectonic plate at a convergent margin. Lithosphere is capped by oceanic crust formed by basaltic magma rising from the asthenosphere. Moving laterally, the lithosphere accumulates a thin layer of marine sediment and eventually starts sinking into the asthenosphere. Along the line of subduction, an oceanic trench is formed, and sediment deposited in the trench, plus sediment from the moving plate, is compressed and deformed to create a mélange. The sinking oceanic crust eventually reaches the temperature where wet partial melting commences and forms andesitic magma, which then rises to form an island arc of stratovolcanoes on the adjacent plate. Behind the island arc, tensional forces lead to the development of a black-arc basin.

of Figure 5.17—a high-pressure, low-temperature metamorphism distinguished by blue schists and eclogites. The blue color comes from a bluish amphibole called glaucophane.

Fore-Arc Ridges and Fore-Arc Basins

Between the trench and the magmatic arc, both of which are prominent topographic features, two less-prominent features are present along many convergent margins: the fore-arc ridge and the fore-arc basin (Fig. 16.13). A **fore-arc ridge** is commonly underlain by mélange and is caused by a local thickening of the crust due to thrust faulting at the edge of the overriding plate. A **fore-arc basin** is a low-lying region between the fore-arc ridge and the magmatic arc. The island of Sumatra in Indonesia is part of a magmatic arc that is flanked by a fore-arc ridge and basin (Fig. 16.14).

Back-Arc Basins

When the sinking rate of a subducting plate is faster than the forward motion of the overriding plate, part of the overriding plate can be subjected to tensional (pulling) stress. The leading edge of the overriding plate must remain in contact with the subduction edge

or else a huge void would open. What happens is that the overriding plate grows slowly larger at a rate equal to the difference in velocities between the two plates. Most commonly, this process is manifested by a thinning of the crust and an opening of an arc-shaped basin behind and parallel to the magmatic arc (Figs. 16.13 and 16.14). Basaltic magma may rise into such a **back-arc basin,** and a small region of new oceanic crust may even form.

Transform Fault Margins

The faults at plate margins are transform faults. As discussed in chapter 15, these are huge, vertical, strike-slip faults cutting down into the lithosphere. They can form when either a divergent margin or a convergent margin fractures the lithosphere (Fig. 16.15). The plates on either side of transform fault margins smash and abrade each other like two strips of sandpaper. As a result, the faults are marked by zones of intensely shattered rock. Where the faults cut oceanic crust, they make elongate zones of narrow ridges and valleys on the seafloor. They also influence topography by forming narrow valleys where they cut

evidence also indicates that about 40 million years ago the Pacific Plate apparently changed its direction of motion.)

Wilson made his suggestion as a way of testing seafloor spreading. However, it was not long before he realized that if long-lived hot spots do exist deep in the mantle, they might provide a series of fixed points against which absolute plate motions can be measured. More than a hundred hot spots have now been identified (Fig. 16.17). Using them for reference, geologists have found that the African Plate is very nearly stationary (evidenced by the fact that volcanoes there seem to be very long lived). Because the African Plate is almost completely surrounded by spreading edges, and because the relative velocities along the encircling ridges are closely matched, we must conclude that the Mid-Atlantic Ridge is moving westward, and that the oceanic ridge that runs up the center of the Indian Ocean is moving to the east. If the absolute motion of the African Plate is zero or nearly so, the Mid-Atlantic Ridge in the southern Atlantic Ocean must be moving westward at the rate of about 2 cm/yr, and the absolute velocity of the South American Plate must be 4 cm/yr.

The Australian–Indian Plate is moving almost directly northward. All other plates, with the exception of the nearly stationary African Plate, are moving approximately eastward or westward. Several plates do not have convergent margins and must therefore be increasing in size. Most of the modern subduction zones are to be found around the Pacific Ocean along the edge of the Pacific Plate, and thus much of the oceanic lithosphere now being destroyed is in the Pacific. It follows then that the Indian Ocean, the Atlantic Ocean, and most other oceans must be growing larger, while the Pacific Ocean must be steadily getting smaller.

CAUSE OF PLATE TECTONICS

Just as Wegener could not explain what made continents drift, we are still unable to say exactly how convection currents make plates of lithosphere move. The situation is analogous to knowing the shape, color, size, and speed of an automobile, and knowing that gasoline supplies the energy needed for movement, but not knowing how the gasoline makes the engine work. Until the driving mechanism is explained, plate tectonics must remain only an approximate description. Meanwhile, we can hypothesize about the causes of the motion and test the hypotheses by making detailed calculations based on the laws of nature.

The lithosphere and asthenosphere are closely bound together. If the asthenosphere moves, the litho-

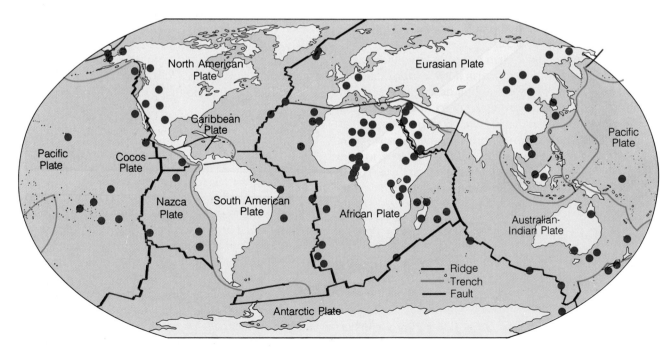

FIGURE 16.17   Long-lived hot spots at the Earth's surface, each a center of volcanism, are believed to lie above deep-seated sources of magma in the mantle. Because the magma sources lie far below the lithosphere and so do not move laterally, hot spots can be used to determine the absolute motions of plates.

sphere must move too, just as the movement of sticky molasses moves a piece of wood floating on the surface of the molasses. Also, movement of the lithosphere causes movement in the asthenosphere. Such is our state of uncertainty, though, that we cannot yet separate the relative importance of the two effects. However, on two points we can be quite certain: (1) the lithosphere must have kinetic energy in order to move, and (2) the source of this kinetic energy is the Earth's internal heat. We know, too, that the heat energy reaches the surface by convection in the mantle. What has not yet been figured out is the precise way convection and plate motions are linked.

Convection in the Mantle

The mantle is solid rock; however, it is hot and apparently weak enough so that, under low strain rates, even small stresses make it flow like a viscous liquid. Like a liquid, too, the mantle is subject to convection currents when a local source of heat causes a mass of rock to become heated to a higher temperature than surrounding rock. The heated mass expands, becomes less dense, and rises very slowly. To compensate for the rising mass, rock that is cooler and denser must sink downward. The rate at which heat reaches the Earth's surface can be accounted for only if convection in the mantle brings heat from the deep interior.

Any discussion about convection is speculative. Evidence from seismic tomography and heat flow indicates that convection of some sort does occur beneath the lithosphere. Even so, it is difficult to see how plate motions can be due entirely to convective motions. For this reason, most scientists believe that lithosphere motion is due to a combination of processes and that convection is only one of the processes. One important thing that convection must do is keep the asthenosphere hot and weak by bringing up heat from the deep mantle and core. In this sense at least, convection is essential for plate tectonics.

Movement of the Lithosphere

Three forces might play a role in moving the lithosphere. The first is a push away from a spreading center. Rising magma at a spreading center creates new lithosphere and in the process pushes the plates sideways (Fig. 16.18A). Once the process is started, it tends to keep itself going. The problem is that pushing involves compression, but the existence of normal faults along a mid-ocean ridge indicates a state of *tension*.

A second way by which lithosphere could be made to move is by dragging. Proponents of the dragging

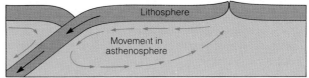

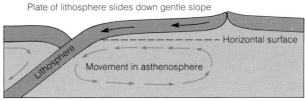

FIGURE 16.18 Three suggested mechanisms by which lithosphere might move over the asthenosphere. A. Magma rising at a spreading edge exerts enough pressure to push the plates of lithosphere apart. B. A tongue of cold, dense lithosphere sinks into the mantle and pulls the rest of the plate behind it. C. A plate of lithosphere slides down a gently inclined surface of asthenosphere.

idea point out that a descending tongue of old, cold lithosphere is more dense than the surrounding hot asthenosphere. Because rock is a poor conductor of heat, they argue, the temperature at the center of a descending slab can be as much as 1000°C cooler than the mantle at depths of 400 to 500 km. The dense slab of lithosphere must therefore sink under its own weight and exert a pull on the entire plate. This is like a heavy weight that hangs over the side of a bed and is tied to the edge of a sheet. The weight falls and pulls the sheet across the bed. To compensate for the descending lithosphere, rock in the asthenosphere must flow slowly back toward the spreading edge (Fig. 16.18B).

However, both the pushing and the dragging mechanism have problems. Plates of lithosphere are brittle, and they are much too weak to transmit large-scale pushing and pulling forces without major deformation occurring in their middle. Deformation is not present, however, and mid-plate seismicity, which would be expected for a plate undergoing deformation, is infrequent.

The third possible mechanism for lithosphere movement is for the plate to slide downhill away from the spreading center. The lithosphere grows cooler and thicker away from a spreading center. As a consequence, the boundary between lithosphere and asthenosphere slopes away from the spreading center. If the slope is as little as 1 part in 3000, its own weight could cause the lithosphere to slide at a rate of several centimeters per year (Fig. 16.18C).

At present, there is no way to choose among the three lithosphere mechanisms. Calculations suggest that each operates to some extent, so that the entire process is possibly more complicated than we now imagine. The prevailing idea at present is that subduction starts when old, cold lithosphere breaks and begins to sink, thereby pulling on the plate and starting the movement, and then the other processes combine to keep the movement going. Only future research will resolve the question.

PLATE TECTONICS, CONTINENTAL CRUST, AND MOUNTAIN BUILDING

In a sense, continental crust is simply a passenger being rafted on large plates of lithosphere. But it is a passenger that is buffeted, stretched, fractured, and altered by the ride.

Someone once characterized continental crust as the product of bump-and-grind tectonics. Each bump between two crust fragments forms a mountain belt, each grind a strike-slip fault, each stretch a rift valley. Scars left in the continental crust by bump-and-grind tectonics are evidence of former plate motions. That this evidence exists is fortunate because the most ancient crust known to exist in the ocean dates only from the mid-Jurassic Period. Indeed, the only direct evidence concerning geological events more ancient than the mid-Jurassic comes from the continental crust.

In order to discuss how the continental crust has been affected by plate tectonics, it is helpful to look first at the large-scale structure of continents.

Regional Structures of Continents

Cratons

On the scale of a continent, two kinds of structural units can be distinguished within the continental crust. The first is a core of very ancient rock called a **craton** (Fig. 16.19). The term is applied to any portion of the Earth's crust that has attained tectonic and isostatic

stability. Rocks within cratons may be deformed, but the deformation is invariably ancient.

Orogens

Draped around cratons are the second kind of crustal building unit, **orogens,** which are elongate regions of crust that have been intensely folded and faulted during continental collisions. Crust in an orogen is commonly thicker than crust in a craton and many orogens—even some very old ones—have not yet attained isostatic equilibrium. Orogens are the eroded roots of ancient mountain ranges that formed as a result of collisions between the cratons. Orogens differ from each other in age, history, size, and details of origin; however, all were once mountainous terrains, and all are younger than the cratons they surround. Only the youngest orogens are mountainous today; ancient orogens, now deeply eroded, reveal their history through the kinds of rock they contain and the kind of deformation present.

Continental Shields

An assemblage of cratons and ancient orogens that has reached isostatic equilibrium is called a **continental shield.** That portion of a continental shield that is covered by a thin layer of little-deformed sediments is call a *stable platform*. North America has a huge continental shield at its core, and around the shield are four young orogens (Fig. 16.19). Because the North American shield crops out in Canada (especially Ontario and Quebec), but is mostly covered by flat-lying sedimentary rocks in the United States, geologists often refer to it as the Canadian Shield.

Through careful mapping and radiometric dating, geologists have identified several ancient cratons and orogens in the Canadian Shield (Fig. 16.19). Within the cratons, all rocks are older than 2.0 billion years. Such rocks can be observed in many places in eastern Canada, but within the United States cratonic rocks crop out only in a small region around Lake Superior. Nevertheless, by drilling through the cover of sedimentary rocks on the stable platform, geologists have discovered that cratons and orogens similar to those that surface in eastern Canada also lie below much of the central United States and part of western Canada.

The small cratons within the Canadian Shield shown in Figure 16.19 were probably minicontinents during the Archean Eon. By about 1.6 billion years ago, these minicontinents had become welded together to form the assemblage of cratons and ancient orogens we see in North America today. Each time two cratonic fragments collided, an orogen was formed between them. The existence of ancient collision belts—orogens—is the best evidence available to sup-

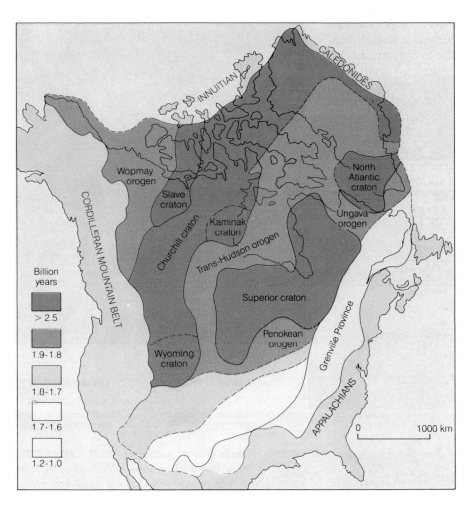

FIGURE 16.19 The North American cratons and associated orogens. The Caledonide, Appalachian, Cordilleran, and Innuitian orogens are each younger than 600 million years. The assemblage of ancient cratons and orogens that is surrounded by the four young orogens is the Canadian Shield.

port the idea that plate tectonics operated at least as far back as 2 billion years ago.

Continental Margins

The fragmentation, drift, and welding together of pieces of continental crust are inevitable consequences of plate tectonics. Various combinations of these processes are responsible for the five types of continental margins we know of today: passive, convergent, collision, transform fault, and accreted terrane. Before we discuss additional evidence for plate tectonics, it will be helpful to review briefly the features associated with each of the continental margins.

Passive Continental Margins

A **passive continental margin** is one that occurs in the stable interior of a plate. The Atlantic Ocean margins of the Americas, Africa, and Europe are each passive continental margins. The eastern coast of North America, for example, is in the stable interior of the North American Plate, far from the plate margins. Passive continental margins develop when a new ocean

basin forms by the rifting of continental crust, as illustrated in Figure 16.20. This process can be seen in the Red Sea, which is a young ocean with an active spreading center running down its axis (Fig. 16.21). New, passive continental margins have formed along both edges of the Red Sea.

Passive continental margins are places where great thicknesses of sediment accumulate. The kinds of sediment deposited are distinctive, and the Red Sea provides an example. Deposition commenced with clastic, non-marine sediments followed by evaporites and then clastic marine shales. The sequence apparently arises in the following manner. Basaltic magma, associated with formation of the new spreading edge that splits the continent, heats and expands the lithosphere so that a plateau forms with an elevation of as much as 2.5 km above sea level. Tensional forces cause normal faults and form a rift so that there is a pronounced topographic relief between the plateau and the floor of the rift. The earliest rifting of the Red Sea must have looked very much the way the African Rift Valley looks today. Before the rift floor sank low enough for seawater to enter, clastic nonmarine sediments, such as conglomerates and sandstones, were shed from the steep

than 2 cm/yr. This suggests that Australia is moving north with an absolute motion of about 5 cm/yr (Fig. B16.2).

The climate of the world in the early Tertiary was much warmer than it is today. At the measured average rate of movement, 40 million years ago the Western Australian Shield was 2000 km south of its present position; in other words, it was in a temperate zone of high rainfall like that of southern Chile today. A warm early Tertiary climate and high rainfall probably explain the origin of the old laterite. Like a ship slowly sailing north, the Western Australia Shield is passing through different climatic zones. Today it lies in the dry, subtropical belt. However, if the present direction and rate of motion continue for another 20 million years, the continent will enter a zone of tropical climate.

SUMMARY

1. The lithosphere is broken into six large and many smaller plates, each about 100 km thick and each slowly moving over the top of the weak asthenosphere beneath it.

2. Three kinds of margins are possible between plates. Divergent margins (spreading centers) are those where new lithosphere forms; plates move away from them. Convergent margins (subduction zones) are lines along which plates compress each other and along which lithosphere capped by oceanic crust is subducted back into the mantle. Transform fault margins are lines where two plates slide past each other.

3. Plate movement can be described in terms of rotation across the surface of a sphere. Each plate rotates around a spreading axis. The spreading axis does not necessarily coincide with the Earth's axis of rotation.

4. Because plate movement is a rotation, the velocity varies from place to place on the plate.

5. Each segment of oceanic ridge that marks a divergent margin between two plates lies on a line of longitude passing through the spreading pole. Each transform fault margin between two plates lies on the line of latitude of the spreading pole.

6. The mechanism that drives a moving plate is not known, but apparently it results from a combination of convection in the mantle plus forces that act on a plate of lithosphere.

7. Two major structural units can be discerned in the continental crust. Cratons are ancient portions of the crust that are tectonically and isostatically stable. Separating and surrounding the cratons are orogens of highly deformed rock, marking the site of mountain ranges.

8. An assemblage of cratons and deeply eroded orogens that forms the core of a continent is a continental shield.

9. There are five kinds of continental margins: passive, convergent, collision, transform fault, and accreted terrane.

10. Passive margins develop by rifting of the continental crust. The Red Sea is an example of a young rift, the Atlantic Ocean is a mature rift.

11. A characteristic sequence of sediments forms along a passive continental margin, starting with clastic nonmarine sediments, followed by marine evaporites, and then marine clastic sediments.

12. Continental convergent margins are the locale of paired metamorphic belts, chains of stratovolcanoes (magmatic arc), and linear belts of granitic batholiths.

13. Collision margins are the locations of fold-and-thrust mountain systems.

14. Transform fault margins occur where the edge of a continent coincides with the transform fault boundary of a plate.

15. Accreted terrane margins arise from the addition of blocks of crust brought in by subduction and transform fault motions.

IMPORTANT TERMS TO REMEMBER

accreted terrane (p. 456)

back-arc basin (p. 445)

continental drift (p. 436)
continental shield
(p. 450)

continental volcanic arc
(p. 443)
craton (p. 450)

décollement (p. 458)
detachment surface (p. 458)

fore-arc basin (p. 445)
fore-arc ridge (p. 445)

hot spot (p. 447)

island arc (p. 443)

magmatic arc (p. 443)
magnetic latitude (p. 436)
mélange (p. 444)

orogen (p. 450)

*passive continental margin
 (p. 451)*
plate triple junction (p. 450)

seafloor spreading (p. 438)
spreading axis (p. 441)
spreading pole (p. 441)

QUESTIONS FOR REVIEW

1. Long ago it was suggested that the location and shape of mountain ranges and other major topographic features were the result of contraction as the Earth cooled. Why is that explanation incorrect?

2. Who was Alfred Wegener and what revolutionary idea did he suggest? Why were scientists reluctant to accept Wegener's idea when it was first proposed?

3. Explain how the apparent wandering of magnetic poles throughout geologic history can be used to help prove continental drift.

4. What are the main features of seafloor spreading? What critical test proved that the seafloor does move?

5. What are the three kinds of margins that bound tectonic plates?

6. How are the velocities of tectonic plates determined? Do magnetic time lines on plates provide relative or absolute plate velocities?

7. What is a spreading pole and how does plate velocity depend on the position of a plate relative to the spreading pole?

8. Why does the radius of curvature differ from one island arc to another?

9. What is a mélange and what kind of metamorphism is associated with mélanges? Where might mélanges be forming today?

10. Draw a cross section through the lithosphere at a convergent plate margin and mark the positions of the fore-arc basin, the back-arc basin, and the magmatic arc.

11. What are cratons and how do they differ from orogens? Name three orogens in North America that are less than a billion years old.

12. Name three fold-and-thrust mountain ranges.

13. What is the tectonic environment of an island arc? Name two examples of modern oceanic island arcs.

14. What is the origin of the Cascade Range; the Sierra Nevada; the Appalachians?

15. What evidence indicates that plate tectonics has been operating for at least the last 2 billion years of Earth history?

16. Name the five kinds of continental margins and describe how they form

17. Describe the sequence of events that leads to the opening of a new ocean basin flanked by two passive continental margins.

18. How does an accreted terrane margin form? Name a continental margin that was modified by terrane accretion.

19. How are the absolute velocities and directions of tectonic plates determined?

20. What is a thermal plume? How might it be connected with plate tectonics?

21. Name three forces that act on lithosphere that might cause it to slide over the asthenosphere.

MINERAL RESOURCES

Can you imagine a world without machines? Our modern world with its 5.2 billion inhabitants couldn't operate without them. Machines are used to produce our food, to make our clothes, to transport us around, and to help us communicate. The metals needed to build machines and the fuels needed to run them are dug from the Earth. Deposits of metallic minerals and fuels are formed by geological processes. Geological processes, therefore, influence the daily lives of each and every one of us.

In many of the previous chapters, we pointed out how geological processes such as weathering, sedimentation, and volcanism can, under suitable conditions, form valuable mineral and energy deposits. The "suitable conditions" are not common, however, and for this reason mineral and energy resources are limited in quantity and hard to find. No geological challenge is more difficult or more rewarding than the search for, and discovery of, new resources of minerals and energy. In this chapter we focus on how and where the different kinds of mineral and energy deposits form.

We turn first to the mineral substances that provide materials from which machines and a myriad other necessary things can be made. The number and diversity of such substances are so great that to make a simple classification covering all of them is almost impossible. Nearly every kind of rock and mineral can be used for something. A society such as ours, which depends so much on energy-intensive industry, not only requires a diverse group of metals for machines, but also demands a host of nonmetallic mineral products, such as shale and limestone for making cement, gypsum for making plaster, salt for making chemical compounds, and calcium phosphate (apatite) for making fertilizer.

Supplies of Minerals

Many industrialized nations are rich in many kinds of **mineral deposits** (any volume of rock containing an enrichment of one or more minerals), which they are exploiting vigorously. Yet no nation is entirely self-sufficient in mineral supplies, and so each must trade with other nations to fulfill its needs (Fig. 17.1).

Mineral resources have three distinctive aspects. First, occurrences of usable minerals are limited in abundance and distinctly localized at places within the Earth's crust. This is the main reason no nation is self-sufficient where mineral supplies are concerned. Because usable minerals are localized, they must be searched out, and the search ranges over the entire globe. The special branch of geology concerned with discovering new supplies of usable minerals is called, appropriately, **exploration geology.**

Second, the quantity of a given material available in any one country is rarely known with accuracy, and the likelihood that new deposits will be discovered is difficult to assess. As a result, production over a period of years can be difficult to predict. Thus, a country that today can supply its need for a given mineral substance may face a future in which it will become an importing nation. As we discussed in the chapter-ending essay for chapter 2, Britain is an example of a country that once supplied most of its own mineral needs but can no longer do so today. A little more than a century ago, Britain was a great mining nation, producing and exporting such materials as tin, copper, tungsten, lead, and iron. Today, the known deposits have been worked out.

Third, unlike plants and animals, which are cropped yearly or seasonally and then replenished, deposits of minerals are depleted by mining and eventually exhausted. This disadvantage can be offset only by finding new occurrences or by using the same material repeatedly—that is, by making use of scrap.

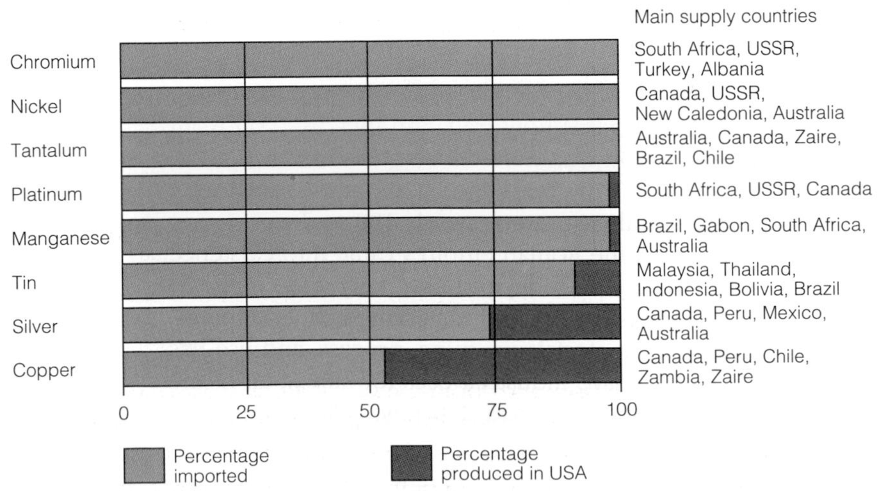

Main supply countries

Mineral	Countries
Chromium	South Africa, USSR, Turkey, Albania
Nickel	Canada, USSR, New Caledonia, Australia
Tantalum	Australia, Canada, Zaire, Brazil, Chile
Platinum	South Africa, USSR, Canada
Manganese	Brazil, Gabon, South Africa, Australia
Tin	Malaysia, Thailand, Indonesia, Bolivia, Brazil
Silver	Canada, Peru, Mexico, Australia
Copper	Canada, Peru, Chile, Zambia, Zaire

Percentage imported / Percentage produced in USA

FIGURE 17.1 Selected mineral substances for which the United States consumption exceeds production. The difference must be supplied by imports. Data are plotted for 1989, but the percentage changes little from year to year.

The peculiarities of the mineral industry place a premium on the skills of exploration geologists and engineers, who play the essential roles in finding and mining mineral deposits. The finding is accomplished through the application of the basic geological principles set forth in this book. Much ingenuity has been expended in bringing the production of minerals to its present state. Because known deposits are being rapidly exploited while demands for minerals continue to grow, we can be sure that even more ingenuity will be needed in the future.

Ore

Minerals for industry are sought in deposits from which the desired substances can be recovered least expensively. The more concentrated the desired minerals, the more valuable the deposit. In some deposits the desired minerals are so highly concentrated that even very rare substances such as gold and platinum can be seen with the naked eye. For every desired mineral substance, a *grade* (level of concentration), exists below which the deposit cannot be worked economically (Fig. 17.2). To distinguish between profitable and unprofitable mineral deposits, the word **ore** is used, meaning an aggregate of minerals from which one or more minerals can be extracted profitably. It is not always possible to say exactly what the grade must be, nor how much of a given mineral must be present, in order to constitute an ore. Two deposits may have the same grade and be the same size, but one is ore and the other is not. There could be many reasons for the difference; for example, the uneconomic deposit could be too deeply buried or located in so remote an area that the costs of mining and transport are so high that the final product is not competitive with the same product from other deposits. Furthermore, as costs and market prices fluctuate, a particular aggregate of minerals may be an ore at one time but not at another.

Gangue

Ore minerals such as sphalerite, galena, and chalcopyrite (chapter 2) from which desired metals can be extracted are usually mixed with other, nonvaluable minerals, collectively termed **gangue** (pronounced gang). Familiar minerals that commonly occur as gangue are quartz, feldspar, mica, calcite, and dolomite.

Hydrothermal Mineral Deposits

Many of the most famous mines in the world contain ores that were formed when their ore minerals were deposited from hydrothermal solutions. As we dis-

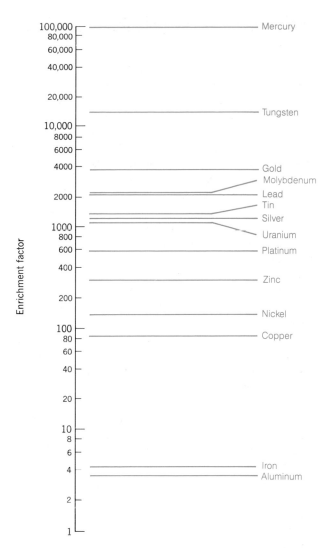

FIGURE 17.2 Before a mineral deposit can be worked profitably, the percentage of valuable metal in the deposit must be greatly enriched above its average percentage in the Earth's crust. The enrichment is greatest for metals that are least abundant in the crust, such as gold and mercury. As mining and mineral processing become more efficient and less expensive, it is possible to work leaner ore, and enrichment factors decline. Note that the scale is a magnitude (logarithmic) scale, in which the major divisions increase by multiples of ten.

cussed in chapter 5, it is probable that more mineral deposits have been formed by deposition from hydrothermal solutions than by any other mechanism. However, the origins of hydrothermal solutions are often difficult to decipher. Some solutions originate when water dissolved in a magma is released as the magma rises and cools. Other solutions are formed from rainwater or seawater that circulate deep in the crust.

An example of the way a hydrothermal solution can form from deeply circulating seawater is shown in

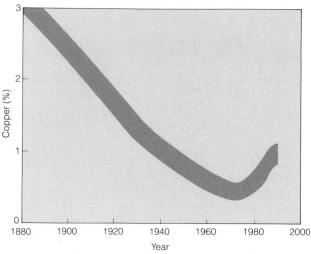

FIGURE 17.3 Declining percentage of copper needed for a mineral deposit to be an ore. The large reduction occurred because large-volume mining with efficient machines led to a steady reduction of mining costs. The reasons for the sharp upturn that started in the 1970s are discussed in the text.

Figure 5.20. Because the heat source for seawater hydrothermal solutions of the kind illustrated in Figure 5.20 is mid-ocean ridge volcanism, and because the ore minerals deposited are always sulfides, mineral deposits formed from such solutions are called **volcanogenic massive sulfide deposits.**

The ore-mineral constituents in volcanogenic massive sulfide deposits come originally from the igneous rocks of the oceanic crust. Heated seawater reacts with the rocks it is in contact with, causing changes in both mineral composition and solution composition. For example, feldspars are changed to clays and epidote, and pyroxenes are changed to chlorites. As the minerals are transformed, trace metals such as copper and zinc, present by atomic substitution, are released and become concentrated in the slowly evolving hydrothermal solution.

The ore challenge is twofold: (1) to find the ores (which all together underlie an infinitesimally small proportion of the Earth's land area), and (2) to mine the ores and get rid of the gangue as cheaply as possible. Both steps are technical problems; engineers have been so successful in solving them that some deposits now considered ore are only one-sixth as rich as were the lowest-grade ores 100 years ago (Fig. 17.3). Notice, however, that the curve in Figure 17.3 has risen up from its lowest value, reached during the 1970s. The reason for this is that overproduction of copper around the world, combined with an economic recession, produced a glut of newly mined copper. This, in turn, drove the price of copper down and led to the closing of many mines, in particular those mines with the lowest grades.

ORIGIN OF MINERAL DEPOSITS

All ores are mineral deposits because each of them is a local enrichment of one or more minerals or mineraloids. The reverse is not true, however. Not all mineral deposits are ores. "Ore" is an economic term, while "mineral deposit" is a geological term. How, where, and why a mineral deposit forms is the result of one or more geological processes. Whether or not a given mineral deposit is an ore is determined by how much we human beings are prepared to pay for its content. Fascinating though the economics of ores and mining is, this topic cannot be explored in this volume. Instead, discussion is limited to the origin of mineral deposits without necessary regard to questions of economics.

In order for a deposit to form, some process or combination of processes must bring about a localized enrichment of one or more minerals. A convenient way to classify mineral deposits is through the principal concentrating process. Minerals become concentrated in five ways:

1. Concentration by hot, aqueous solutions flowing through fractures and pore spaces in crustal rock to form **hydrothermal mineral deposits** (chapter 5).
2. Concentration by magmatic processes within a body of igneous rock to form **magmatic mineral deposits** (chapter 3).
3. Concentration by precipitation from lake water or seawater to form **sedimentary mineral deposits** (chapter 4).
4. Concentration by flowing surface water in streams or along the shore, to form *placers* (chapter 9).
5. Concentration by weathering processes to form **residual mineral deposits** (chapter 7).

Hydrothermal solutions having similar compositions can form in many different ways. Which ore constituents are carried in solution depends on the kinds of rocks involved in the formation of the solution. For example, copper and zinc are present in pyroxenes by atomic substitution, so that the pyroxene-rich rocks of the oceanic crust yield solutions charged with copper and zinc. Most volcanogenic massive sulfide deposits are copper and zinc rich as a result.

The most important question concerning hydrothermal solutions is not *where* the water and dissolved mineral constituents came from, but rather what made the solutions precipitate their soluble mineral load and form a mineral deposit?

Causes of Precipitation

When a hydrothermal solution moves slowly upward, as with groundwater percolating through an aquifer, the solution cools very slowly. If dissolved minerals were precipitated from such a slow-moving solution, they would be spread over great distances and would not be sufficiently concentrated to form an ore. But when a solution flows rapidly, as in an open fracture through a mass of shattered rock, or through a layer of porous tephra where flow is less restricted, cooling can be sudden and can happen over short distances. Rapid precipitation and a concentrated mineral deposit are the result. Other effects—such as boiling, a rapid decrease in pressure, composition changes of the solution caused by reactions with adjacent rock, and cooling as a result of mixing with seawater—can also cause rapid precipitation and form concentrated deposits. When valuable minerals are present, an ore can be the result.

Examples of Precipitation

Veins form when hydrothermal solutions deposit minerals in open fractures, and many such veins are found in regions of volcanic activity (Fig. 17.4). The famous gold deposits at Cripple Creek, Colorado, were formed in fractures associated with a small caldera, and the huge tin and silver deposits in Bolivia are in fractures that are localized in and around stratovolcanoes. In each case the fractures formed as a result of volcanic activity, and the magma chambers that fed the volcanoes served as the sources of the hydrothermal solutions that rose up and formed the mineralized veins.

A cooling granitic stock or batholith is a source of heat just as the magma chamber beneath a volcano is—and it can also be a source of hydrothermal solutions. Such solutions move outward from a cooling stock and will flow through any fracture or channel, metasomatically altering the surrounding rock in the process and commonly depositing valuable minerals. Many famous ore bodies are associated with intrusive igneous rocks. The tin deposits of Cornwall, England, and the copper deposits at Butte, Montana; Bingham, Utah; and Bisbee, Arizona, are examples.

Magmatic Mineral Deposits

The processes of partial melting and fractional crystallization discussed in chapter 3 are two ways of separating some minerals from others. Fractional crystallization, especially, can lead to the creation of valuable mineral deposits. The processes involved are entirely magmatic, and so such deposits are referred to as magmatic mineral deposits.

Pegmatites formed by fractional crystallization of granitic magma (chapter 3) commonly contain rich concentrations of such elements as lithium, beryllium, cesium, and niobium. Much of the world's lithium is mined from pegmatites such as those at King's Mountain, North Carolina, and Bikita in Zimbabwe. The great Tanco pegmatite in Manitoba, Canada, produces much of the world's cesium, and pegmatites in many countries yield beryl, one of the main ore minerals of beryllium.

Crystal setting, another process of fractional crystallization, can also form valuable mineral deposits. The process is especially important in low-viscosity basaltic magma. When a large chamber of basaltic magma crystallizes, one of the first minerals to form is chromite, the main ore mineral of chromium. As shown in Figure 3.40, settling of the dense chromite crystals to the bottom of the magma chamber can produce almost pure layers of chromite. The world's principal ores of chromite, in the Bushveld Igneous Complex in South Africa and the Great Dike of Zimbabwe, were all formed as a result of crystal setting.

FIGURE 17.4 A rich vein in Potosi, Bolivia, containing chalcopyrite, sphalerite, and galena cutting andesite. The andesite has been altered metasomatically by the hydrothermal solution that deposited the ore minerals.

Sedimentary Mineral Deposits

The term *sedimentary mineral deposit* is applied to any local concentration of minerals formed through processes of sedimentation. Any process of sedimentation can form localized concentrations of minerals, but it has become common practice to restrict use of the term sedimentary to those mineral deposits formed through precipitation of substances carried in solution.

Evaporite Deposits

The most direct way in which sedimentary mineral deposits form is by evaporation of lake water or seawater. The layers of salts that precipitate as a consequence of evaporation are called evaporite deposits (chapter 4).

Examples of salts that precipitate from lake waters of suitable composition are sodium carbonate (Na_2CO_3), sodium sulfate (Na_2SO_4), and borax ($Na_2B_4O_7 \cdot 10H_2O$). Huge evaporite deposits of sodium car-

bonate were laid down in the Green River Basin of Wyoming during the Eocene Epoch. This is the same lake in which the rich Green River oil shales were deposited (Fig. 4.18). Borax and other boron-containing minerals are mined from evaporite lake deposits in Death Valley and Searles and Borax lakes, all in California, and in Argentina, Bolivia, Turkey, and China.

Much more common and important than lake-water evaporites are the marine evaporites formed by evaporation of seawater. The most important salts that precipitate from seawater are gypsum ($CaSO_4 \cdot 2H_2O$), halite (NaCl), and carnallite ($KCl \cdot MgCl_2 \cdot 6H_2O$). Low-grade metamorphism of marine evaporite deposits causes another important mineral, sylvite (KCl), to form from carnallite. Marine evaporite deposits are widespread; in North America, for example, strata of marine evaporites underlie as much as 30 percent of the land area (Fig. 17.5). Most of the salt that we use, plus the gypsum used for plaster and the potassium used in plant fertilizers, is recovered from marine evaporites.

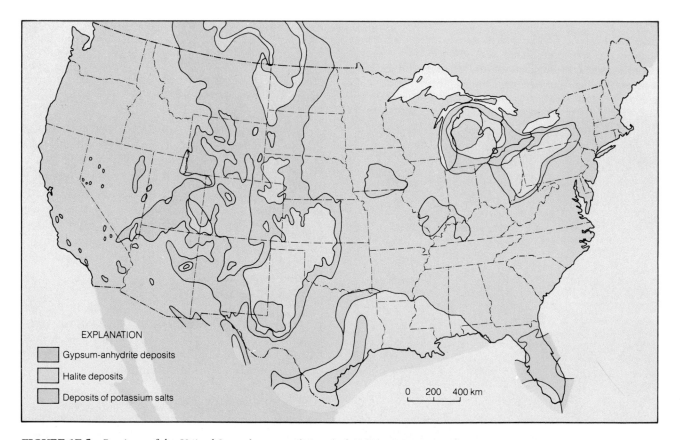

EXPLANATION

☐ Gypsum-anhydrite deposits

☐ Halite deposits

☐ Deposits of potassium salts

0   200   400 km

FIGURE 17.5   Portions of the United States known to be underlain by marine evaporite deposits. The areas underlain by gypsum and anhydrite do not contain halite. The areas underlain by halite are also underlain by gypsum and anhydrite. The areas underlain by potassium salts are also underlain by halite and by gypsum and anhydrite.

Iron Deposits

Sedimentary deposits of iron minerals are widespread, but the amount of iron in average seawater is so small that such deposits cannot have formed from seawater that is the same as today's seawater.

All sedimentary iron deposits are tiny by comparison with the class of deposits characterized by the *Lake Superior-type iron deposits*. These remarkable deposits, mined principally in Michigan and Minnesota, were long the mainstay of the United States steel industry but are declining in importance today as imported ores replace them. The deposits are of early Proterozoic age (about 2 billion years or older) and are found in sedimentary basins on every craton, particularly in Labrador, Venezuela, Brazil, the U.S.S.R., India, South Africa, and Australia. Every aspect of the Lake Superior deposits indicates chemical precipitation, as we discussed in chapter 4. The deposits are interbedded layers of chert and several different kinds of iron minerals. Because the deposits are so large, it is inferred that the iron and silica must have been transported in surface water, but the cause of precipitation remains unknown. Many experts suspect that Lake Superior-type deposits may be ancient evaporites

FIGURE 17.6 Leaching of silica during weathering of a Lake Superior-type iron deposit leads to formation of a secondarily enriched mass of iron minerals that is rich enough to be an ore. This sample, from the Hamersley Range in Western Australia, was developed by secondary enrichment from the kind of iron-rich sediment shown in Figure 4.14.

that formed from seawater of a different composition than today's seawater.

Lake Superior-type iron deposits are not ores. The grades of the deposits range from 15 to 30 percent Fe by weight, and the deposits are so fine grained that the iron minerals cannot be easily separated from the gangue. Two additional processes can form ore. First, as discussed in chapter 7, leaching of silica during weathering can lead to secondary enrichment and can produce ores containing as much as 66 percent Fe. Compare Figure 4.14, which is a Lake Superior-type iron deposit in the Hamersley Range, Western Australia, with Figure 17.6, a sample of ore developed by secondary enrichment in the Hamersley Range. The rocks in Figure 4.14 contain about 25 percent Fe, while those in Figure 17.6 have had most of the silica leached out and contain about 60 percent Fe.

The second way a Lake Superior-type iron deposit can become an ore is through metamorphism. Two changes occur as a result of metamorphism. First, grain sizes increase so that separating ore minerals from the gangue becomes easier and therefore cheaper. Second, new mineral assemblages form, and iron silicate and iron carbonate minerals originally present can be replaced by magnetite or hematite, both of which are desireable ore minerals. The grade is not increased by metamorphism. It is the changes in grain size and mineralogy that change the sedimentary rock into an ore. Iron ores formed as a result of metamorphism are called *taconites,* and they are now the main kind of ore mined in the Lake Superior region.

Stratabound Deposits

Some of the world's most important ores of lead, zinc, and copper occur in sedimentary rocks. The ore minerals—galena, sphalerite, chalcopyrite and pyrite—occur in such regular, fine layers that they look like sediments (Fig. 17.7). The sulfide mineral layers are enclosed by and parallel to the sedimentary strata in which they occur, and for this reason such deposits are called *stratabound mineral deposits.* They look like sediments but are not sediments in the truest sense of the term. It is more correct to consider most stratabound deposits as being diagenetic in origin.

Stratabound deposits form when a hydrothermal solution invades and reacts with a muddy sediment. Reactions between sediment grains and the solution cause deposition of the ore minerals. Deposition commonly occurs before the sediment has become a sedimentary rock.

The famous copper deposits of Zambia, in central Africa, are stratabound ores, as are the great Kupferschiefer deposits of Germany and Poland. The world's largest and richest lead and zinc deposits, at Broken Hill and Mount Isa in Australia and at Kimberley in British Columbia, are also stratabound ores.

FIGURE 17.7 Stratabound ore of lead and zinc from Kimberley, British Columbia. The layers of pyrite (yellow), sphalerite (brown), and galena (grey) are parallel to the layering of the sedimentary rock in which they occur. The specimen is 4 cm across.

Placers

The way in which a mineral with a high specific gravity can become concentrated by flowing water was discussed in chapters 9 and 13. Deposits of minerals having high specific gravities are *placers*. The most important minerals concentrated in placers are gold, platinum, cassiterite (SnO_2), and diamond. Typical locations of placers are illustrated in Figure 17.8.

Gold is the most valuable mineral recovered from placers; more than half of the gold recovered throughout all of human history has come from placers. This is the result of the huge gold production from South Africa, which has come from placers.

The South African gold deposits are really fossil placers, and they have many unusual features. Most placers are found in stream gravels that are geologically young. The South African fossil placers are a series of gold-bearing conglomerates (Fig. 17.9) that

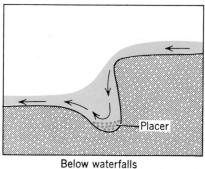

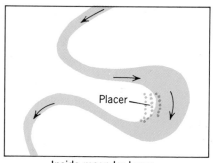

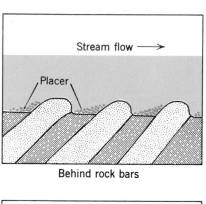

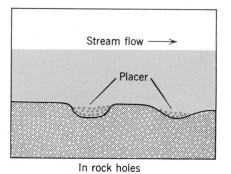

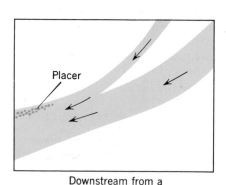

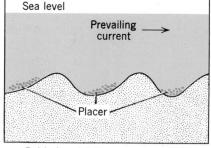

FIGURE 17.8 Placers occur where barriers allow flowing water to carry away the suspended load of lightweight particles while trapping particles carried in the bed load. Placers can form whenever water moves but are most commonly associated with streams or long-shore currents.

FIGURE 17.9 Gold is recovered from ancient fossil deposits of the Witwatersrand, South Africa. The gold is found at the base of conglomerate layers interbedded with finer-grained sandstone, here seen in weathered outcrop at the site where gold was first discovered in 1886.

were laid down 2.7 billion years ago as gravels in the shallow marginal waters of a marine basin. Associated with the gold are grains of pyrite and uranium minerals. So far as size and richness are concerned, nothing like the deposits in the Witwatersrand Basin has been discovered anywhere else. Nor has the original source of all the placer gold been discovered, and so it is not possible to say why so much of the world's minable gold should be concentrated in this one sedimentary basin.

Mining in the Witwatersrand Basin has reached a depth of 3600 m (11,800 ft). This is the deepest mining in the world, and there are plans to continue mining to depths as great as 4500 m. Despite such ambitious plans, the heyday of gold mining in South Africa has probably passed because the deposits are running out of ore.

Through the middle years of the 1980s, the price of gold fluctuated between about $14 and $16 a gram. This led to a boom in gold prospecting and to the discovery of a large number of new ore deposits in the United States, Canada, Australia, the Pacific islands, and elsewhere. Most of the new discoveries are hydrothermal deposits. Despite all the new discoveries, South Africa with its huge fossil placers continues to dominate the world's gold production. In 1991, South Africa still supplied about 40 percent of all the gold produced in the world, but the production rate is dropping steadily.

Residual Mineral Deposits

Weathering occurs because newly exposed rock is not chemically stable when it is in contact with rainwater and the atmosphere. Chemical weathering, in particu-

lar, leads to mineral concentration through the removal of soluble materials in solution and the concentration of a less soluble residue.

A common example of a deposit formed through residual concentration is laterite (chapter 7). Limonite is among the least soluble of the many minerals formed during chemical weathering. Under conditions of high rainfall in a warm, tropical climate, other minerals are slowly leached out of a soil, leaving an iron-rich limonitic crust of laterite at the surface (Fig. 17.10). In a few places, laterites can even be mined for iron.

A.

B.

FIGURE 17.10 Residual mineral deposits rich in iron and aluminum are typically formed under tropical or semitropical conditions. A. Red laterite enriched in iron, near Djenné, Mali. The upper two meters of the laterite consist of rounded concretions of limonite cemented by more limonite to create a hardened mass. Residual iron ores have been mined in the past, but no large mining activity is occurring today. B. Bauxite from Weipa in Queensland, Australia. Long-continued leaching of clastic sedimentary rocks under tropical conditions has removed all original constituents, such as silica, calcium, and magnesium, leaving a rich bauxite consisting largely of the mineral gibbsite ($Al(OH)_3$). Nodules of gibbsite form by repeated solution and redeposition. The Weipa bauxite deposits are among the largest and richest in the world.

While iron-rich laterite is by far the most common kind of residual mineral deposit, the most important deposits so far as human exploitation are concerned are the aluminous laterites called bauxites (chapter 7). Bauxites are the source of the world's aluminum.

Bauxites are widespread in the world, but they are concentrated in the tropics because that is where lateritic weathering occurs. Where bauxites are found in present-day temperate conditions, such as France, China, Hungary, and Arkansas, it is clear that the climate was tropical when the bauxites formed.

All bauxites are vulnerable to erosion. They are not found in glaciated regions, for example, because overriding glaciers scrape off the soft surface materials. The vulnerability of bauxites means that most deposits are geologically young. More than 90 percent of all known deposits formed during the last 60 million years, and all of the very large deposits formed less than 25 million years ago.

Metallogenic Provinces

Many kinds of mineral deposits tend to occur in groups and to form what exploration geologists call **metallogenic provinces.** These are defined as limited regions of the crust within which mineral deposits occur in unusually large numbers. A striking example is the metallogenic province shown in Figure 5.23, which runs along the western side of the Americas. Within the province is the world's greatest concentration of large hydrothermal copper deposits. These deposits are associated with intrusive igneous rocks that are invariably porphyritic (chapter 3), and they are therefore called **porphyry copper deposits.** The intrusive igneous rocks, and therefore the deposits themselves, were formed as a consequence of subduc-

tion because they are in, or adjacent to, old stratovolcanoes.

Metallogenic provinces form as a result of either climatic control (as in the formation of bauxite deposits in the tropics) or plate tectonics. Magmatic, hydrothermal, and stratabound deposits all form near present or past plate boundaries (Fig. 17.11). This is hardly surprising as the deposits are related directly or indirectly to igneous activity, and most igneous activity we now know is related to plate tectonics.

USEFUL MINERAL SUBSTANCES

It is convenient for purposes of discussion to group mineral products on the basis of the way they are used rather than the way they occur. Excluding substances used for energy, there are two broad groups: (1) minerals from which metals such as iron, copper, and gold can be recovered, and (2) nonmetallic minerals, such as salt, gypsum, and clay, used not for the metals they contain but for their properties as chemical compounds. The nonmetallic substances can be further subdivided on the basis of more specialized uses (Table 17.1).

Without exception, the useful metals are present in the crust in such small amounts that we can mine and recover them only when rich mineral deposits can be located.

Geochemically Abundant Metals

Metals can be usefully subdivided on the basis of their average percentage in the crust. Those present in such abundance that they make up 0.1 percent by weight

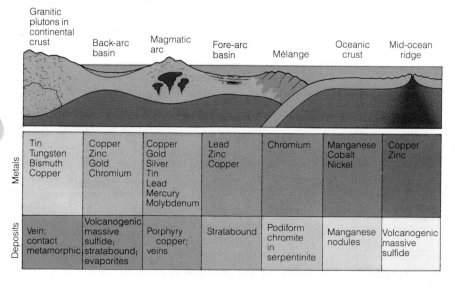

	Granitic plutons in continental crust	Back-arc basin	Magmatic arc	Fore-arc basin	Mélange	Oceanic crust	Mid-ocean ridge
Metals	Tin Tungsten Bismuth Copper	Copper Zinc Gold Chromium	Copper Gold Silver Tin Lead Mercury Molybdenum	Lead Zinc Copper	Chromium	Manganese Cobalt Nickel	Copper Zinc
Deposits	Vein; contact metamorphic	Volcanogenic massive sulfide; stratabound; evaporites	Porphyry copper; veins	Stratabound	Podiform chromite in serpentinite	Manganese nodules	Volcanogenic massive sulfide

FIGURE 17.11 Locations of certain kinds of mineral deposits in terms of plate structures.

TABLE 17.1 Principal mineral substances, grouped according to use

Metals
  Geochemically abundant metals
    Iron, aluminum, magnesium, manganese, titanium
  Geochemically scarce metals
    Copper, lead, zinc, nickel, chromium, gold, silver, tin, tungsten, mercury, molybdenum, uranium, platinum, palladium, and many others
Nonmetallic substances
  Used for chemicals
    Sodium chloride (halite), sodium carbonate, sulfur, borax, fluorite
  Used for fertilizers
    Calcium phosphate (apatite), potassium chloride (sylvite), sulfur, calcium carbonate (limestone), sodium nitrate
  Used for building
    Gypsum (for plaster), limestone, clay (for brick and tile), asbestos, sand gravel, crushed rock of various kinds, shale (for cement)
  Used for ceramics and abrasives
    Ceramics: Clay, feldspar, quartz
    Abrasives: Diamond, garnet, corundum, pumice, quartz

or more of the crust are considered to be **geochemically abundant.** These metals are iron, aluminum, manganese, magnesium, and titanium.

Geochemically abundant metals require comparatively small enrichment factors to form extremely large deposits. The minerals that are concentrated in deposits of the geochemically abundant metals tend to be oxides and hydroxides. The most important kinds of deposits are residual, sedimentary, and magmatic deposits.

Geochemically Scarce Metals

Metals that make up less than 0.1 percent by weight of the crust are said to be **geochemically scarce.**

With the exception of copper, zinc, and chromium, minerals of the geochemically scarce metals are not found in common rocks. However, chemical analysis of any rock will reveal that even though the minerals are absent, geochemically scarce metals are certainly present. Further research will reveal that the scarce metals are present exclusively as a result of atomic substitution (chapter 2). Atoms of the scarce metals (such as nickel, cobalt, and copper) can readily substitute for more common atoms (such as magnesium and calcium). In order for a mineral deposit to form, therefore, some gathering and concentrating agent, such as a hydrothermal solution, must react with the rock-forming minerals and leach the scarce metals from them. The solution must then transport the metals in solution and deposit them as separate minerals in a localized place. With such a complicated chain of events, it is hardly surprising that deposits of geochemically scarce metals are rarer and very much smaller than deposits of the abundant metals.

Most ore minerals of the scarce metals are sulfides; a few, such as the ore minerals of tin and tungsten, are oxides. In the case of gold, platinum, palladium, and a few less common elements, the metal itself is the most important mineral. Most scarce metal deposits form as hydrothermal or magmatic mineral deposits. In the case of gold and platinum, placer concentration is also important.

ENERGY RESOURCES

A healthy, hard-working person can produce just enough muscle energy to keep a single 100-watt light bulb burning for 8 hours a day. It costs about 10 cents to purchase the same amount of energy from the local electrical utility. Viewed strictly as machines, humans aren't worth much. By comparison, the amount of mechanical and electrical energy used each 8-hour working day in North America could keep three hundred 100-watt bulbs burning for every person living there.

To see where all this energy is used, it is necessary to sum up all the energy employed to grow and transport food, make clothes, cut lumber for new homes, light streets, heat and cool office buildings, and do myriad other things. The uses can be grouped into three categories: transportation, home and commerce, and industry (meaning all manufacturing and raw material processing plus the growing of foodstuffs). The present-day uses and sources of energy in the United States are summarized in Figure 17.12.

How much energy do all the people of the world use? The total is enormous. The energy drawn annually from the major fuels—coal, oil, and natural gas—plus that from nuclear power plants, is 2.6×10^{20} J. Nobody keeps accurate accounts of all the wood and animal dung burned in the cooking fires of Africa and Asia, but the amount has been estimated to be so large that when it is added to the 2.6×10^{20} J figure, the world's total energy consumption rises to about 3.0×10^{20} J annually. This is equivalent to the burning of 2 metric tons of coal or 10 barrels of oil for every living man, woman, and child each year! Energy consumption around the world is very uneven, however. In less developed countries such as India and Tanzania, energy use is equivalent to burning only 3 or 4 barrels of oil per person per year, while in a devel-

Sources of energy

Uses of
energy

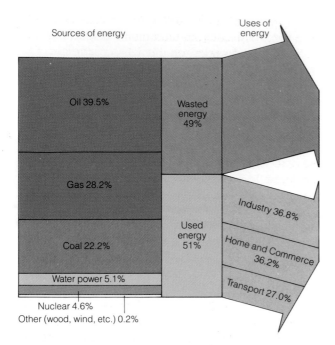

FIGURE 17.12   Uses and sources of energy in the United States. Wasted energy arises both from inefficiencies of use and from the fact that the laws of thermodynamics impose a limit to the efficiency of any engine and therefore a limit on the fraction of available energy that can be usefully employed.

oped country such as the United States, energy use is equivalent to burning more than 50 barrels of oil per person per year.

Supplies of Energy

The chief sources of the energy consumed in highly industrialized nations are few: the fossil fuels (coal, oil, natural gas), hydroelectric and nuclear power, wood, wind, and a very small amount of muscle energy. As recently as a century ago, wood was an important fuel in industrial societies, but now it is used mainly for space heating in some dwellings.

FOSSIL FUELS

The term *fossil fuel* refers, as we discussed in chapter 4, to the remains of plants and animals trapped in sediment that can be used for fuel. The kind of sediment, the kind of organic matter trapped, and the changes in the organic matter as a result of burial and diagenesis determine the kind of fossil fuel that forms.

In the ocean, microscopic photosynthetic phytoplankton and bacteria are the principal sources of

trapped organic matter. Shales do most of the trapping. Once bacteria and phytoplankton are trapped in the shale, the organic compounds they contain—*proteins, lipids,* and *carbohydrates*—become part of the shale, and it is these compounds that are transformed (mainly by heat) to oil and natural gas.

On land, it is trees, bushes, and grasses that contribute most of the trapped organic matter to shales; these large land plants are rich in resins, waxes, and lignins, which tend to remain solid and form coals rather than oil or natural gas.

In many marine and lake shales, burial temperatures never reach the levels at which the original organic molecules are converted to the organic molecules found in oil and natural gas. Instead, what happens is that an alteration process occurs in which waxlike substances with large molecules are formed. This material, which remains solid, is called *kerogen,* and it is the substance in so-called *oil shales.* Kerogen can be converted to oil and gas by mining the shale and heating it in a retort.

Coal

The combustible sedimentary rock we call coal is the most abundant fossil fuel. Most of the coal mined is eventually burned either under boilers to make steam for electrical generators, or else converted to coke, an essential ingredient in the smelting of iron ore and the making of steel. In addition to its use as a fuel, coal is a raw material for nylon and many other plastics, plus a multitude of other organic chemicals. The conditions under which organic matter accumulates in swamps as peat, then during burial and diagenesis is converted to coal, is discussed in chapter 4. Coalification involves the loss of volatile materials such as H_2O, CO_2, and CH_4 (methane). As the volatiles escape, the remaining coal is increasingly enriched in carbon. Through coalification, peat is converted successively to *lignite* (one type of coal), *subbituminous coal,* and *bituminous coal* (Fig. 4.17). These coals are sedimentary rocks. However, *anthracite,* a still later phase in the coalification process, is a metamorphic rock.

Because of its low volatile content, anthracite is hard to ignite, but once alight it burns with almost no smoke. In contrast, lignite is rich in volatiles, burns smokily, and ignites so easily that it is dangerously subject to spontaneous ignition.

In regions where metamorphism has been intense, coal has been changed so thoroughly that it has been converted to graphite, in which all volatiles have been lost. Graphite will not burn in an ordinary fire.

Occurrence of Coal

A coal seam is a flat, lens-shaped body having the same surface area as the swamp in which it originally accumulated. Most coal seams tend to occur in groups. In western Pennsylvania, for example, 60 seams of bituminous coal are found. This clustering indicates that the coal must have formed in a slowly subsiding site of sedimentation.

Coal swamps seem to have formed in many sedimentary environments, of which two types predominate. One consists of slowly subsiding basins in continental interiors and the swampy margins of shallow inland seas formed at times of high sea level. This is the home environment of the bituminous and subbituminous coal seams in Utah, Montana, Wyoming, and the Dakotas. The second sedimentary environment consists of continental margins with wide continental shelves (that is, continental margins in plate interiors) that were flooded at times of high sea level. This is the environment of the bituminous coals of the Appalachian region.

Coal-Forming Periods

Although peat can form under even subartic conditions, it is clear that the luxuriant plant growth needed to form thick and extensive coal seams developed most readily in a tropical or semitropical climate. The Great Dismal Swamp in Virginia and North Carolina is one of the largest modern peat swamps. It contains an average thickness of 2 m of peat. However, unless this swamp lasts for millions of years, even that dense growth is insufficient to produce a coal seam as thick as some of the seams in Pennsylvania.

Peat formation has been widespread and more or less continuous from the time land plants first appeared about 450 million years ago, during the Silurian Period. The size of peat swamps has varied greatly, however, and so, as a consequence, has the amount of coal formed. By far the greatest period of coal swamp formation occurred during the Carboniferous and Permian Periods, when Pangaea existed. The great coal beds of Europe and the eastern United States formed at this time, when the plants of coal swamps were giant ferns and scale trees (gymnosperms). The second great period of coal deposition peaked during the Cretaceous Period but commenced in the early Jurassic and continued until the mid-Tertiary. The plants of the coal swamps during this period were flowering plants (angiosperms) much like flowering plants today.

Petroleum: Oil and Natural Gas

As was mentioned in the introductory essay to this chapter, rock oil is one of the earliest resources our ancestors learned to use. However, the major use of oil really started about 1847 when a merchant in Pittsburgh, Pennsylvania, started bottling and selling rock oil from natural seeps to be used as a lubricant. Five years later, in 1852, a Canadian chemist discovered that heating and distillation of rock oil yielded kerosene, a liquid that could be used in lamps. This discovery spelled doom for candles and whale-oil lamps. Wells were soon being dug by hand near Oil Springs, Ontario, in order to produce oil. In Romania in 1856, using the same hand-digging process, workers were producing 2000 barrels a year.* In 1859, the first oil well was drilled in Titusville, Pennsylvania. On August 27, 1859, at a depth of 21.2 m, oil-bearing strata were encountered and up to 35 barrels of oil a day were pumped out. Oil was soon discovered in West Virginia (1860), Colorado (1862), Texas (1866), California (1875), and many other places.

The earliest known use of natural gas was about 3000 years ago in China, where gas seeping out of the ground was collected and transmitted through bamboo pipes to be ignited and used to evaporate salt water in order to recover salt. It wasn't long before the Chinese were drilling wells to increase the flow of gas. Modern use of gas started in the early seventeenth century in Europe, where gas made from wood and coal was used for illumination. Commercial gas companies were founded as early as 1812 in London and 1816 in Baltimore. The stage was set for the exploitation of an accidental discovery at Fredonia, New York, in 1821. A water well drilled in that year produced not only water but also bubbles of a mysterious gas. The gas was accidentally ignited and produced such a spectacular flame that a new well was drilled on the same site and wooden pipes were installed to carry the gas to a nearby hotel, where 66 gas lights were installed. By 1872, natural gas was being piped as far as 40 km from its source.

Origin of Petroleum

Petroleum is a term used for both oil and natural gas. That petroleum is a product of the decomposition of organic matter trapped in sediment was discussed in chapter 4. That petroleum migrates through aquifers and becomes trapped in reservoirs was discussed in the essay at the end of chapter 14.

The migration of petroleum deserves further discussion. The sediment in which organic matter is accumulating today is rich in clay minerals, whereas most of the strata that constitute oil or gas pools are sandstones (consisting of quartz grains), limestones and dolostones (consisting of carbonate minerals), and much-fractured rock of other kinds. Long ago, geologists realized that oil and gas form in one kind of

---

*A barrel is equal to 42 U.S. gal and is the volume generally used when commercial production of oil is discussed.

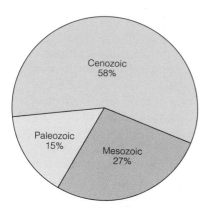

FIGURE 17.13 Percentage of world's total oil production from strata of different ages.

material (shale) and at some later time migrate to another (sandstone or limestone).

Petroleum migration is analogous to groundwater migration. When oil and gas are squeezed out of the shale in which they originated and enter a body of sandstone or limestone, they can migrate more easily than before because most sandstones and limestones are more permeable than any shale. The force of molecular attraction between oil and quartz or carbonate minerals is weaker than that between water and quartz or carbonate minerals. Hence, because oil and water do not mix, water remains fastened to the quartz or carbonate grains while oil occupies the central parts of the larger openings in the porous sandstone or limestone. Because it is lighter than water, the oil tends to glide upward past the carbonate- and quartz-held water. In this way it becomes segregated from the water; when it encounters a trap, it can form a pool.

Most of the petroleum that forms in sediments does not find a suitable trap and eventually makes its way, along with groundwater, to the surface. It is estimated that no more than 0.1 percent of all the organic matter originally buried in a sediment is eventually trapped in an oil or gas pool. It is not surprising, therefore, that the highest ratio of oil and gas pools to volume of sediment is found in rock no older than 2.5 million years, and that nearly 60 percent of all the oil and gas discovered so far has been found in strata of Cenozoic age (Fig. 17.13). This does not mean that older rocks produced less petroleum. It simply means that oil in older rocks has had a longer time in which to escape.

Distribution of Oil

Petroleum deposits, like coal, are frequent but are distributed unevenly. The reasons for uneven petroleum distribution are not as obvious as they are with coal. Suitable source sediments for petroleum are very widespread and seem as likely to form in subarctic

waters as in tropical regions. The critical controls seem to be a supply of heat to effect the conversion of solid organic matter to oil and gas and the formation of a suitable trap before the petroleum has leaked away.

Conversion of solid organic matter to oil and gas happens within a specific range of depth and temperature defined by the geothermal gradients shown in Figure 17.14. If a thermal gradient is too low (less than 1.8 C/100 m) conversion does not occur to either oil or gas. If the gradient is above 5.5 C/100 m, conversion to gas starts at such shallow depths that very little trapping occurs. Once oil and gas have been formed, they will accumulate in pools only if suitable traps are present. Most oil and gas pools are found beneath anticlines; the timing of the folding event is therefore a critical part of the trapping process. If folding occurs after petroleum has formed and migrated, pools cannot form. The great oil pools in the Middle East arose through the fortunate coincidence of the right thermal gradient and the development of anticlinal traps during the collision of Europe and Asia with Africa.

How much oil is there in the world? This is an extremely controversial question. Approximately 600 billion barrels have been already been pumped out of the ground. A lot of additional oil has been located by drilling but is still waiting to be pumped. Probably

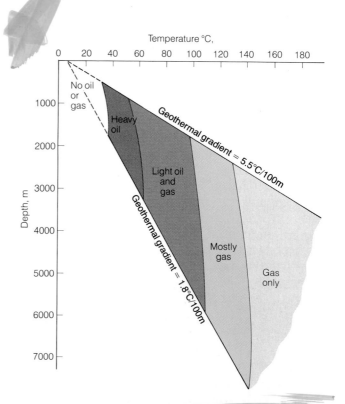

FIGURE 17.14 Regions of depth and temperature within which oil and gas are generated and trapped.

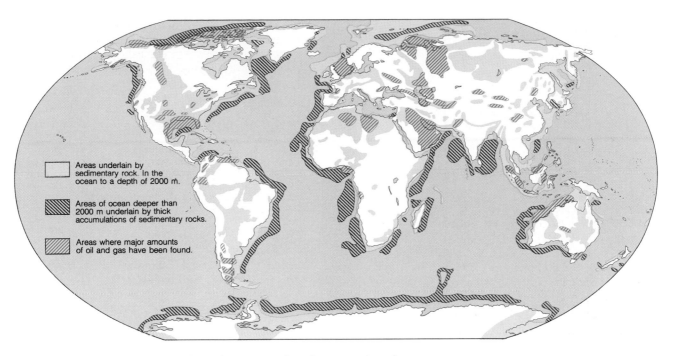

FIGURE 17.15 Areas underlain by sedimentary rock and regions where large accumulations of oil and gas have been located. Where the ocean is deeper than 2000 m, sedimentary rock has yet to be tested for its oil and gas potential.

a great deal more oil remains to be found by drilling. Unlike coal, for which the volume of strata in a basin of sediment can be accurately estimated even before drilling, the volume of undiscovered oil can only be guessed at. The way guesses are made is to use the accumulated experience of a century of drilling. Knowing how much oil has been found in an intensively drilled area, such as eastern Texas, experts make estimates of probable oil volumes in other regions where rock types and structures are similar to what we find in eastern Texas. Using this approach, and considering all the sedimentary basins of the world (Fig. 17.15), experts estimate that somewhere between 1500 and 3000 billion barrels of oil will eventually be discovered.

Tars

Oil that is exceedingly viscous will not flow easily and cannot be pumped. Colloquially called **tar,** heavy, viscous oil acts as a cementing agent between mineral grains in an oil pool. The tar can be recovered only if the sandstone is mined and heated enough to make the tar flow. The resulting tar must then be processed to recover the valuable gasoline fraction. The cost of mining and treating "tar sands," as heavy, viscous oil deposits are called, is high, but it is technologically possible and someday tar sands may be an important source of fuel. The largest known occurrence of tar sands is in Alberta, Canada, where the *Athabasca Tar Sand* covers an area of 5000 km² and reaches a thickness of 60 m (Fig 17.16). Similar deposits almost as large are known in Venezuela and in the U.S.S.R.

Oil Shale

Another potential source of petroleum is kerogen in shale. If the kerogen is heated, it breaks down and forms liquid and gaseous hydrocarbons similar to those in oil and gas. All shales contain some kerogen, but to be considered an energy resource the kerogen must yield more energy than is required to mine and heat it. Only those shales that yield 40 or more liters of distillate per ton can be considered because the energy needed to mine and process a ton of shale is equivalent to that created by burning 40 liters of oil.

The world's largest deposit of rich oil shale is in the United States. During the Eocene Epoch, many large, shallow lakes existed in basins in Colorado, Wyoming, and Utah; in three of them a series of rich organic sediments was deposited that are now the Green River Oil Shales (Fig. 4.18). The richest shales were deposited in the lake in Colorado now called the Piceance Basin. These shales are capable of producing as much as 240 liters of oil per ton. Scientists of the U.S. Geological Survey estimate that, in the

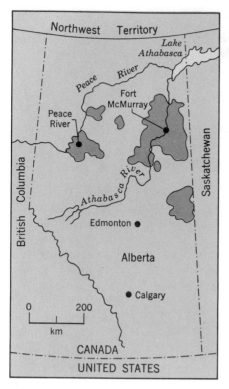

Area underlain
by tar sand

FIGURE 17.16 The part of the Province of Alberta underlain by the Athabasca Tar Sand.

TABLE 17.2 Amounts of fossil fuels possibly recoverable worldwide (unit of comparison is a barrel of oil)

Fossil Fuel	Total Amount in Ground (billions of barrels)	Amount Possibly Recoverable (billions of barrels)
Coal	About 100,000	62,730[a]
Oil and gas (flowing)	1500–3000	1500–3000
Trapped oil in pumped-out pools	1500–3000	0–?
Viscous oil (tar sands)	3000–6000	500–?
Oil shale	Total unknown; much greater than coal	1000–?

[a]0.22 ton of coal = 1 barrel of oil.

Green River Oil Shale alone, oil-shale resources capable of producing 50 liters or more of oil per ton of shale can ultimately yield about 2000 billion barrels of oil.

Rich deposits of oil shale in other parts of the world have not been adequately explored, but there is a huge deposit in Brazil called the Irati Shale. Another very large deposit is known in Queensland, Australia, and others have been reported in such widely dispersed places as South Africa and China. Although oil shales have been mined and processed in an experimental fashion in the United States, the only countries where extensive commercial production has been tried are the U.S.S.R. and China. Production expenses today make exploitation of oil shales in all countries unattractive by comparison with oil and gas. Most experts believe, however, that large-scale mining and processing of oil shale will eventually happen.

How Much Fossil Fuel?

Are supplies of fossil fuels adequate to meet future demands? If we use a barrel of oil as our unit of measurement, we can compare quantities of all fossil fuels

directly. Approximately 0.22 ton of coal produces the same amount of heat energy as one barrel of oil. Thus, the world's recoverable coal reserves of $13,800 \times 10^9$ tons are equivalent to about 63,000 billion barrels of oil.

Considering the approximate world-use rate of oil (30 billion barrels a year), and comparing the estimated recoverable amounts of fossil fuels (Table 17.2), it is apparent that only coal seems to have the capacity to meet our long-term demands.

OTHER SOURCES OF ENERGY

Three sources of energy other than fossil fuels have already been developed to some extent: the Earth's plant life (so-called biomass energy), hydroelectric energy, and nuclear energy. Five others—the Sun's heat, winds, waves, tides, and the Earth's internal heat—have been tested and developed on a limited basis but none has yet been shown to be developable on a large scale. The day may not be far off, however, when one or more of the four could become locally important.

Biomass Energy

Scientists working for the United Nations estimate that wood and animal dung used for cooking and heating fires now amounts to energy production of 4×10^{19} J annually. This is approximately 14 percent of the world's total energy use. The greatest use of wood as a fuel occurs in developing countries, where the cost of fossil fuel is very high in relation to income.

Measurements made on living plant matter indicate that new plant growth on land equals 1.5×10^{11} metric tons of dry plant matter each year. If all of this were burned, or used in some other way as a biomass energy source, it would produce almost nine times more energy than the world uses each year. Obviously, this is a ridiculous suggestion because in order to do so all the forests would have to be destroyed, plants could not be eaten, and agricultural soils would be devastated. Nevertheless, controlled harvesting of fuel plants could probably increase the fraction of the biomass now used for fuel without serious disruption to forests or to food supplies. In several parts of the world, such as Brazil, China, and the United States, experiments are already under way to develop this energy source.

Hydroelectric Power

Hydroelectric power is recovered from the potential energy of stream water as it flows to the sea. As discussed in chapter 9, in order to convert the power of flowing water to electricity efficiently, it is necessary to dam streams. Unfortunately, reservoirs behind dams fill with silt, and so even though water power is continuous, dams and reservoirs have limited lifetimes.

Water power has been used in small ways for thousands of years, but only in the twentieth century has it been used to any significant extent for generating electricity. All the water flowing in the streams of the world has a total recoverable energy estimated at 9.2×10^{19} J/yr, an amount equivalent to burning 15 billion barrels of oil per year. Thus, even if all the possible hydropower in the world were developed, we could satisfy only about one third of the present world energy needs. We have to conclude that, for those fortunate countries with large rivers and suitable dam sites, hydropower is very important, but for most countries hydropower holds limited potential for development.

Nuclear Energy

Nuclear energy is the heat energy produced during controlled transformation of suitable radioactive isotopes (a process called **fission.**) Three of the radioactive atoms that keep the Earth hot by spontaneous decay—^{238}U, ^{235}U, and ^{232}Th—can be mined and used to obtain nuclear energy. Fission is accomplished by bombarding the radioactive atoms with neutrons, thus accelerating the rate of decay and the release of heat energy. The device in which this operation is carried out is called a **pile.**

When ^{235}U fissions, it not only releases heat and forms new elements but also ejects some neutrons from its nucleus. These neutrons can then be used to induce more ^{235}U atoms to fission, and a continuous chain reaction occurs. The function of a pile is to control the flux of neutrons so that the rate of fission can be controlled. When a chain reaction proceeds without control, an atomic explosion occurs. Controlled fission, therefore, is the method used by nuclear power plants, and a tremendous amount of energy can be obtained in the process. The fissioning of one gram of ^{235}U produces as much heat as the burning of 13.7 barrels of oil. Unfortunately, however, ^{235}U is the only natural radioactive isotope that will maintain a chain reaction, and it is the least abundant of the three radioactive isotopes that are mined for nuclear energy. Only one atom of each 138.8 atoms of uranium in nature is ^{235}U. The remaining atoms are ^{238}U, which will not sustain a chain reaction. However, if ^{238}U is placed in a pile with ^{235}U that is undergoing a chain reaction, some of the neutrons will bombard the ^{238}U and convert it to plutonium-239(^{239}Pu). This new isotope can, under suitable conditions, sustain a chain reaction of its own. The pile in which the conversion of ^{238}U takes place is called a **breeder reactor.** The same kind of device can be used to convert ^{232}Th into ^{233}U, which also will sustain a chain reaction. Unfortunately, breeder reactors and nuclear power plants based on them are more complex and less safe than ^{235}U plants, so all the present nuclear power plants use ^{235}U.

Already there are more than 300 piles in nuclear power plants operating around the world. They utilize the heat energy from fission to produce steam that drives turbines and generates electricity. Approximately 7.6 percent of the world's electrical power is derived from nuclear power plants. In France, more than half of all the electrical power comes from nuclear plants; the fraction is rising sharply in some other European countries and in Japan, too. The reason for the increase is obvious. Japan and most European countries do not have adequate supplies of fossil fuels in order to be self-sufficient.

Many problems are associated with nuclear energy. The isotopes used in power plants are the same isotopes used in atomic weapons, and so a security problem exists. The possibility of a power plant failing in some unexpected way creates a safety problem. The dreadful Chernobyl disaster in 1986 in the U.S.S.R. is an example of such an event. Finally, the problem of safe burial of dangerous radioactive waste matter must be faced. Some of the waste matter will retain dangerous levels of radioactivity for thousands of years.

Geothermal Power

Geothermal power, as the Earth's internal heat flux is called, has been used for more than 50 years in New Zealand, Italy, and Iceland and more recently in other parts of the world, including the United States. How this is done was discussed in the essay on geothermal energy at the end of chapter 3.

Most of the world's geothermal steam reservoirs are close to plate margins because plate margins are where most recent volcanic activity has occurred. A depth of 3 km seems to be a rough lower limit for big geothermal steam and hot-water pools. It is estimated that the world's geothermal reservoirs could yield about 8×10^{19} J—equivalent to burning 13 billion barrels of oil. This estimate incorporates the observation that, in New Zealand and Italy, only about 1 percent of the energy in a geothermal reservoir is recoverable. If the recovery efficiency were to rise, the estimate of recoverable geothermal resources would also rise. But even if the efficiency rose to 50 percent, geothermal power, like hydropower, could satisfy only a small part of human energy needs. For this reason, as we discussed in the chapter-closing essay for chapter 3, a good deal of attention is being given to creating artificial geothermal steam fields. So far experiments have been only partially successful.

Energy from Winds, Waves, Tides, and Sunlight

The most obvious source of energy is the Sun. The amount of energy reaching the Earth each year from the Sun is approximately 4×10^{24} J—that is, ten thousand times more than we humans use. We already put some of the Sun's energy to work in greenhouses and in solar homes, but the amount so used is tiny. The major challenge is to convert solar energy directly to electricity. Devices that effect such a conversion, called photovoltaic devices, have been invented. So far their costs are too high and efficiencies too low for most uses, although they are already widely used in small calculators, radios, and other devices that use very little power.

Winds and waves are both secondary expressions of solar energy. Winds, in particular, have been used as an energy source for thousands of years through sails on ships and windmills. Today, huge farms of windmills are being erected in suitably windy places (Fig. 17.17). Although there are problems and high costs with windmills, it seems very likely that by the year 2000 or sooner, windmills will be cost-competitive with coal-burning electrical power plants. Unfortunately, much of the wind energy is in very-high-altitude winds. Steady surface winds have only about 10 percent of the energy the human race now uses. As with hydro- and geothermal power, therefore, wind power may become locally significant but will probably not be globally important.

Waves, which arise from winds blowing over the ocean, contain an enormous amount of energy. We can see how powerful waves are along any coastline during a storm. Wave power has been used to ring bells and blow whistles as navigational aids for centuries, but so far no one has discovered how to tap wave energy on a large scale. Devices that have been designed to do this tend to fail because of corrosion or storm damage.

Tides arise from the gravitational forces exerted on the Earth by the Moon and the Sun. If a dam is put across the mouth of a bay so water can be trapped at high tide, the outward flowing water at low tide can drive a turbine. Unfortunately the efficiency of the process is low, and few places around the world have tides high enough to make tidal energy feasible.

It is clear that there are numerous sources of energy and that there is far more energy available than we can use. What is not yet clear is when, or even whether, we will be clever enough to learn how to tap the different energy sources in nonhazardous ways that don't disrupt the environment.

FIGURE 17.17 A field of windmills near Palm Springs, California, designed to generate electricity using the kinetic energy of wind.

MINERAL DEPOSITS FORMING TODAY

Three extraordinary discoveries over a 15-year period have changed our thinking about mineral deposits. The first discovery, in 1962, was accidental. Until that year no one was sure where to look for modern hydrothermal solutions or even how to recognize one when it was found. Drillers seeking oil and gas in the Imperial Valley of southern California were astonished when they struck a 320°C brine at a depth of 1.5 km. As the brine flowed upward, it cooled and precipitated minerals it had been carrying in solution. Over 3 months, the well deposited 8 tons of siliceous scale containing 20 percent copper and 8 percent silver by weight. The drillers had found a hydrothermal solution that could, under suitable flow conditions, form a rich mineral deposit.

The Imperial Valley is a sediment-filled graben covering the join between the Pacific and North American plates, where the East Pacific Rise passes under North America (Fig. B17.1). Volcanism is the source of heat for the brine solution discovered in 1962. These brines provided the first unambiguous evidence that hydrothermal solutions can leach metals from ordinary sediments.

Before geologists had a chance to fully absorb the significance of the Imperial Valley discovery, a second remarkable find was announced. In 1964, oceanographers discovered a series of hot, dense, brine pools at the bottom of the Red Sea. The brines are trapped in the graben formed by the spreading center between the Arabian and African plates (Fig. B17.2), and they are so much more salty, and therefore more dense, than seawater that they remain ponded in the graben even though they are as hot as 60°C. Many such brine pools have now been discovered.

The Red Sea brines rise up the normal faults associated with the central rift of a spreading center and, like the Imperial Valley brines, have evolved to their present compositions through reactions with the enclosing rocks. The Red Sea brine discovery was surprising, but even more surprising was the discovery that sediments at the bottom of the pools contained ore minerals such as chalcopyrite, galena, and sphalerite. In other words, the oceanographers had discovered modern stratabound mineral deposits in the process of formation.

The third remarkable discovery was really a series of discoveries that commenced in 1978. Scientists using deep-diving submarines made a series of dives on the East Pacific Rise at 21°N

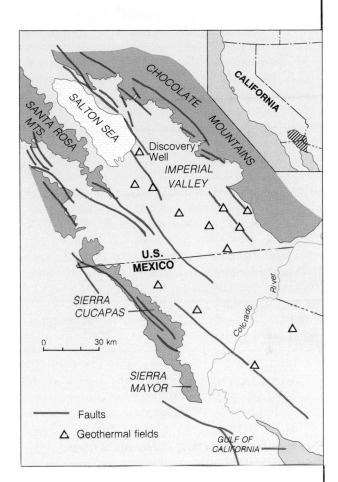

FIGURE B17.1 The Imperial Valley graben (also known as the Salton Trough). The graben is bounded by the Chocolate Mountains on the east and the Santa Rosa Mountains on the west. Hydrothermal solutions were discovered in a well drilled on the southern end of the Salton Sea. Places where geothermal activity is known, and where other hydrothermal solutions may be present at depth, are marked with triangles.

latitude. To their amazement, they found 300°C hot springs emerging from the seafloor 2500 m below sealevel. Around the hot springs lay a blanket of sulfide minerals. The submariners watched a modern volcanogenic sulfide deposit forming before their eyes.

Each of the discovery sites—Imperial Valley, Red Sea, and 21°N—is on a spreading center, and so there is no doubt that the deposits are forming as a result of plate tectonics. Soon the hunt was on to see if seafloor deposits could be found above subduction zones. In 1989, a joint German–Japanese oceanographic expedition to the western Pacific discovered the first modern

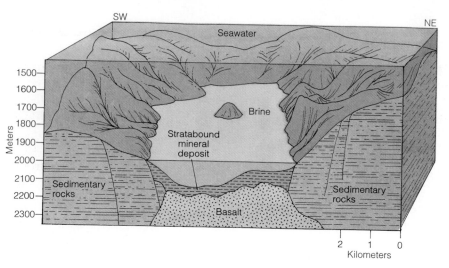

FIGURE B17.2 Topography of the Red Sea graben near the Atlantis II brine pool. Hot, dense brines rise up normal faults, pond on the floor of the graben, and form stratabound deposits rich in copper and zinc.

subduction-related deposits. No longer are geologists limited to speculating about how certain mineral deposits *might* have formed. Today we can study them as they grow.

SUMMARY

1. When a mineral deposit can be worked profitably, it is called an ore. The waste material mixed with ore minerals is gangue.

2. Mineral deposits form when minerals become concentrated in one of five ways: (1) precipitation from hydrothermal solutions to form hydrothermal mineral deposits; (2) concentration through crystallization to form magmatic mineral deposits; (3) concentration from lake water or seawater to form sedimentary mineral deposits; (4) concentration in flowing water to form placers; and (5) concentration through weathering to form residual deposits.

3. Hydrothermal solutions are brines, and they can either be given off by cooling magma or else form when either groundwater or seawater penetrate the crust, become heated, and react with the enclosing rocks.

4. Hydrothermal mineral deposits form when hydrothermal solutions deposit dissolved minerals because of cooling, boiling, pressure drop, mixing with colder water, or through chemical reactions with enclosing rocks.

5. Chromite, the main ore mineral of chromium, is the most important mineral concentrated by fractional crystallization.

6. Sedimentary mineral deposits are varied. The largest and most important are evaporites. Marine evaporite deposits supply most of the world's gypsum, halite, and potassium minerals.

7. Gold, platinum, cassiterite, diamonds, and other minerals are commonly found mechanically concentrated in placers.

8. Bauxite, the main ore of aluminum, is the most important kind of residual mineral deposit. Bauxite forms as a result of tropical weathering.

9. The distribution of many kinds of mineral deposits is controlled by plate tectonics because most magmas and most sedimentary basins are where they are because of plate tectonics.

10. Geochemically abundant metals, which make up 0.1 percent or more of the crust, tend to form residual, sedimentary, or magmatic mineral deposits. The amounts available for exploitation in such deposits are enormous.

11. Deposits of geochemically scarce metals, present in the crust in amounts less than 0.1 percent, form mainly as hydrothermal, magmatic, and placer deposits. Amounts of scarce metals available for exploitation are limited and geographically restricted.

12. Nonmetallic substances are used mainly as chemicals, fertilizers, building materials, and ceramics and abrasives.

13. Coal originated as plant matter in ancient swamps and is both abundant and widely distributed.

14. Oil and gas originated as organic matter trapped in shales and decomposed chemically due to

486

heat and pressure following burial. Later, these fluids moved through reservoir rocks and were caught in geologic traps to form pools.

15. When heated, part of the solid organic matter found in shale—called kerogen—will convert to oil and gas. Oil from shales is the world's largest resource of fossil fuel. Unfortunately, most shale contains so little kerogen that more oil must be burned to heat the shale than is produced by the conversion process.

16. Nuclear energy is derived from atomic nuclei of radioactive isotopes, chiefly uranium. The nuclear energy available from naturally occurring radioactive elements is the single largest energy resource now available.

17. Other sources of energy currently used to some extent are geothermal heat, energy from flowing streams, winds, waves, tides, and the Sun's heat.

IMPORTANT TERMS TO REMEMBER

breeder reactor (p. 483)

exploration geology (p. 468)

fission (p. 483)

gangue (p. 469)
geochemically abundant metals (p. 477)
geochemically scarce metals (p. 477)

hydrothermal mineral deposit (p. 470)

magmatic mineral deposit (p. 470)
metallogenic province (p. 476)
mineral deposit (p. 468)

ore (p. 467)

pile (p. 483)

porphyry copper deposit (p. 476)

residual mineral deposit (p. 470)

sedimentary mineral deposit (p. 470)

tar (p. 481)

volcanogenic massive sulfide deposit (p. 470)

QUESTIONS FOR REVIEW

1. What are mineral deposits? Describe five ways by which a mineral deposit can form.

2. If there are any mineral deposits in the area where you live or study, what kind of deposits are they and how did they form?

3. What factors determine whether a mineral deposit is ore or not?

4. How do hydrothermal solutions form and how do they form mineral deposits?

5. Briefly describe the formation of three kinds of sedimentary mineral deposit.

6. What factors control the concentration of minerals in placers? Name four minerals mined from placers.

7. Compare and contrast mineral deposits of the geochemically abundant and geochemically scarce metals.

8. What is a fossil fuel? Name four kinds of fossil fuel.

9. During what two periods in the Earth's history was most coal formed? Explain why coal formed when and where it did and why the coal is now found where it is.

10. What kind of rocks serve as source rocks for petroleum? In what kinds of rocks does petroleum tend to be trapped? Why?

11. Oil drillers find more petroleum per unit volume of rock in Cenozoic rocks than in Paleozoic rocks of the same kind. Explain.

12. Oil shales are rich in organic matter. Explain why such shales have not served as source rock for petroleum.

13. Discuss the relative amounts of energy available from the different fossil fuels. What is your opinion about how fossil fuels will be used in the future?

14. What is nuclear energy? How is it used to make electricity and what possible dangers are there in developing nuclear energy?

15. What limitations are there to the development of hydroelectric power? Of wave and wind power?

analyses before they were overcome by the high temperatures. In each case, the analyses suggest that the rocks analyzed were basaltic in composition. A few analyses indicated that some rocks on Venus have relatively high potassium contents. This observation suggests that Earthlike magmatic differentiation, probably following Bowen's reaction series, may have been operating.

As this book was nearing completion, a spacecraft called *Magellan* reached Venus. *Magellan* went into orbit around Venus and started examining the surface topography using radar. The results have provided scientists with an unprecedented view of Venus. It will take years before all of the results have been studied, but already some important issues have been resolved.

Volcanism

The dominant process on the surface of Venus is volcanism. Vast volcanic plains cover 80 percent of the surface, and scattered across the plains are thousands of shield volcanoes ranging from 2 to 10 km in diameter. The volcanic plains and shield volcanoes were clearly formed from low-viscosity, basaltic magma.

Not all of the volcanism on Venus has been basaltic, however. A cluster of seven steep-sided volcanic domes have been discovered (Fig. 18.15). Each dome is about 25 km across, and except for the fact of their large size, the domes look very like the rhyolitic and dacitic lava domes that occur on the Earth. The lava domes indicate that high-viscosity magma occurs on Venus. This observation confirms a suspicion suggested by the chemical analyses performed by the Russian spacecraft: magmatic differentiation has occurred on Venus.

Tectonics

Sensitive radar measurements of Venus's shape show that the total relief from the bottom of the deepest chasm to the top of the highest peak is 13 km. By comparison, the total relief on the Earth is 20 km. The radar measurements also show that Venus does not have a bimodal distribution of its surface topography the way the Earth does (Fig. 1.10). The surface of Venus is unimodal. Because the Earth's bimodal surface topography is a result of plate tectonics, it seems unlikely that plate tectonics operate on Venus.

Two elongate mountain ranges, Ishtar Terra and Aphrodite Terra (Fig. 18.16) have been discovered on Venus. The ranges seem to have formed by a combination of both extension and compression. One popular hypothesis for their formation is that mantle plumes thrust the surface upward and produced extensional features, then the highest areas slid downward along thrust faults as a result of gravity, thereby producing compressional features. The origin of Venus's mountains, the reasons for the lava plains being crisscrossed by giant faults, and the origin of huge grabens are all

FIGURE 18.15   Steep-sided lava domes on Venus seen by the *Magellan* spacecraft early in 1991. Each dome is about 25 km in diameter. The shapes and steep sides resemble rhyolitic lava domes on the Earth, and this suggests that high viscosity magma might be produced on Venus by some form of magmatic differentiation.

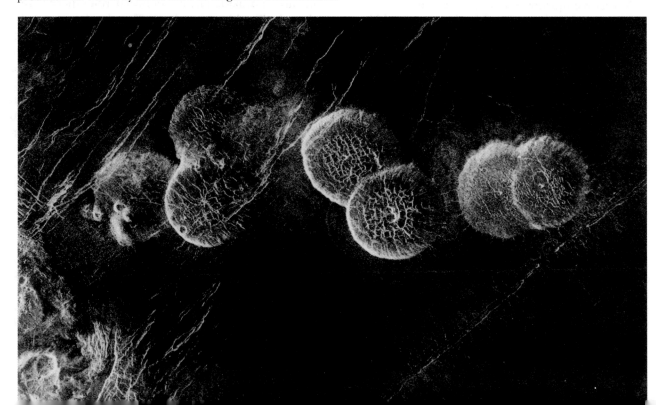

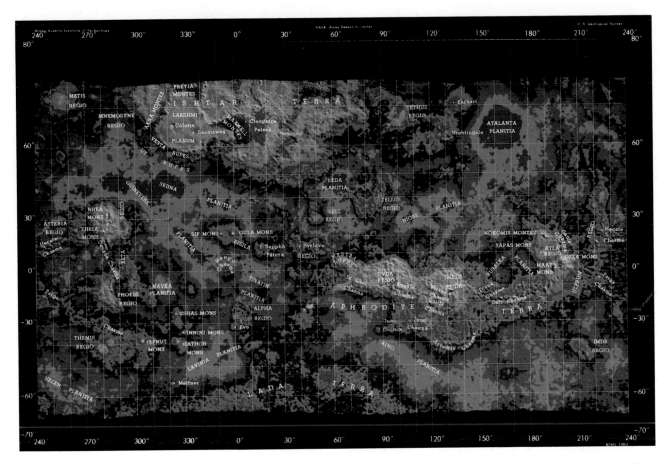

FIGURE 18.16 Topography of Venus as determined by radar. Blue areas are lowlands; green, yellow, and red, in that order, represent increasingly high topography.

topics for further research. Venus may not have plate tectonics, but it certainly seems to have mantle convection and to be tectonically active. The more we learn about Venus, the more apparent it becomes that Venus and the Earth are more similar than the Earth and any other planet.

Comparison of the Terrestrial Planets

Despite the differences between the terrestrial planets, they also share many features in common. All the terrestrial planets seem to have metallic cores, all have undergone partial melting to form basaltic magma, and all have been modified by impact cratering. The differences between the terrestrial planets apparently arise through the interaction of several factors. First, the planet size controls not only the atmosphere, but also the thermal properties. Small planets cool rapidly, and magmatic activity soon ceases. Large planets cool more slowly and remain magmatically active for much longer. Second, the distance of a planet from the Sun determines whether or not H_2O can exist as a liquid.

The third factor is the presence or absence of life. The Earth's atmosphere is the way it is because living plants and animals play essential roles in the geochemical cycles that control the atmosphere's composition. If life had developed on Venus, that planet would probably have an atmosphere like the Earth's. However, life apparently did not develop, so all of the CO_2 is still in the atmosphere. On the Earth, plants and animals have been the means whereby carbon dioxide has been removed from the atmosphere, and the carbon is locked up in rocks as fossil organic matter and as calcium carbonate.

THE JOVIAN PLANETS AND THEIR MOONS

Jupiter

The giant, gassy planets tell us little about the evolution of the Earth, but they provide the best-preserved samples of the gases from which all of the planets are believed to have formed. Thus, they reveal much about how the solar system may have formed. Jupiter is the largest and best studied of the giant planets. It has an atmosphere composed of hydrogen, helium, ammonia, and methane, plus other trace constituents. It is inferred that a rocky core exists inside the dense atmosphere.

Jupiter has about twice the mass of all the other planets combined. Had it been slightly larger, it would have reached an internal temperature high enough for nuclear burning to start and as a result it would have been a sun. Jupiter is unusual in many ways. One feature is that it gives off twice as much energy as it receives from the Sun. The reason for this seems to be that Jupiter is still undergoing gravitational contraction that gives off heat energy.

The Jovian Moons

One of the most interesting things about Jupiter is its moons, four of which are as large as, or larger, than the Earth's Moon. The moons closest to Jupiter, Io and Europa, have densities of 3.5 and 3.2 g/cm^3, respectively, indicating that they are rocky bodies.

The moon closest to Jupiter, Io, is extraordinary. It is a highly colored body with shades of yellow and orange predominating, suggesting that it is covered by sulfur and sulfurous compounds (Fig. 18.17). Impact craters are absent, and the reason is not hard to find—Io is volcanically active. Not only is Io volcanically active, it is by far the most volcanically active body in the solar system. Impacts by bolides certainly must occur, but the craters are quickly covered up by volcanic debris.

Io's volcanism seems to be of two kinds. The first is the familiar basaltic volcanism found so widely throughout the solar system. Lava plains and shield volcanoes are the result. One of the shield volcanoes, Ra Patera, is almost as large as Olympus Mons on Mars. Fresh lava flows can be seen on its slopes. The second kind of volcanism seems to involve sulfur and sulfur dioxide (SO_2). Huge orange-yellow flows of what is presumed to be molten sulfur have been seen, some as much as 700 km long. Most striking, however, are active volcanic plumes that throw sprays of sulfurous gases and entrained solid particles as high as 300 km above the surface of Io. Nine active plumes were observed by the two *Voyager* spacecraft as they flew by Io (Fig. 1.5A). The volcanic plumes seem to be geyserlike in origin, but the fluid that boils and erupts is SO_2, not H_2O. It has been estimated that the plumes eject 10^{16}g of fine, solid particles each year. This quantity is sufficient to bury the surface of Io with a layer of pyroclastic debris 100 m thick in a million years; it is no wonder no impact craters are to be seen. The

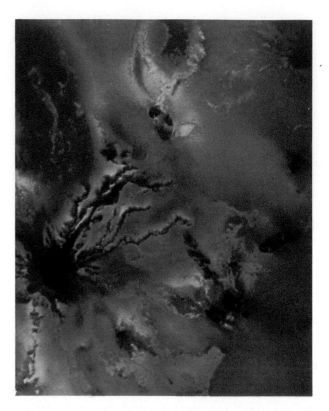

FIGURE 18.17   The bright colors on the surface of Io are caused by sulfur and sulfurous compounds given off during volcanism. The feature in the lower left of the image is a volcanic cone; numerous lava flows, basaltic in composition, radiate out from the volcano. The width of the field of view is 1000 km. The image was taken by *Voyager I* in 1979 at a distance of 128,500 km.

process of surface renewal is much faster than it is on the Earth.

The amount of heat energy needed to drive Io's volcanoes is much greater than the heat that could be produced through radioactive decay in a stony planet. Io's volcanic heat comes from a different source—the gravitational pull exerted by the huge mass of Jupiter. As Io moves around Jupiter in an elliptical orbit it is periodically stretched more or less by the gravitational pull—more during a close approach, less when far away. The bending and stretching due to the fluctuating gravitational pull generate heat, just as a copper wire becomes hot if it is bent back and forth. No other object in the solar system demonstrates the effect of tidal stresses so dramatically as Io.

Farther away from Jupiter than Io are the three large moons Europa, Ganymede, and Callisto. Their densities decrease the farther they are away from Jupiter. Europa has a density of 3.2 g/cm^3, Ganymede, 2.0 g/cm^3, and Callisto, 1.8 g/cm^3. The reason for the lowered densities was discovered during the *Voyager* missions: each of the moons is sheathed with a layer of ice. The outer moons, and especially Europa, may have small metallic cores, but their densities suggest that their main masses lie in thick mantles consisting of ice and silicate minerals. Above the mantle are crusts of nearly pure ice 100 km or more thick.

Craters are rare on Europa; however, the surface is split and crisscrossed by an intricate network of fractures (Fig. 1.5B). Presumably some tectonic process renews Europa's ice surface through fracture and upwelling. The upwelling may involve melting of the ice. The source of energy that results in melting is probably tidal, caused, as with Io, by the gravitational pull of Jupiter.

Ganymede has a much thicker sheath of ice than does Europa. Ganymede is too far from Jupiter to be influenced by tides, and it is pitted by craters. However, on Ganymede some slow-acting process is apparently renewing and reworking the ice surface because some regions have few if any impact craters. The surface is divided into dark and light areas (Fig. 18.18). The dark areas are more heavily cratered than the light areas and are presumably older. Within the lighter (younger) areas there are striking grooves, fractures, and graben-like structures. It is possible that a unique kind of plate tectonics may be operating on Ganymede. The dark regions are ancient ice continents, the lighter areas are the places where ice rises convectively from the depths to create new ice crust.

The most distant moon, Callisto, must contain the greatest proportion of ice because its density is the least of the jovian moons. The surface is again icy, but it must be very ancient ice as the number of impact craters is great. No evidence has yet been found to suggest that the surface of Callisto is being renewed and reworked.

Saturn and its Moons

Saturn has a composition like that of Jupiter but it is not as large. Like Jupiter, Saturn radiates more energy than it receives from the Sun. The most striking feature of Saturn is its immense ring system, discussed in the opening essay. Before the *Voyager* mission to Saturn in 1980, little was known of Saturn's moons. Less is known about them still than is known about the moons of Jupiter. Most of the moons are small and have low densities. In composition and structure they seem to resemble Ganymede and Callisto. They are ice-covered, and cratering is extensive. Two of the moons display evidence, like that on Ganymede, suggesting that ice flow may be renewing the surfaces.

The most distinctive among the Saturnian moons is Titan, a body larger than Mercury. It is the only moon in the solar system large enough to retain a substantial atmosphere. Unfortunately, the atmosphere is an opaque, orange-colored smog that shrouds the surface from view. The composition of Titan's atmosphere is mostly nitrogen; however, ethane, acetylene, ethylene, hydrogen cyanide, and other unpleasant substances are also present.

The density of Titan is 1.9 g/cm³, which suggests that it contains about 45 percent ice and 55 percent rocky matter. Sunlight working on the atmosphere has caused ethylene and acetylene to form through photochemical reactions, and it is these compounds that produce the smog that covers Titan.

Because sunlight cannot penetrate the smog, the surface of Titan must be a cold and unfriendly place. Scientists who have studied the data sent back by *Voyager II* suggest that the surface temperature is −180°C, that Titan is covered by an ocean of liquid ethane and methane, and that continents of ice rise up from the ocean floor. Titan is stranger even than science fiction. Who could have imagined a planetary body with oceans of liquid hydrocarbons and continents of ice?

FIGURE 18.18 The surface of Ganymede, largest of Jupiter's moons, viewed from a distance of 312,000 km by *Voyager 2* in 1979. Ganymede is covered by a thick crust of ice. The dark surface is ancient ice, presumably covered by dust and impact debris. It is split into continent-sized fragments that are separated by light-colored, grooved terrains of younger ice. Ganymede is apparently tectonically active, and the grooved terrains seem to be the places where new ice rises from below, but how this happens, and what causes the grooves, is not known. The field of view is approximately 1300 km across.

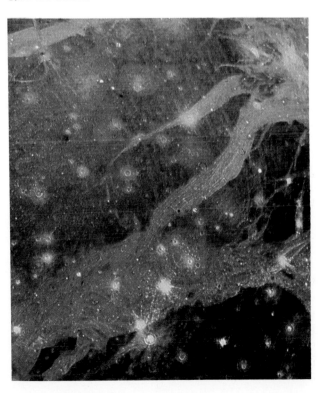

IS THERE LIFE OUT THERE?

So far as we know there is only one planet in the solar system—the Earth—that is home to carbon-based life. Mercury, the Moon, and the satellites of the jovian planets with the exception of Titan lack atmospheres. Venus is so hot nothing could live on its surface, and the fact that so much CO_2 is in Venus's atmosphere makes it unlikely life has ever existed there. The one possible exception is Mars. Mars has ice, and we know at times in the past water has been present because we can see water-cut channels.

Mars today is very dry, and its atmosphere is about 0.007th of an Earth atmosphere. That is far too low for life. But Mars may once have had a denser atmosphere and may once have been warmer. If that is the case, perhaps life existed on Mars in the past. The two *Viking* spacecraft that landed on Mars carried a very sensitive instrument called a mass spectrometer. The instrument was designed to test samples of Martian soil to see if any organic compounds might be present. On the Earth, rocks that are more than 2 billion years old still contain traces of organic compounds that indicate life forms. If similar organic molecules could be found in Martian rocks and soil they would provide strong evidence that life once existed on Mars. Unfortunately nothing suggestive of past life was found by the *Viking* tests.

Does life exist on planets that circle other suns out there in space? There are billions upon billions of suns, so it is almost inevitable that there are billions of planets too. Surely a few of the planets must be earthlike and therefore might support life. The real problem is how we might ever find out about life on other planets. If we sent out radio messages and hoped to get a reply, we might have to wait hundreds of years because the distances to even the closer suns are so great. However, consider the possibility that a somewhat more advanced civilization than ours actually does exist somewhere out in space. Presumably they would already have sent messages. So far we haven't heard or seen any messages because they would ask the same kind of questions we do. Such evidence is negative and not very good, but so far it does suggest that the Earth might well be unique and we might very well be alone.

SUMMARY

1. Planets close to the Sun, such as Mercury, Mars, the Earth, and Venus, are small, dense, rocky bodies. Planets farther away, such as Saturn and Jupiter, are large, low-density bodies.

2. The Moon has a layered structure probably formed by differentiation.

3. The Moon probably has a small core surrounded by a thick mantle and is capped by a crust 65 km thick.

4. On the Moon, magma was formed early but is no longer generated. The Moon is a magmatically dead planet.

5. The highlands of the Moon are remnants of ancient crust built by magmatic differentiation more than 4 billion years ago.

6. The maria (lunar lowlands) are vast basins created by the impacts of giant meteorites and later filled in by lava flows.

7. Each of the terrestrial planets went through a period of internal radioactive heating that led to generation of basaltic magma.

8. The Earth and possibly Mars and Venus are still producing magma from radioactive heating. Mars and Venus both appear to be one-plate planets, not multiplate planets like the Earth.

9. Mars seems to be magmatically active. Olympus Mons, a shield volcano on Mars, is the largest volcano yet found in the solar system.

10. The principal eroding agent on Mars is wind-driven sand and dust. Water or some other flowing liquid cut stream channels at some time in the past.

11. Venus has about the same size and density as the Earth. It has a dense atmosphere of carbon dioxide and a surface temperature of about 500°C.

12. None of the other terrestrial planets has a climate hospitable to human life.

13. The large outer planets, Jupiter, Saturn, Neptune, and Uranus, have thick atmospheres that obscure their rocky cores. Pluto, the outermost planet, is probably covered by ice.

14. The moons of Jupiter have progressively thicker outer layers of ice the farther away they are from Jupiter.

IMPORTANT TERMS TO REMEMBER

asteroid (p. 491)

bolide (p. 491)

ecliptic (p. 490)

impact cratering (p. 491)

lunar highlands (p. 494)

magma ocean (p. 497)

mare (p. 495)
mare basins (p. 497)
maria (p. 495)

planetology (p. 490)

QUESTIONS FOR REVIEW

1. What is planetology? How is planetology related to geology?

2. Name the four terrestrial planets in order of their positions out from the Sun.

3. Besides the terrestrial planets, what other solid objects are to be found in the inner reaches of the solar system?

4. What is the most important process for shaping planetary surfaces in the solar system?

5. Briefly describe the kinds of magmatic activity that have been observed on planets and moons other than the Earth.

6. What evidence leads scientists to conclude that in the earliest days of the Moon's history it had a molten outer layer and a solid interior?

7. What is the origin of the lunar maria?

8. How could it be possible for evidence of water erosion to exist on Mars without lakes, streams, or seas being present?

9. What is the source of the energy that made Io, one of Jupiter's moons, the most tectonically active object in the solar system?

10. Compare the structure of Ganymede, a moon of Jupiter, with the Earth's moon. What explanation can you offer for the differences?

(GCMs) that attempt to link processes in the atmosphere, the hydrosphere, and the biosphere. The sheer complexity of these natural systems means that such models, of necessity, are greatly simplified representations of the real world. Furthermore, many of the linkages and processes in the climate system are still poorly understood and therefore difficult to model. The models do not yet adequately portray the dynamics of ocean circulation or cloud formation, two of the most important elements of the climate system. Also absent from the models are many of the complex biogeochemical processes that link climate to the biosphere. Despite these limitations, GCMs have been very successful in simulating the general character of present-day climates and have greatly improved weathering forecasting. This success encourages us to use these models to gain a general global picture of future climates as the Earth's physical and chemical balance changes.

To run a modeling experiment that simulates the climate, a set of **boundary conditions** is specified. Boundary conditions are mathematical expressions of the physical state of the Earth's climate system at the period of interest for the experiment. Thus, an experiment designed to simulate the present climate would prescribe as boundary conditions the orbitally determined solar radiation reaching the Earth, the geographic distribution of land and ocean, the position and heights of mountains and plateaus, the concentrations of atmospheric trace gases, sea-surface temperatures, the limits of sea ice, the snow and ice cover on land, the *albedo* (reflectivity) of land, ice, and water surfaces, and the effective soil moisture (the sum of water input and water loss).

Because the solution of the complex mathematical equations of a GCM requires considerable amounts of computer time, the three-dimensional grid spacing (the distance between points on and above the globe for which the solutions are calculated) in the model experiments is large in order to keep costs manageable. As a result, the resolution of the models is rather coarse: grid points commonly are separated by 4 or 5 degrees of latitude, or 450–550 km. Therefore, although these models can generate a reasonable picture of global and hemispheric climatic conditions, they are poor at resolving conditions at the scale of small countries, states, or counties. Until more powerful computers are built, and/or the cost of running a model experiment decreases, the spacial resolution of GCMs is likely to remain relatively coarse.

Model Estimates of Greenhouse Warming

Predictions of climatic change related to greenhouse warming are based mainly on the results of five GCMs,

four developed in the United States and one in the United Kingdom. The models differ in detail, as well as in the assumptions they employ. Nevertheless, they all predict that the anthropogenically generated greenhouse gases already in the atmosphere will lead to an average global temperature increase of at least 0.5 to 1.5°C. This prediction is consistent with the 0.5°C rise in temperature inferred from the instrumental record. They further predict that if the greenhouse gases continue to build up until their combined effect is equivalent to a doubling of the preindustrial CO_2 concentration, then average global temperatures will rise between 1 and 5°C. This does not mean, however, that the temperature will increase uniformly all over the Earth. Instead, the projected temperature change varies geographically, with the greatest change occurring in the polar regions (Fig. 19.8).

The rate at which the projected warming will occur depends on a number of basic uncertainties: How rapidly will concentrations of the greenhouse gases increase? How rapidly will the oceans, a major reservoir of heat and a fundamental element in the climate system (Fig. 19.1), respond to changing climate? How will changing climate affect ice sheets and cloud cover? What is the range of natural variations in the climate system on the century time scale? The potential complexity is well illustrated by clouds. If the temperature of the lower atmosphere increases, more water will evaporate from the oceans. The increased atmospheric moisture will create more clouds, but clouds reflect solar energy back into space, which will have a cooling effect on the surface air, thereby having a result opposite to that of the greenhouse effect.

Because of such uncertainties about the climate system, scientists tend to be cautious in their predictions. Nevertheless, there is a general consensus that (1) human activities have led to increasing atmospheric concentrations of carbon dioxide and other trace gases that have enhanced the greenhouse effect; (2) global mean surface air temperature has increased by 0.3 to 0.6°C during the last 100 years, and this increase may be the direct result of the enhanced greenhouse effect; and (3) during the next century global average temperature will likely increase at about 0.3°C per decade, assuming emission rates do not change. This projected increase will lead to a global average temperature about 1°C warmer than present by the year 2025 and as much as 3°C warmer by the end of the next century. If governmental controls lead to lower emission rates, the decadal rise in temperature may be only 0.1–0.2°C. Nevertheless, the temperature increase related to the continued release of greenhouse gases will be larger and more rapid than any experienced in human history. In effect, we may be about to experience a "super interglaciation" warmer than any interglaciation of the past 2 million years (Fig. 19.9).

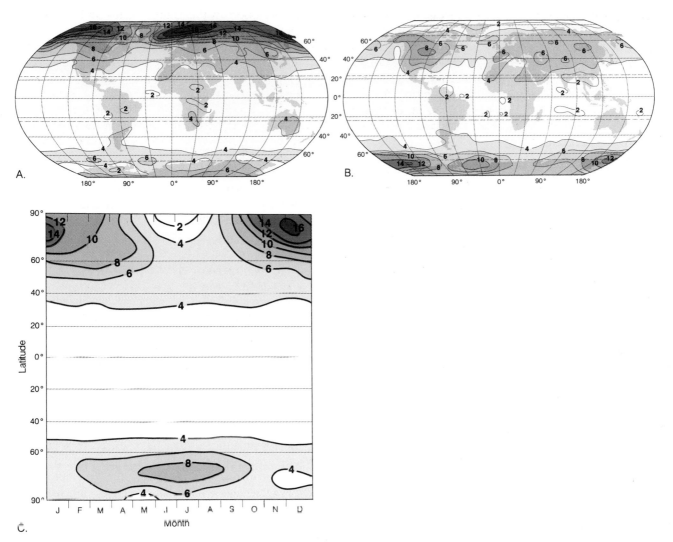

FIGURE 19.8 A forecast of future changes in surface air temperature (in °C) that would result from an effective doubling of atmospheric CO_2 concentration relative to that of the present. A. Temperature increases for Winter (December, January, February). For example, along the lines labeled 4, the projected temperature increase is everywhere 4°C. B. Temperature increases for Summer (June, July, August). C. A latitudinal cross section showing changes in zonal average air temperature through the year. This graph is a summary of the map patterns shown in A and B, but includes the spring and autumn months as well. Greenhouse warming is greatest at high latitudes, where temperature increases as great as 16°C are forecast by the model for the Northern Hemisphere winter.

Environmental Effects of Global Warming

An increase in global surface air temperature by a few degrees does not sound like much. Surely, you say, we can put up with this rather insignificant change. However, if we stop and consider that the difference in average global temperature between the present and the coldest part of the last ice age was only about 5°C, we can begin to see how even a slight temperature change of a degree or two could well have global repercussions.

Global warming is just one result of our great geochemical "experiment." There are many physical and biological side effects that are of considerable interest and concern.

Global Precipitation Changes.

A warmer atmosphere will lead to increased evaporation from oceans, lakes, and streams, and to greater precipitation. However, the distribution of precipitation will be uneven. Climate models suggest that the equatorial regions will receive more rainfall, in part because

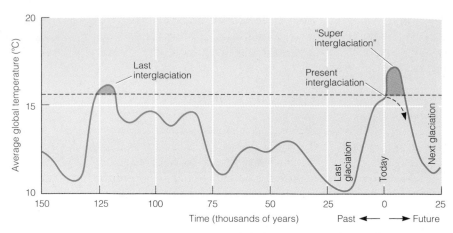

FIGURE 19.9 The course of the average global temperature during the past 150,000 years and 25,000 years into the future. The natural course of climate (dashed black line) would be declining temperatures leading to the next glacial maximum, about 23,000 years from now. With greenhouse warming, a continuing rise of temperature may lead to a "super interglaciation" within the next several centuries. The temperature may then be warmer than during the last interglaciation, and warmer than at any time in human history. The decline toward the next glaciation would thereby be delayed a millennium or more. The dashed black line just above 15°C marks the present average global temperature.

warmer temperatures will increase rates of evaporation over the tropical oceans and promote the formation of rainclouds, whereas some interior portions of large continents, which are distant from precipitation sources, will become both warmer and drier.

Changes in Vegetation. Shifts in precipitation patterns are likely to upset ecosystems, causing vegetation communities, and animals dependent on them, to adjust to new conditions. Forest boundaries may shift during coming centuries in response to altered temperature and precipitation regimes. Some prime midcontinental agricultural regions are likely to face increased droughts and substantially reduced soil moisture that will have a negative impact on crops. Higher-latitude regions with short, cool growing seasons may see increased agricultural production as summer temperatures increase.

Increased Storminess. The shift to a warmer, wetter atmosphere favors an increase in tropical storm activity. Regions that now are impacted by hurricanes and typhoons could see an increase in the size and frequency of these devastating storms.

Melting (and Growing?) Glaciers. Because warmer summers favor increased ablation, worldwide recession of low- and middle-latitude mountain glaciers is likely in a warmer world. On the other hand,

warmer air in high latitudes can evaporate and transport more moisture from the oceans to adjacent ice sheets, which may cause them to grow larger.

Reduction of Sea Ice. The greatly enhanced heating projected for high northern latitudes (Fig. 19.8) favors the shrinkage of sea ice. A reduction in polar sea ice, which has a high albedo, should enhance the greenhouse effect by reducing the amount of solar radiation reflected back into space. Models show much less heating in the high-latitude Southern Hemisphere, suggesting little change in sea ice cover there.

Thawing of Frozen Ground. Rising summer air temperatures will begin to thaw vast regions of perennially frozen ground at high latitudes. The thawing will likely affect natural ecosystems as well as cities and engineering works built on frozen ground.

Rise of Sea Level. As the temperature of ocean water rises, its volume will expand, causing world sea level to rise. This rise in sea level, supplemented by meltwater from shrinking mountain glaciers, is likely to increase calving along the margins of tidewater glaciers and ice sheets, thereby leading to additional sea level rise. The rising sea will inundate coastal regions where millions of people live and will make the tropical regions even more vulnerable to larger and more frequent cyclonic storms.

Changes in the Hydrologic Cycle. Shifting patterns of precipitation and warmer temperatures will likely lead to some significant local and regional changes in stream runoff and groundwater levels.

Decomposition of Soil Organic Matter. As the temperature rises, the rate of decomposition of organic matter in soil will increase. Soil decomposition releases CO_2 to the atmosphere, thereby further enhancing the greenhouse effect. If world temperature rises by 0.3°C per decade, during the next 60 years soils will release an amount of CO_2 equal to nearly 20 percent of the projected CO_2 release due to combustion of fossil fuels, assuming the present rate of fuel consumption continues.

Breakdown of Gas Hydrates. Gas hydrates are ice-like solids in which gas molecules, mainly methane, are locked in the structure of water. They are found in ocean sediments and beneath frozen ground. By one estimate, gas hydrates worldwide may hold 10,000 billion metric tons of carbon, twice the carbon in all the known coal, gas, and oil reserves on land. They accumulate in ocean sediments beneath a water depth of 500 m where the temperature is low enough and the pressure high enough to permit their formation. They also accumulate beneath frozen ground, which acts as a seal to prevent upward migration and escape of the gas. When gas hydrates break down, they release methane. Global warming at high latitudes will result in thawing of frozen ground that may well destabilize the hydrates there, release large volumes of methane, and thus amplify the greenhouse effect.

While our present knowledge of how the Earth works, coupled with computer modeling, enables us to make educated projections of surface environmental changes that will result from greenhouse warming, most scientists are reluctant to make firm forecasts. Instead, they hedge their bets with qualifying adjectives like "possible," "probable," and "uncertain." Their understandable caution emphasizes the gap between what we know about the Earth and what we would like to know and points to the many challenges that still face scientists studying global change.

THE PAST AS A KEY TO THE FUTURE

To increase the likelihood that our predictions about the changing global environment are correct, we can invoke Ayer's Law. This useful tenet of geology states: "Anything that did happen, can happen." In other words, we can use the geologic record, which provides an invaluable archive of the past history of natural environmental changes on the Earth, as a key to understanding what the future may hold in store.

Lessons from the Past

Examining changes to physical and biological systems that occurred when human influence was absent or minimal allows us to see how these systems responded to sudden changes of climate. We can use this information to help us anticipate the character of environmental changes that may happen on a warming Earth.

One example is a very rapid warming event that occurred in the North Atlantic region at the end of the last glaciation. Oxygen-isotope data derived from Swiss lake sediment and from a Greenland ice core can be plotted as curves that can be read as temperature curves (Fig. 19.10). The curves are similar and show an abrupt warming close to 10,000 years ago (arrows in Fig. 19.10). Detailed studies of the ice core indicate that the Greenland climate warmed 7°C in only 40 years, a rate that exceeds even the very rapid average global rate of temperature rise we can expect during the next century. As we learn more about this remarkable past natural climatic event, and see how the Earth's physical and biological systems reacted to it, we should gain important insights about future environmental changes that could happen in a rapidly warming world.

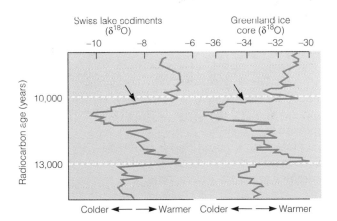

FIGURE 19.10 Measurements of oxygen isotopes in the sediments of a Swiss lake and an ice core from the Greenland Ice Sheet show an abrupt and rapid change of climate at the end of the last glaciation (arrows). The curves, which can be viewed as recording changes in temperature, show a sudden shift from cold glacial climate to warmer postglacial climate. Detailed studies of the ice core indicate that average temperature in Greenland abruptly climbed about 7°C in only 40 years.

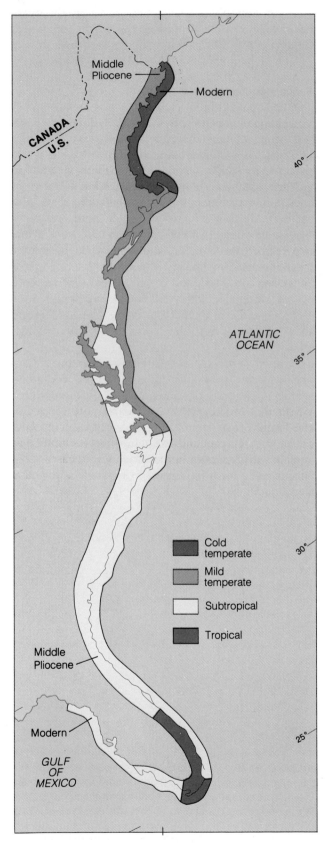

Cold temperate

Mild temperate

Subtropical

Tropical

We can also explore the geologic record for information about times when the climate was warmer than now. Of particular interest are the early Holocene, from 10,000 to about 6000 years ago, when average temperatures were 0.5 to 1°C warmer; the warmest part of the last interglaciation, about 120,000 years ago, when global temperatures were about 1 to 2°C higher; and the middle Pliocene, about 4.5 to 3 million years ago, when average temperatures may have been 3 to 4°C warmer than present. These periods do not provide perfect analogs for present global warming because the distribution of solar radiation reaching the Earth's surface was different then (e.g., Figs. 11.32 and 11.33). Nevertheless, these intervals enable us to see how plants and animals responded to climatic conditions that may have been broadly similar to those we may experience in the near future.

During the warmest parts of the Pliocene Epoch, for example, world sea level was tens of meters higher than now, pointing to reduced global ice cover. Isotopic measurements of North Atlantic deep-sea cores show that winter sea-surface temperatures were at least 3°C warmer than at present. In response to warmer ocean waters, temperature-sensitive marine organisms had different distributions. The boundaries between tropical, subtropical, and temperate assemblages of ostracodes, a type of crustacean, shifted northward along the east coast of the United States during the Middle Pliocene in response to warmer coastal waters (Fig. 19.11). Whereas the northern limit of the tropical assemblage now lies mid-way up the east coast of Florida, Mid-Pliocene fossils of tropical species are found as far north as central North Carolina. Similarly, the cold-temperate assemblage, which is now distributed from Cape Cod, Massachusetts northward into Canada, was absent from New England waters in the Middle Pliocene.

FIGURE 19.11 Distribution of ostracode assemblages along the continental margin of the eastern United States at present (seaward of present coastline) and during the Middle Pliocene (landward of present coastline). Because ostracode distribution is closely related to water temperature, the Pliocene range of these organisms can be used to show how ocean temperatures at that time differed from modern temperatures. During the Middle Pliocene, relative sea level was higher than now, the tropical ostracode assemblage ranged halfway up the east coast of Florida, the subtropical assemblage extended well north of its present limit, and the mild-temperate assemblage reached north into Canada.

Modeling Past Global Changes

Paleoclimatic reconstructions offer a means of testing the accuracy of climate models. Our confidence in using GCMs for predicting the future will be strengthened if the models not only can accurately reproduce the present climate, but also can reproduce past climates. We can test, or "validate," these climate simulations by comparing the model results against independent geologic evidence of the conditions then prevailing.

The same GCMs that are used to model the present global weather are also used to model ancient climates. The main difference is in the boundary conditions. For example, one set of experiments has attempted to simulate the Earth's climates at 3000-year intervals since the last glacial maximum about 18,000 years ago. The specified boundary conditions changed substantially over this time interval (Fig. 19.12): solar radiation varied by as much as 8 percent as the Earth went through one precessional cycle; the area of glacier ice shrank, and airborne dust decreased as the ice-age came to an end; and both sea-surface temperature and atmospheric CO_2 increased as the climate moved toward its present interglacial state. During each of the seven time periods modeled, the boundary conditions were different from those of the other periods. As a result, the successive simulations change as the model Earth passes from a glacial age to an interglacial age.

As an example of such a simulation, we can examine the Eastern Hemisphere results of an experiment focusing on 9000 years ago (Fig. 19.13A). You can see what the specified boundary conditions are by noting the values at 9000 years in Figure 19.12. The model "predicts" that 9000 years ago the climate in a broad belt across northern Africa and southern Asia experienced warmer and wetter summers and cooler winters; the Mediterranean region at that time had warmer, drier summers, southern Africa had colder, drier winters, and Australia had colder winters and warmer summers. The model also indicates that the increase in summer moisture was largely due to an increase in the strength of the summer monsoon (moisture-bearing winds) in western Africa and southern Asia. Figure 19.12 shows us that the boundary conditions for 9000 years ago were different from those of today mainly with respect to Northern Hemisphere solar radiation, which was about 8 percent greater in summer and 8 percent less in winter. This difference was a major factor in strengthening the summer monsoon.

To see how reasonable these model predictions are, we can examine geologic evidence of conditions about 9000 years ago (Fig. 19.13B). The evidence points to a broad belt of land across northern Africa, the Middle East, and China that had greater effective moisture (i.e., precipitation minus evaporation) than now. In Africa and western China the increased moisture is evidenced by high water levels in closed-basin

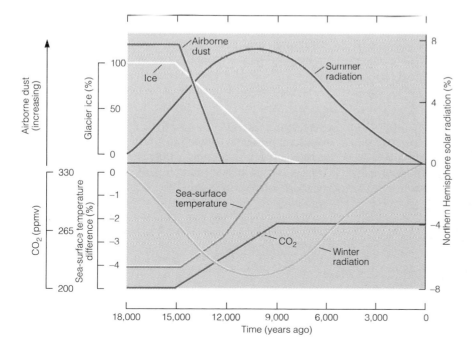

FIGURE 19.12 Boundary conditions used in one set of GCM simulations of world climate between the last glacial maximum (18,000 years ago) and the present. Solar radiation values are based on known astronomical motions of the Earth (Fig. 11.33), while ice, dust, CO_2, and sea-surface temperature values are based on geological data.

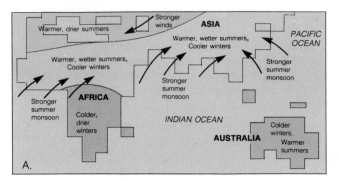 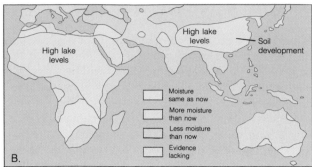

FIGURE 19.13 Climates of Africa and southern Asia during the early Holocene (9000 years ago). A. GCM simulation indicates that a belt crossing northern Africa and southeastern Asia had warmer, wetter summers and cooler winters 9000 years ago. These conditions resulted from a strengthening of summer monsoon winds, which bring precipitation to the continents from the warm tropical oceans. At that time, the Mediterranean region and Australia had warmer, drier summers, while southern Africa and Australia had colder winters. B. Geologic evidence generally bears out the GCM results: a belt running from northern Africa through the Middle East and China received more moisture 9000 years ago, as shown by high lake levels and paleosols that indicate increased soil moisture. Evidence also shows that southern Africa and Madagascar, as well as the northeastern Mediterranean region, were drier than now.

lakes and the widespread occurrence in China of an early Holocene soil that indicates an increase in monsoon precipitation. In this example, the results of the model simulation are generally consistent with the paleoclimatic data assembled by geologists.

PERSPECTIVES OF GLOBAL CHANGE

Most of us would find it fascinating, and no doubt convenient, if we could look into a crystal ball and see the future. It certainly would be helpful if we had a clear vision of our climatic future. At present, the best we can conclude is that the force of scientific evidence and theory makes it very probable that the climate is warming up and will continue to warm as we add greenhouse gases to the atmosphere. There also is a high probability that average global temperatures ultimately will increase by 2 to 4°C, leading to widespread environmental changes.

It is less probable that the temperature will increase steadily, for there are natural, and as yet largely unpredictable, modulations of the climate system on the time scale of years to decades. As an example, the huge eruption of Krakatau in 1883 led to a 0.5°C drop in average global temperature (chapters 3 and 11), and the even larger Tambora eruption of 1815 was

followed by the "year without a summer," during which mid-summer snow and frost caused severe hardships in Europe and New England. The huge eruption of the Philippine volcano Pinatubo in June 1991 introduced a vast quantity of fine ash and sulfurous gas into the stratosphere, where it quickly began to spread into the northern and southern hemispheres. Current estimates are that as the veil of dust and gas spreads throughout the atmosphere and reflects incoming solar radiation, surface temperatures will be lowered for at least several years. The volcanic effect may well reverse temporarily any upward trend in average global temperature attributable to continued emission of greenhouse gases.

While the short-term prospect (on the scale of human generations) is for a warmer world, if we stand back and look at our great geochemical "experiment" from a geological perspective, we will perceive that it is only a brief, very rapid, yet nonrepeatable perturbation in the Earth's climatic history. It is nonrepeatable because once the Earth's store of easily extractable fossil fuels is used up, most likely within the next several hundred years, the human impact on the atmosphere will inevitably decline, and the climate system should revert to a more natural state. The greenhouse perturbation may well last a thousand years, and perhaps more, but ultimately the changing geometry of the Earth's orbit will propel the climate system inexorably into the next glacial age (Fig. 19.9).

WHY WAS THE MIDDLE CRETACEOUS CLIMATE SO WARM?

It's probably a good thing we did not live during the Middle Cretaceous Period. Not only was the world inhabited by huge carnivorous dinosaurs, but the climate was one of the warmest in the Earth's history. Evidence that the world was much warmer than today is compelling (Fig. B19.1A). Warm-water marine faunas were widespread, coral reefs grew 5 to 15° closer to the poles than they do now, and vegetation zones were displaced about 15° poleward of their present positions. Peat deposits that would give rise to widespread coal formations formed at high latitudes, and dinosaurs, which are generally thought to have preferred warm climates, ranged north of the arctic circle. Sea level was 100 to 200 m higher, implying the absence of polar ice sheets, and isotopic measurements of deep-sea deposits indicate that intermediate and deep waters in the oceans were 15 to 20° warmer than now.

GCM simulations of the Middle Cretaceous world suggest that several factors likely were involved in producing such warm conditions: geography, ocean circulation, and atmospheric composition. The modeling simulations show that the Middle Cretaceous arrangement of continents and oceans (Fig. B19.1A), which influenced ocean circulation and planetary albedo, could account for nearly 5°C of warming; of this amount, about a third is attributable to the absence of polar ice sheets. However, geography alone is inadequate to explain warmer year-round temperatures at high latitudes. Could the poleward transfer of heat be the answer? The oceans now account for about a third of the present poleward heat transfer, but modeling shows that even with the geography and ocean circulation rearranged as they were in the Middle Cretaceous, oceanic heat transfer cannot explain the greater high-latitude warmth. If the geologic data have been correctly interpreted, and the modeling results are reliable, some other factor must be involved. This factor appears to be CO_2, the major greenhouse trace gas.

Can an enhanced greenhouse effect be the key to explaining the exceptionally warm Middle Cretaceous climate? GCM experiments show that by rearranging the geography and also increasing carbon dioxide 6 to 8 times above present concentrations, the warmer temperatures can be explained. Geochemical reconstructions of changing atmospheric CO_2 levels over the past

100 million years point to at least a tenfold increase in CO_2 during the Middle Cretaceous, leading to average temperatures as much as 8°C higher than now (Fig. B19.1B). Compared to the

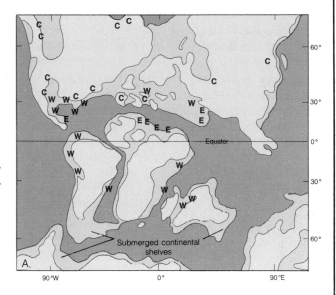

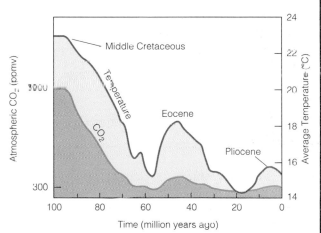

FIGURE B19.1 A. During the Middle Cretaceous, sea level was 100–200 m higher than now and flooded large areas of the continents, producing shallow seas. Warm-water faunas (W) and evaporites (E) were present at low to middle latitudes, while coal deposits (C) developed in arctic regions, implying warmer year-around temperatures. B. Geochemical reconstruction of changing atmospheric CO_2 concentration and average global temperature over the past 100 million years. High CO_2 values and temperatures in the Middle Cretaceous contrast with much lower values of the present. Other intervals of higher temperature and CO_2 occurred during the Eocene and the Mid-Pliocene.

average conditions forecast for the twenty-first century (2–4°C hotter), the Middle Cretaceous climate must have been a scorcher!

If CO_2 was an important factor in Middle Cretaceous warming, we still are faced with explaining how this gas increased so substantially. Unlike the modern world, combustion of fossil fuels cannot provide the answer. The most likely source is volcanic activity, which today constitutes a major source of CO_2 entering the atmosphere. Prior to the industrial revolution, volcanism probably replenished the atmosphere with as much as 65 percent of the carbon lost to sedimentation, mostly by slow, noneruptive degassing of CO_2 from magmas in the upper crust.

Geologic evidence points to an unusually high rate of volcanic activity in the Middle Cretaceous. Rates of continental drift were then about three times as great as now, implying increased extrusion rates at spreading ridges. In addition, vast outpourings of lava created a succession of great undersea plateaus across the southern Pacific Ocean between 135 and 115 million years ago, the time of maximum Cretaceous warmth. One of these—the Ontong-Java Plateau in the southwest Pacific—has more than twice the area of Alaska and reaches a thickness of 40 km. Such a massive outpouring of lava likely released massive amounts of CO_2. Could this gas emission have been sufficient to warm the climate to unprecedented levels? By one calculation, the eruptions could have released enough CO_2 to raise the atmospheric concentration to 20 times the preindustrial value, in the process raising average global temperature as much as 10°C. Other estimates range from 8 to 12 times the preindustrial value.

Recently, geologists have proposed that these vast lava outpourings are the result of super hot spots. Hot spots are thought to be due to rising plumes of hot rock (chapter 16). A plume like that responsible for the Hawaiian hot spot, which is about 200–300 km across, must originate somewhere in the mesosphere, possibly at a depth of 670 km, where a distinct seismic discontinuity occurs (chapter 15). A super hot spot is thought to arise from a *superplume,* which is conceived as being a plumelike mass of unusually hot rock that rises from the base of the mantle at a rate of 10–20 cm/yr and spreads out in a mushroom shape as it reaches shallower depths where confining pressures are lower. Modeling studies suggest that the size of a plume head depends on the depth at which the plume originates (Fig. B19.2) and that the Hawaiian plume must originate at a depth of less than 1000 km.

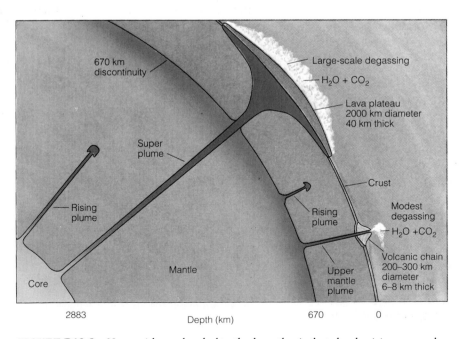

FIGURE B19.2   New evidence has led to the hypothesis that slowly rising superplumes from the core-mantle boundary build huge lava plateaus when they reach the lithosphere and may give rise to large-scale degassing of CO_2 that greatly enhances the greenhouse effect. By contrast, plumes rising from the base of the upper mantle at 670 km produce much smaller hot spots that generate series of volcanoes like those of the Hawaiian-Emperor chain.

The huge Middle Cretaceous lava plateaus, however, are thousands of kilometers across, implying that they could only have formed from superplumes rising from the core-mantle boundary, nearly 2900 km deep. Such a plume would be about 1000 km wide when it reached the upper mantle and would flatten to two or three times this width as it approached the base of the lithosphere. A superplume would be an efficient mechanism for allowing heat to escape from the Earth's core. If this hypothesis is correct, then the plate-tectonic cycle cools the mantle both by heat loss at spreading ridges and by the downward plunge of plates of cool lithosphere, while superplumes cool the core. By this reasoning, the core and atmosphere are linked dynamically, and the warm Middle Cretaceous climate was a direct consequence of the cooling of the Earth's deep interior.

SUMMARY

1. Synthetic chlorofluorocarbon (CFC) gases entering the upper atmosphere break down to chlorine, which destroys the protective ozone layer. Discovery of a vast and recurring ozone hole over Antarctica has led to international efforts to eliminate CFC production by the end of the century.

2. The Earth's climate system involves the atmosphere, hydrosphere, lithosphere, and biosphere. Changes affecting it operate on time scales ranging from decades to millions of years.

3. Using information from the geologic record, geologists can measure the magnitude and geographic extent of past climatic changes, determine the range of climatic variability on different time scales, and test the accuracy of computer models that simulate past climatic conditions.

4. The carbon cycle is among the most important of the Earth's biogeochemical cycles. Carbon resides in the atmosphere, the biosphere, the hydrosphere, and in the crust and cycles through these reservoirs at different rates.

5. The anthropogenic extraction and burning of fossil fuels perturbs the natural carbon cycle and has led to an increase in atmospheric CO_2 since the start of the Industrial Revolution.

6. The greenhouse effect, caused by the trapping of long-wave infrared radiation by trace gases in the atmosphere, makes the Earth a habitable planet.

7. The increase in atmospheric trace gases (CO_2, CH_4, O_3, N_2O, and the CFCs) due to human activities is projected to warm the lower atmosphere by 2° to 4°C by the end of the next century.

8. A probable 0.5°C increase in average global temperature since the mid-nineteenth century may reflect the initial part of this warming. The rate of warming is likely to reach 0.3°C per decade and may lead to a "super interglaciation," making the Earth warmer than at any time in human history.

9. Potential physical and biological consequences of global warming include global changes in precipitation and vegetation patterns, increased storminess, melting of glaciers, sea ice, and frozen ground, a worldwide rise of sea level, local and regional changes in the hydrologic cycle, and increased rates of organic decomposition in soils.

10. Evidence of past intervals of rapid environmental change in the geologic record and reconstructions of past warmer intervals can provide insights into physical and biological responses to global warming. Such reconstructions also permit evaluation of climate model simulations by general circulation models.

11. Although the Earth's surface environments may change substantially during the next several centuries in response to greenhouse warming, viewed from the geologic perspective, this interval will appear as only a brief perturbation in the Earth's climatic history.

IMPORTANT TERMS TO REMEMBER

biogeochemical cycles (p. 512)
boundary conditions (p. 518)

general circulation model (GCM) (p. 517)

greenhouse effect (p. 513)

QUESTIONS FOR REVIEW

1. Why does chlorine have such an adverse affect on the ozone layer, despite the fact it is released into the atmosphere in very small amounts?

2. Describe the carbon cycle and indicate why we regard it as one of the most important biogeochemical cycles.

3. In what ways can carbon be trapped in the Earth and become part of the rock cycle? How can such stored carbon once again find its way into the atmosphere?

4. If atmospheric CO_2 can be dissolved in streams, lakes, groundwater, and the oceans, and also efficiently absorbed by vegetation, why is the burning of fossil fuels causing the CO_2 content of the atmosphere to increase?

5. How is the Earth's atmosphere similar to a garden greenhouse, and why?

6. What are the anthropogenic sources of the principle greenhouse gases?

7. What geologic evidence indicates that the present concentrations of carbon dioxide and methane in the atmosphere are exceptional compared to those of the last several hundred thousand years?

8. What are the major boundary conditions that must be specified in GCM climate simulation experiments? How have these conditions changed since the maximum of the last glaciation?

9. Why do we remain uncertain about the extent to which average global temperature will rise in the next century as a result of greenhouse warming?

10. Give an example of an environmental effect arising from global warming that could enhance the greenhouse effect and lead to additional warming.

11. What factors are likely to cause world sea level to rise in a warming climate? In what ways is rising sea level likely to impact the human population?

12. Why is the geologic record important in helping predict the environmental effects of greenhouse warming? Give two examples.

APPENDIX A

Units and Their Conversions

ABOUT SI UNITS

Regardless of the field of specialization, all scientists use the same units and scales of measurement. They do so to avoid confusion and the possibility that mistakes can creep in when data are converted from one system of units, or one scale, to another. By international agreement the SI units are used by all, and they are the units used in this text. SI is the abbreviation of Système International d'Unités (in English, the International System of Units).

Some of the SI units are likely to be familiar, some unfamiliar. The SI unit of length is the meter (m), of area the square meter (m^2), and of volume the cubic meter (m^3). The SI unit of mass is the kilogram (kg), and of time the second (s). The other SI units used in this book can be defined in terms of these basic units. Three important ones are:

The newton (N), a unit of force defined as that force needed to accelerate a mass of 1 kg by 1 m/s^2; hence 1 N = 1 kg·m/s^2. (The period between kg and m indicates multiplication.)

The Joule (J), a unit of energy or work, defined as the work done when a force of 1 newton is displaced a distance of 1 meter; hence 1 J = 1 N·m. One important form of energy so far as the Earth is concerned is heat. The outward flow of the Earth's internal heat is measured in terms of the number of joules flowing outward from each square centimeter each second; thus, the unit of heat flow is $J/cm^2/s$.

The pascal (Pa), a unit of pressure defined as a force of 1 newton applied across an area of 1 square meter; hence 1 Pa = 1 N/m^2. The pascal is a numerically small unit. Atmospheric pressure, for example (15 lb/in^2), is 101,300 Pa. Pressure within the Earth reaches millions or billions of pascals. For convenience, earth scientists sometimes use 1 million pascals (megapascal, or MPa) as a unit.

Temperature is a measure of the internal kinetic energy (expressed as movement) of the atoms and molecules in a body. In the SI system, temperature is measured on the Kelvin scale (K). The temperature intervals on the Kelvin scale are arbitrary, and they are the same as the intervals on the more familiar Celsius scale (°C). The difference between the two scales is that the Celsius scale selects 100°C as the temperature at which water boils at sea level, and 0°C as the freezing temperature of water at sea level. Zero degrees Kelvin, on the other hand, is absolute zero, the temperature at which all atomic and molecular motions cease. Thus, 0°C is equal to 273.15 K, and 100°C is 373.15 K. The temperatures of processes on and within the Earth tend to be at or above 273.15 K.

Despite the inconsistency, earth scientists still use the Celsius scale when geological processes are discussed.

Appendix A provides a table of conversion from older units to Standard International (SI) units.

PREFIXES FOR MULTIPLES AND SUBMULTIPLES

When very large or very small numbers have to be expressed, a standard set of prefixes is used in conjunction with the SI units. Some prefixes are probably already familiar; an example is the centimeter (which is one hundredth of a meter, or 10^{-2} m). The standard prefixes are

giga	1,000,000,000 =	10^9
mega	1,000,000 =	10^6
kilo	1,000 =	10^3
hecto	100 =	10^2
deka	10 =	10
deci	0.1 =	10^{-1}
centi	0.01 =	10^{-2}
milli	0.001 =	10^{-3}
micro	0.000001 =	10^{-6}
nano	0.000000001 =	10^{-9}
pico	0.000000000001 =	10^{-12}

One measure used commonly in geology is the nanometer (nm), a unit by which the sizes of atoms are measured; 1 nanometer is equal to 10^{-9} meter.

COMMONLY USED UNITS OF MEASURE

Length

Metric Measure

1 kilometer (km)	= 1000 meters (m)
1 meter (m)	= 100 centimeters (cm)
1 centimeter (cm)	= 10 millimeters (mm)
1 millimeter (mm)	= 1000 micrometers (μm) (formerly called microns)
1 micrometer (μm)	= 0.001 millimeter (mm)
1 angstrom (Å)	= 10^{-8} centimeters (cm)

Nonmetric Measure

1 mile (mi)	= 5280 feet (ft) = 1760 yards (yd)

1 yard (yd)	= 3 feet (ft)
1 fathom (fath)	= 6 feet (ft)

Conversions

1 kilometer (km)	= 0.6214 mile (mi)
1 meter (m)	= 1.094 yards (yd) = 3.281 feet (ft)
1 centimeter (cm)	= 0.3937 inch (in)
1 millimeter (mm)	= 0.0394 inch (in)
1 mile (mi)	= 1.609 kilometers (km)
1 yard (yd)	= 0.9144 meter (m)
1 foot (ft)	= 0.3048 meter (m)
1 inch (in)	= 2.54 centimeters (cm)
1 inch (in)	= 25.4 millimeters (mm)
1 fathom (fath)	= 1.8288 meters (m)

Area

Metric Measure

1 square kilometer (km^2)	= 1,000,000 square meters (m^2) = 100 hectares (ha)
1 square meter (m^2)	= 10,000 square centimeters (cm^2)
1 hectare (ha)	= 10,000 square meters (m^2)

Nonmetric Measure

1 square mile (mi^2)	= 640 acres (ac)
1 acre (ac)	= 4840 square yards (yd^2)
1 square foot (ft^2)	= 144 square inches (in^2)

Conversions

1 square kilometer (km^2)	= 0.386 square mile (mi^2)
1 hectare (ha)	= 2.471 acres (ac)
1 square meter (m^2)	= 1.196 square yards (yd^2) = 10.764 square feet (ft^2)
1 square centimeter (cm^2)	= 0.155 square inch (in^2)
1 square mile (mi^2)	= 2.59 square kilometers (km^2)
1 acre (ac)	= 0.4047 hectare (ha)
1 square yard (yd^2)	= 0.836 square meter (m^2)
1 square foot (ft^2)	= 0.0929 square meter (m^2)
1 square inch (in^2)	= 6.4516 square centimeter (cm^2)

Volume

Metric Measure

1 cubic meter (m^3)	= 1,000,000 cubic centimeters (cm^3)
1 liter (l)	= 1000 milliliters (ml) = 0.001 cubic meter (m^3)
1 centiliter (cl)	= 10 milliliters (ml)
1 milliliter (ml)	= 1 cubic centimeter (cm^3)

Nonmetric Measure

1 cubic yard (yd^3)	= 27 cubic feet (ft^3)
1 cubic foot (ft^3)	= 1728 cubic inches (in^3)
1 barrel (oil) (bbl)	= 42 gallons (U.S.) (gal)

Conversions

1 cubic kilometer (km^3)	= 0.24 cubic miles (mi^3)
1 cubic meter (m^3)	= 264.2 gallons (U.S.) (gal) = 35.314 cubic feet (ft^3)
1 liter (l)	= 1.057 quarts (U.S.) (qt) = 33.815 ounces (U.S. fluid) (fl. oz)
1 cubic centimeter (cm^3)	= 0.0610 cubic inch (in^3)
1 cubic mile (mi^3)	= 4.168 cubic kilometers (km^3)
1 acre-foot (ac-ft)	= 1233.46 cubic meters (m^3)
1 cubic yard (yd^3)	= 0.7646 cubic meter (m^3)
1 cubic foot (ft^3)	= 0.0283 cubic meter (m^3)
1 cubic inch (in^3)	= 16.39 cubic centimeters (cm^3)
1 gallon (gal)	= 3.784 liters (l)

Mass

Metric Measure

1000 kilograms (kg)	= 1 metric ton (also called a tonne) (m.t)
1 kilogram (kg)	= 1000 grams (g)

Nonmetric Measure

1 short ton (sh.t)	= 2000 pounds (lb)
1 long ton (l.t)	= 2240 pounds (lb)
1 pound (avoirdupois (lb)	= 16 ounces (avoirdupois) (oz) = 7000 grains (gr)

1 ounce (avoirdupois) (oz)	= 437.5 grains (gr)
1 pound (Troy) (Tr. lb)	= 12 ounces (Troy) (Tr. oz)
1 ounce (Troy) (Tr. oz)	= 20 pennyweight (dwt)

Conversions

1 metric ton (m.t)	= 2205 pounds (avoirdupois) (lb)
1 kilogram (kg)	= 2.205 pounds (avoirdupois) (lb)
1 gram (g)	= 0.03527 ounce (avoirdupois) (oz) = 0.03215 ounce (Troy) (Tr. oz) = 15,432 grains (gr)
1 pound (lb)	= 0.4536 kilogram (kg)
1 ounce (avoirdupois) (oz)	= 28.35 grams (g)
1 ounce (avoirdupois) (oz)	= 1.097 ounces (Troy) (Tr. oz)

Pressure

1 pascal (Pa)	= 1 newton/square meter (N/m^2)
1 kilogram/square centimeter (kg/cm^2)	= 0.96784 atmosphere (atm) = 14.2233 pounds/square inch (lb/in^2) = 0.98067 bar
1 bar	= 0.98692 atmosphere (atm) = 10^5 pascals (Pa) = 1.02 kilograms/square centimeter (kg/cm^2)

Energy and Power

Energy

1 joule (J)	= 1 newton meter (N.m)
	= 2.390×10^{-1} calorie (cal)

	= 9.47×10^{-4} British thermal unit (Btu)
	= 2.78×10^{-7} kilowatt-hour (kWh)
1 calorie (cal)	= 4.184 joule (J)
	= 3.968×10^{-3} British thermal unit (Btu)
	= 1.16×10^{-6} kilowatt-hour (kWh)
1 British thermal unit (Btu)	= 1055.87 joules (J)
	= 252.19 calories (cal)
	= 2.928×10^{-4} kilowatt-hour (kWh)
1 kilowatt hour	= 3.6×10^6 joules (J)
	= 8.60×10^5 calories (cal)
	= 3.41×10^3 British thermal units (Btu)

Power (energy per unit time)

1 watt (W)	= 1 joule per second (J/s)
	= 3.4129 Btu/h
	= 1.341×10^{-3} horsepower (hp)
	= 14.34 calories per minute (cal/min)
1 horsepower (hp)	= 7.46×10^2 watts (W)

Temperature

To change from Fahrenheit (F) to Celsius (C)

$$°C = \frac{(°F - 32°)}{1.8}$$

To change from Celsius (C) to Fahrenheit (F)

$$°F = (°C \times 1.8) + 32°$$

APPENDIX B

Tables of the Chemical Elements and Naturally Occurring Isotopes

TABLE B.1 *Alphabetical List of the Elements*

Element	Symbol	Atomic Number	Crustal Abundance, Weight Percent	Element	Symbol	Atomic Number	Crustal Abundance, Weight Percent
Actinium	Ac	89	Man-made	Mercury	Hg	80	0.000002
Aluminum	Al	13	8.00	Molybdenum	Mo	42	0.00012
Americium	Am	95	Man-made	Neodymium	Nd	60	0.0044
Antimony	Sb	51	0.00002	Neon	Ne	10	Not known
Argon	Ar	18	Not known	Neptunium	Np	93	Man-made
Arsenic	As	33	0.00020	Nickel	Ni	28	0.0072
Astatine	At	85	Man-made	Niobium	Nb	41	0.0020
Barium	Ba	56	0.0380	Nitrogen	N	7	0.0020
Berkelium	Bk	97	Man-made	Nobelium	No	102	Man-made
Beryllium	Be	4	0.00020	Osmium	Os	76	0.00000002
Bismuth	Bi	83	0.0000004	Oxygen[b]	O	8	45.2
Boron	B	5	0.0007	Palladium	Pd	46	0.0000003
Bromine	Br	35	0.00040	Phosphorus	P	15	0.1010
Cadmium	Cd	48	0.000018	Platinum	Pt	78	0.0000005
Calcium	Ca	20	5.06	Plutonium	Pu	94	Man-made
Californium	Cf	98	Man-made	Polonium	Po	84	Footnote[d]
Carbon[a]	C	6	0.02	Potassium	K	19	1.68
Cerium	Ce	58	0.0083	Praseodymium	Pr	59	0.0013
Cesium	Cs	55	0.00016	Promethium	Pm	61	Man-made
Chlorine	Cl	17	0.0190	Protactinium	Pa	91	Footnote[d]
Chromium	Cr	24	0.0096	Radium	Ra	88	Footnote[d]
Cobalt	Co	27	0.0028	Radon	Rn	86	Footnote[d]
Copper	Cu	29	0.0058	Rhenium	Re	75	0.00000004
Curium	Cm	96	Man-made	Rhodium[c]	Rh	45	0.00000001
Dysprosium	Dy	66	0.00085	Rubidium	Rb	37	0.0070
Einsteinium	Es	99	Man-made	Ruthenium[c]	Ru	44	0.00000001
Erbium	Er	68	0.00036	Samarium	Sm	62	0.00077
Europium	Eu	63	0.00022	Scandium	Sc	21	0.0022
Fermium	Fm	100	Man-made	Selenium	Se	34	0.000005
Fluorine	F	9	0.0460	Silicon	Si	14	27.20
Francium	Fr	87	Man-made	Silver	Ag	47	0.000008
Gadolinium	Gd	64	0.00063	Sodium	Na	11	2.32
Gallium	Ga	31	0.0017	Strontium	Sr	38	0.0450
Germanium	Ge	32	0.00013	Sulfur	S	16	0.030
Gold	Au	79	0.0000002	Tantalum	Ta	73	0.00024
Hafnium	Hf	72	0.0004	Technetium	Tc	43	Man-made
Helium	He	2	Not known	Tellurium[c]	Te	52	0.000001
Holmium	Ho	67	0.00016	Terbium	Tb	65	0.00010
Hydrogen[b]	H	1	0.14	Thallium	Tl	81	0.000047
Indium	In	49	0.00002	Thorium	Th	90	0.00058
Iodine	I	53	0.00005	Thulium	Tm	69	0.000052
Iridium	Ir	77	0.00000002	Tin	Sn	50	0.00015
Iron	Fe	26	5.80	Titanium	Ti	22	0.86
Krypton	Kr	36	Not known	Tungsten	W	74	0.00010
Lanthanum	La	57	0.0050	Uranium	U	92	0.00016
Lawrencium	Lw	103	Man-made	Vanadium	V	23	0.0170
Lead	Pb	82	0.0010	Xenon	Xe	54	Not known
Lithium	Li	3	0.0020	Ytterbium	Yb	70	0.00034
Lutetium	Lu	71	0.000080	Yttrium	Y	39	0.0035
Magnesium	Mg	12	2.77	Zinc	Zn	30	0.0082
Manganese	Mn	25	0.100	Zirconium	Zr	40	0.0140
Mendelevium	Md	101	Man-made				

Source: After K. K. Turekian, 1969.

[a] Estimate from S. R. Taylor (1964).

[b] Analyses of crustal rocks do not usually include separate determinations for hydrogen and oxygen. Both combine in essentially constant proportions with other elements, so abundances can be calculated.

[c] Estimates are uncertain and have a very low reliability.

[d] Elements formed by decay of uranium and thorium. The daughter products are radioactive with such short half-lives that crustal accumulations are too low to be measured accurately.

TABLE B.2 *Naturally Occurring Elements Listed in Order of Atomic Numbers, Together with the Naturally Occurring Isotopes of Each Element, Listed in Order of Mass Numbers*

Atomic Number	Name	Symbol	Mass Numbers[b] of Natural Isotopes
1	Hydrogen	H	1,2, $\boxed{3}$[c]
2	Helium	He	3, 4
3	Lithium	Li	6, 7
4	Beryllium	Be	9, $\boxed{10}$
5	Boron	B	10, 11
6	Carbon	C	12, 13, $\boxed{14}$
7	Nitrogen	N	14, 15
8	Oxygen	O	16, 17, 18
9	Fluorine	F	19
10	Neon	Ne	20, 21, 22
11	Sodium	Na	23
12	Magnesium	Mg	24, 25, 26
13	Aluminum	Al	27
14	Silicon	Si	28, 29, 30
15	Phosphorus	P	31
16	Sulfur	S	32, 33, 34, 36
17	Chlorine	Cl	35, 37
18	Argon	A	36, 38, 40
19	Potassium	K	39, $\boxed{40}$, 41
20	Calcium	Ca	40, 42, 43, 44, 46, $\boxed{48}$
21	Scandium	Sc	45
22	Titanium	Ti	46, 47, 48, 49, 50
23	Vanadium	V	$\boxed{50}$, 51
24	Chromium	Cr	50, 52, 53, 54
25	Manganese	Mn	55
26	Iron	Fe	54, 56, 57, 58
27	Cobalt	Co	59
28	Nickel	Ni	58, 60, 61, 62, 64
29	Copper	Cu	63, 65
30	Zinc	Zn	64, 66, 67, 68, 70
31	Gallium	Ga	69, 71
32	Germanium	Ge	70, 72, 73, 74, 76
33	Arsenic	As	75
34	Selenium	Se	74, 76, 77, 80, 82
35	Bromine	Br	79, 81
36	Krypton	Kr	78, 80, 82, 83, 84, 86
37	Rubidium	Rb	85, $\boxed{87}$
38	Strontium	Sr	84, 86, 87, 88
39	Yttrium	Y	89
40	Zirconium	Zr	90, 91, 92, 94, 96
41	Niobium	Nb	93
42	Molybdenum	Mo	92, 94, 95, 96, 97, 98, 100
44	Ruthenium	Ru	96, 98, 99, 100, 101, 102, 104
45	Rhodium	Rh	103
46	Palladium	Pd	102, 104, 105, 106, 108, 110
47	Silver	Ag	107, 109
48	Cadmium	Cd	106, 108, 110, 111, 112, 113, 114, 116
49	Indium	In	113, $\boxed{115}$
50	Tin	Sn	112, 114, 115, 116, 117, 118, 119, 120, 122, 124
51	Antimony	Sb	121, 123
52	Tellurium	Te	120, 122, 123, 124, 125, 126, 128, 130
53	Iodine	I	127
54	Xenon	Xe	124, 126, 128, 129, 130, 131, 132, 134, 136
55	Cesium	Cs	133
56	Barium	Ba	130, 132, 134, 135, 136, 137, 138
57	Lanthanum	La	$\boxed{138}$, 139
58	Cerium	Ce	136, 138, 140, $\boxed{142}$
59	Praseodymium	Pr	141
60	Neodymium	Nd	142, 143, $\boxed{144}$, 145, 146, 148, 150

TABLE B.2 *(Continued)*

Atomic Number[a]	Name	Symbol	Mass Numbers[b] of Natural Isotopes
62	Samarium	Sm	144, 147 , 148 , 149 , 150, 152, 154
63	Europium	Eu	151, 153
64	Gadolinium	Gd	152 , 154, 155, 156, 157, 158, 160
65	Terbium	Tb	159
66	Dysprosium	Dy	156, 158, 160, 161, 162, 163, 164
67	Holmium	Ho	165
68	Erbium	Er	162, 166, 167, 168, 170
69	Thulium	Tm	169
70	Ytterbium	Yb	168, 170, 171, 172, 173, 174, 176
71	Lutetium	Lu	175, 176
72	Hafnium	Hf	174, 176, 177, 178, 179, 180
73	Tantalum	Ta	180, 181
74	Tungsten	W	180, 182, 183, 184, 186
75	Rhenium	Re	185, 187
76	Osmium	Os	184, 186, 187, 188, 189, 190, 192
77	Iridium	Ir	191, 193
78	Platinum	Pt	190, 192 , 195, 196, 198
79	Gold	Au	197
80	Mercury	Hg	196, 198, 199, 200, 201, 202, 204
81	Thallium	Tl	203, 205
82	Lead	Pb	204, 206, 207, 208
83	Bismuth	Bi	209
84	Polonium	Po	210
86	Radon	Rn	222
88	Radium	Ra	226
90	Thorium	Th	232
91	Protactinium	Pa	231
92	Uranium	U	234 , 235 , 238

[a] Atomic number = number of protons.

[b] Mass number = protons + neutrons.

[c] □ indicates isotope is radioactive.

APPENDIX C

Tables of the Properties of Common Minerals

TABLE C.1 *Properties of the Common Minerals with Metallic Luster*

Mineral	Chemical Composition	Form and Habit	Cleavage	Hardness	Specific Gravity	Other Properties	Most Distinctive Properties
Bornite	Cu_5FeS_4	Massive. Crystals very rare.	None. Uneven fracture.	3	5	Brownish bronze on fresh surface. Tarnishes purple, blue, and black. Grayish-black streak.	Color, streak.
Chalcocite	Cu_2S	Massive. Crystals very rare.	None. Conchoidal fracture.	2.5	5.7	Steel-gray to black. Dark gray streak.	Streak.
Chalcopyrite	$CuFeS_2$	Massive or granular.	None. Uneven fracture.	3.5 – 4	4.2	Golden yellow to brassy yellow. Dark green to black streak.	Streak. Hardness distinguishes from pyrite.
Chromite	$FeCr_2O_4$	Massive or granular.	None. Uneven fracture.	5.5	4.6	Iron black to brownish black. Dark brown streak.	Streak and lack of magnetism distinguishes from ilmenite and magnetite.
Copper	Cu	Massive, twisted leaves and wires.	None. Can be cut with a knife.	2.5 – 3	9	Copper color but commonly stained green.	Color, specific gravity, malleable.
Galena	PbS	Cubic crystals, coarse or fine-grained granular masses.	Perfect in three directions at right angles.	2.5	7.6	Lead-gray color. Gray to gray-black streak.	Cleavage and streak.
Gold	Au	Small irregular grains.	None. Malleable.	2.5	19.3	Gold color. Can be flattened without breakage.	Color, specific gravity, malleability.
Hematite	Fe_2O_3	Massive, granular, micaceous	Uneven fracture.	5–6	5	Reddish-brown, gray to black. Reddish-brown streak.	Streak, hardness.
Ilmenite	$FeTiO_3$	Massive or irregular grains.	Uneven fracture.	5.5 – 6	4.7	Iron-black. Brown-reddish streak differing from hematite.	Streak distinguishes hematite. Lack of magnetism distinguishes magnetite.
Limonite (*Goethite* is most common.)	A complex mixture of minerals, mainly hydrous iron oxides.	Massive, coatings, botryoidal crusts, earthy masses.	None.	1 – 5.5	3.5 – 4	Yellow, brown, black, yellowish-brown streak.	Streak.

TABLE C.1 *(Continued)*

Mineral	Chemical Composition	Form and Habit	Cleavage	Hardness / Specific Gravity	Other Properties	Most Distinctive Properties
Magnetite	Fe_3O_4	Massive, granular. Crystals have octahedral shape.	None. Uneven fracture.	5.5–6.5 / 5	Black. Black streak. Strongly attracted to a magnet.	Streak, magnetism
Pyrite ("Fool's gold")	FeS_2	Cubic crystals with striated faces. Massive.	None. Uneven fracture.	6–6.5 / 5.2	Pale brass-yellow, darker if tarnished. Greenish-black streak.	Streak. Hardness distinguishes from chalcopyrite. Not malleable, which distinguishes from gold.
Pyrolusite	MnO_2	Crystals rare. Massive, coatings on fracture surfaces.	Crystals have a perfect cleavage. Massive, breaks unevenly.	2–6.5 / 5	Dark gray, black or bluish black. Black streak.	Color, streak.
Pyrrhotite	FeS	Crystals rare. Massive or granular.	None. Conchoidal fracture.	4 / 4.6	Brownish-bronze. Black streak. Magnetic.	Color and hardness distinguish from pyrite, magnetism from chalcopyrite.
Rutile	TiO_2	Slender, prismatic crystals or granular masses.	Good in one direction. Conchoidal fracture in others.	6–6.5 / 4.2	Reddish-brown (common), black (rare). Brownish streak. Adamantine luster.	Luster, habit, hardness.
Sphalerite (zinc blende)	ZnS	Fine to coarse granular masses. Tetrahedron shaped crystals.	Perfect in six directions.	3.5–4 / 4	Yellowish-brown to black. White to yellowish-brown streak. Resinous luster.	Cleavage, hardness, luster.
Uraninite	UO_2 to U_3O_8	Massive, with botryoidal forms. Rare crystals with cubic shapes.	None. Uneven fracture.	5–6 / 6.5–10	Black to dark brown. Streak black to dark brown. Dull luster.	Luster and specific gravity distinguish from magnetite. Streak distinguishes from ilmenite and hematite.

TABLE C.2 *Properties of Rock-Forming Minerals with Nonmetallic Luster*

Mineral	Chemical Composition	Form and Habit	Cleavage	Hardness / Specific Gravity		Other Properties	Most Distinctive Properties
Amphiboles. (A complex family of minerals, *Hornblende* is most common.)	$X_2Y_5Si_8O_{22}(OH)_2$ where X = Ca, Na; Y = Mg, Fe, Al.	Long, six-sided crystals; also fibers and irregular grains	Two; intersecting at 56° and 124°	5–6	2.9–3.8	Common in metamorphic and igneous rocks. *Hornblende* is dark green to black; *actinolite*, green; *tremolite*, white.	Cleavage, habit.
Andalusite	Al_2SiO_5	Long crystals, often square in cross-section.	Weak, parallel to length of crystal.	7.5	3.2	Found in metamorphic rocks. Often flesh-colored.	Hardness, form.
Anhydrite	$CaSO_4$	Crystals are rare. Irregular grains or fibers.	Three, at right angles.	3	2.9	Alters to gypsum. Pearly luster, white or colorless.	Cleavage, hardness.
Apatite	$Ca_5(PO_4)_3(F, OH, Cl)$	Granular masses. Perfect six-sided crystals.	Poor. One direction.	5	3.2	Green, brown, blue, or white. Common in many kinds of rocks in small amounts.	Hardness, form.
Aragonite	$CaCO_3$	Massive, or slender, needle-like crystals.	Poor. Two directions.	3.5	2.9	Colorless or white. Effervesces with dilute HCl.	Effervescence with acid. Poor cleavage distinguishes from calcite.
Asbestos			See Serpentine				
Augite			See Pyroxene				
Biotite			See Mica				
Calcite	$CaCO_3$	Tapering crystals and granular masses.	Three perfect; at oblique angles to give a rhomb-shaped fragment.	3	2.7	Colorless or white. Effervesces with dilute HCL.	Cleavage, effervescence with acid.
Chlorite	$(Mg, Fe)_5(Al, Fe)_2 Si_3O_{10}(OH)_8$	Flaky masses of minute scales.	One perfect; parallel to flakes.	2–2.5	2.6–2.9	Common in metamorphic rocks. Light to dark green. Greasy luster.	Cleavage—flakes not elastic, distinguishes from mica. Color.
Dolomite	$CaMg(CO_3)_2$	Crystals with rhomb-shaped faces. Granular masses.	Perfect in three directions as in calcite.	3.5	2.8	White or gray. Does not effervesce in cold, dilute HCl unless powdered. Pearly luster.	Cleavage. Lack of effervescence with acid.

TABLE C.2 (*Continued*)

Mineral	Chemical Composition	Form and Habit	Cleavage	Hardness / Specific Gravity		Other Properties	Most Distinctive Properties
Epidote	Complex silicate of Ca, Fe and Al	Small elongate crystals. Fibrous.	One perfect, one poor.	6–7	3.4	Yellowish-green to dark green. Common in metamorphic rocks.	Habit, color. Hardness distinguishes from chlorite.
Feldspars: Potassium feldspar (*orthoclase* is a common variety)	$KAlSi_3O_8$	Prism-shaped crystals, granular masses.	Two perfect, at right angles.	6	2.6	Common mineral. Flesh-colored, pink, white, or gray.	Color, cleavage.
Feldspar	$NaAlSi_3O_8$ (albite) and $CaAl_2Si_2O_8$ (anorthite) and all compositions between.	Irregular grains, cleavable masses. Tabular crystals.	Two perfect, not quite at right angles.	6–6.5	2.6–2.7	White to dark gray. Cleavage planes may show fine parallel striations.	Cleavage. Striations on cleavage planes will distinguish from potassium feldspar.
Fluorite	CaF_2	Cubic crystals, granular masses.	Perfect in four directions.	4	3.2	Colorless, bluish green. Always an accessory mineral.	Hardness, cleavage, does not effervesce with acid.
Garnets	$X_3Y_2(SiO_4)_3$; X = Ca, Mg, Fe, Mn; Y = Al, Fe, Ti, Cr.	Perfect crystals with 12 or 24 sides. Granular masses.	None. Uneven fracture.	6.5–7.5	3.5–4.3	Common in metamorphic rocks. Red, brown, yellowish-green, black.	Crystals, hardness, no cleavage.
Graphite	C	Scaly masses.	One, perfect. Forms slippery flakes.	1–2	2.2	Metamorphic rocks. Black with metallic to dull luster.	Cleavage, color. Marks paper.
Gypsum	$CaSO_4 \cdot 2H_2O$	Elongate or tabular crystals. Fibrous and earthy masses.	One, perfect. Flakes bend but are not elastic.	2	2.3	Vitreous to pearly luster. Colorless.	Hardness, cleavage.
Halite	NaCl	Cubic crystals.	Perfect to give cubes.	2.5	2.2	Tastes salty. Colorless, blue.	Taste, cleavage.
Hornblende			See Amphibole				
Kaolinite	$Al_2Si_2O_5(OH)_4$	Soft, earthy masses. Submicroscopic crystals.	One, perfect.	2–2.5	2.6	White, yellowish. Plastic when wet; emits clayey odor. Dull luster.	Feel, plasticity, odor.

TABLE C.2 (*Continued*)

Mineral	Chemical Composition	Form and Habit	Cleavage	Hardness / Specific Gravity	Other Properties	Most Distinctive Properties
Kyanite	Al_2SiO_5	Bladed crystals.	One perfect. One imperfect.	4.5 parallel to blade, 7 across blade / 3.6	Blue, white, gray. Common in metamorphic rocks.	Variable hardness, distinguishes from sillimanite. Color.
Mica: *Biotite*	$K(Mg, Fe)_3$-$AlSi_3O_{10}$-$(OH)_2$	Irregular masses of flakes.	One, perfect.	2.5–3 / 2.8–3.2	Common in igneous and metamorphic rocks. Black, brown, dark green.	Cleavage, color. Flakes are elastic.
Muscovite	$KAl_3Si_3O_{10}(OH)_2$	Thin flakes.	One, perfect.	2–2.5 / 2.7	Common in igneous and metamorphic rocks. Colorless, pale green or brown.	Cleavage, color. Flakes are elastic.
Olivine	$(Mg, Fe)_2SiO_4$	Small grains, granular masses.	None. Conchoidal fracture.	6.5–7 / 3.2–4.3	Igneous rocks. Olive green to yellowish-green.	Color, fracture, habit.
Orthoclase			See Feldspar			
Plagioclase			See Feldspar			
Pyroxene (A complex family of minerals. *Augite* is most common.)	$XY(SiO_3)_2$ $X = Y = Ca$, Mg. Fe	8-sided stubby crystals. Granular masses.	Two, perfect, nearly at right angles.	5–6 / 3.2–3.9	Igneous and metamorphic rocks. *Augite*, dark green to black; other varieties white to green.	Cleavage
Quartz	SiO_2	6-sided crystals, granular masses.	None. Conchoidal fracture.	7 / 2.6	Colorless, white, gray, but may have any color, depending on impurities. Vitreous to greasy luster.	Form, fracture, striations across crystal faces at right angles to long dimension.
Serpentine (Fibrous variety is *asbestos*)	$Mg_3Si_2O_5(OH)_4$	Platy or fibrous.	One, perfect.	2.5–5 / 2.2–2.6	Light to dark green. Smooth, greasy feel.	Habit, hardness.
Sillimanite	Al_2SiO_5	Long needle-like crystals, fibers.	Breaks irregularly, except in fibrous variety.	6–7 / 3.2	White, gray. Metamorphic rocks.	Hardness distinguishes from kyanite. Habit.
Talc	$Mg_3Si_4O_{10}(OH)_2$	Small scales, compact masses.	One, perfect.	1 / 2.6–2.8	Feels slippery. Pearly luster. White to greenish.	Hardness, luster, feel, cleavage.

TABLE C.2 *(Continued)*

Mineral	Chemical Composition	Form and Habit	Cleavage	Hardness / Specific Gravity		Other Properties	Most Distinctive Properties
Tourmaline	Complex silicate of B, Al, Na, Ca, Fe, Li and Mg.	Elongate crystals, commonly with triangular cross section.	None.	7–7.5	3–3.3	Black, brown, red, pink, green, blue, and yellow. An accessory mineral in many rocks.	Habit.
Wollastonite	CaSiO₃	Fibrous or bladed aggregates of crystals.	Two, perfect.	4.5–5	2.8–2.9	Colorless, white, yellowish. Metamorphic rocks. Soluble in HCl.	Habit. Solubility in HCl and hardness distinguish amphiboles, kyanite, sillimanite.

TABLE C.3 *Properties of Some Common Gemstones*

Mineral and Variety	Composition	Form and Habit	Cleavage	Hardness / Specific Gravity		Other Properties	Most Distinctive Properties
Beryl: *Aquamarine* (blue) *Emerald* (green) *Golden beryl* (golden-yellow)	Be₃Al₂Si₆O₁₈	Six-sided, elongate crystals common.	Weak.	7.5–8	2.75	Bluish green, green, yellow, white, colorless. Common in pegmatites	Form. Distinguished from apatite by its hardness.
Corundum: *Ruby* (red) *Sapphire* (blue)	Al₂O₃	Six-sided, barrel-shaped crystals.	None, but breaks easily across its crystal.	9	4	Brown, pink, red, blue, colorless. Common in metamorphic rocks. Star sapphire is opalescent with a six-sided light spot showing.	Hardness.
Diamond	C	Octahedron-shaped crystals.	Perfect, parallel to faces of octahedron.	10	3.5	Colorless, yellow; rarely red, orange, green, blue or black.	Hardness, cleavage.
Garnet: *Almandite* (red) *Grossularite* (green, cinnamon-brown) *Andradite* (variety *demantoid* is green)	A rock-forming mineral—See Table C.2						

TABLE C.3 *(Continued)*

Mineral and Variety	Composition	Form and Habit	Cleavage	Hardness / Specific Gravity		Other Properties	Most Distinctive Properties
Opal (A mineraloid)	$SiO_2 \cdot nH_2O$	Massive, thin coating. Amorphous.	None. Conchoidal fracture.	5–6	2–2.2	Colorless, white, yellow, red, brown, green, gray, opalescent.	Hardness, color, form.
Quartz: (1) Coarse crystals *Amethyst* (violet) *Cairngorm* (brown) *Citrine* (yellow) *Rock crystal* (colorless) *Rose quartz* (pink) (2) Fine-grained *Agate* (banded, many colors) *Chalcedony* brown, gray) *Heliotrope* (green) *Jasper* (red)	A rock-forming mineral—See Table C.2						
Topaz	$Al_2SiO_4(OH, F)_2$	Prism-shaped crystals, granular masses.	One, perfect.	8	3.5	Colorless, yellow, blue, brown.	Hardness, form, color.
Tourmaline	A rock-forming mineral—See Table C.2						
Zircon	$ZrSiO_4$	Four-sided elongate crystals, square in cross-section.	None.	7.5	4.7	Brown, red, green, blue, black.	Habit, hardness.

Glossary

Some definitions are not included in the glossary; *units of measurement* can be found in Appendix A, *chemical elements* are listed in Appendix B, and *mineral names* are given in Appendix C.

A horizon. A soil horizon that either underlies the O horizon or is the uppermost horizon. Generally dark colored and characterized by an accumulation of organic matter.

Aa. A rubbly, rough-looking form of lava, usually basaltic in composition.

Ablation. The loss of mass from a glacier.

Ablation area. A region of net loss on a glacier characterized by a surface of bare ice and old snow from which the last winter's snowcover has melted away.

Absolute velocity (of a plate). The velocity of a plate of lithosphere measured against a fixed, external frame of reference. Compare with *relative velocity.*

Abyssal plain. A large flat area of the deep seafloor having slopes less than about 1 m/km, and ranging in depth below sealevel from 3 to 6 km.

Accreted terrane. Block of crust moved laterally by strike-slip faulting or by a combination of strike-slip faulting and subduction, then accreted to a larger mass of continental crust. Also called a *suspect terrane.*

Accreted terrane continental margin. Continental margin modified by the addition of island arcs and other rafted-in, exotic fragments of crust.

Accretion. The process by which solid bodies gather together to form a planet or a continent.

Accumulation. The addition of mass to a glacier.

Accumulation area. An upper zone on a glacier, covered by remnants of the previous winter's snowfall and representing an area of net gain in mass.

Active zone. The margin of a tectonic plate where deformation is occurring.

Agglomerate. A pyroclastic rock consisting of bomb-sized tephra, i.e., tephra in which the average particle diameter is greater than 64 mm.

Aggradation. Depositional upbuilding, as by a stream.

Albedo. The reflectivity of the surface of a planet.

Alluvial fan. A fan-shaped body of alluvium typically built where a stream leaves a steep mountain valley.

Alluvium. Sediment deposited by streams in nonmarine environments.

Alpha particle (α-particle). An atomic particle expelled from an atomic nucleus during certain radioactive transformations, equivalent to an $^4_2$He nucleus stripped of its electrons.

Amorphous. A term applied to solids that lack internal atomic order.

Amphibolite. A metamorphic rock of intermediate grade, generally coarse grained, containing abundant amphibole.

Amygdule. A vesicle filled by secondary minerals such as calcite and quartz deposited by groundwater.

Andesite. A fine-grained igneous rock with the composition of a diorite.

Andesite line. A line on a map that roughly surrounds the Pacific Ocean basin and inside of which andesite is not found.

Andesitic magma. One of the three common magma types; a magma with an SiO_2 content of about 60 percent by weight.

Angle of repose. The steepest angle, measured from the horizontal, at which rock debris remains stable.

Angular unconformity. An unconformity marked by angular discordance between older and younger rocks.

Anhydrous. A term applied to a substance that is H_2O free. Opposite of hydrous.

Anion. An ion with a negative electrical charge.

Anorthosite. A coarse-grained igneous rock consisting largely of plagioclase.

Antecedent stream. A stream that has maintained its course across an area of the crust that was raised across its path by folding or faulting.

Anthracite. A metamorphic rock derived from coal by heat and pressure.

Anticline. An upfold in the form of an arch.

Apparent polar wandering. The apparent motions of the magnetic poles derived from measurements of pole positions using paleomagnetism.

Aquiclude. A body of impermeable or distinctly less permeable rock adjacent to an *aquifer.*

Aquifer. A body of permeable rock or regolith saturated with water and through which groundwater moves.

Archean. The eon that follows the Hadean Eon.

Arête. A jagged, knife-edged ridge crest created where glaciers have eroded back into a ridge.

Arkose (-arkosic sandstone). A sandstone in which feldspar is a major mineral component.

Artesian aquifer. An aquifer in which water is under hydraulic pressure.

Artesian spring. A natural spring that draws its supply of water from an artesian aquifer.

Artesian system. An inclined aquifer that permits water confined in it to rise to the surface in a well or along a fissure.

Artesian well. A well in which water rises above the aquifer.

Ash. Tephra in which particles have an average diameter less than 2 mm. Also called *volcanic ash.*

Ash tuff. Pyroclastic rock in which the tephra particles are less than 2 mm in diameter.

Asphalt. See *tar.*

Asteroids. Irregularly shaped rocky bodies that have orbits lying between the orbits of Mars and Jupiter.

Asthenosphere. The region of the mantle where rocks become ductile, have little strength, and are easily deformed. It lies at a depth of 100 to 350 km below the surface.

Asymmetric fold. A fold in which one limb dips more steeply than the other.

Atmosphere. The mixture of gases, predominantly nitrogen, oxygen, carbon dioxide, and water vapor, that surrounds the Earth.

Atoll. A coral reef, often roughly circular in plan, that encloses a shallow lagoon.

Atom. The smallest individual particle that retains all the properties of a given chemical element.

Atomic number. The number of protons in the nucleus of an atom.

Atomic substitution. See *ionic substitution*.

Axial plane. An imaginary plane that divides a fold as symmetrically as possible, and that passes through the axis.

Axis (of a fold). The median line between the limbs, along the crest of an anticline or the trough of a syncline.

B horizon. A soil horizon generally lying below an A horizon, usually brownish or reddish in color, and commonly enriched in clay and iron oxides.

Back-arc basin. An arc-shaped basin formed by crustal thinning behind a magmatic arc.

Backshore. A zone extending inland from a berm to the farthest point reached by waves.

Backwash. The seaward return of water down a beach following the swash of a wave.

Bajada. A broad alluvial apron composed of coalescing adjacent fans.

Bar. An accumulation of alluvium formed in a channel where a decrease in stream velocity causes deposition.

Barchan dune. A crescent-shaped sand dune with horns pointing downwind.

Barrier island. A long island built of sand, lying offshore and parallel to the coast.

Barrier reef. A reef separated from the land by a lagoon.

Basalt. A fine-grained igneous rock with the composition of a gabbro.

Basaltic magma. One of the three common types of magma; contains about 50 percent SiO_2 by weight.

Base level. The limiting level below which a stream cannot erode the land.

Batholith. The largest kind of pluton. A very large, igneous body of irregular shape that cuts across the layering of the rock it intrudes.

Bauxite. An aluminous laterite formed by tropical weathering. The preferred ore of aluminum.

Bay. A wide, open, curving indentation or inlet of a sea or lake into an adjacent land mass.

Bay barrier. A ridge of sand or gravel that completely blocks the mouth of a bay.

Beach. Wave-washed sediment along a coast, extending throughout the surf zone.

Beach drift. The irregular movement of particles along a beach as they travel obliquely up the slope of a beach with the swash and directly down this slope by the backwash.

Beach ridge. A low ridge of sand parallel to and on the landward side of a beach.

Bed. The smallest formal unit of a body of sediment or sedimentary rock.

Bedding. The layered arrangement of strata in a body of sediment or sedimentary rock.

Bedding plane. The top or bottom of a bed.

Bed load. Coarse particles that move along the bottom of a stream channel.

Bedrock. The continous mass of solid rock that makes up the crust.

Benioff strain seismograph. See *strain seismograph*.

Benioff zone. A narrow, well-defined zone of deep earthquake foci beneath a seafloor trench.

Berm. A nearly horizontal or landward-sloping bench formed of sediment deposited by waves.

Beta particle (β-particle). An electron expelled from an atomic nucleus during certain radioactive transformations.

Biogenic rock. Rock formed by lithification of biogenic sediment.

Biogenic sediment. Sediment composed mainly of fossil remains.

Biogeochemical cycle. A natural cycle describing the movements and interactions through the Earth's spheres of the chemicals essential to life.

Biomass energy. The energy obtained through burning plant matter.

Biosphere. The totality of the Earth's organisms and, in addition, organic matter that has not yet been completely decomposed.

Bituminous coal. The highest grade of coal.

Blueschist. A metamorphic rock formed under conditions of high pressure and low temperature containing blue-colored amphiboles.

Body waves. Seismic waves that travel outward from an earthquake focus and pass through the Earth.

Bolide. An impacting body, either a meteorite, an asteroid, or a comet.

Bombs. Tephra particles having average diameters greater than 64 mm.

Bond (chemical). The electrostatic force that holds atoms together to form compounds by sharing and transfer of electrons. See *covalent bond, ionic bond, metallic bond,* and *van der Waal's bond.*

Bottomset layer. A gently sloping, fine, thin part of each layer in a delta.

Boundary conditions. A mathematical expression of the physical state of the Earth's climate system at the period of interest for a climate-simulation experiment.

Bowen's reaction series. A schematic description of the order in which different minerals crystallize during the cooling and progressive crystallization of a magma. See *continuous* and *discontinuous reaction series.*

Braid delta. A delta composed of coarse-grained sediment built by a braided stream into a standing body of water.

Braided stream. A channel system consisting of a tangled network of two or more smaller branching and reuniting channels that are separated by islands or bars.

Breaker. An oversteepened wave that collapses in a mass of turbulent water against a shore or reef.

Breakwater. An offshore barrier built to protect a beach or anchorage from incoming waves.

Breccia. A coarse-grained rock composed of cemented angular fragments of broken rock.

Breeder reactor. A nuclear reactor in which nonfissionable isotopes such as ^{238}U are converted to fissionable isotopes.

Brittle fracture. Rupture of a solid body that is stressed beyond its elastic limit.

Burial metamorphism. Metamorphism caused solely by the burial of sedimentary or pyroclastic rocks.

Butte. Isolated, often flat-topped, steep-sided, desert hill. Smaller than a *mesa*.

C horizon. The deepest soil horizon, lying beneath the A horizon and/or B horizon of a soil profile; often yellowish-brown in color and consisting of weathered parent rock or sediment.

Calcareous ooze. A deep-sea pelagic sediment composed largely of calcareous skeletal remains. See *deep-sea ooze.*

Caldera. A roughly circular, steep-walled volcanic basin several kilometers or more in diameter.

Caliche. A solid, almost impervious layer of whitish calcium carbonate in a soil profile.

Calving. The progressive breaking off of icebergs from a glacier that terminates in deep water.

Capillary attraction. The adhesive force between a liquid, such as water, and a solid.

Carbonate shelf. A shallow marine shelf where sedimentation is dominated by carbonate-secreting organisms.

Carbonic acid. A weak acid resulting from the solution of small quantities of carbon dioxide in rain or groundwater.

Cataclastic metamorphism. Metamorphism that involves change of texture caused by mechanical effects such as crushing and shearing, but no change in mineral assemblage.

Catastrophism. The concept that all of the Earth's major features, such as mountains, valleys, and oceans, have been produced by a few great catastrophic events.

Cation. A positive ion.

Cave. A natural underground opening, generally connected to the surface and large enough for a person to enter.

Cavern. A large cave or system of interconnected cave chambers.

Celsius scale. A temperature scale in which the boiling point of water is 100°, and the freezing point is 0°.

Cementation. The diagenetic process by which clastic sediments are converted to rock through deposition or precipitation of minerals in the spaces between the grains.

Cenozoic. The youngest era of the Phanerozoic Eon.

Central rift valley. A long, narrow valley at the crest of a midocean ridge.

Chalk. Compacted carbonate shells of minute floating organisms.

Channel. The passageway in which a stream flows.

Chemical elements. The most fundamental substances into which matter can be separated by chemical means.

Chemical sediment. Sediment formed by precipitation of minerals from solution in water.

Chemical weathering. The decomposition of rocks through chemical reactions such as hydration and oxidation.

Chrons. See *magnetic chrons*.

Cirque. A bowl-shaped hollow on a mountainside open downstream, bounded upstream by a steep slope (headwall), and excavated mainly by frost wedging and by glacial abrasion and plucking.

Cirque glacier. A glacier that occupies a bowl-shaped hollow on the side of a mountain.

Chloroflurocarbons. Synthetic industrial gases that destroy ozone in the upper atmosphere and contribute to the greenhouse effect. Also called CFCs.

Clast. Any individual particle of clastic sediment.

Clastic sediment. See *detritus*.

Cleavage. The tendency of a mineral to break in preferred directions along bright, reflective plane surfaces.

Climate. The average weather conditions of a place or area over a period of years.

Coal. A black, combustible, sedimentary or metamorphic rock consisting chiefly of decomposed plant matter and containing more than 50 percent organic matter.

Coalification. The stages by which plant matter is converted first to peat, then lignite, subbituminous coal, and bituminous coal.

Col. A gap or pass in a mountain crest where the headwalls of two cirques intersect.

Collision zone. A convergent plate margin where two plates collide.

Colluvium. Loose, incoherent deposits on or at the base of slopes and moving mainly by creep.

Column. A stalactite joined with a stalagmite, forming a connection between the floor and roof of a cave.

Columnar joints. Joints that split igneous rocks into long prisms or columns.

Comet. A small celestial body that circles the Sun with a highly elliptical orbit.

Compaction. A decrease in porosity and bulk of a body of sediment as additional sediment is deposited above it, or due to pressures resulting from deformation.

Complex ion. A strongly bonded pair of ions that act in the same way as a single ion, forming compounds by bonding with other elements.

Composition (of a mineral). The proportions of the various chemical elements in a mineral.

Compound. A combination of atoms of different elements bonded together.

Compressive stress. Differential stress that squeezes and compresses a body.

Compressional waves. See *P waves*.

Conchoidal fracture. Breakage resulting in smooth, curved surfaces.

Concretion. A hard, localized body, having distinct boundaries, enclosed in sedimentary rock, and consisting of a substance precipitated from solution, commonly around a nucleus.

Conduction. The means by which heat is transmitted through solids without deforming the solid.

Cone of depression. A conical depression in the water table immediately surrounding a well.

Confined aquifer. An aquifer bounded by aquicludes.

Confining stress. See *uniform stress*.

Conformable strata. Strata that have been deposited layer upon layer without interruption.

Conglomerate. A sedimentary rock composed of clasts of rounded gravel set in a finer-grained matrix.

Consequent stream. A stream whose pattern is determined solely by the direction of slope of the land.

Contact metamorphism (also called *thermal metamorphism*). Metamorphism adjacent to an intrusive igneous rock.

Continental collision margin. A plate margin along which two continental masses collide.

Continental convergent margin. The margin of a continent that is adjacent to a subduction zone.

Continental crust. The part of the Earth's crust that comprises the continents, which has an average thickness of 45 km.

Continental drift. The slow, lateral move-

ments of continents across the surface of the Earth.

Continental rise. A region of gently changing slope where the floor of the ocean basin meets the margin of a continent.

Continental shelf. A submerged platform of variable width that forms a fringe around a continent.

Continental shield. An assemblage of cratons and orogens that has reached isostatic equilibrium.

Continental slope. A pronounced slope beyond the seaward margin of the continental shelf.

Continental volcanic arc. An arcuate chain of andesitic volcanoes on the continental crust formed as a result of subduction.

Continuous reaction series. The continuous change of mineral composition, through ionic substitution, as a magma crystallizes. See *discontinuous reaction series*.

Convection. The process by which hot, less-dense materials rise upward, being replaced by cold, dense, downward flowing material to create a convection current.

Convection current. The flow of material as a result of convection.

Convergent margin. The zone where plates meet as they move toward each other. See *subduction zone*.

Coquina. A limestone composed solely or chiefly of loosely aggregated shells and shell fragments.

Core. The spherical mass, largely metallic iron, at the center of the Earth.

Coriolis effect. An effect that causes any body that moves freely with respect to the rotating solid Earth to veer toward the right in the Northern Hemisphere and toward the left in the Southern Hemisphere, regardless of the initial direction of the moving body.

Correlation. Determination in time-stratigraphic age of the succession of strata found in two or more different areas.

Covalent bond. The force between two atoms that have filled their energy-level shells by sharing one or more electrons.

Crater. A funnel-shaped depression, opening upward, at the top of a volcano from which gases, fragments of rock, and lava are ejected.

Craton. A core of ancient rock in the continental crust that has attained tectonic and isostatic stability.

Creep. The imperceptibly slow downslope movement of regolith.

Creep of glacier ice. Slow deformation of glacier ice, with movement occurring along the internal planes of ice crystals.

Crevasse. A deep, gaping fissure in the upper surface of a glacier.

Cross section. See *geologic cross section*.

Cross bedding. Beds that are inclined with respect to a thicker stratum within which they occur.

Crust. The outermost and thinnest of the Earth's compositional layers, which consists of rocky matter that is less dense than the rocks of the mantle below.

Crystal. A solid compound composed of ordered, three-dimensional arrays of atoms or ions chemically bonded together and displaying crystal form.

Crystal faces. The planar surfaces that bound a crystal.

Crystal form. The geometric arrangement of crystal faces.

Crystalline. See *crystal structure.*

Crystal settling. The process of fractional crystallization by which dense minerals sink and form segregated layers of one or more minerals in a magma chamber.

Crystal structure. The geometric pattern that atoms assume in a solid. Any solid that has a crystal structure is said to be *crystalline.*

Curie point. A temperature above which permanent magnetism is not possible.

Dacite. A fine-grained igneous rock with the composition of a granodiorite.

Daughter atom (-daughter). The product arising from radioactive decay. Compare *parent.*

Debris avalanche. A granular flow of regolith moving at a high velocity (≥ 10 m/s).

Debris fall. The relatively free fall or collapse of regolith from a steep cliff or slope.

Debris flow. The downslope movement of a mass of unconsolidated regolith more than half of which is coarser than sand.

Debris slide. The slow to rapid downslope movement of regolith across an inclined surface. Compare *rockslide.*

Décollement. A body of rock above the detachment surface of a thrust fault.

Decomposition (of rocks). Chemical weathering.

Deep-sea fan. Huge fan-shaped body of sediment at the base of the continental slope that spreads downward and outward to the deep seafloor.

Deep-sea ooze. A muddy marine sediment composed mainly of the remains of microscopic marine organisms. See also *calcareous ooze* and *siliceous ooze.*

Deflation. The picking up and removal of loose particles by wind.

Delta. A body of sediment deposited by a stream where it flows into standing water.

Density. The average mass per unit volume.

Denudation. The sum of the weathering, mass-wasting, and erosional processes that result in the progressive lowering of the Earth's surface.

Desert. Arid land, whether "deserted" or not, in which annual rainfall is less than 250 mm (10 in) or in which the evaporation rate exceeds the precipitation rate.

Desertification. The invasion of desert into nondesert areas.

Desert pavement. A surface layer of coarse particles concentrated chiefly by deflation.

Desert varnish. A thin, dark, shiny coating consisting mainly of manganese and iron oxides, formed on the surfaces of stones and rock outcrops in desert regions after long exposure.

Detachment surface. The surface along which a large-scale thrust fault moves.

Detritus (also called *clastic sediment* and *detrital sediment*). The loose fragmented debris produced by the mechanical breakdown of older rocks.

Diagenesis. Chemical, physical, and biological changes that affect sediment after its initial deposition and during and after its slow transformation into sedimentary rock.

Diatomite. A sedimentary rock formed by lithification of *siliceous ooze.*

Differential stress. Stress in a solid that is not equal in all directions.

Differential weathering. Weathering that occurs at different rates or intensity as a result of variations in the composition and structure of rocks.

Dikes. Tabular, parallel-sided sheets of intrusive igneous rock that cut across the layering of the intruded rock.

Diorite. A coarse-grained igneous rock consisting mainly of plagioclase and ferromagnesian minerals. Quartz is sparse or absent.

Dip. The angle in degrees between a horizontal plane and an inclined plane, measured down from horizontal in a plane perpendicular to the strike.

Dip-slip fault. A normal or reverse fault on which the only component of movement lies in a plane normal to the strike of the fault surface.

Discharge. The quantity of water that passes a given point in a stream channel per unit time.

Discharge area. Area where subsurface water is discharged to streams or to bodies of surface water.

Disconformity. An irregular surface of erosion between parallel strata.

Discontinuous reaction series. The discontinuous sequence of reactions by which early formed minerals in a crystallizing magma react with residual liquid to form new minerals. See *continuous reaction series.*

Dispersion. Waves of different wavelengths traveling at different velocities.

Dissolution. The chemical weathering process whereby minerals and rock material pass directly into solution.

Dissolved load. Matter dissolved in stream water.

Divergent margin (of a plate). A fracture in the lithosphere where two plates move apart. Also called a *spreading center.*

Divide. The line that separates adjacent drainage basins.

Dolostone. A sedimentary rock composed chiefly of the mineral dolomite.

Drainage basin. The total area that contributes water to a stream.

Drift. See *glacial drift.*

Dripstone. A deposit chemically precipitated by dripping water in an air-filled cavity.

Dropstone. A stone released from a melting iceberg that plunges into unconsolidated sediment on the seafloor or a lake bottom.

Drumlin. A streamlined hill consisting of glacially deposited sediment and elongated parallel with the direction of ice flow.

Ductile deformation. The irreversible deformation induced in a solid that is stressed beyond its elastic limit but before rupture occurs.

Dune. A mound or ridge of sand deposited by wind.

E horizon. A soil horizon, sometimes present below the A horizon, that is grayish or whitish in color.

Earthflow. A granular flow of regolith with velocities ranging from 10^{-5} to 10^{-1} m/s.

Earthquake focus. The point of the first release of energy that causes an earthquake.

Earthquake magnitude. See *Richter magnitude scale.*

Earth's gravity. An inward-acting force with which the Earth tends to pull all objects toward its center.

Eccentricity (of Earth's orbit). The degree to which the shape of the Earth's orbit departs from perfect circularity.

Ecliptic. Plane of the Earth's orbit around the Sun.

Eclogite. A metamorphic rock containing garnet and jadeitic pyroxene.

Ejecta blanket. Layer of broken rock surrounding an impact crater.

Elastic deformation. The reversible or nonpermanent deformation that occurs when an elastic solid is stretched and squeezed and the force is then removed.

Elastic limit. The limiting stress beyond which a body suffers irreversible deformation.

Elastic rebound theory. The theory that earthquakes result from the release of stored elastic energy by slippage on faults.

Electrons. Negatively charged atomic particles.

Emergence. An increase in the area of land exposed above sea level resulting from uplift of the land and/or fall of sea level.

End moraine. A ridgelike accumulation of drift deposited along the margin of a glacier.

Energy-level shell. The specific energy level of electrons as they orbit the nucleus of an atom.

Eolian. Pertaining to the wind, especially erosional and depositional processes, as well as landforms and sediments resulting from wind action.

Eon. The largest interval of geologic time. We are now in the fourth eon.

Epicenter. That point on the Earth's surface that lies vertically above the focus of an earthquake.

Epidote amphibolite. A metamorphic rock containing both amphibole and epidote as major constituents.

Epoch. The time during which a geologic series accumulates.

Equilibrium line. A line that marks the level on a glacier where net mass loss equals net gain.

Era. The primary time division of eons.

Erosion. The complex group of related processes by which rock is broken down physically and chemically and the products moved.

Erratic. A glacially deposited rock fragment whose composition differs from that of the bedrock beneath it.

Eruption column. A mixture of ash and hot gases that rises upward as a column above an erupting volcano.

Esker. A long narrow ridge, often sinuous, composed of stratified drift.

Estuary. A semienclosed body of coastal water within which seawater is diluted with fresh water.

Evaporite. Sedimentary rock composed chiefly of minerals precipitated from a saline solution through evaporation.

Evaporite deposits. Layers of salts that precipitate as a consequence of evaporation.

Exfoliation. The spalling off of successive shells, like the "skins" of an onion, around a solid rock core.

Exploration geology. The special branch of geology concerned with discovering new supplies of usable minerals.

Exposure (also called an *outcrop*). A place where rock or sediment is exposed at the Earth's surface.

External processes. All the activities involved in erosion, and also in the transport and deposition of the eroded materials.

Extraterrestrial material. Material originating outside the Earth.

Extrusive igneous rock. Rock formed by the solidification of magma poured out onto the Earth's surface.

Facies. A distinctive group of characteristics, within a rock unit, that differs as a group from those elsewhere in the same unit. See also *sedimentary facies* and *metamorphic facies.*

Fan. See *alluvial fan.*

Fan delta. A gravel-rich delta formed where an alluvial fan builds outward into a standing body of water.

Fault. A fracture in a rock along which movement occurs.

Fault breccia. Crushed and broken rock adjacent to a fault.

Ferromagnesian minerals. The common rock-forming minerals that contain iron and/or magnesium as essential constituents.

Fiord. See *fjord.*

Fission. Controlled radioactive transformation.

Fissure eruption. Extrusion of lava or pyroclasts and associated gases along an extended fracture.

Fjord. A deep glacially carved valley submerged by the sea. Also spelled *fiord.*

Fjord glacier. A glacier that occupies a fjord.

Flash flood. A local and sudden flood of water through a stream channel, generally of relatively great volume and short duration.

Flood. A discharge great enough to cause a stream to overflow its banks.

Floodplain. The part of any stream valley that is inundated during floods.

Flowstone. A deposit chemically precipitated from flowing water in the open air or in an air-filled cavity.

Fluvial. Of, or pertaining to, streams or rivers, especially erosional and depositional processes of streams and the sediments and landforms resulting from them.

Foliation. The planar texture of mineral grains, principally micas, produced by metamorphism.

Fold. An individual bend or warp in layered rock.

Folding. The bending of rocks or sediments.

Footwall block. The block of rock below an inclined fault.

Fore-arc basin. A basin parallel to a deep-sea trench and separated from the trench by a *fore-arc ridge.*

Fore-arc ridge. See *fore-arc basin.*

Foreset layer. The coarse, thick, steeply sloping part of each layer in a delta.

Foreshore. A zone extending from the level of lowest tide to the average high-tide level.

Formation. A body of rock distinctive enough on the basis of physical properties to constitute a basic unit for geologic mapping. The basic unit of rock stratigraphy.

Fossil. The naturally preserved remains or traces of an animal or a plant.

Fossil fuel. Remains of plants and animals trapped in sediment that may be used for fuel.

Fringing reef. A coral reef attached to or bordering the adjacent land.

Frost-heaving. The lifting of regolith by the freezing of water contained within it.

Frost-wedging. The formation of ice in a confined opening within rock, thereby causing the rock to be forced apart.

Fumarole. A volcanic vent that emits only gases.

Gabbro. A coarse-grained igneous rock in which olivine and pyroxene are the predominant minerals and plagioclase is the feldspar present. Quartz is absent.

Gamma rays (γ-rays). Very short wavelength electromagnetic radiation given off by an atomic nucleus during certain radioactive transformations.

Gangue. The nonvaluable minerals of an ore.

Garnet peridotite. A coarse-grained igneous rock consisting largely of olivine, garnet, and pyroxene.

Gelifluction. Downslope movement of the thawed surface layer of regolith in a region of perennially frozen ground.

General Circulation Model (GCM). A mathematical model used to simulate present and past climate conditions on the Earth.

Geochemically abundant elements. Those chemical elements that individually comprise 0.1 percent or more by weight of the crust.

Geochemically scarce elements. Those chemical elements that individually comprise less than 0.1 percent by weight of the crust.

Geologic column. A composite diagram combining in chronological order the succession of known strata, fitted together on the basis of their fossils or other evidence of relative or actual age.

Geologic cross section. A diagram showing the arrangement of rocks in a vertical plane.

Geologic map. A map that shows the distribution, at the surface, of rocks of various kinds or of various ages.

Geologic time scale. A sequential arrangement of geologic time units, as currently understood.

Geology. The science of the Earth.

Geologists. Scientists who study the Earth.

Geosyncline. A great trough that has received thick deposits of sediment during its slow subsidence through long geologic periods.

Geothermal gradient. The rate of increase of temperature downward in the Earth.

Geothermal power. Heat energy drawn from the Earth's internal heat.

Geyser. A thermal spring equipped with a system of plumbing and heating that causes intermittent eruptions of water and steam.

Glacial drift. Sediment deposited directly by glaciers or indirectly by meltwater in streams, in lakes, and in the sea. Also called *drift.*

Glacial grooves. See *glacial striations.*

Glacialmarine drift (also referred to as *glacialmarine sediment*). Terrigenous sediment dropped onto the seafloor from floating ice shelves or from icebergs.

Glacial striations. Subparallel scratches inscribed on a clast or a bedrock surface by rock debris embedded in the base of a glacier. Wider and deeper markings on bedrock are *glacial grooves.*

Glaciation. The modification of the land surface by the action of glacier ice.

Glacier. A permanent body of ice, consisting largely of recrystallized snow, that shows evidence of downslope or outward movement due to the stress of its own weight.

Gneiss. A high-grade metamorphic rock, always coarse-grained and foliated, with marked compositional layering but with imperfect cleavage.

Gondwanaland. The southern half of Pangaea, consisting of present-day Australia, India, Madagascar, Africa, and South America.

Graben (also called a *rift*). A trenchlike structure bounded by parallel normal faults. See *half-graben*.

Grade. A term for the level of concentration of a metal in an ore. Usually expressed as a percentage.

Graded bed. A layer in which the size of the sedimentary particles grades upward from coarse to finer.

Graded stream. A stream in which the slope has become so adjusted, under conditions of available discharge and prevailing channel characteristics, that the stream is just able to transport the sediment load available to it.

Gradient. A measure of the vertical drop over a given horizontal distance.

Grain flow. Mass wasting of dry or nearly dry granular sediment with air filling the pore space.

Granite. A coarse-grained igneous rock containing quartz and feldspar, with potassium feldspar being more abundant than plagioclase.

Granitic. Any coarse-grained igneous or metamorphic rock having a texture and composition resembling that of a granite.

Granodiorite. A coarse-grained igneous rock resembling a granite, in which plagioclase is more abundant than potassium feldspar.

Granular flow. A type of flow in which the weight of the flowing mass is supported by grain-to-grain contact or repeated collision between grains.

Granular texture. The interlocking arrangements of mineral grains in granitic rocks.

Granulite. A high-grade metamorphic rock, usually coarse-grained and indistinctly foliated, containing pyroxenes as a major mineral.

Gravimeter (also called a *gravity meter*). A sensitive device for measuring the pull of gravity at any locality.

Gravity anomaly. Variations in the pull of gravity after correction for latitutde and altitude.

Gravity meter. See *gravimeter*.

Greenhouse effect. The property of the Earth's atmosphere by which long wavelength heat rays from the Earth's surface are trapped or reflected back by the atmosphere.

Greenhouse gases. The gases in the atmosphere, mainly H_2O, CO_2, CFCs, and CH_4, that cause the greenhouse effect.

Greenschist. A low-grade metamorphic rock rich in chlorite.

Greywacke. See *lithic sandstone*.

Groin. A low wall, built on a beach, that crosses the shoreline at a right angle.

Groundmass. The fine-grained matrix of a porphyry.

Ground moraine. Widespread drift with a relatively smooth surface topography consisting of gently undulating knolls and shallow closed depressions.

Groundwater. All the water contained in the spaces within bedrock and regolith.

Growth habit. A characteristic growth form of a mineral.

Hadean. The oldest eon.

Half-graben. A trenchlike structure formed when the hanging-wall block moves downward on a curved fault surface. See *graben*.

Half-life. The time required to reduce the number of parent atoms by one-half as a result of radioactive decay.

Hand specimen. A rock sample of convenient size to hold in the hand for study.

Hanging valley. A glacial valley whose mouth is at a relatively high level on the steep side of a larger glaciated valley.

Hanging-wall block. The block of rock above an inclined fault.

Hardness. Relative resistance of a mineral to scratching.

Hard water. Water containing an unusually high amount of calcium carbonate.

Headwall. The steep cliff that bounds the upslope side of a cirque.

Heat. The energy a body has due to the motions of its atoms.

Heat energy. The energy of a hot body.

Heat flow. The outward flow of heat from the Earth's interior.

High grade of metamorphism. Metamorphism under conditions of high temperature and high pressure.

Hinge fault. A fault on which displacement dies out perceptibly along strike and ends at a definite point.

Highlands. See *lunar highlands*.

Historical geology. Study of the chronology of the Earth's past events, both physical and biological.

Homogeneous stress. See *uniform stress*.

Horn. A sharp-pointed peak bounded by the intersecting walls of three or more cirques.

Hornfels. A hard, fine-grained rock developed during contact metamorphism of a shale.

Horst. An elevated elongate block of crust bounded by parallel normal faults. See *graben*.

Hot spot. A fixed point on the Earth's surface defined by long-lived volcanism.

Humus. The decomposed residue of plant and animal tissues.

Hurricane. A tropical cyclonic storm having winds that exceed 120 km/h. See *typhoon*.

Hydration. The incorporation of water into a crystal structure.

Hydraulic gradient. The slope of the water table.

Hydrocarbon. Any organic compound (gaseous, liquid, or solid) consisting wholly of carbon and hydrogen.

Hydroelectric power. Energy recovered from the potential energy of rivers as they flow downward to the sea.

Hydrologic cycle. The day-to-day and long-term cyclic changes in the hydrosphere.

Hydrolysis. A chemical reaction in which the H^+ or OH^- ions of water replace ions of a mineral.

Hydrosphere. The totality of the Earth's water, including the oceans, lakes, streams, water underground, and all the snow and ice, including glaciers.

Hydrothermal mineral deposit. Any local concentration of minerals formed by deposition from a hydrothermal solution.

Hydrothermal solutions. Hot brines either given off by cooling magmas, or produced by reactions between hot rock and circulating water, that concentrate ore minerals.

Hydrous. A term applied to substances that contain H_2O or (OH).

Iapetus. The name given to the ocean that disappeared when North America and Europe collided during the Paleozoic Era.

Ice cap. A dome-shaped body of ice and snow that covers a mountain highland, or lower-lying land at high latitude, and that displays generally radial outward flow.

Ice-contact stratified drift. Stratified sediment deposited in contact with supporting glacier ice.

Ice field. A broad, nearly level area of glacier ice in a mountainous region consisting of many interconnected mountain glaciers.

Ice sheet. A continent-sized mass of ice that overwhelms nearly all the land surface within its margin.

Ice shelf. Thick glacier ice, connected to glaciers on land, that floats on the sea and commonly is located in large coastal embayments at high latitudes.

Igneous rock. Rock formed by the cooling and consolidation of magma.

Ignimbrite. The poorly sorted mass of tephra deposited by a pyroclastic flow.

Impact cratering. The process by which a planetary surface is deformed as a result of a transfer of energy from a bolide to the planetary surface.

Index fossil. A fossil that can be used to identify and date the strata in which it is found and is useful for local correlation of rock units.

Index mineral. A mineral whose first appearance marks the outer limits of a specific zone of metamorphism.

Inertia. The resistance a large mass has to sudden movement.

Inertial seismograph. A device for measuring earthquake waves based on inertia of a mass suspended on a sensitive spring.

Inner core. The central, solid portion of the Earth's core.

Inorganic compound. Chemical compounds that do not consist largely of the elements carbon and hydrogen.

Inselberg. Steep-sided mountain, ridge, or isolated hill rising abruptly from adjoining monotonously flat plains.

Intergranular fluids. The fluids, both liquid and gas, that fill the tiny pore spaces in a rock.

Intermediate grade of metamorphism. Metamorphism under conditions of intermediate pressures and temperatures.

Internal processes. All activities involved in movement or chemical and physical change of rocks in the Earth's interior.

Intrusive igneous rock. Any igneous rock formed by solidification of magma below the Earth's surface.

Ion. An atom that has excess positive or negative charges caused by electron transfer.

Ionic bond. The electrostatic attraction between negatively and positively charged ions.

Ionic radius. The distance from the center of the nucleus to the outermost shell of the orbiting electrons.

Ionic substitution (also called *atomic substitution*). The substitution of one ion for another in a random fashion throughout a crystal structure.

Inertia. The resistance of a stationary mass to sudden movement.

Inertial seismograph. A seismograph that uses the inertia of a large mass against which to measure seismic vibrations.

Island arc. An arcuate chain of andesitic stratovolcanoes sitting on oceanic crust, parallel to a sea-floor trench, and separated from it by a distance of 150–300 km.

Isoclinal fold. A fold in which both limbs are parallel.

Isograd. A line on a map connecting points of first occurrence of a given mineral in metamorphic rocks.

Isostasy. The ideal property of flotational balance among segments of the lithosphere.

Isotopes. Atoms of an element having the same atomic number but differing mass numbers.

Joint. A fracture in a rock on which no observable movement has occurred.

Jovian planets. Giant planets in the outer regions of the solar system that are characterized by great masses, low densities, and thick atmospheres consisting primarily of hydrogen and helium.

K horizon. A horizon, present in some arid-zone soils beneath the B horizon, that is impregnated with calcium carbonate.

Kame. A short, steep-sided knoll of stratified drift.

Kame terrace. A terrace of ice-contact stratified drift along a valley side.

Karst topography. An assemblage of topographic forms resulting from dissolution of carbonate bedrock and consisting primarily of closely spaced sinks.

Kerogen. Insoluble, waxlike organic matter found in sedimentary rocks, especially shales.

Kettle. A basin within a body of drift created by melting out of a mass of underlying ice.

Key bed. A thin and generally widespread bed with sedimentary characteristics so distinctive that it can be easily recognized but not confused with any other bed.

Kimberlite pipes. Narrow, pipelike masses of igneous rocks, sometimes containing diamonds, that intrude the crust but originate deep in the mantle.

Kinetic energy. The energy possessed by a moving body.

Laccolith. A lenticular pluton intruded parallel to the layering of the intruded rock, above which the layers of the invaded country rock have been bent upward to form a dome.

Lacustrine. Pertaining to, produced by, or formed in a lake.

Lagoon. A bay inshore from an enclosing reef or island paralleling a coast.

Lake Superior-type iron deposit. Iron-rich chemical-sedimentary rocks in which chert and iron-rich layers are interbedded on a fine scale. All known deposits are early Proterozoic in age, or older.

Laminar flow. A pattern of flow in which fluid particles move in parallel layers.

Landslide. Any perceptible downslope movement of a mass of bedrock or regolith, or a mixture of the two.

Lapilli. Tephra with particles having an average diameter between 2 and 64 mm.

Lapilli tuff. Pyroclastic rock in which the average diameter of tephra particles ranges between 2 and 64 mm.

Lateral moraine. An end moraine built along the side of a valley glacier.

Laterite. A hardened soil horizon characterized by extreme weathering that has led to concentration of secondary oxides of iron and aluminum.

Latitude. Part of a grid used for describing positions on the Earth's surface, consisting of parallel circles concentric to the poles. The circles are called *parallels of latitude*.

Laurasia. The northern half of Pangaea, consisting of present-day Asia, Europe, and North America.

Lava. Magma that reaches the Earth's surface through a volcanic vent.

Lava dome. A dome-shaped mass of sticky, gas-poor lava erupted from a volcanic vent following a major eruption.

Law of faunal succession. Fossil faunas and floras succeed one another in a definite, recognizable order.

Law of original horizontality. See *original horizontality*.

Leaching. The continued removal, by water solutions, of soluble matter from bedrock or regolith.

Left-lateral fault. A strike-slip fault in which relative motion is such that to an observer looking directly at the fault, the motion of the block on the opposite side of the fault is to the left. A *right-lateral fault* has right-handed movement.

Levee. See *natural levee*.

Lignite. A low-grade coal with a calorific value between that of peat and bituminous coal.

Liquification. The rapid fluidization of sediment as a result of an abrupt shock such as earthquakes.

Limbs. The sides of a fold.

Limestone. A sedimentary rock consisting chiefly of calcium carbonate, mainly in the form of the mineral calcite.

Linear dune. A long, straight, ridge-shaped dune paralleling the wind direction.

Lithic sandstone. A dark-colored sandstone containing quartz, feldspar, and a large amount of tiny rock fragments. Also called *greywacke*.

Lithification. The process that converts a sediment into a sedimentary rock.

Lithology. The systematic description of rocks in terms of mineral assemblage and texture.

Lithosphere. The outer 100 km of the solid Earth, where rocks are harder and more rigid than those in the plastic asthenosphere.

Little Ice Age. The interval of generally cool climate between the middle thirteenth and middle nineteenth centuries, during which mountain glaciers expanded worldwide.

Load. The material that is moved or carried by a natural transporting agent, such as a stream, the wind, a glacier, or waves, tides, and currents.

Local base level. Any base level, other than sea level, below which a stream controlled by that base level cannot erode the land.

Loess. Wind-deposited silt, sometimes accompanied by some clay and fine sand.

Long profile. A line drawn along the surface of a stream from its source to its mouth.

Longitude. Part of a grid used for describing positions on the Earth's surface, consisting of half circles joining the poles. The half circles are called *meridians*.

Longshore current. A current, within the surf zone, that flows parallel to the coast.

Low grade of metamorphism. Metamorphism under conditions of low temperature and low pressure.

Low-velocity zone. A region in the mantle, approximately between a depth of 100 and 350 km, where seismic-wave velocities decrease.

Lunar highlands. Mountainous regions on the Moon, believed to consist of anorthosite and gabbro.

Luster. The quality and intensity of light reflected from a mineral.

Magma. Molten rock, together with any suspended mineral grains and dissolved gases, that forms when temperatures rise and melting occurs in the mantle or crust.

Magma ocean. Molten outer layer of the Moon formed as a result of intense rain of bolides.

Magmatic arc. An arcuate chain of magmatic activity lying above a subduction zone, parallel to and separated from the seafloor trench by 100 to 400 km.

Magmatic differentiation by fractional crystallization. Compositional changes that occur in magmas by the separation of early formed minerals from residual liquids.

Magmatic differentiation by partial melting. The process of forming magmas with differing compositions by the incomplete melting of rocks.

Magmatic mineral deposit. Any local concentration of minerals formed by magmatic processes in an igneous rock.

Magnetic chrons. Periods of predominantly normal polarity (as at present), or predominantly reversed polarity.

Magnetic declination. The clockwise angle from true north assumed by a magnetic needle.

Magnetic field. Magnetic lines of force surrounding the Earth.

Magnetic inclination. The angle with the horizontal assumed by a freely swinging bar magnet.

Magnetic latitude. The latitude of a place on the Earth with respect to the magnetic poles.

Magnitude (of an earthquake). See *Richter magnitude scale*.

Mantle. The thick shell of dense, rocky matter that surrounds the core.

Marble. A metamorphic rock derived from limestone and consisting largely of calcite.

Mare. Dark-colored lowland region of the Moon underlain by basalt.

Mare basins. Huge circular impact structures on the Moon that were later filled by basaltic flows.

Maria. Plural of *mare*.

Mass balance (of a glacier). The sum of the accumulation and ablation on a glacier during a year.

Mass number. The sum of the protons and neutrons in the nucleus of an atom.

Mass-wasting. The movement of regolith downslope by gravity without the aid of a transporting medium.

M-discontinuity. See *Mohorovičić discontinuity*.

Migmatite. A composite rock containing both igneous and metamorphic portions.

Mineral. Any naturally formed, crystalline solid with a definite chemical composition and a characteristic crystal structure.

Mineral assemblage. The variety and abundance of minerals present in a rock.

Mineral deposit. Any volume of rock containing an enrichment of one or more minerals.

Mineral group. A mineral that displays extensive ionic substitution without changing the cation–anion ratio.

Mineralogy. The special branch of geology that deals with the classification and properties of minerals.

Mineraloid. A naturally occurring mineral-like solid that lacks either a crystal structure or a definite composition, or both.

Modified Mercalli Scale. A scale used to compare earthquakes based on the intensity of damage caused by the quake.

Moho. See *Mohorovičić discontinuity*.

Mohorovičić discontinuity (also called *M-discontinuity* and *Moho*). The seismic discontinuity that marks the base of the crust.

Moh's relative hardness scale. A scale of relative mineral hardness determined by scratching, divided into 10 steps, each marked by a common mineral.

Molecule. The smallest unit that retains all the properties of a compound.

Monocline. A local steepening in an otherwise uniformly dipping pile of strata.

Moraine. An accumulation of drift deposited beneath or at the margin of a glacier and having a surface form that is unrelated to the underlying bedrock.

Mountain chain. A large scale, elongate geologic feature consisting of numerous ranges or systems, regardless of similarity in form or equivalence in ages.

Mountain range. An elongate series of mountains forming a single geologic feature.

Mountain system. A group of ranges similar in general form, structure, and alignment, and presumably owing their origin to the same general causes.

Meander. A looplike bend of a stream channel.

Mechanical deformation. The changes in texture of a rock due to grinding, crushing, and development of foliation during metamorphism.

Mechanical weathering. Disintegration of rocks by mechanical processes, such as frost-wedging.

Megascopic. Features that can be seen by the unaided eye, or by the eye assisted by a simple lens that magnifies up to 10 times.

Mélange. A chaotic mixture of broken, jumbled, and thrust-faulted rock above a subduction zone.

Mercalli Scale. See *Modified Mercalli Scale*.

Mesa. An isolated, flat-topped, steep-sided, desert landform larger than a *butte*.

Mesosphere. The region between the base of the asthenosphere and the core/mantle boundary.

Mesozoic. The middle era of the Phanerozoic Eon.

Metallic bond. A form of covalent bond between atoms in which electron sharing occurs with inner energy-level shells rather than the outermost shells.

Metallogenic provinces. Limited regions of the crust within which mineral deposits occur in unusually large numbers.

Metamorphic aureole. A shell of metamorphic rock, produced by contact metamorphism, surrounding an igneous intrusion.

Metamorphic facies. Contrasting assemblages of minerals that reach equilibrium during metamorphism within a specific range of physical conditions belonging to the same metamorphic facies.

Metamorphic rock. Rock whose original compounds or textures, or both, have been transformed to new compounds and new textures by reactions in the solid state as a result of high temperature, high pressure, or both.

Metamorphic zones. The regions on a map between isograds.

Metamorphism. All changes in mineral assemblage and rock texture, or both, that take place in sedimentary and igneous rocks in the solid state within the Earth's crust as a result of changes in temperature and pressure.

Metasomatism. The process by which rocks have their composition distinctly altered by the addition or removal of ions in solution.

Meteorites. Small stony or metallic objects from interplanetary space that impact a planetary surface.

Microscopic. Those features of rocks that require high magnification in order to be viewed.

Mid-ocean ridges. Continuous rocky ridges on the ocean floor, many hundreds to a few thousand kilometers wide with a relief of more than 0.6 km. Also called *oceanic ridges* and *oceanic rises*.

Mudcracks. Cracks caused by shrinkage of wet mud as its surface dries.

Mudflow. A flowing mass of predominantly fine-grained rock debris that generally has a high enough water content to make it highly fluid; a rapidly moving type of *debris flow.*

Mudstone. A clastic sedimentary rock composed of mineral fragments finer than those in a siltstone.

Natural gas. The gaseous component of petroleum. Chiefly methane.

Natural levee. A broad, low ridge of fine alluvium built along the side of a stream channel by water that spreads out of the channel during floods.

Neutron. An electrically neutral particle with a mass 1833 times greater than that of the electron.

Nonconformity. Stratified rocks that unconformably overlie igneous or metamorphic rocks.

Normal fault. A fault, generally steeply inclined, along which the hanging-wall block has moved relatively downward.

Nuclear energy. The heat energy produced during controlled fission or fusion of atoms.

Nucleus (of an atom). The assemblage of protons and neutrons in the core of an atom.

O horizon. An accumulated layer of humus that is the uppermost horizon in many soil profiles.

Oblique slip fault. A fault on which movement includes both horizontal and vertical components.

Obsidian. An extrusive igneous rock that is wholly or largely glass.

Oceanic crust. The crust beneath the oceans.

Oceanic ridges. See *mid-ocean ridge.*

Oceanic rise. See *mid-ocean ridge.*

Oil. The liquid component of petroleum.

Oilfield. A group of oil pools, usually of similar type, or a single pool in an isolated position.

Oil pool. An underground accumulation of oil and gas in a reservoir limited by geologic barriers.

Oil shale. A shale containing wax-like substances that will break down to liquid and gaseous hydrocarbons when heated.

Oolitic limestone. A sedimentary rock composed of accumulations of tiny, round, calcareous bodies called *oolites.*

Open fold. A fold in which the two limbs dip gently and equally, and away from the axis.

Ore. An aggregate of minerals from which one or more minerals can be extracted profitably.

Organic compound. Chemical compounds made from carbon and hydrogen, with or without other elements such as nitrogen and oxygen.

Original horizontality (law of). Water-laid sediments are deposited in strata that are horizontal, or nearly horizontal, and parallel or nearly parallel to the Earth's surface.

Orogenic belts. See *orogens.*

Orogens. Elongate regions of the crust that have been intensively folded, faulted, and thickened as a result of continental collisions.

Orogeny. The process by which large regions of the crust are deformed and uplifted to form mountains.

Outcrop. See *exposure.*

Outcrop area. The area on a geologic map shown as occupied by a particular rock unit.

Outer core. The outer portion of the Earth's core, which is molten.

Outwash. Stratified drift deposited by melt-water streams.

Outwash plain. A body of outwash that forms a broad plain.

Outwash terrace. A terrace formed by dissection of an outwash plain or valley train.

Overland flow. The movement of runoff in broad sheets or groups of small, interconnecting rills.

Overturned fold. A fold in which the strata in one limb have been tilted beyond vertical.

Oxbow lake. A crescent-shaped, shallow lake occupying the abandoned channel of a meandering stream.

Oxidizing environment. A sedimentary environment in which oxygen is present and organic remains are readily converted by oxidation into carbon dioxide and water.

Pahoehoe. A smooth, ropy-surfaced lava flow, usually basaltic in composition.

Paleomagnetism. Remanent magnetism in ancient rock recording the direction of the magnetic poles at some time in the past.

Paleosol. A soil that formed at the ground surface and subsequently was buried and preserved.

Paleozoic. The oldest era of the Phanerozoic Eon.

Pangaea. The name given to a supercontinent that formed by collision of all the continental crust during the late Paleozoic.

Parabolic dune. A sand dune of U-shape with the open end of the U facing upwind.

Parallel of latitude. See *latitude.*

Parallel strata. Strata whose individual layers are parallel.

Parent. An atomic nucleus undergoing radioactive decay. Compare *daughter product.*

Parent material (of a soil). The regolith from which a soil develops.

Passive continental margin. A continental margin in a plate interior.

Peat. An unconsolidated deposit of plant remains that is the first stage in the conversion of plant matter to coal.

Pediment. A sloping surface cut across bedrock and thinly or discontinuously veneered with alluvium that slopes away from the base of a highland in an arid or semiarid environment.

Pegmatite. An exceptionally coarse-grained intrusive igneous rock, commonly granitic in composition and texture.

Pelagic sediment. Sediment consisting of the remains of marine organisms living in the open ocean.

Perched water body. A water body perched atop an aquiclude that lies above the main water table.

Percolation. The movement of groundwater in the saturated zone.

Peridotite. A coarse-grained igneous rock consisting largely of olivine, with or without pyroxene.

Period. The time during which a geologic system accumulated.

Permeability. A measure of how easily a solid allows a fluid to pass through it.

Petroleum. Gaseous, liquid, and semisolid substances occurring naturally and consisting chiefly of chemical compounds of carbon and hydrogen.

Petrology. The special branch of geology that deals with the occurrence, origin, and history of rocks.

Phanerozoic. The eon that follows the Proterozoic Eon.

Phase transition. Atomic repacking caused by changes in pressure and temperature.

Phenocrysts. The isolated large mineral grains in a porphyry.

Photosynthesis. The process by which plants combine water and carbon dioxide to make carbohydrates and oxygen.

Phyllite. A well-foliated metamorphic rock in which the component platy minerals are just visible.

Physical geology. The study of the processes that operate at or beneath the surface of the Earth, and the materials on which those processes operate.

Piedmont glacier. A broad glacier that terminates on a piedmont slope beyond confining mountain valleys and is fed by one or more large valley glaciers.

Pile. A device in which nuclear fission can be controlled.

Pillow basalt. Discontinuous, pillow-shaped masses of basalt, ranging in size from a few centimeters to a meter or more in greatest dimension.

Placer. A deposit of heavy minerals concentrated mechanically.

Plane of the ecliptic. See *ecliptic.*

Planetary accretion. The process by which bits of condensed solid matter were gathered to form the planets.

Planetary nebula. A flattened, rotating disk of gas surrounding a proto-sun.

Planet. A large celestial body that revolves around the Sun in an elliptical orbit.

Planetology. A comparative study of the Earth with the Moon and with the other planets.

Plateau basalt. Flat plains of lava formed as a result of a fissure eruption of basalt.

Plate tectonics. The special branch of tectonics that deals with the processes by which the lithosphere is moved laterally over the asthenosphere.

Plate triple junction. Junction between three plate spreading edges. The angle between any two edges is 120°.

Playa. A dry lake bed in a desert basin.

Plunge (of a fold). The angle between a fold axis and the horizontal.

Plunging fold. A fold with an inclined axis.

Pluton. Any body of intrusive igneous rock, regardless of shape or size.

Point bar. An arcuate deposit of sand or gravel along the inside of the bend of a meander loop.

Polar easterlies. Globe-encircling belts of easterly winds in the high latitudes of both hemispheres.

Polar front. The region where equatorward-moving polar easterlies meet poleward-moving westerlies.

Polar (cold) glacier. A glacier in which the ice is below the pressure melting point throughout, and the ice is frozen to its bed.

Polarity reversals. Changes of the Earth's magnetic field to the opposite polarity.

Polymerization. The process of linking silicate tetrahedra into large anion groups.

Polymorph. A compound that occurs in more than one crystal structure.

Pores (*-pore space*). The innumerable tiny openings in rock and regolith that can be filled by water or other fluids.

Porosity. The proportion (in percent) of the total volume of a given body of bedrock or regolith that consists of pore spaces.

Porphyry. Any igneous rock consisting of coarse mineral grains scattered through a mixture of fine mineral grains.

Porphyry copper deposit. A class of hydrothermal mineral deposit associated with intrusions of porphyritic igneous rocks.

Potential energy. Stored energy.

Precession of the equinoxes. A progressive change in the Earth–Sun distance for a given date.

Pressure melting point. The temperature at which ice can melt at a given pressure.

Primary waves. See *P waves.*

Principle of stratigraphic superposition. See *stratigraphic superposition.*

Principle of Uniformitarianism. The same external and internal processes we recognize in action today have been operating unchanged, though at different rates, throughout most of the Earth's history.

Progradation. The outward extension of a shoreline into the sea or a lake due to sedimentation.

Prograde metamorphic effects. The metamorphic changes that occur while temperatures and pressures are rising.

Proterozoic. The eon that follows the Archean.

Proton. A positively charged particle with a mass 1832 times greater than the mass of an electron.

Pumice. A natural glassy froth made by gases escaping through a viscous magma.

P waves. Seismic body waves transmitted by alternating pulses of compression and expansion. *P* waves pass through solids, liquids, and gases.

Pyroclast. A fragment of rock ejected during a volcanic eruption.

Pyroclastic flow. A hot, highly mobile flow of tephra that rushes down the flank of a volcano during an eruption.

Pyroclastic rocks. Rocks formed from pyroclasts.

Pyrometer. An optical device for measuring temperature.

Quartzite. A metamorphic rock consisting largely of quartz, and derived from a sandstone.

Radiation. Transmission of heat energy through the passage of electromagnetic waves.

Radioactivity. The process by which isotopes of one element transform spontaneously to other isotopes of the same or different elements.

Radiometric age. The length of time a mineral has contained its built-in radioactivity clock.

Rainshadow. A dry region on the downwind side of a mountain range where precipitation is noticeably less than on the windward side.

Recharge. The addition of water to the saturated zone of a groundwater system.

Recharge area. Area where water is added to the saturated zone.

Recrystallization. The formation of new crystalline minerals within a rock.

Recumbent fold. A fold in which the axial plane is horizontal.

Reducing environment. An environment in which oxygen is lacking and organic matter does not decay, but instead is slowly transformed into solid carbon.

Reef. A generally ridgelike structure composed chiefly of the calcareous remains of sedentary marine organisms (e.g., corals, algae).

Reef limestone. A carbonate sedimentary rock formed of fossil reef organisms.

Reflection. The bouncing of a wave off the surface between two media.

Refraction. The change in velocity when a wave passes from one medium to another.

Regional metamorphism. Metamorphism affecting large volumes of crust and involving both mechanical and chemical changes.

Regolith. The irregular blanket of loose, noncemented rock particles that covers the Earth.

Relative velocity (of a plate). The apparent velocity of one plate relative to another.

Relief. The range in altitude of a land surface.

Replacement. The process by which a fluid dissolves matter already present and at the same time deposits from solution an equal volume of a different substance.

Reservoir rock. A permeable body of rock in which petroleum accumulates.

Residual mineral deposit. Any local concentration of minerals formed as a result of weathering.

Resurgent dome. The uplifting of the collapsed floor of a caldera to form a structural dome.

Retrograde metamorphic effects. Metamorphic changes that occur as temperature and pressure are declining.

Reverse fault. A fault, generally steeply inclined, along which the hanging-wall block has moved relatively upward.

Rhyolite. A fine-grained igneous rock with the composition of a granite.

Rhyolite magma. One of the three common magma types. A magma with an SiO_2 content of about 70 percent by weight.

Richter magnitude scale. A scale, based on the recorded amplitudes of seismic body waves, for comparing the amounts of energy released by earthquakes.

Rift. See *graben* and *half-graben.*

Right-lateral fault. See *left-lateral fault.*

Rind. See *weathering rind.*

Rip current. A high-velocity current flowing seaward from the shore as part of the backwash from a wave.

Ripple mark. One of a series of small, fairly regular, subparallel ridges preserved in rock and representing a former rippled sedimentary surface.

Rock. Any naturally formed, nonliving, firm and coherent aggregate mass of mineral matter that constitutes part of a planet.

Rock cleavage (also called *slaty cleavage*). The property by which a rock breaks into platelike fragments along flat planes.

Rock cycle. The cyclic movement of rock material, in the course of which rock is created, destroyed, and altered through the operation of internal and external Earth processes.

Rockfall. The free falling of detached bodies of bedrock from a cliff or steep slope.

Rock flour. Fine rock particles produced by glacial crushing and grinding.

Rock glacier. A lobe of ice-cemented rock

debris that moves slowly downslope in a manner similar to glaciers.

Rockslide. The sudden and rapid downslope movement of detached masses of bedrock across an inclined surface.

Rock-stratigraphic unit. Any distinctive rock unit that can be distinguished from other strata on the basis of composition and physical properties.

Roof rock. A rock, such as shale, that is impermeable and caps a petroleum reservoir.

Runoff. The fraction of precipitation that flows over the land surface.

Saltation. The progressive forward movement of a sediment particle in a series of short intermittent jumps along arcing paths.

Sand sea. Vast tract of shifting sand.

Sand ripples. A series of small and rather regular ridges on the surface of a body of sand, such as a dune.

Sandstone. A medium-grained clastic sedimentary rock composed chiefly of sand-sized grains.

Saturated zone. The groundwater zone in which all openings are filled with water.

Scale (of a map). The proportion between a unit of distance on a map and the unit it represents on the Earth's surface.

Schist. A well-foliated metamorphic rock in which the component platy minerals are clearly visible.

Schistosity. The parallel arrangement of coarse grains of the sheet-structure minerals, like mica and chlorite, formed during metamorphism under conditions of differential stress.

Sea arch. An opening through a headland, generally produced by wave erosion, that forms a bridge of rock over water.

Sea cave. A cave at the base of a seacliff produced by wave erosion.

Seafloor spreading (theory of). A theory proposed during the early 1960s in which lateral movement of the oceanic crust away from midocean ridges was postulated.

Seamount. An isolated submerged volcanic mountain standing more than 1000 m above the seafloor.

Secondary enrichment. The process by which a sulfide mineral deposit is chemically weathered and enriched in its metal content as a result.

Secondary mineral. A mineral formed later than the rock enclosing it, usually at the expense of an earlier formed primary mineral.

Secondary waves. See *S waves*.

Sediment. Regolith that has been transported by any of the external processes.

Sediment drifts. Huge bodies of sediment, up to hundreds of kilometers long, deposited and shaped by deep ocean currents along a continental margin.

Sediment flows. Mass wasting of mixtures of sediment, water, and air.

Sedimentary facies. A distinctive group of characteristics within a sedimentary unit that differs, as a group, from those elsewhere in the same unit.

Sedimentary mineral deposit. Any local concentration of minerals formed through processes of sedimentation.

Sedimentary rock. Any rock formed by chemical precipitation or by sedimentation and cementation of mineral grains transported to a site of deposition by water, wind, ice, or gravity.

Seismic belts. Large tracts of the Earth's surface that are subject to frequent earthquake shocks.

Seismic sea waves (also called *tsunami*). Long wavelength ocean waves produced by sudden movement of the seafloor following an earthquake. Incorrectly called tidal waves.

Seismic tomography. A way of revealing inhomogeneities in the mantle by measuring slight differences in the frequencies and velocities of seismic waves.

Seismic waves. Elastic disturbances spreading outward from an earthquake focus.

Seismograph. The device used to study the shocks and vibrations caused by earthquakes.

Seismology. The study of earthquakes.

Serpentinite. A rock composed largely of the mineral serpentine.

Setting time. The moment a mineral starts accumulating a daughter product produced by radioactive decay.

Shale. A fine-grained, clastic sedimentary rock.

Shard. See *volcanic shard*.

Shear strength. The internal resistance of a body to movement.

Shear stress (on a free-standing body). The force acting on a body that causes slippage or translation.

Shear waves. See *S waves*.

Sheet erosion. The erosion performed by overland flow.

Shield volcano. A volcano that emits fluid lava and builds up a broad dome-shaped edifice with a surface slope of only a few degrees.

Shore profile. A vertical section along a line perpendicular to a shore.

Silicate (-*silicate mineral*). A mineral that contains the silicate anion.

Silicate anion. A complex ion $(SiO_4)^{-4}$, that is present in all silicate minerals.

Silicate mineral. See *silicate*.

Siliceous ooze. Any pelagic deep-sea sediment of which at least 30 percent consists of siliceous skeletal remains. See *deep-sea ooze*.

Sills. Tabular, parallel-sided sheets of intrusive igneous rock that are parallel to the layering of the intruded rock.

Siltstone. A sedimentary rock composed mainly of silt-sized mineral fragments.

Sinkhole. A large solution cavity open to the sky.

Slate. A low-grade metamorphic rock with a pronounced slaty cleavage.

Slaty cleavage. See *rock cleavage*.

Slickensides. Striated or highly polished surfaces on hard rocks abraded by movement along a fault.

Slip face. The straight, lee slope of a dune.

Slump. A type of slope failure in which a downward and outward rotational movement of rock or regolith occurs along a concave-up slip surface.

Slurry flow. A moving mass of sediment that is saturated with water trapped among the grains and transported with the flowing mass.

Snowline. The lower limit of perennial snow.

Soft water. Groundwater that contains little dissolved matter and no appreciable calcium.

Soil. The part of the regolith that can support rooted plants.

Soil horizons. The subhorizontal weathered zones formed as a soil develops.

Soil profile. A vertical section through a soil that displays its component horizons.

Sole marks. Irregularities formed by currents together with tracks and other markings preserved on the bedding plane of sandstone or siltstone.

Solifluction. The very slow downslope movement of waterlogged soil and surficial debris.

Source rock. A sedimentary rock containing organic matter that is a source of petroleum.

Spatter cone. A cone-shaped pile of bits of lava surrounding a volcanic vent.

Spatter rampart. A linear pile of bits of lava erupted along a fissure.

Specific gravity. A number stating the ratio of the weight of a substance to the weight of an equal volume of pure water. A dimensionless number numerically equal to the density.

Spheroidal weathering. The successive loosening of concentric shells of decayed rock from a solid rock mass as a result of chemical weathering.

Spit. An elongate ridge of sand or gravel that projects from land and ends in open water.

Spreading axis. The axis of rotation of a plate of lithosphere.

Spreading center (also called a *divergent margin*). The new, growing edge of a plate. Coincident with a midocean ridge.

Spreading pole. The point where a spreading axis reaches the Earth's surface.

Spring. A flow of groundwater emerging naturally at the ground surface.

Stable platform. That portion of a craton that is covered by a thin layer of little-deformed sediments.

Stable zone. The interior part of a tectonic plate.

Stack. An isolated rocky island or steep rock mass near a cliffy shore, detached from a headland by wave erosion.

Stalactite. An iciclelike form of dripstone and flowstone, hanging from cave ceilings.

Stalagmite. An "icicle" of dripstone and flowstone projecting upward from cave floors.

Star dune. An isolated hill of sand having a base that resembles a star in plan.

Steady state. A condition in which the rate of arrival of material or energy equals the rate of escape.

Stock. A small, irregular body of intrusive igneous rock, smaller than a batholith, that cuts across the layering of the intruded rock.

Stoping. The process by which a rising body of magma wedges off fragments of overlying rock that then sink through the magma chamber.

Strain. The measure of the changes in length, volume, and shape in a stressed material.

Strain rate. The rate at which a rock is forced to change its shape or volume.

Strain seismograph. A device for recording earthquake waves based on the flexure of a long, rigid rod.

Strata. See *stratum.*

Stratabound mineral deposits. Ores of lead, zinc, copper, and other metals enclosed in sedimentary rocks in such a way that they closely resemble primary sediments.

Stratification. The layered arrangement of sediments, sedimentary rocks, or extrusive igneous rocks.

Stratified drift. Glacial drift that is both sorted and stratified.

Stratigraphic superposition (principle of). In a sequence of strata, not later overturned, the order in which they were deposited is from bottom to top.

Stratigraphy. The study of strata.

Stratovolcanoes. Volcanoes that emit both tephra and viscous lava, and that build up steep conical mounds.

Stratum (plural = *strata*). A distinct layer of sediment that accumulated at the Earth's surface.

Streak. A thin layer of powdered mineral made by rubbing a specimen on a nonglazed porcelain plate.

Stream. A body of water that carries detrital particles and dissolved substances and flows down a slope in a definite channel.

Streamflow. The flow of surface water in a well-defined channel.

Stress. The magnitude and direction of a deforming force.

Striations. See *glacial striations.*

Strike. The compass direction of a horizontal line in an inclined plane.

Strike-slip fault. A fault on which displace-

ment has been horizontal and parallel to the strike of the fault.

Structural geology. The branch of geology devoted to the study of rock deformation.

Structure (of minerals). See *crystal structure.*

Subatomic particles. The small particles that combine to form an atom—electrons, protons, and neutrons.

Subduction. The process by which old, cold lithosphere sinks into the asthenosphere.

Subduction zone (also called a *convergent margin*). The linear zone along which a plate of lithosphere sinks down into the asthenosphere.

Submarine canyon. A steep-sided valley on the continental shelf or slope resembling a river-cut canyon on land.

Submergence. A rise of water level relative to the land so that areas formerly dry are inundated.

Subsequent stream. A stream whose course has become adjusted so that it occupies belts of weak rock or other geologic structures.

Superposed stream. A stream that was let down, or superposed, from overlying strata onto buried bedrock having composition or structure unlike that of the covering strata.

Surf. Wave activity between the line of breakers and the shore.

Surface waves. Seismic waves that are guided by the Earth's surface and do not pass through the body of the Earth.

Surge. An unusually rapid movement of a glacier marked by dramatic changes in glacier flow and form.

Suspect terrane. See *accreted terrane.*

Suspended load. Fine particles suspended in a stream.

Swash. The surge of water up a beach caused by waves moving against a coast.

S waves. Seismic body waves transmitted by an alternating series of sideways (shear) movements in a solid. *S* waves cause a change of shape and cannot be transmitted through liquids and gases.

Symmetrical fold. A fold in which both limbs dip equally away from the axial plane.

Syncline. A downfold with a troughlike form.

System. The primary unit in a time-stratigraphic sequence of rocks.

Taconite. Iron ore found as a result of metamorphism of Lake-Superior-type iron deposits.

Talus. The apron of rock waste sloping outward from the cliff that supplies it.

Tar (also called *asphalt*). An oil that is viscous and so thick it will not flow.

Tarn. A small, generally deep mountain lake occupying a cirque.

Tectonics. The study of movement and deformation of the lithosphere.

Temperate (warm) glacier. A glacier in which the ice is at the pressure-melting point and water and ice coexist in equilibrium.

Tensional stress. The differential stress on a body that causes stretching and elongation.

Tephra. A loose assemblage of pyroclasts.

Tephra cone. A cone-shaped pile of tephra deposited around a volcanic vent.

Terminal moraine. An end moraine deposited at the front of a glacier.

Terminus. The outer, lower margin of a glacier.

Terrace. An abandoned floodplain formed when a stream flowed at a level above the level of its present channel and floodplain.

Terrane. A large piece of crust with a distinctive geological character.

Terrestrial planets. The innermost planets of the solar system (Mercury, Venus, Earth, and Mars), which have high densities and rocky compositions.

Tethys. The name of a narrow sea separating Gondwanaland from Laurasia.

Texture. The overall appearance that a rock has because of the size, shape, and arrangement of its constituent mineral grains.

Thermal metamorphism. See *contact metamorphism.*

Thermal plume. A vertically rising mass of heated rock in the mantle.

Thermal spring. A natural spring that emits hot water.

Thin section. A thin slice of rock glued to a glass slide and used for microscopic examination.

Thrust faults (also called *thrusts*). Low-angle reverse faults with dips less than 15°.

Tidal bore. A large, turbulent, wall-like wave of water caused by the meeting of two tides or by the rush of tide up a narrowing inlet, river, estuary, or bay.

Tidal bulge. A bulge in bodies of marine and fresh water, produced by the gravitational attraction of the Moon and Sun, that moves around the Earth as it rotates.

Tide. The twice-daily rise and fall of the ocean surface resulting from the gravitational attraction of the Moon and Sun.

Till. A nonsorted sediment deposited directly from glacier ice.

Tillite. A nonsorted sedimentary rock of glacial origin (i.e., a lithified till).

Tilt (of axis). The angle of the Earth's rotational axis with respect to the plane of the Earth's orbit.

Time-stratigraphic unit. All the rocks or sediments that formed during a specific interval of geologic time. Compare *rock-stratigraphic unit.*

Titius–Bode rule. The distance to each planet is approximately twice as far as the next inner one, measuring from the Sun.

Tombolo. A ridge of sand or gravel that connects an island to the mainland or to another island.

Topography. The relief and form of the land.

Topset layer. A layer of stream sediment that overlies the foreset layers in a delta.

Trade wind. A globe-encircling belt of winds in the low latitudes. They blow from the northeast in the northern hemisphere and the southeast in the southern hemisphere.

Transform. The junction point where one of the major deformation features—a mid-ocean ridge, a seafloor trench, or a strike-slip fault–meets another.

Transform fault. The special class of strike-slip fault that links major structural features.

Transform fault continental margin. The margin of a continent that coincides with a transform fault.

Transform fault margin (of a plate). A fracture in the lithosphere along which two plates slide past each other.

Transverse dune. A sand dune forming a wavelike ridge transverse to wind direction.

Trap. A reservoir rock plus a roof rock that serve to accumulate petroleum.

Trenches. Long, narrow, very deep, and arcuate basins in the seafloor.

Tributary. A stream that joins a larger stream.

Triple junction. See *plate triple junction*.

Tsunami. See *seismic sea wave*.

Tuff. A pyroclastic rock consisting of ash- or lapilli-sized tephra, hence *ash tuff* and *lapilli tuff*.

Turbidite. A graded layer of sediment deposited by a turbidity current.

Turbidity current. A gravity-driven current consisting of a dilute mixture of sediment and water having a density greater than the surrounding water.

Turbulent flow. A pattern of flow in which particles of fluid move in swirls and eddies.

Typhoon. A term used in the western Pacific Ocean for a tropical cyclonic storm. See *hurricane*.

Unconfined aquifer. An aquifer with an upper surface that coincides with the water table.

Unconformity. A substantial break or gap in a stratigraphic sequence that marks the absence of part of the rock record.

Unconformity-bounded sequence. A grouping of strata that is bounded at its base and top by unconformities of regional or interregional extent.

Uniform layer. A layer of sediment or sedimentary rock that consists of particles of about the same diameter.

Uniform stress. Stress that is equal in all directions. Also called *confining stress* or *homogeneous stress*.

Uniformitarianism. See *Principle of Uniformitarianism*.

Unsaturated zone (zone of aeration). The groundwater zone in which open spaces in regolith or bedrock are filled mainly with air.

Valley glacier. A glacier that flows from a cirque or cirques onto and along the floor of a valley.

Valley train. A body of outwash that partly fills a valley.

van der Waals bond. A weak electrostatic attraction that arises because certain ions and atoms are restored from a spherical shape.

Varve. A pair of sedimentary layers deposited during the seasonal cycle of a single year.

Ventifact. Any bedrock surface or stone that has been abraded and shaped by wind-blown sediment.

Vesicle. A small opening, in extrusive igneous rock, made by escaping gas originally held in solution under high pressure while the parent magma was underground.

Viscosity. The internal property of a substance that offers resistance to flow.

Volcanic ash. See *ash*.

Volcanic breccia. A breccia formed as a result of explosive volcanic activity.

Volcanic neck. The approximately cylindrical conduit of igneous rock forming the feeder pipe of a volcanic vent that has been stripped of its surrounding rock by erosion.

Volcanic sediment (in the ocean). Sediment from submarine volcanoes, together with ash from oceanic and nonoceanic volcanic eruptions.

Volcanic shard. A particle of ash-sized, glassy tephra.

Volcano. The vent from which igneous matter, solid rock, debris, and gases are erupted.

Volcanogenic massive sulfide deposit. A mineral deposit formed by deposition of sulfide minerals from a submarine hot spring.

Wall-rock alteration. Changes produced in the mineral assemblages of rocks lining the flow channel of a hydrothermal solution.

Water quality. The fitness of water for human use, as affected by physical, chemical, and biological factors.

Water table. The upper surface of the saturated zone of groundwater.

Wave. An oscillatory movement of water characterized by an alternate rise and fall of the water surface.

Wave base. The effective lower limit of wave motion, which is half of the wavelength.

Wave-cut bench. A bench or platform cut across bedrock by surf.

Wave-cut cliff. A coastal cliff cut by surf.

Wavelength. The distance between the crests or troughs of adjacent waves.

Wave refraction. The process by which the direction of a series of waves, moving into shallow water at an angle to the shoreline, is changed.

Weathering. The chemical alteration and mechanical breakdown of rock materials during exposure to air, moisture, and organic matter.

Weathering rind. A discolored rim of weathered rock surrounding an unweathered core.

Welded tuff (also called *ignimbrite*). Pyroclastic rocks, the glassy fragments of which were plastic and so hot when deposited that they fused to form a glassy rock.

Well. An excavation in the ground designed to tap a supply of underground liquid, especially water or petroleum.

Westerlies. Globe encircling belts of winds centered at about 45° latitude in both hemispheres.

Xenoliths. Fragments of country rock still enclosed in a magmatic body when it solidifies.

Yardang. An elongate, streamlined, wind-eroded ridge.

Zone of aeration. See *unsaturated zone*.